Plant Diversity and Evolution

Genotypic and Phenotypic Variation in Higher Plants

Plant Diversity and Evolution

Genotypic and Phenotypic Variation in Higher Plants

Edited by

Robert J. Henry

Centre for Plant Conservation Genetics
Southern Cross University
Lismore, Australia

CABI Publishing

CABI Publishing is a division of CAB International

CABI Publishing
CAB International
Wallingford
Oxfordshire OX10 8DE
UK

Tel: +44 (0)1491 832111
Fax: +44 (0)1491 833508
E-mail: cabi@cabi.org
Website: www.cabi-publishing.org

CABI Publishing
875 Massachusetts Avenue
7th Floor
Cambridge, MA 02139
USA

Tel: +1 617 395 4056
Fax: +1 617 354 6875
E-mail: cabi-nao@cabi.org

A catalogue record for this book is available from the British Library, London, UK.

Library of Congress Cataloging-in-Publication Data
Henry, Robert J.
  Plant diversity and evolution : genotypic and phenotypic variation in higher plants / Robert J Henry.
    p. cm.
    Includes bibliographical references (p.   ).
    ISBN 0-85199-904-2 (alk. paper)
    1. Plant diversity. 2. Plants--Evolution. I. Title.

  QK46.5.D58H46 2005
  581.7--dc22

                        2004008213

ISBN 0 85199 904 2

Typeset in 9/11pt Baskerville by Columns Design Ltd, Reading.
Printed and bound in the UK by Cromwell Press, Trowbridge.

Contents

Contributors

Victor A. Albert, The Natural History Museums and Botanical Garden, University of Oslo, NO-0318 Oslo, Norway

Ihsan A. Al-Shehbaz, Missouri Botanical Gardens, PO Box 299, St Louis, MO 63166-0299, USA, Email: ihsan.al-shehbaz@mobot.org

Carl Beierkuhnlein, University Bayreuth, Lehrstuhl fur Biogeografie, D-95440 Bayreuth, Germany, Email: Carl.Beierkuhnlein@uni-bayreuth.de

Margaret Byrne, Science Division, Department of Conservation and Land Management, Locked Bag 104, Bentley Delivery Centre, WA 6983, Australia, Email: margaretb@calm.wa.gov.au

Mark Chase, Royal Botanic Gardens, Kew, Richmond, Surrey TW9 3DS, UK, Email: m.chase@kew.org

David J. Coates, Science Division, Department of Conservation and Land Management, Locked Bag 104, Bentley Delivery Centre, WA 6983, Australia, Email: davidc@calm.wa.gov.au

Claude dePamphilis, Department of Biology, The Huck Institutes of the Life Sciences and Institute of Molecular Evolutionary Genetics, The Pennsylvania State University, University Park, PA 16802, USA

Jeff Doyle, Department of Plant Biology, 228 Plant Science Building, Cornell University, Ithaca, NY 14853–4301, USA

Michael W. Frohlich, Department of Botany, Natural History Museum, London SW7 5BD, UK

Philip J. Harris, School of Biological Sciences, The University of Auckland, Private Bag 92019, Auckland, New Zealand, Email: p.harris@auckland.ac.nz

Robert J. Henry, Centre for Plant Conservation Genetics, Southern Cross University, PO Box 157, Lismore, NSW 2480, Australia, Email: rhenry@scu.edu.au

Ken Hill, Royal Botanic Gardens, Mrs Macquaries Road, Sydney NSW 2000, Australia, Email: ken.hill@rbgsyd.nsw.gov.au

Robert K. Jansen, Integrative Biology, University of Texas, Austin, TX 78712-0253, USA, Email: jansen@mail.utexas.edu

Anke Jentsch, UFZ Centre for Environmental Research Leipzig, Conservation Biology and Ecological Modelling, Permoserstr. 15, D-04318 Leipzig, Germany

Sangtae Kim, Department of Botany and the Genetics Institute, University of Florida, Gainesville, FL 32611, USA

Marcus A. Koch, Heidelberg Institute of Plant Sciences, Biodiversity and Plant Systematics, Im Neuenheimer Feld 345, D69129, Heidelberg, Germany, Email: marcus.koch@urz.uni-heidelberg.de

Hongzhi Kong, Laboratory of Systematic and Evolutionary Botany, Institute of Botany, The Chinese Academy of Sciences, Beijing 100093, China and Department of Biology, The Huck Institutes of the Life Sciences and Institute of Molecular Evolutionary Genetics, The Pennsylvania State University, University Park, PA 16802, USA

James Leebens-Mack, Department of Biology, The Huck Institutes of the Life Sciences and Institute of Molecular Evolutionary Genetics, The Pennsylvania State University, University Park, PA 16802, USA

Hong Ma, The Huck Institutes of the Life Sciences and Institute of Molecular Evolutionary Genetics, The Pennsylvania State University, University Park, PA 16802, USA

Sally A. Mackenzie, Plant Science Initiative, N305 Beadle Center for Genetics Research, University of Nebraska, Lincoln, NE 68588-0660, USA, Email: smackenzie2@unl.edu

Gay McKinnon, School of Plant Science, University of Tasmania, Private Bag 55, Hobart, TAS 7001, Australia, Email: Gay.McKinnon@utas.edu.au

Thomas Mitchell-Olds, Department of Genetics and Evolution, Max Planck Institute of Chemical Ecology, Hans-Knoll Strasse 8, 07745, Jena, Germany, Email: tmo@ice.mpg.de

Eviatar Nevo, Institute of Evolution, University of Haifa, Mt Carmel, Haifa, Israel, Email: nevo@research.haifa.ac.il

Linda A. Raubeson, Department of Biological Sciences, Central Washington University, Ellensburg, WA 98926-7537, Email: raubeson@cwu.edu

Tim F. Sharbel, Laboratoire IFREMER de Genetique et Pathologie, 17390 La Tremblade, France, Email: Tim.Sharbel@ifremer.fr

Douglas E. Soltis, Department of Botany and the Genetics Institute, University of Florida, Gainesville, FL 32611, USA, Email: dsoltis@botany.ufl.edu

Pamela S. Soltis, Florida Museum of Natural History and the Genetics Institute, University of Florida, Gainesville, FL 32611, USA

Kerr Wall, Department of Biology, The Huck Institutes of the Life Sciences and Institute of Molecular Evolutionary Genetics, The Pennsylvania State University, University Park, PA 16802, USA

Peter G. Waterman, Centre for Phytochemistry, Southern Cross University, Lismore, NSW 2480, Australia, Email: waterman@nor.com.au, pwaterma@scu.edu.au

Jonathan Wendel, Department of Ecology, Evolution and Organismal Biology, Iowa State University, Ames, IA 50011, USA, Email: jfw@iastate.edu

Mi-Jeong Yoo, Department of Botany and the Genetics Institute, University of Florida, Gainesville, FL 32611, USA

1 Importance of plant diversity

Robert J. Henry

Centre for Plant Conservation Genetics, Southern Cross University, PO Box 157, Lismore, NSW 2480, Australia

Introduction

Plants are fundamental to life, providing the basic and immediate needs of humans for food and shelter and acting as an essential component of the biosphere maintaining life on the planet. Higher plant species occupy a wide variety of habitats over most of the land surface except for the most extreme environments and extend to fresh water and marine habitats. Plant diversity is important for the environment in the most general sense and is an essential economic and social resource. The seed plants (including the flowering plants) are the major focus of this book and are related to the ferns and other plant groups as shown in Fig. 1.1.

Types of Plant Diversity

Plant diversity can be considered at many different levels and using many different criteria. Phenotypic variation is important in the role of plants in the environment and in practical use. Analysis of genotypic variation provides a basis for understanding the genetic basis of this variation. Modern biological research allows consideration of variation at all levels from the DNA to the plant characteristic (Table 1.1). Genomics studies the organism at the level of the genome

(DNA). Analysis of expressed genes (transcriptome), proteins (proteome), metabolites (metabolome) and ultimately phenotypes (phenome) provides a range of related layers for investigation of plant diversity.

Diversity of Plant Species

More than a quarter of a million higher plant species have been described. Continual analysis identifies new, previously undescribed species and may group more than one species together (lumping) or divide species into more than one (splitting). The use of DNA-based analysis has begun to provide more objective evidence for such reclassifications. Evolutionary relationships may be deduced using these approaches. The analysis of plant diversity at higher taxonomic levels allows identification of genetic relationships between different groups of plants. The family is the most useful and important of these classification levels. A knowledge of evolutionary relationships is important in ensuring that management of plant populations is conducted to allow continuation of effective plant evolution, allowing longer-term plant diversity and survival to be maintained. The use of DNA analysis has greatly improved the reliability and likely stability of such classifications. Chase presents an

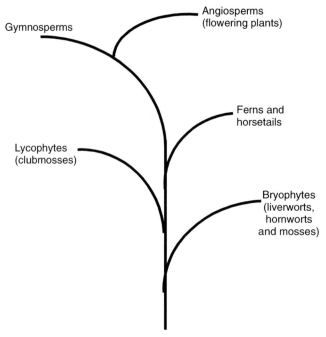

Fig. 1.1. Phylogenetic relationships between higher plants (based upon Pryer *et al.*, 2001).

Table 1.1. Levels of analysis of diversity in plants.

Level	Whole system	Study of whole system
DNA	Genome	Genomics
RNA	Transcriptome	Transcriptomics
Protein	Proteome	Proteomics
Metabolite	Metabolome	Metabolomics
Phenotype	Phenome	Phenomics

updated review of the relationships between the major groups of flowering plants in Chapter 2. This analysis draws together recent evidence from plant DNA sequence analysis. The rate of evolution of new species varies widely in different plant groups (Klak *et al.*, 2004). The factors determining these differences are likely to be important determinants of evolutionary processes.

Evolutionary relationships are important in plant conservation and also in plant improvement. Plant breeders increasingly look to source genes from wild relatives for use in the introduction of novel traits or the development of durable pest and disease resistance (Godwin, 2003).

Diversity within Plant Species

Diversity within a population of plants of the same species may be considered a primary level of variation. Coates and Byrne present an analysis of the causes and patterns of variation within plant species in Chapter 9. Principles of population genetics can be used to analyse and understand the variation within and between populations of a species. Reproductive mechanisms are a key determinant of plant diversity. Plants may reproduce by either sexual or asexual means. Clonal or vegetative propagation usually results in relatively little genetic variation except that arising from somatic mutations. Sexual reproduction can involve many different reproductive mechanisms that produce different levels of variation within the population. Outbreeding species are generally much more variable than inbreeding or self-pollinating species. Some species use more than one of these methods of reproduction. Examples include a mix of vegetative variants, mixed outcrossing and mechanisms such as apomixis. Morphological and other phenotypic variation within species can be extreme. Variation in one or a small number of genes can result in very large morphological differences in the plant. Maize was domesticated from teosinte, a very different plant in appearance. However, a mutation in a single gene has been shown to explain the major morphological differences (Wang *et al.*, 1999). This emphasizes the importance of DNA analysis in determination of plant diversity.

Factors determining diversity within species are also being better defined by the use of DNA analysis methods. The influence of environmental factors in driving adaptive selection relative to other factors of evolutionary history in determining genetic structure of plant populations can now be examined experimentally. Nevo explores these issues in Chapter 14. Habitat fragmentation may limit gene flow in wild plant populations (Rossetto, 2004). This has become an important issue in managing the impact of human activities on plant diversity and evolutionary capacity.

Plant Diversity at the Community and Ecosystem Level

Diversity can also be considered at the plant community level. Indeed this is probably what most people think of when they consider plant diversity. This diversity of species within any given plant community is often termed the species richness. The number of species is one measure of this diversity but the frequency of different species in the population is another. Populations may contain only one or a few dominant species and very small numbers of individuals from a large number of species or they may be composed of much more equal numbers of different species.

The diversity of different plant communities that make up the wider ecosystem is another level to be considered. Plant communities may extend over very wide geographic ranges while in others a complex mosaic of different plant communities can exist in close proximity. This is usually determined by the uniformity of the environment, which, in turn, is determined by differences in substrate or microenvironment. This is an important level of analysis of plant diversity for use in the conservation of plant and more general biodiversity.

Plant Diversity Enriching and Sustaining Life

Plants and plant diversity contribute directly and indirectly to the enrichment of life experiences for humans. A world in which few other life forms existed would in a narrow sense limit opportunities for ecotourism, but this is a much wider issue. A key driver for support for nature conservation is the human perception that diversity of life forms has a value beyond that associated with the importance, however great that might be, of diversity for environmental sustainability and economic reasons.

Human food is sourced directly or indirectly from plants. The role of plants in the food chain is dominant for all animal life. This provides immediate and everyday examples of the importance of plant diversity in contributing to a diversity of foods. A small number of plant species account for a relatively large proportion of the calories and protein in human diets. Most human diets include smaller amounts of a larger number of plant species. Many more plant species are regionally important as human food. Chapter 15 (Henry) expands on these issues.

Environmental Importance of Plant Diversity

Plant diversity is a key contributor to environmental sustainability on a global scale. Studies of species richness demonstrate the greater productivity of more diverse plant communities. The mechanisms that promote the co-existence of large numbers of species may include the ability of competitors to thrive at different times and places (Clark and McLauchlan, 2003). More research is needed in this area because of the scale of the potential environmental importance of this issue. This topic is reviewed by Beierkuhnlein and Jentsch in Chapter 13.

Social and Economic Importance of Plant Diversity

Social uses of plants may include ceremonial and other specific social applications. However, the greatest social use of plants probably relates to their use as ornamentals. Ornamental plants often reflect social status or identity. Foods from some plants have a social value extending beyond that contributed by their nutritional value.

Agriculture and forestry are primary industries of great economic importance. The food industry as an extension of agriculture can be considered to depend upon plant diversity. Ornamental plants are also of considerable economic importance. Fibre crops (such as cotton and hemp) provide a major source of materials for clothing. Forest species are key sources of building materials for shelter for many human populations. Plants remain the source of many medicinal compounds. All of these uses have social and economic importance.

Overview of *Plant Diversity and Evolution*

This book brings together a wide range of issues and perspectives on plant diversity and evolution. Diversity at the genome (gene) and phenome (trait) level is considered. A contemporary analysis of diversity and relationships in the flowering plants is provided for angiosperms in Chapter 2 and the gymnosperms in Chapter 3. Diversity in non-nuclear genomes is analysed for the chloroplast in Chapter 4 and the mitochondria in Chapter 5. The complication of reticulate evolution in the interpretation of plant relationships is evaluated by McKinnon in Chapter 6. The evolution and role of polyploidy in plants is reviewed by Wendel and Doyle in Chapter 7. In Chapter 8, Mitchell-Olds *et al.* provide an analysis of a plant family, the Brassicaceae, which includes *Arabidopsis*, the first plant for which a complete genome sequence was determined. Patterns of variation in plant populations and their basis are explored by Coates and Byrne in Chapter 9. The evolution of the key organ, the flower, is reviewed by Soltis *et al.* in Chapter 10. Two key features of plants – the cell wall and diverse secondary metabolism – are described in an evolutionary context by Harris and Waterman in Chapters 11 and 12, respectively. The plant cell is characterized by the presence of a cell wall essential to the structure of plants. The cell wall is not only of biological significance. The chemistry of cell walls is the basis of wood and paper chemistry. The secondary metabolites in plants play a major role in the defence of the plant. These compounds are also of use to humans in many applications, including use as drugs or drug precursors in medicine. The ecological significance of plant diversity is the subject of Chapter 13. Nevo explores the impact of domestication on plant diversity in Chapter 14 and Henry describes conservation of diversity in plants of environmental, social and economic importance in Chapter 15.

This compilation brings together information on plant diversity and evolution in a general sense and provides essential background for an understanding of plant biology and plant use in industry.

References

Clark, J.S. and McLauchlan, J.S. (2003) Stability of forest biodiversity. *Nature* 423, 635–638.
Godwin, I. (2003) Plant germplasm collections as sources of useful genes. In: Newbury, H.J. (ed.) *Plant Molecular Breeding*. Blackwell, Oxford, pp. 134–151.

Klak, C., Reeves, G. and Hedderson, T. (2004) Unmatched tempo of evolution in Southern African semi-desert ice plants. *Nature* 427, 63–65.

Pryer, K.M., Schnelder, H., Smith, A.R., Cranfill, R., Wolf, P.G., Hunt, J.S. and Sipes, S.D. (2001) Horsetails and ferns are the monophyletic group and the closest living relatives of seed plants. *Nature* 409, 618–622.

Rossetto, M. (2004) Impact of habitat fragmentation on plant populations. In: Henry, R.J. (ed.) *Plant Conservation*. Haworth Press, New York.

Wang, R.L., Stec, A., Hey, J., Lukens, L. and Dooebley, J. (1999) The limits of selection during maize domestication. *Nature* 398, 236–239.

2 Relationships between the families of flowering plants

Mark W. Chase

Royal Botanic Gardens, Kew, Richmond, Surrey TW9 3DS, UK

Introduction

In the past 10 years, enormous improvements have been made to our ideas of angiosperm classification, which have involved new sources of information as well as new approaches for handling of systematic data. The former is the topic of this chapter, but a few comments on the latter are appropriate. Before the Angiosperm Phylogeny Group classification (APG, 1998), the process of assessing relationships was mired in the use of gross morphology and a largely intuitive understanding of which characters should be emphasized (effectively a method of character weighting). Morphological features and other non-molecular traits (such as development, biosynthetic pathways and physiology) are worthy of study, but their use in phylogenetic analyses is limited by the prior information possessed by the researcher through which the acquisition of new data is filtered and the inherently complex and largely unknown genetic basis of nearly all traits. It has become increasingly clear that morphology and other phenotypic data are not appropriate for phylogenetic studies (Chase *et al.*, 2000a), but instead should be interpreted in the light of phylogenetic trees produced by analysis of DNA data, preferably DNA sequences.

It is clear that an improved understanding of all phenotypic patterns is important, but it is equally clear that assessments of phylogenetic patterns should involve as few interpretations and as many data points as is possible. Other forms of DNA data (e.g. gene order and restriction endonuclease data) suffer from limitations similar to those of morphology, and thus also should be abandoned as appropriate data for phylogenetic analyses. Prior to the APG effort (1998), there was no single, widely accepted phylogenetic classification of the angiosperms, regardless of the data type upon which a classification was based. Instead, classifications were established largely on the authority of the author; choice of which of the many in simultaneous existence should be used depended to a large degree on geography, such that in the USA the system of Cronquist (1981) was predominant, whereas in Europe those of Dahlgren (1980) or Takhtajan (1997) were more likely to be used. To a large degree, these competing systems agreed on most issues, but in the end they disagreed on many points, including the relationships of some of the largest families, such as Asteraceae, Fabaceae, Orchidaceae and Poaceae. When trying to establish why these differences existed, it soon becomes evident that the authors of these classifications were using the same data but interpreting them differently, usually in line with their intuitive assessments of which suites of characters were most informative.

The issue of ranks and authority

Other differences between morphologically based classifications (e.g. Cronquist, 1981; Thorne, 1992; Takhtajan, 1997) have to do with the hierarchical ranks given to the same groups of lower taxa. For example, Platanaceae (one genus, *Platanus*) were placed in the order Hamamelidales by Cronquist (1981), the order Platanales by Thorne (1992), and the subclass Platanidae by Takhtajan (1997), but only in the first case was it associated with any other families. In APG (1998, 2003), Platanaceae were included in Proteales along with Nelumbonaceae and Proteaceae and were listed as an optional synonym of Proteaceae (APG, 2003). Higher categories composed of single taxa are a redundancy in classification and make them less informative than systems with many taxa in each higher category. All clades in a cladogram should not be named, and lumping to an extreme degree can also make the system less informative, but monogeneric families, such as Platanaceae, should not then be the sole component of yet higher taxa unless such a taxon is sister to a larger clade composed of many higher taxa. Thus recognition of Zygophyllales composed of only Zygophyllaceae was included in APG (2003) for exactly this reason, but had Zygophyllaceae been shown to be sister to any single order, they would have been included there so that redundancy of the classification could have been reduced.

Regardless of these considerations, all classifications prior to APG (1998) could only be revised or improved by the originating author; if an author made changes (usually viewed as 'improvements') to the classification of another author, then what resulted was viewed as the second author's classification, not merely as a revision of the first. The long succession of major classifications of the angiosperms was the result of the fact that these were not composed of sets of falsifiable hypotheses. They were indisputably hypotheses of relationships, but their highly intuitive basis meant that they were not subject to improvement through evaluation of emerging new data. The only way changes could be incorporated was by the original author changing his or her mind. This intuitive basis made researchers in other fields of science view classification as more akin to philosophy than science. Thus, in spite of many years of careful study and syntheses of many data, plant taxonomy came to be viewed as an outmoded field of research. It was clear that all of the different ideas of relationships for a given family, Fabaceae for example, observed in competing modern classifications could not be simultaneously correct, and if selection of one over the others was based on an assessment of which author was the most authoritative, then perhaps framing a research programme around a classification was unwise. It would perhaps be better to think that predictivity should not be an attribute of classification and to ignore the evolutionary implications for research in other fields. Although it is immediately clear to researchers in other areas of science that classifications should be subject to modification on the basis of being demonstrated to put together unrelated taxa, this did not appear to matter to many taxonomists.

The APG classification is not the work of a single author, and the data are analysed phylogenetically, that is, without any influence of preconceived ideas of which characters are more reliable or informative, other than that DNA sequences from all three genetic compartments that agree about patterns of relationships (Soltis *et al.*, 2000) are likely to produce a predictive classification. If new data emerge that demonstrate that any component of the APG system places together unrelated taxa, then the system will be modified to take these data into account. There is no longer a need for competing classifications, and over time the APG system should be improved by more study and the addition of more data.

Monophyly and classification

The concept of monophyly has had a long and problematic history, and some have claimed that the phylogeneticists have twisted its original meaning. It is not worthwhile to include these arguments here, but it is appropriate to mention that the APG system follows the priorities for making decisions about which families to recognize that were proposed by Backlund and Bremer (1998), which means that the first priority is that all taxa are monophyletic in the phylogenetic sense of the word, i.e. that all members of a taxon must be more closely related to all other members of that taxon than they are to the members of any other taxon. This is in contrast to what an evolutionary taxonomist would propose; in such an evolutionary system, if some of the members of a group had developed one or more major novel traits then that group could be segregated into a separate family, leaving behind in another family the closest relatives of the removed group (the phylogenetic taxonomist would term the remnant group as being paraphyletic to the removed group, which is not permitted in a phylogenetic classification). Aside from the philosophical considerations, which have been debated extensively, there is a practical reason for eliminating paraphyletic groups: it is impossible to get two evolutionary taxonomists to agree on when to split a monophyletic group in this manner. Is one major novel trait enough or should there be two or more? How do we define a 'major trait' such that everyone understands when to split a monophyletic group? This problem is similar to that of falsifying hypotheses that are based on someone else's intuition. If given the same set of taxa, how likely is it that two evolutionary taxonomists would split them in the same manner and how would either be able to prove the other wrong? Therefore, the practical solution is to avoid the use of paraphyly, which is what the APG system did. It is simply impractical to include paraphyletic taxon in a system, because to do so forces the process of classification back into the hands of authority and incorporates intuition in the process, which is not only undesirable but also unscientific.

From the standpoint of the genetics, use of paraphyly is also unwise. This is because there are few traits for which we know the genetic basis, and what may appear to be a 'major trait' could in fact be a genetically simple change. Therefore, recognition of paraphyletic taxa does not involve an appreciation of how 'major' underlying genetic change might be and assumes that the taxonomist can determine this simply by appearances, which we know to be incorrect. The use of paraphyly in classification therefore decreases predictivity of the system and on this basis should also be avoided.

What follows in this chapter is compatible with the use of monophyly in what has come to be known as 'Hennigian monophyly', after the German taxonomist, Hennig, whose ideas formed the basis for phylogenetic (cladistic) classification. It is of no importance that an earlier definition of 'monophyly' may or may not have existed. The term as used in this sense has been widely accepted as of prime importance in the construction of a predictive system of classification, and classification should be as practical as possible and as devoid of historical and philosophical concepts as possible because this makes classification subject to change simply because new generations develop new philosophies, which inevitably means that classification must change. Change of classification is undesirable on this basis, and therefore the tenets under which a classification is formulated should be as far removed from historical and philosophical frameworks as possible because if a classification is to be used by scientists in other fields, it should change as little as possible.

Angiosperm Relationships

The overall framework of extant angiosperm relationships (Fig. 2.1) has become clear only since the use of DNA sequences to elucidate phylogenetic patterns, beginning with Chase et al. (1993). Analyses using up to 15 genes from all three genomic compartments of plant cells (nucleus, mitochondrion and plastid) have yielded consistent and well-supported estimates of relationships (Qiu et al., 2000; Zanis et al., 2002). Studies of genes have placed the previously poorly known monogeneric

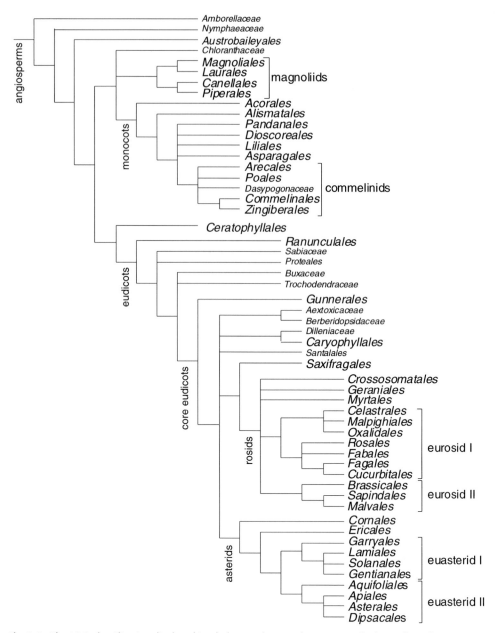

Fig. 2.1. The APG classification displayed in cladogram format. The patterns of relationships shown are those that were well supported in Soltis *et al.* (2000) or other studies; the data analysed in these studies included at least plastid *rbcL* and *atpB* and nuclear 18S rDNA sequences. Rosid and asterid families not yet placed in one of the established orders are not shown (modified from APG, 2003).

family Amborellaceae as sister to the rest of the angiosperms. *Amborella*, restricted to New Caledonia, has, since the three-gene analysis of Soltis *et al.* (1999, 2000), been the subject of a great number of other studies and has been shown to have a number of not particularly primitive traits, such as separately sexed plants. One study (Barkman *et al.*, 2000) used a technique to 'reduce' noise in DNA sequences, which resulted in *Amborella* being placed sister to Nymphaeaceae (the waterlilies). It is not clear how the subject of noise in DNA sequences should be identified, but several other techniques were employed by Zanis *et al.* (2002), and they found that the rooting at the node with *Amborella* alone could not be rejected by any partition of the data (e.g. codons, transitions/tranversions, synonymous/non-synonymous). Thus it seems reasonable to conclude that the rooting issue was resolved in favour of that of *Amborella*, but more study is required. Following *Amborella*, the next node splits Nymphaeaceae from the rest, followed by a clade composed of Austrobaileyaceae, Schisandraceae and Trimeniaceae. This arrangement of families (the ANITA grade of Qiu *et al.*, 1999) results in each being given ordinal status: Amborellales, Nymphaeales and Austrobaileyales. None of these families is large (Nymphaeaceae is the largest with eight genera and 64 species), and were it not for their phylogenetic placement, they would probably receive little attention. They are critical in terms of understanding patterns of morphological and genomic change within the angiosperms, and thus no study purporting to present a comprehensive overview can ignore them. They have thus been studied extensively but are problematic none the less because it is clear that they are the last remnants of their lineages. As such they are unlikely to represent adequately the traits of these lineages, so their use in the study of how morphological characters have changed must be qualified by an appreciation of the instability caused by having so few representatives of these earliest lineages to diverge from the rest of the angiosperms. It could well be that the traits ancestral for the angiosperms are not to be found in the families of the ANITA grade, but rather in the descendants of the other line, the bulk of the families of angiosperms. 'Basal' families in a phylogenetic sense are not necessarily primitive (the concept of heterobathmy applies here: most plants are mixtures of advanced and primitive traits, for example dioecy and vesselless wood, respectively, in *Amborella*).

The remainder of the angiosperms fall into two large groups, the monocots and eudicots (dicots with triaperturate pollen), and a number of smaller clades: Canellales, Laurales, Magnoliales, Piperales (these four orders collectively known as the 'eumagnoliids' or simply 'magnoliids'), Ceratophyllaceae (monogeneric) and Chloranthaceae (four genera). These smaller groups were in previous systems typically included with the eudicots in the 'dicots' because, like the eudicots, they have two cotyledons. None the less, they share with the monocots uniaperturate pollen, and it would appear that the magnoliids are collectively sister to the monocots (Duvall *et al.*, 2005). The relationships of Ceratophyllum and Chloranthaceae have been difficult to establish, but it would appear that the former are related to the monocots and the latter perhaps sister to the monocots plus magnoliids. More study is required before these issues can be settled.

As stated above, the monocots were considered one of the two groups of angiosperms, but they share with the primitive dicots pollen with a single germination pore. In this respect, they are not an obvious group on their own, but they deviate substantially from the primitive dicots in having scattered vascular bundles in their stems (as opposed to having them arranged in a ring) and leaves generally with parallel venation (as opposed to a net-like reticulum). Their flowers are generally composed of whorls of three parts, typically two whorls each of perianth parts and stamens and a single whorl of carpels, but there are numerous exceptions to this format.

Within the monocots, the relationships of nearly all families are well established as well as the general branching order of the orders *sensu* APG (1998, 2003). Monogeneric Acoraceae (Acorales) are sister to the rest (Chase *et al.*, 1993, 2000b; Duvall *et al.*, 1993a,b); the sole genus, *Acorus*, in most systems of classification was included in Araceae (the aroids), but most morphologists had concluded that it did not belong there (Grayum, 1987). The issue of what is the most primitive monocot family was not settled by the position of *Acorus* because most of the characters judged to be primitive in the monocots are found in Alismatales (Dahlgren *et al.*, 1985). Alismatales (13 families), which include Araceae, Tofieldiaceae and the alismatid families (Alismataceae, Aponogetonaceae, Butomaceae, Cymodoceaceae, Hydrocharitaceae, Juncaginaceae, Limnocharitaceae, Posidoniaceae, Potamogetonaceae, Ruppiaceae and Zosteraceae), are then the next successive sister to the rest of the monocots. The alismatid families were previously the only components of Alismatales, but analyses of DNA data have indicated a close relationship of these to Araceae and Tofieldiaceae, the former being considered either an isolated family or related to Areceae (the palms) and the latter a part of Melanthiaceae, all of which have now been proven to be erroneous placements.

Alismatales include a large number of aquatic taxa, both freshwater and marine. The flowering rush family (Butomaceae) and water plantain family (Alismataceae) include mostly emergent species, whereas others, such as the pondweed family (Potamogetonaceae) and frog's bit family (Hydrocharitaceae), have species that are submerged, with perhaps only their flowers reaching the surface. Yet others, such as Najadaceae, have underwater pollination. The eel grass family (Zosteraceae) and the sea grass families (Cymodoceaceae and Posidoniaceae) are all marine and ecologically important; they are also among the relatively small number of angiosperms that have conquered marine habitats.

The next several orders have typically been considered the 'lilioid' monocots because they were by and large included in the heterogeneous broad concept of Liliaceae by most authors (Hutchinson, 1934, 1967; Cronquist, 1981). Liliaceae in this expansive circumscription included all monocots with six showy tepals (in which the sepals looked like petals), six stamens and three fused carpels. If the plants were either arborescent (e.g. *Agave*, *Dracaena*) or had broad leaves with net-like venation (e.g. *Dioscorea*, *Trillium*), they were placed in segregate families, but we now know that these distinctions are not reliable for the purposes of family delimitation. Instead of one large family, we now have five orders, Asparagales, Dioscoreales, Liliales, Pandanales and Petrosaviales (Chase *et al.*, 2000b).

Asparagales (14 families) is the largest order of the monocots and contains the largest family, Orchidaceae (the orchids, 750 genera, 20,000 species; one of the two largest families of the angiosperms, the other being Asteraceae). The onion and daffodil family (Alliaceae) and the asparagus and hyacinth family (Asparagaceae) are the enlarged optional concepts of these families proposed by APG (2003). Up to 30 smaller families have sometimes been recognized in Asparagales, but this large number of mostly small families makes learning the families of the order difficult and trivializes the concept of family. Therefore, I favour the optional fewer/larger families recommended by APG (2003). For example, APG II proposed to lump the following in Asparagaceae: Agavaceae (already including Anemarrhneaceae, Anthericaceae, Behniaceae and Hostaceae), Aphyllanthaceae, Hyacinthaceae, Laxmanniaceae, Ruscaceae (already including Convallariaceae, Dracaenaceae, Eriospermaceae and Nolinaceae) and Themidaceae. Hesperocallidaceae have recently been shown to be embedded in Agavaceae, thus further reducing the number of families in Asparagales. Asparagales include a number of genera that can produce a form of secondary growth, which permits them to become tree-like; these include the Joshua tree (*Yucca*), aloes (*Aloe*) and the grass trees of Australia (*Xanthorrhoea*).

Orchidaceae are famous for their extravagant flowers and bizarre pollination biology, but only one, the vanilla orchid (*Vanilla*), is of agricultural value. Many are important in the cut flower and pot plant trade worldwide. Other well-known members of Asparagales include *Iris*, *Crocus* and *Gladiolus* (Iridaceae), *Aloe*, *Phormium* and *Hemerocallis* (Xanthorrhoeaceae), *Allium* (onion), *Narcissus* (daffodils), *Hippeastrum* (amaryllis) and *Galanthus* (snowdrops; all Alliaceae), *Asparagus*, *Hyacinthus* (hyacinth), *Agave* (century plant), *Hosta* and *Yucca*, *Convallaria* (lily of the valley), *Dracaena*, *Cordyline* and *Triteleia* (all Asparagaceae). There are many of these that are of minor horticultural importance. Asparagus, onion and agave (fibre and tequila) are the only agriculturally exploited species.

Dioscoreales are composed of three families, but only Dioscoreaceae, which are large forest understorey plants or vines, are large and well known. Species of *Dioscorea* (yams) are a source of starch in some parts of the world, as well as of medicines (e.g. birth control compounds). A few species are grown as ornamentals (e.g. bat flower, *Tacca*). Burmanniaceae are all peculiar mycoparasitic herbs, some of which are without chlorophyll, but these are not common and have no commercial uses.

Liliales have 11 families, including the well-known Liliaceae (in the narrow sense) and the cat-briars, Smilacaceae (another group of vines with a nearly worldwide distribution). Like a number of genera in Asparagales (e.g. *Narcissus*, *Allium*), many members of Liliaceae have bulbs; *Lilium* and *Tulipa* (tulips) are horticulturally important. Colchicaceae also have many species with bulbs, but unlike Liliaceae, which has a north temperate distribution, Colchicaceae are primarily found in the southern hemisphere, although the autumn crocus (*Colchicum*) is found in Europe and is the source of colchicine, an alkaloid that interferes with meiosis and causes chromosome doubling (polyploidy). Alstroemeriaceae, Peruvian lily, is also used in horticulture.

Pandanales are a tropical order containing the screw pines, Pandanaceae, and the Panama hat family, Cyclanthaceae. Screw pines, *Pandanus*, are immense herbs without secondary growth; the leaves are used as thatch, and the fruits are eaten. Cyclanthaceae are straggling vines that look similar to palms (but they are distantly related); they are local sources of fibre and of course are used for Panama hats.

The remaining monocots were recognized as a group, the commelinids, before the advent of DNA phylogenetics because of their shared possession of silica bodies and UV-fluorescent compounds in their epidermal cells. They are otherwise a diverse group of plants and include small herbs, a few vines and tree-like herbs such as the palms and bananas. Arecales include only the palms, Arecaceae (or the more traditionally used Palmae), which are important throughout the tropics as sources of food, beverage and building materials.

Commelinales include the bloodroots (Haemodoraceae), pickerelweed and water hyacinths (Pontederiaceae) and the large spiderwort family, Commelinaceae. The gingers, Zingiberaceae, and bananas, Musaceae, are members of Zingiberales, whereas the largest commelinid order, Poales, contains the wind-pollinated grasses, Poaceae (Graminae), and sedges, Cyperaceae, which dominate regions where woody plants cannot grow, as well as the Spanish mosses, Bromeliaceae, which like the orchids (Orchidaceae; Asparagales) are epiphytes. In addition to being ecologically important, grasses are the foundation of agriculture worldwide and include maize (*Zea*), rice (*Oryza*) and wheat (*Triticum*), as well as a number of minor grains, such as barley (*Hordeum*) and oats (*Avena*).

Eudicots

Eudicots are composed of three major groups: caryophyllids (a single order, Caryophyllales), rosids (13 orders) and asterids (nine orders). In addition to these (the core eudicots), there are a number of smaller families and orders that form a grade with respect to the core eudicots. The largest of these are Ranunculales, which include the buttercups (Ranunculaceae) and poppies (Papaveraceae), and Proteales, which include the plane tree (Platanaceae), lotus (Nelumbonaceae) and protea (Proteaceae) families. The last is an important family in South Africa and Australia where they are one of the dominant groups of plants. The placement of the lotus (*Nelumbo*) in this order was one of the most controversial aspects of the early phylogenetic studies based on DNA sequences, but subsequent studies have demonstrated that this is a robust result. The lotus is a 'waterlily' (an herbaceous plant with rhizome and round leaves attached to the stem in their middle), but its similarities to the true waterlilies are due to convergence.

The so-called 'basal' eudicots (i.e. Ranunculales and Proteales) have flowers that lack the organization typical for the larger group. The strict breakdown into sepals, petals, stamens and carpels is not obvious in many of these taxa. Some have what appears to be a regular organization, but upon closer inspection this breaks down. For example, some Ranunculaceae have whorls of typical appearance, but the sepals are instead bracts and the petals are most likely derived from either sepals or stamens. Numbers of parts are also not regular, and fusion within whorls or between whorls is rare, whereas in the core eudicots flowers take on a characteristic 'synorganization' in which numbers are regular and whorls of adjacent parts are often fused or otherwise interdependent. This is not to say that there are not complicated flowers in these basal lineages because there are some rather extraordinary ones: for example, in Ranunculaceae, there are *Delphinium* species with highly zygomorphic flowers in which the parts are highly organized. None the less, synorganization is typically the hallmark of the core eudicots.

Caryophyllids

The flowers of Caryophyllales (29 families; APG, 2003) often look like those of other core eudicot families, and thus some of the members of this order were previously thought to be rosids (e.g. the sundews, Droseraceae, which were thought to be related to Saxifragaceae) or asterids (e.g. the leadworts, Plumbaginaceae, which many authors thought were related to Primulaceae because of their similar pollen and breeding systems with stamens of different lengths). The core Caryophyllales have a long history of recognition, and in the past they have been called the Centrospermae because of their capsules with seeds arranged on centrally located placenta. This group was clearly identified in the first DNA studies (Chase *et al.*, 1993), so previous workers were correct in recognizing this group, but the DNA analyses placed a number of additional families with the core Caryophyllales. In addition to their fruit characters, Centrospermae also have betalain floral pigments that have replaced the anthocyanins typically found in angiosperms. Another common characteristic is anomalous secondary growth; such plants are woody

and often small trees or shrubs, but the way in which they make wood does not follow the typical pattern for angiosperms, which is probably an indication that these plants are derived from herbs that lost the ability to make woody growth. None the less, some of these groups do make wood that appears to be typical, so it is not yet clear whether or not Caryophyllales are ancestrally herbaceous. Good examples of this anomalous woodiness are the cacti (Cactaceae). Well-known examples of core Caryophyllales families include Amaranthaceae (which include spinach and beets), Caryophyllaceae (carnations), Cactaceae and Portulacaceae (pusley and spring beauty). Cactaceae and several other families adapted to arid zones are known to be closely related to various members of Portulacaceae, but a formal transfer of these families to the last has not yet been proposed (although it will almost certainly be treated this way in a future update of the APG system).

In the DNA studies, Centrospermae (core caryophyllids) were found to have a number of previously undetected relatives. Many of these have chemical and pollen similarities to the core group, and some have anomalous secondary growth as well. The core set of families are well known for their abilities to adapt to harsh environments, particularly deserts and salty sites, and their newly discovered relatives are similarly adapted. For example, the tamarisks (Tamaricaceae) and frankenias (Frankeniaceae) have salt-secreting glands, and jojoba (Simondsiaceae) grows in the arid zones of western North America along with cacti. The leadworts (Plumbaginaceae) and jewelweeds (Polygonaceae) also include a number of plants adapted to dry and salty conditions. The ecological diversity displayed by these plants was increased by the recognition that several families of carnivorous plants are members of Caryophyllales. These are the sundews and Venus fly trap (Droseraceae) and the Asian pitcher plants (Nepenthaceae).

Carnivory evolved several times in the angiosperms, and there are members in each of the major groups: *Brochinnia* (Bromeliaceae) in the monocots, the Australian pitcher plants (*Cephalotus*,

Cephalotaceae) in the rosids and the bladderworts (Lentibulariaceae) and New World pitcher plants (Sarraceniaceae), each related to different groups of the asterids. Botanists had debated the affinities of each of these groups of carnivorous plants for many years, and most had proposed multiple origins. However, there was little agreement about which of the carnivorous plants might be closely related and with which other families they shared a common history. DNA data were crucial to establish patterns of relationships (Albert *et al.*, 1992) because the highly modified morphology of these plants as well as the diversity of floral types made assessments of their relationships largely a matter of intuitive weighting of the reliability of these characters.

Santalales

Before turning to the rosids, I would like to mention briefly two APG orders of core eudicots that have not been placed in the three major groups because they have yet to obtain a clear position in the results of the DNA studies. The first of these are Santalales (six families), which include a large number of parasitic plants, all of which are photosynthetic but none the less obligate parasites. Some, like the sandalwood family (Santalaceae), attach to their hosts via underground haustoria, whereas others, like the mistletoes (Loranthaceae), grow directly on the branches of their woody host plants. Although most are parasites on woody species, some, such as the Western Australian Christmas tree (*Nuytsia*), attack herbaceous plants (they are one of the few trees in the areas where they grow). Santalales have a long history of recognition as a group, and nearly all proposed classifications have included them, more or less with the same circumscription as in APG (1998, 2003). Like other core eudicots, species in Santalales have organized flowers, but they have unusual numbers of whorls. Rosids and caryophyllids generally have one whorl each of calyx (sepals), corolla (petals) and carpels, whereas there are two whorls of stamens (sometimes with an amplification of

these). Asterids are similar except that there is a single whorl of stamens. Santalales have typically many whorls of some parts, particularly stamens (up to as many as 16 in some cases), so they clearly deviate from the main themes of the core eudicots. It is likely that Santalales evolved before the number of whorls became fixed or that they have simply retained a degree of developmental flexibility that was lost in the other major groups.

Saxifragales

Unlike Santalales, Saxifragales (12 families) is a novel order in the APG system (1998, 2003). The name has been used previously by some authors, but the circumscription of the order is different. Some of the families are woody and wind-pollinated, for example the witch hazel family (Hamamelidaceae, although some genera are pollinated by insects) and the sweet gum family (Altingiaceae), and these were previously considered to be related to the other wind-pollinated families (see Hamamelidae below). Others are woody and insect-pollinated, for example the gooseberry and currant family (Grossulariaceae), and yet others are herbaceous and insect-pollinated, for example the stonecrops (Crassulaceae), peonies (Paeoniaceae) and saxifrages (Saxifragaceae). The order has many species with a particular type of vein endings in their leaves, but in general they are diverse in most traits. If not thought to be related to Hamamelidae, then they were thought to be related to the rosids in Rosales and clustered near Saxifragaceae. New results have shown that a small tropical family, Peridiscaceae, are also related (Davis and Chase, 2004).

Dilleniaceae

This small family is only mentioned here because, although it is an unplaced-to-order core eudicot, it is the namesake of subclass Dilleniidae, which figured importantly in many previous systems of angiosperm classification (e.g. Cronquist, 1981). They occupy a potentially critical position within the core eudicots as sister to one of the other major groups (i.e. asterids, caryophyllids or rosids) or perhaps to a pair or all three, so, when they are placed, an understanding of their floral organization might be key to understanding floral evolution of the eudicots in general. In the three-gene analysis of Soltis *et al.* (2000), they were sister to Caryophyllales but this was not a clear result. If additional gene data also place them in this position, they will be included in Caryophyllales.

Rosids

Like Carophyllales, rosids and asterids have a long history of recognition, and similarly the DNA sequence studies have considerably enlarged the number of groups associated with them (see below). In contrast to the Caryophyllales and the asterids, many groups of plants long thought to be rosids have been demonstrated to have relationships to the first two groups, and thus the rosids have somewhat fewer families than in many systems of classification. The additional families have come mostly from the group called by many previous authors the dilleniids (e.g. in Cronquist, 1981, subclass Dilleniidae) and hamamelids (subclass Hamamelidae, *sensu* Cronquist). Before discussing the rosids, it is appropriate to first discuss these two groups that are not present in the APG system.

Hamamelidae (Cronquist, 1981) contained nearly all of the families of wind-pollinated trees, including such well-known families as the beeches and oaks (Fagaceae), birches (Betulaceae) and plane tree (Platanaceae). They were often split into 'lower' and 'higher' Hamamelidae, in recognition of their degree of advancement. The syndrome of wing pollination is highly constraining of floral morphology on a mechanical basis, and convergence in distantly related families was always suspected. Nevertheless, since the syndrome is one associated with either great modification or loss of many floral organs (e.g. petals are nearly always absent and stamens are held

on long filaments so that they can dangle in the wind), determination of other relationships was made difficult, leading most workers to place them together. DNA studies have been of major significance in sorting out the diverse patterns of relationships; some families are now placed among the non-core eudicots (e.g. Platanaceae in Proteales; Trochodendraceae, unplaced to order), Saxifragales (e.g. Daphniphyllaceae and Hamamelidaceae), rosids (most of the 'higher' Hamamelidae such as Betulaceae and Fagaceae in Fagales, see below) or even asterids (e.g. Eucommiaceae in Garryales).

At least in the case of Hamamelidae, botanists had the characters associated with wind pollination as the basis for placing the families in one taxonomic category, but the basis for Dilleniidae was always much weaker and less consistent among the authors who recognized the group. Basically (and explaining their characters in APG terminology), they were core eudicots that tended to have many petals and stamens, with the latter maturing centrifugally. In all other respects, they were diverse and difficult to place. With respect to the APG system (1998, 2003), families of this subclass are now placed in either the rosids (e.g. Brassicaceae, Clusiaceae, Cucurbitaceae, Malvaceae and Passifloraceae) or asterids (Ericaceae, Primulaceae and Theaceae). The only exceptions to this are Paeoniaceae and Dilleniaceae, which are Saxifragales and unplaced in the core eudicots thus far, respectively. Thus with respect to all previous systems of angiosperm classification, that of APG (1998, 2003) does not contain in any form two of the previously recognized major taxa, which have been shown by DNA studies to be polyphyletic (Chase *et al.*, 1993; Savolainen *et al.*, 2000; Soltis *et al.*, 2000).

Within the rosids, there are still several orders not yet placed to either of the two larger groups, eurosid I and II: Crossosomatales, Geraniales and Myrtales. Crossosomatales are a small order, with three families, none of which is well known. It is another of the APG orders that no one had predicted. Geraniales have four families, of which only Geraniaceae are well known (the temperate genera *Geranium* and largely South

African *Pelargonium*, the 'geranium' of commerce, which are both important horticulturally). On the other hand, Myrtales (13 families) have several important families, including the combretum family (Combretaceae), the melastome family (Melastomataceae), the myrtle and guava family (Myrtaceae) and the fuchsia and evening primrose family (Onagraceae). Melastomataceae and Myrtaceae are both large and ecologically important in the tropics, whereas Onagraceae are horticulturally important. Onagraceae have been studied for many years by several American botanists and have become a minor model family.

The remainder of the rosids are split into two major clades, which have been referred to as eurosid I and eurosid II. Alternative names, fabids and malvids, have also been suggested for these two clades, respectively. Celastrales (three families) are another order unique to the APG classification (in the sense of their circumscription). The reasonably large spindle family, Celastraceae, is the only one of any particular note in this order, which is sister to one of the larger orders, Malpighiales (28 families). Malpighiales and Celastrales share a particular seed type with a fibrous middle layer. Seed characters appear to be significant taxonomic characters in the angiosperms as a whole, but unfortunately they are relatively poorly studied. Within Malpighiales, the most important families are the mangosteen family (Clusiaceae or Guttiferae), a large tropical family with several species important for their fruit or timber, the spurge family (Euphorbiaceae), the passionfruit family (Passifloraceae) and the violet family (Violaceae). Also related to these two orders are Oxalidales (six families), in which the oxalis (Oxalidaceae) and elaeocarp (Elaeocarpaceae) families are placed. Both of these are sources of ornamentals, and some species of oxalis are important weeds. The southern hemisphere cunon family (Cunoniaceae) includes some important tree species.

The rest of the families make up a clade that has been termed the 'nitrogen-fixing clade' (Soltis *et al.*, 1995) because at least some members of each order are known to harbour nitrogen-fixing bacteria in root nodules. This trait is important because these plants can thereby grow on poorer soil and enrich it (e.g. farmers alternate crops so that in some years they plant legumes, one of the major nitrogen-fixing families). It has been hypothesized (Soltis *et al.*, 1995) that this trait evolved in the common ancestor of this clade and then was lost in many of the genera, although the reasons why such a valuable trait would be lost is not clear. The alternative hypothesis, and perhaps the more likely one, is that there are some preconditions that are required for the trait to evolve and these were present in the common ancestor; possession of the preconditions then made it more likely that the trait would evolve. If nitrogen fixation can be engineered in plants that currently are not capable of this, then it is more likely that this will be possible in non-fixing species in this clade than those in other clades.

Cucurbitales (seven families) contain the familiar cucumber and melon family (Cucurbitaceae) as well as the begonia family (Begoniceae), which is common in our gardens. They are sister to Fagales (seven families), which are important (mostly) north temperate forest trees. These include the birch family (Betulaceae), the she-oak family (Casuarinaceae, one of the tropical members of this order), the beech and oak family (Fagaceae, with some tropical genera), the walnut, pecan and hickory nut family (Juglandaceae) and the southern beeches (Nothofagaceae). These families are well known for their timbers as well as their fruits (nuts), and they are dominant members of many temperate and tropical ecosystems.

Fabales (four families) are important because they include the legume family, which, as mentioned above, are capable of fixing nitrogen and thus enriching many of the soils in which they grow. They are also important as food-producing plants and are grown as crops throughout the world. Beans and pulses are a good source of protein; soybean is widely grown and soya is a widely used meat substitute. Many legumes, both herbs and woody species, are also common ornamentals, and some of the tropical genera are timber species. Most are ecologically important throughout the world. The other large family of Fabales is the milkwort family, Polygalaceae. Both of these families have highly characteristic zygomorphic flowers of similar general construction, although no previous author had suggested that they were closely related. DNA studies were the first to place these families in one clade (Chase *et al.*, 1993). Milkworts curiously are unable to fix nitrogen.

Rosales (nine families) in the APG circumscription (2003) are radically different from those of most previous systems (e.g. Cronquist, 1981). Among the important families are the rose family (Rosaceae), which include many ornamentals as well as fruit-bearing species, such as apples, cherries, peaches, plums, raspberries and strawberries. A few are also important timbers (e.g. cherry and white beam). Both Rosaceae and Rhamnaceae include a number of nitrogen-fixing genera, and the latter include a number of timber species as well as some minor fruit-bearing genera (e.g. jujube). Circumscription of the last set of families in Rosales is in flux, but these have long been recognized as a natural group. Relative to their limits as used in APG (2003), the marijuana and hops family (Cannabaceae) should now include the hop-hornbeam family (Celtidaceae), which has been split from Ulmaceae. The nettle family (Urticaceae) have a number of temperate herbs of minor importance and a larger number of tropical trees that are timber species; many are sources of fibres. The fig and mulberry family (Moraceae) are a mostly tropical group, which are important ecologically and as a source of fruits. Figs are well known for their symbiotic relationships with their pollinators, fig-wasps, each species of which generally has a one-to-one relationship with a species of fig. This relationship is one of the longest enduring known; it probably dates back to 90 million years ago, when the first fig-wasp fossils are known.

In the second major clade of the rosids, there are only three orders, Brassicales, Malvales and Sapindales. Brassicales (15 families, most of them small) include all of the families that produce mustard oils, but their morphological traits were so diverse that only one author ever previously included

them in a single order (Dahlgren, 1980). This circumscription was so highly criticized by other taxonomists that in this next classification, he split them again into several unrelated orders. The basis for including them in a single order was simply due to the presence of mustards oils, which involve a complicated biosynthetic pathway for their synthesis; chemists interested in plant natural chemistry had long believed that it was highly unlikely that such a process could have evolved so many times in distantly related groups (up to six times if you consider the placement of these families in the system of Cronquist, 1981). Thus DNA data figured importantly in the recognition of this circumscription of the order. The largest family in the order is the mustard family, Brassicaceae (Cruciferae), which include the well-known broccoli, Brussels sprouts, cabbage and cauliflower, all of which are selected forms of the same species. In APG (2003), the circumscription of Brassicaceae included the caper family (Capparaceae), but recent studies have shown that by segregating a third family, Cleomaceae, it would then be appropriate to reinstate Capparaceae as a recognized family. Other commonly encountered families of Brassicales are the papaya (pawpaw) family (Caricaceae) and the nasturtium family (Tropaeolaceae).

Malvales (nine families) are well known for their production (in various parts of the plants) of mucilaginous compounds (e.g. the original source of marshmallow is the marsh mallow, a species in Malvaceae; sugar mixed with these polysaccharides is what was originally used to make the candy, but it is now artificially synthesized). Nearly all of the nine families produce at least some of these compounds. The best-known family of the order is the mallow and hibiscus family (Malvaceae), which before the application of DNA data was typically split into four families, Bombacaceae, Malvaceae, Sterculiaceae and Tiliaceae. Chocolate is also a commercial product from a species in the family, and okra is an edible fruit of a species of hibiscus. A number of ornamentals are found in the next largest family in Malvales, the thymelea family (Thymelaeaceae), which include the daphne, whereas the next largest family is the dipterocarps (Dipterocarpaceae), which is the most important family of the Old World tropical forests and produces timbers.

The last order of the eurosid II clade is Sapindales (nine families), which are nearly all woody species, whereas Brassicales and to a lesser degree Malvales have many herbaceous species. The largest family of the order is that of the maple and litchi (Sapindaceae), which is a largely tropical group; the well-known north temperate maples and horse chestnuts (buckeyes) are two exceptions to this distribution. Another important family of tropical forest trees is the mahogany family (Meliaceae), from which also comes an important insecticide, neem. The citrus or rue family (Rutaceae) is also an important woody group, but there are some herbaceous species, such as rue itself, which is a temperate genus. Grapefruit, lemons, limes and oranges, as well as a number of minor fruits, are important commercially. The poison ivy and cashew family (Anacardiaceae) is another largely tropical group; the family is well known for its highly allergenic oils, which cause severe and sometimes fatal reactions in many people. Cashews, mangoes and pistachios are important commercial members of the family.

Asterids

The second major group of eudicots is the asterids, which are subdivided into three major subgroups, only the last two of which have typically been considered to be members of formally recognized asterid taxa. Asterids differ in a number of technical and chemical characters from the rosids, but their flowers differ in having fused petals to which a single whorl of stamens is typically attached. This sympetalous corolla fused to the stamens is sometimes modified late in floral development, such that when these flowers open they appear to have free petals, but in terms of their development they are none the less derived from a fused condition (this situation has been termed 'early sympetaly' by Erbar and Leins (1996)). Some rosids can also be sympetalous (e.g. the papaya family,

Caricaceae), but this is rarely encountered; rosids of most orders have two whorls of stamens that are rarely attached to the petals (Celastraceae and Rhamnaceae are two exceptions, but they have lost different whorls). The caryophyllids are more similar to the asterids in some ways (seed and pollen characters), but to rosids in others (e.g. lack of fused petals).

Cornales (six families) were previously associated with the rosids because of their unfused petals. The dogwood family (Cornaceae) is the best known of the order and is largely north temperate. The hydrangea family (Hydrangeaceae) is well known for its ornamental species; it had been previously associated with Saxifragaceae by nearly all authors. The loasa family (Loasaceae) had been frequently placed near the passionflower family (Passifloraceae); this family includes a number of plants with stinging hairs (such as the nettles, Urticaceae).

Ericales (23 families) have previously been split into as many as seven orders by some authors (e.g. Cronquist, 1981; Diapensiales, Ebenales, Lecythidales, Polemoniales, Primulales, Theales and Sarraceniales), but DNA data do not discriminate among these clearly so the order has been broadly defined in APG (1998). Well-known families among Ericales include the heath and rhododendron family (Ericaceae), ebony family (Ebenaceae), phlox family (Polemoniaceae), primula family (Primulaceae), North American pitcher plants (Sarraceniaceae), zapote family (Sapotaceae) and tea family (Theaceae). Commercially important timber families include the ebonies and zapotes, and Ericaceae include a number of ornamentals (azaleas, ericas and rhododendrons).

In the first of the two core or euasterid clades (four orders), which some authors have termed the lamiids (Bremer *et al.*, 2001), there are four families unplaced to order, the most important of which is the borage family (Boraginaceae) and in which a number of ornamentals are included (forget-me-not, etc.). Garryales is a small order with two small families: Garryaceae include the ornamentals *Garrya* and *Aucuba*. *Eucommia* (Eucommoniaceae) is a wind-pollinated genus formerly placed in subclass Hamamelidae (Cronquist, 1981; see above). Gentianales (five families) include the milkweeds (Apocynaceae), gentians (Gentianaceae) and the fifth largest family in the angiosperms, the madders (Rubiaceae). Milkweeds are a largely tropical family of vines and trees, some of which are locally important timbers; the succulent milkweeds of South Africa are common in cultivation and include the carrion flowers that attract flies to pollinate them and which deceive the female flies so well that they lay eggs on what they think is a rotting animal carcass. Gentians are common herbs, some ornamental, in temperate zones but include as well some tropical trees. Rubiaceae are largely tropical woody plants, but in the temperate zones there are some herbs; many are important timber species, such as teak (*Tectonia*), as well as medicinal plants and coffee (*Coffea* species).

The largest order of the lamiids is Lamiales (21 families), which include the acanths (Acanthaceae), *Catalpa* and bignon family (Bignoniaceae), African violet family (Gesneriaceae), mints (Lamiaceae or Labiatae), olive and lilac family (Oleaceae), veronica family (Plantaginaceae), snapdragon family (Scrophulariaceae), broomrape family (Orobanchaceae) and verbena and teak family (Verbenaceae). Scrophulariaceae have been much studied and remain problematic in their circumscription. Orobanchaceae include the obligate, non-photosynthetic genera that most previous authors assigned there, but along with these the former 'hemiparasitic' genera, such as the Indian paint brush (*Castileja*) and lousewort (*Pedicularis*), which had been included in Scrophulariaceae, have been transferred to Orobanchaceae. The genera related to *Veronica*, such as foxglove (*Digitalis*, the source of the heart medicine, digitalin), are now considered to be Plataginaceae, which had formerly been a monogeneric family. A number of other segregates from Scrophulariaceae have recently been proposed as well, such as the pocketbook plant (Calceolariaceae). Further changes are likely as more studies are completed. The mints (Lamiaceae) are the sources of many herbs, such as basil (*Ocimum*),

lavender (*Lavendula*), rosemary (*Rosmarinus*), marjoram and oregano (*Oreganum*) and sage (*Salvia*), the last of which also has a number of ornamentals.

The last of the lamiid orders is Solanales (five families), which include the morning glory family (Convolvulaceae) and the potato and tomato family (Solanaceae). Convolvulaceae also contain the sweet potato (*Ipomoea*), which is of major importance as a staple (starch) crop in some tropical regions (e.g. New Guinea). In addition to potato and tomato (*Solanum*), Solanaceae also include aubergine (also *Solanum*), sweet and hot peppers (*Capsicum*) and tomatillo (*Physalis*), as well as many ornamentals, such as petunia (*Petunia*), poor man's orchid (*Schizanthus*) and devil's trumpet (*Brugmannsia*). Solanaceae are also well known for their drug plants, including belladonna (*Atropa*) and tobacco (*Nicotiana*), the most widely used drug plant of all.

The last clade of euasterids is the lobeliids, which includes four orders. There are still a number of small families that are not yet placed in one of these orders (e.g. the escallonia family, Escalloniaceae, and brunia family, Bruniaceae). Two of the orders were previously not considered asterids at all by most previous authors (e.g. Cronquist, 1981; Thorne, 1992). Apiales and Aquifoliales were usually allied to rosid families, although authors such as Cronquist (1981) admitted that at least the former was transitional between his subclasses Rosidae and Asteridae. Apiales (ten families) include the carrot family (Apiaceae or Umbelliferae), ivy family (Araliaceae) and pittosporum family (Pittosporaceae). Apiaceae include mostly herbaceous plants, and in addition to carrots (*Daucus*), they produce parsnips (*Pastinaca*) and fennel (*Foeniculum*, which is both a vegetable and a herb). Other genera provide herbs, such as dill (*Anethum*), parsley (*Petroselinum*) and chervil (*Anthriscus*). Araliaceae include the common English ivy (*Hedera*; other types of ivy, such as Virginia creeper and poison ivy, are included in Vitaceae and Anacardiaceae, respectively). Other well-known members of Araliaceae include aralia (*Aralia*) and ginseng (*Panax*), the latter of which is considered an important tonic in the Far East.

Aquifoliales (five families, none of them large) include the holly family (Aquifoliaceae) and the stemonura family (Stemonuraceae, which is largely tropical). Two small families, Helwingiaceae and Phyllonomaceae, include shrubs with flowers borne in the middle of their leaves. Holly (*Ilex*) was important in the religious rituals of the pre-Christians in Europe and later became identified with Christmas because its leaves persisted through the winter. One species of *Ilex* is commonly used as a tea in southern South America, principally Argentina, and others are frequently used as ornamentals.

Asterales (11 families) have the greatest number of species in the asterids because they contain the daisy and sunflower family (Asteraceae or Compositae), which is one of the two largest families of flowering plants (the other is the orchids, Orchidaceae). Asteraceae are economically and ecologically important and contain herbaceous plants as well as woody genera. *Helianthus* is the sunflower, which is cultivated for its seeds that are rich in proteins, as well as the Jerusalem artichoke (Jerusalem in this case is a corruption of *hira sol*, sunflower in Spanish); *Cynara* is the true artichoke, which is the large flower head that is harvested before it opens; *Lactuca* is lettuce, and *Chicorium* is chicory. Other species are important weeds, such as dandelion (*Taraxacum*), sticktight (*Bidens*), English daisy (*Bellis*) and ragweed (ironically named *Ambrosia*). Many cultivated ornamentals are also members of Asteraceae, including marigold (*Calendula*), African marigold (*Tagetes*, which is native to Mexico, in spite of its common name), dahlia (*Dahlia*), cosmos (*Cosmos*), batchelor's button (*Centaurea*), daisy and chrysanthemum (*Chrysanthemum*) and aster (*Aster* and several segregate genera). Other important families in Asterales include the bluebell family (Campanulaceae), the goodenia family (Goodeniaceae) and the bog-bean family (Menyanthaceae). Campanulaceae include several ornamentals, such as lobelia and cardinal flower (both *Lobelia*), Canterbury bells (*Campanula*) and bellflowers (*Platycodon*). Goodeniaceae is an Australasian family that has produced some ornamentals, such as *Scaevola*.

Dipsacales (two families) is the last order of asterids. Caprifoliaceae and Adoxaceae are the only included families, the former often treated as five more narrowly circumscribed families. The former includes elder (*Sambucus*, used as a fruit; the flowers as a drink) and snowball bush (*Viburnum*). Caprifoliaceae include a number of ornamentals, such as honeysuckle (*Lonicera*), abelia (*Abelia*), morina (*Morina*) and scabious (*Scabiosa*). *Dipsacus*, teasel, has in the past been used to card wool, but is now an introduced weed in many parts of the world.

The Future

One of the major questions regarding the APG system of classification is its stability. As is obvious when comparing the original APG classification with APG II, only a few changes have been made. The orders described in the original have been uniformly retained, and even family circumscription has changed relatively little. The evidence produced by 10 years of phylogenetic studies on the angiosperms has been remarkably consistent, and thus a system built upon such a base is likely to be stable. Changes anticipated are continued refinement of familial circumscriptions and likely recognition of some small orders for families that consistently fall outside the major clades, such as Berberidopsidales (composed of just Berberidosidaceae and Aextoxicaceae).

There are also some small families and genera that should be placed once suitable material becomes available for DNA work, but in the general scheme of things these are trivial matters. From time to time a genus misplaced within a family is also discovered. For example, *Aphanopetalum* was considered a member of Cunoniaceae by most authors, but it does not fall into either Cunoniaceae or even Oxalidales, and instead is related to Saxifragales, in which APG II placed it. None the less, these sorts of change have little effect on the overall system and do not complicate matters greatly (most people did not know *Aphanopetalum* so such changes have little effect on users of the APG classification).

The major improvement that is needed in the APG system is for there to be greater confidence in the higher-level relationships (above orders) so that a formal nomenclature can be adopted for superorders or sub-classes, but to achieve this will require additional data collected in an organized manner. At present, the relationships among the major groups of eudicots (e.g. asterids, caryophyllids and rosids) and the basal clades of angiosperms (Chloranthaceae, magnoliids, monocots, eudicots and probably Ceratophyllaceae) are not clear. Collaborative efforts are under way to address these uncertainties, and in the future we can expect clarification of these issues.

References

Albert, V.A., Williams, S.E. and Chase, M.W. (1992) Carnivorous plants: phylogeny and structural evolution. *Science* 257, 1491–1495.

APG (Angiosperm Phylogeny Group) (1998) An ordinal classification of the families of flowering plants. *Annals of the Missouri Botanical Garden* 85, 531–553.

APG (Angiosperm Phylogeny Group) (2003) An update of the Angiosperm Phylogeny Group classification for the orders and families of flowering plants: APG II. *Botanical Journal of the Linnaean Society* 141, 399–436.

Backlund, A. and Bremer, K. (1998) To be or not to be: principles of classification and monotypic plant families. *Taxon* 47, 391–400.

Barkman, T.J., Chenery, G., McNeal, J.R., Lyons-Weiler, J., Ellisens, W.J., Moore, M., Wolfe, A.D. and dePamphilis, C.W. (2000) Independent and combined analyses of sequences from all three genome compartments converge on the root of flowering plant phylogeny. *Proceedings of the National Academy of Sciences USA* 97, 13166–13171.

Bremer, K., Backlund, A., Sennblad, B., Swenson, U., Andreasen, K., Hjertson, M., Lundberg, J., Backlund, M. and Bremer, B. (2001) A phylogenetic analysis of 100+ genera and 50+ families of euasterids based on morphological and molecular data with notes on possible higher level morphological synapomorphies. *Plant Systematics and Evolution* 229, 137–169.

Chase, M.W., Soltis, D.E., Olmstead, R.G., Morgan, D., Les, D.H., Mishler, B.D., Duvall, M.R., Price, R.A., Hills, H.G., Qiu, Y.-L., Kron, K.A., Rettig, J.H., Conti, E., Palmer, J.D., Manhart, J.R., Sytsma, K.J., Michael, H.J., Kress, W.J., Karol, K.G., Clark, W.D., Hedrén, M., Gaut, B.S., Jansen, R.K., Kim, K.J., Wimpee, C.F., Smith, J.F., Furnier, G.R., Strauss, S.H., Xiang, Q.Y., Plunkett, G.M., Soltis, P.S., Swensen, S.M., Williams, S.E., Gadek, P.A., Quinn, C.J., Eguiarte, L.E., Golenberg, E., Learn, G.H. Jr, Graham, S.W., Barrett, S.C.H., Dayanandan, S. and Albert, V.A. (1993) Phylogenetics of seed plants: an analysis of nucleotide sequences from the plastid gene *rbcL*. *Annals of the Missouri Botanical Garden* 80, 528–580.

Chase, M.W., Fay, M.F. and Savolainen, V. (2000a) Higher-level classification in the angiosperms: new insights from the perspective of DNA sequence data. *Taxon* 49, 685–704.

Chase, M.W., Soltis, D.E., Soltis, P.S., Rudall, P.J., Fay, M.F., Hahn, W.H., Sullivan, S., Joseph, J., Givnish, T., Sytsma, K.J. and Pires, J.C. (2000b) Higher-level systematics of the monocotyledons: an assessment of current knowledge and a new classification. In: Wilson, K.L. and Morrison, D.A. (eds) *Monocots: Systematics and Evolution*. CSIRO, Melbourne, pp. 3–16.

Cronquist, A. (1981) *An Integrated System of Classification of Flowering Plants*. Columbia University Press, New York.

Dahlgren, R.M.T. (1980) A revised system of classification of the angiosperms. *Botanical Journal of the Linnaean Society* 80, 91–124.

Dahlgren, R.M.T., Clifford, H.T. and Yeo, P.F. (1985) *The Families of the Monocotyledons: Structure, Evolution and Taxonomy*. Springer, Berlin.

Davis, C.C. and Chase, M.W. (2004) Elatinaceae are sister to Malpighiaceae, and Peridiscaceae are members of Saxifragales. *American Journal of Botany* 91, 149–157.

Duvall, M.R., Clegg, M.T., Chase, M.W., Lark, W.D., Kress, W.J., Hills, H.G., Eguiarte, L.E., Smith, J.F., Gaut, B.S., Zimmer, E.A. and Learn, G.H. Jr (1993a) Phylogenetic hypotheses for the monocotyledons constructed from *rbcL* sequences. *Annals of Missouri Botanical Garden* 80, 607–619.

Duvall, M.R., Learn, G.H. Jr, Eguiarte, L.E. and Clegg, M.T. (1993b) Phylogenetic analysis of *rbcL* sequences identifies *Acorus calamus* as the primal extant monocotyledon. *Proceedings of the National Academy of Sciences USA* 90, 4611–4644.

Duvall, M., Mathews, S., Mohammad, N. and Russell, T. (2005) Placing the monocots: conflicting signal from trigenomic analyses. In: Columbus, J.T. (ed.) *Proceedings of the Third International Conference on Monocots*. Aliso Press, Los Angeles, California.

Erbar, C. and Leins, P. (1996) Distribution of the character state 'early sympetaly' and 'late sympetaly' within the 'sympetalae tetracyclicae' and presumably allied groups. *Botanica Acta* 109, 427–440.

Grayum, M.H. (1987) A summary of evidence and arguments supporting the removal of *Acorus* from the Araceae. *Taxon* 36, 723–729.

Hutchinson, J. (1934) *The Families of Flowering Plants*. Oxford University Press, Oxford.

Hutchinson, R. (1967) *The Genera of Flowering Plants*. Clarendon Press, Oxford.

Qiu, Y.-L., Lee, J., Bernasconi-Quadroni, F., Soltis, D.E., Soltis, P.S., Zanis, M., Chen, Z., Savolainen, V. and Chase, M.W. (1999) The earliest angiosperms: evidence from mitochondrial, plastid and nuclear genomes. *Nature* 402, 404–407.

Qiu, Y.-L., Lee, J., Bernasconi-Quadroni, F., Soltis, D.E., Soltis, P.S., Zanis, M., Chen, Z., Savolainen, V. and Chase, M.W. (2000) Phylogeny of basal angiosperms: analysis of five genes from three genomes. *International Journal of Plant Sciences* 161, S3–S27.

Savolainen, V., Chase, M.W., Hoot, S.B., Morton, C.M., Soltis, D.E., Bayer, C., Fay, M.F., de Bruijn, A.Y., Sullivan, S. and Qiu, Y.-L. (2000) Phylogenetics of flowering plants based upon a combined analysis of plastid *atpB* and *rbcL* gene sequences. *Systematic Biology* 49, 306–362.

Soltis, D.E., Soltis, P.S., Morgan, D.R., Swensen, S.M., Mullin, B.C., Dowd, J.M. and Martin, P.G. (1995) Chloroplast gene sequence data suggest a single origin of the predisposition for symbiotic nitrogen fixation in angiosperms. *Proceedings of the National Academy of Sciences USA* 92, 2647–2651.

Soltis, D.E., Soltis, P.S., Chase, M.W., Mort, M.E., Albach, D.C., Zanis, M., Savolainen, V., Hahn, W.H., Hoot, S.B., Fay, M.F., Axtell, M., Swensen, S.M., Nixon, K.C. and Farris, J.S. (2000) Angiosperm phylogeny inferred from a combined data set of 18S rDNA, *rbcL*, and *atpB* sequences. *Botanical Journal of the Linnaean Society* 133, 381–461.

Soltis, P.S., Soltis, D.E. and Chase, M.W. (1999) Angiosperm phylogeny inferred from multiple genes as a tool for comparative biology. *Nature* 402, 402–404.

Takhtajan, A. (1997) *Diversity and Classification of Flowering Plants*. Columbia University Press, New York.

Thorne, R.F. (1992) An updated phylogenetic classification of the flowering plants. *Aliso* 13, 365–389.

Zanis, M.J., Soltis, D.E., Soltis, P.S., Mathews, S. and Donoghue, M.J. (2002) The root of the angiosperms revisited. *Proceedings of the National Academy of Sciences USA* 99, 6848–6853.

3 Diversity and evolution of gymnosperms

Ken Hill

Royal Botanic Gardens, Mrs Macquaries Road, Sydney, NSW 2000, Australia

Introduction

Lindley (1830) introduced the class Gymnospermae for that group of seed plants possessing exposed or uncovered ovules as one of four classes of seed plants. Two were the monocots and the dicots, with the Gymnospermae placed between these, and a fourth group (the Rhizanths) was added for a number of highly modified hemiparasites such as *Rafflesia* and *Balanophora*. Within the class Gymnospermae, Lindley recognized five 'natural orders', Gnetaceae, Cycadaceae, Coniferae, Taxaceae and Equisetaceae. Equisetaceae have since been shown not to be seed plants, Coniferae and Taxaceae have been combined and an additional group has been introduced for the then unknown ginkgo. This gives us the four divisions of gymnosperms recognized today, Cycadophyta, Ginkgophyta, Pinophyta and Gnetophyta, with all of the flowering plants treated as the fifth division of seed plants, the Magnoliophyta (Judd *et al.*, 2002).

Nomenclature

The gymnosperms have been variously placed in a class Gymnospermae or a division Gymnospermophyta. Within this group, the subgroups have been recognized at the rank of orders, classes, subclasses, divisions, subdivisions or phyla, giving rise to the different spellings often seen in the literature (e.g Cycadales, Cycadae, Cycadinae, Cycadophyta, Cycadophytina). All subgroups including the flowering plants or angiosperms are treated here as divisions with the termination '-phyta'.

A Monophyletic Group?

The extant seed plants (the Spermatophyta) have been shown to be a monophyletic group; that is, the entire group arose from a single common ancestor, with initial radiation in the Late Palaeozoic (Stewart and Rothwell, 1993). The five lineages recognized within the seed plants have been shown to be monophyletic by most studies (e.g. Crane, 1988; Loconte and Stevenson, 1990; Qiu and Palmer, 1999), although the status of the Gnetophyta and Pinophyta has been questioned by some recent molecular studies (Bowe *et al.*, 2000; Chaw *et al.*, 2000; Rydin *et al.*, 2002; Soltis *et al.*, 2002). However, exact relationships among these lineages and the pattern and chronology of divergence remain unclear. A number of morphological and molecular cladistic studies published over the past 10 years on all or part of the Spermatophyta differ in details of divergence, and no consensus is yet available (Fig. 3.1).

© CAB International 2005. *Plant Diversity and Evolution: Genotypic and Phenotypic Variation in Higher Plants* (ed. R.J. Henry)

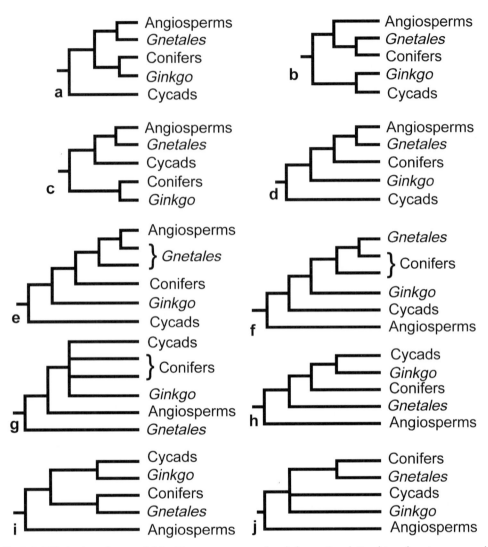

Fig. 3.1. Differing hypotheses published in recent years on the phylogenetic relationships of angiosperms and the four clades constituting the gymnosperms. (a) Parenti, 1980; (b) Hill and Crane, 1982; (c) Crane, 1985a; Doyle, 1998a; (d) Doyle and Donoghue, 1986; (e) Loconte and Stevenson, 1990; (f) Qiu *et al.*, 1999; (g) Soltis *et al.*, 2002; (h) Goremykin *et al.*, 1996; (i) Chaw *et al.*, 1997; Bowe *et al.*, 2000; (j) Rydin *et al.*, 2002.

Although regarded as a natural group for many years, more recent morphological studies have suggested that the Gymnospermae may be paraphyletic (Parenti, 1980; Hill and Crane, 1982; Crane, 1985a,b; Doyle and Donoghue, 1986; Bremer *et al.*, 1987; Loconte and Stevenson, 1990; Nixon *et al.*, 1994; Rothwell and Serbet, 1994; Doyle, 1996, 1998a,b). Central to these conclusions was the recognition of the 'Anthophyte' clade,

placing the Gnetophyta on the stem lineage of the Magnoliophyta (Crane, 1985a,b; Doyle and Donoghue, 1986, 1992; Friedman, 1992; Donoghue, 1994; Doyle, 1996; Frohlich and Meyerwitz, 1997; Nickrent *et al.*, 2000). Still more recently, molecular phylogenetic studies have failed to corroborate the Anthophyte clade and in many cases have supported a monophyletic Gymnospermae (Hasebe *et al.*, 1992; Goremykin *et al.*, 1996; Chaw *et al.*,

1997, 2000; Stefanovic *et al.*, 1998; Bowe *et al.*, 2000; Pryer *et al.*, 2001; Soltis *et al.*, 2002). Other recent studies support a close phylogenetic relationship between ginkgo and cycads and between Gnetales and conifers (Raubeson, 1998). However, support for the deeper phylogenetic structure in all of these studies has been low and results have been inconsistent, and the monophyly of the Gymnospermae must still be regarded as an open question (Rydin *et al.*, 2002). The four divisions will be discussed separately below.

Origins of the Gymnosperms

Primitive plants (bryophytes and pteridophytes and some of their even more primitive algal progenitors) disperse by means of haploid spores, which establish a free-living haploid gametophyte generation. These reproduce using motile flagellated sperm, which must swim through free water to find and then fertilize ova (Raven *et al.*, 1992). This limits habitat to sites with free water. Their gametophyte generations are free-living and also lack conductive vascular tissues, are not differentiated into true organs such as leaves and roots, have fixed stomates that cannot close and have poorly developed cuticles. They are consequently sensitive to environmental conditions, and in particular cannot withstand desiccation. Evolution of seed plants represents a major step in surviving different and varying environmental conditions. The advent of pollen eliminates dependency on water for fertilization, and the seed allows wider and more successful dispersal.

The earliest known seed plants have been reported from the Late Devonian (Famennian) of West Virginia (Rothwell *et al.*, 1989). A number of other seed structures have been reported from the latest Devonian and Early Carboniferous. Many have unusual morphologies and show no similarities to extant seed plants.

It has been suggested that extensive morphological variability often seen early in the history of lineages occurs because the new organisms are moving into new 'adaptive spaces' (vacant or underutilized ecological niches) where they are suffering little competition (Lewin, 1988). This allows many, sometimes impractical, forms to develop and coexist. Later, when members of the new lineage begin to compete, many of the early morphologies are removed by selection. In this case the adaptive space is the dry land made accessible by the acquisition of 'key adaptations' allowing the plants to resist desiccation. The 'key adaptations' allowing this movement were the protection of the fragile gametophyte stage of the life cycle by the evolution of pollen (which also eliminates dependency on water for fertilization) and the evolution of seeds (which also enable the transport and protection of plant embryos). These features also eliminate the fragile free-living gametophyte stage from the life cycle.

This 'experimental' period lasted less than 40 million years in the case of the seed plants, and was largely over by the middle of the Carboniferous. A few basic designs became established and common. At this point, lineages assigned to modern-day Cycadophyta, Ginkgophyta and Pinophyta were in existence, and seed plants became more species rich. Although three of the five extant lineages were in existence by the end of the Carboniferous, these progenitors differed in many ways from their living descendents (Florin, 1939; Miller, 1982).

Seed plants have many anatomical and reproductive features in common (Foster and Gifford, 1989). The primary vascular structure is a eustele (vascular bundles are organized into bundles of xylem internally flanked by bundles of phloem on the outside). All have secondary growth occurring from a bifacial vascular cambium (producing cells on both sides; in seed plants phloem is produced on the outside and xylem on the inside). Gametophytes are wholly developed within the spore mother cell walls with the exception of the sperm cell transfer by haustorial growth of the microgametophyte as the pollen tube. There is thus no free-living gametophyte generation in seed plants; the entire gametophyte generation is sustained by or parasitic on the sporophyte generation.

Gymnosperm Characteristics

Gymnosperms and angiosperms are differentiated by several features (also see Foster and Gifford, 1989; Raven *et al.*, 1992).

1. Gametophytes. Angiosperms are characterized by simplified megagametophytes reduced to only embryos in seeds, with endosperm not derived from megagametophyte tissue. However, gymnosperm gametophytes vary in structure between the divisions, and may represent a gradational reduction in complexity to the angiosperm condition.

2. Integuments. Most but not all angiosperms have ovules completely surrounded by two integuments (bitegmic). The exceptions apparently represent secondary loss of one integument. Most gymnosperms are unitegmic except in the gnetophytes, where the second integument may not be homologous with that of the angiosperms.

3. Pollen wall morphology. The pollen of angiosperms is different from that of all other seed plants in having a tectate–columellate structure, in which the outer layer of the pollen wall (exine) is differentiated into two layers separated by columns. In all other seed plants, pollen has a two-layer structure (exine and intine) but within these layers structure is homogeneous.

4. Vessels. Xylem vessels characterize most angiosperms. However, several basal angiosperm families lack vessels, while some ferns, *Selaginella*, *Equisetum* and *Gnetum*, all have vessel-like cells. Developmental morphology can differentiate vessels as probable independent developments in these groups.

5. Sieve elements with companion cells in phloem. Companion cells also occur in gnetophytes, but apparently again through a different developmental pathway.

6. Carpels. Angiosperms have ovules enclosed in carpels, whereas gymnosperms have exposed ovules (hence the name). However, the nature of the carpel varies widely and is not completely enclosing in some cases.

The origin and ancestry of the flowering plants remains a mystery. No clear ancestral lineage has been identified beyond the complex of Mesozoic seed plants clearly recognizable as flowering plants from which all of the major lineages branched. The first unequivocally true angiosperms appear in the fossil record as both pollen and macrofossils in the Early Cretaceous (Beck, 1976; Friis *et al.*, 1987). It has been suggested that the earlier angiosperms lived in upland, possibly arid, regions where they were unlikely to enter the fossil record (Cleal, 1989). The lack of any trace (including pollen, which can be widely transported) of these pre-Cretaceous angiosperms through the Carboniferous to Cretaceous time gap makes this hypothesis unlikely. If the gymnosperms are indeed monophyletic, their sister group the angiosperms must date from the same period, the Carboniferous. This leaves a gap of over 150 million years with no fossil record of angiosperms – a period longer than their entire known fossil history. This could be either because the gymnosperms are not a natural group, or because the stem lineage of the angiosperms lacked distinguishing angiosperm synapomorphies.

Reproductive Features Common to All Gymnosperms

A feature of all gymnosperms is the aggregation of reproductive structures into separate male and female cones or strobili (Chamberlain, 1935).

The male cone or microstrobilus in cycads, conifers and ginkgo consists of a cone axis with spirally arranged modified leaves (microsporophylls), each bearing two or several pollen sacs (microsporangia) abaxially (on the underside). Gnetophyta have a more complex microstrobilus structure that varies from family to family (Foster and Gifford, 1989).

Diploid cells inside the microsporangium (microsporocytes) undergo meiosis to produce four haploid cells (microspores). Each microspore divides mitotically to produce a microgametophyte, which becomes a pollen grain. The development of the microgametophyte occurs inside the microspore wall, and this all occurs inside the microsporangium. The outer wall of the microspore forms the pollen wall. The microgametophyte thus consists of four nuclei: two prothallial nuclei, one tube nucleus and one generative nucleus. The tube nucleus forms a pollen tube that digests its way thorough the megasporangium. The generative nucleus divides mitotically to produce two sperm cells.

The female cone or megastrobilus differs in different gymnosperm groups (Foster and Gifford, 1989). Cycads have a simple structure consisting of a cone axis, modified leaves (sporophylls) and two or several ovules on the underside of the sporophylls.

In conifers the female cone consists of a cone axis and cone scales (modified branches because they are subtended by a bract (a type of a leaf)). On the surface of a cone scale, there are two or several ovules. Gnetophyta also have a compound cone scale structure, but in both male and female cones. Female reproductive structures in ginkgo are highly reduced, and homologies with either the cycad megasporophyll or conifer cone scale have been disputed (Florin, 1951; Meyen, 1981).

An ovule consists of an integument, a megasporangium, and a diploid megaspore mother cell (megasporocyte). The megasporocyte divides meiotically to produce four haploid megaspores. Three of these degenerate leaving one megaspore, which divides mitotically many times to produce a megagametophyte. Specialized regions of the megagametophyte will differentiate into two archegonia, and each of these archegonia will produce a single egg cell. Development of the megagametophyte occurs inside the megaspore (all inside the megasporangium and the integument).

The pollen is carried by wind or insects to a mucilaginous droplet, which exudes from the micropyles of the ovules. The drop retracts (or evaporates) bringing the pollen into the pollen chamber where a haustorial pollen tube forms and the final stages of male gametophyte development take place. At fertilization, one of the sperm cells unites with the egg cell to produce a 2n zygote that will divide mitotically to produce an embryo.

The seed is made up of the embryo ($2n$), the endosperm or food source that comes from the megagametophyte ($1n$), and the seed coat that is derived from the integument (from the $2n$ parent sporophyte). There is no double fertilization as in angiosperms, although gnetophytes show a double fertilization of a different kind (see below).

Cycadophyta

The cycads are a distinct monophyletic group, defined by the presence of cycasin, girdling leaf traces, simple megasporophylls, the absence of axillary buds and the primary thickening meristem, which gives rise to the pachycaul habit (Stevenson, 1981, 1990).

Present-day occurrence

The modern cycads comprise two families with ten genera and about 300 species distributed across the warm, subtropical environments of the Americas, Africa, eastern Asia and Australasia. Most individual genera, however, have more limited geographical ranges. Many extant cycads show relictual distributions, although other groups are clearly actively evolving (Gregory and Chemnick, 2004).

Vegetative morphology

All living cycads are dioecious, long-lived, slow-growing woody perennials (also see Foster and Gifford, 1989; Norstog and Nichols, 1997).

Stems are pachycaul (short and thick) with a broad pith and cortex and manoxylic wood (a wood type that contains abundant parenchyma, typical of cycads), and may be subterranean or aerial. Internal stem structure is characterized by a eustele with endarch protoxylem, where a small amount of manoxylic wood is produced from a bifacial vascular cambium. Leaf vasculature traces in the stem are girdling; that is, traces arise from the stele at a point opposite the point of leaf attachment and dichotomously branch, with the two branches girdling the stele as they transverse the cortex and meeting again before entering the petiole. This condition is known in no other extant group of seed plants. Axillary buds are absent, and vegetative branching is either dichotomous or adventitious.

Cycad roots are heteromorphic, with contractile and coralloid roots in addition to normally functioning roots. Contractile tissue is present in roots and, to a lesser extent, stems, especially in juvenile plants (Stevenson, 1980). Coralloid roots are highly modified roots, with apogeotropic growth and extensive dichotomous branching, with the branches shortened, thickened and modified to internally accommodate symbiotic cyanobacteria (Nathanielsz and Staff, 1975).

Leaves are large, spirally arranged, pinnate, bipinnate or bipinnatifid, exstipulate or stipulate or with a stipular hood, loosely pubescent at least when young, and usually arranged in crowns on the stem-apex. The leaves are often scleromorphic, owing to the strong fibres, thick cuticle and thick hypodermis. Leaf-bases may be persistent or abscisent, depending on species. Leaves are interspersed with scale-leaves (cataphylls), except in *Stangeria* and *Bowenia*.

The pachycaul habit of modern-day cycads is thought by some to be a Tertiary development, many Mesozoic cycads having dense wood and a leptocaul habit (Delevoryas, 1993).

Reproductive morphology

Sporophylls of both sexes are simple and spirally arranged in determinate strobilate structures (except in Cycadaceae) carried on stem apices. The strobilate structure is lacking in Cycadaceae, with flushes of sporophylls developing at the stem apex in the same manner as flushes of leaves. The abaxial surfaces of male sporophylls carry numerous sporangia in two 'patches' that open by slits. Pollen is cymbiform, monosulcate and bilaterally symmetrical.

Female sporophylls are simple and entire (dissected in Cycadaceae), and carry naked, unitegmic ovules. Seeds are large, with a two-layered testa: a fleshy and distinctly coloured outer layer, and a woody inner layer. The embryo is straight, with two cotyledons, which are usually united at the tips; germination is cryptocotular.

Although widely accepted in the past to be wind pollinated (Chamberlain, 1935), recent studies in several regions indicate that cycads are mostly insect pollinated, often by closely commensal beetles (Norstog *et al.*, 1986; Tang, 1987; Donaldson *et al.*, 1995; Stevenson *et al.*, 1998). This contrasts with both *Ginkgo* and the conifers, all of which are wind pollinated (Page, 1990). Chemistry of the pollinator-attractants in cycads is markedly different from that of any flowering plants (Pellmyr *et al.*, 1991), suggesting an independent origin for this pollination syndrome.

Male gametophytes produce large, multiflagellate and motile sperm cells, sharing some similarities with those of *Ginkgo* but otherwise unlike those of any other seed plants.

Cycad seeds are large, with a fleshy outer coat (sarcotesta) over a hard, stony layer (sclerotesta), and copious haploid, maternally derived endosperm. The fertilized embryo develops slowly but continuously until germination, with short-term chemical inhibition of germination by the sarcotesta but no real dormancy (Dehgan and Yuen, 1983). This makes seeds relatively short-lived and subject to damage by desiccation.

Dispersal

The fleshy sarcotesta attracts animals, mainly birds, rodents, small marsupials and fruit-eating bats, which serve as dispersal agents (Burbidge and Whelan, 1982; Tang, 1987). In most cases, the fleshy coat is eaten off the seed and the entire seed is not consumed. Dispersal is consequently limited to the usually short distance that the animals can carry the seed.

Cycas subsection *Rumphiae* has seeds with a spongy endocarp not seen elsewhere among the cycads (Guppy, 1906; Dehgan and Yuen, 1983; Hill, 1994), which gives a potential for oceanic dispersal, and it has been demonstrated that seeds maintain viability after prolonged immersion in sea water (Dehgan and Yuen, 1983). Subsection Rumphiae is the only subgroup of the genus to occur on oceanic islands, and is widely distributed through the Indian and western Pacific oceans, as well as all non-mainland parts of South-east Asia (Hill, 1994).

Distribution and ecology

Cycad plants are long-lived and slow growing, with slow recruitment and population turnover. The fleshy and starch-rich stems are highly susceptible to fungal attack, and almost all species grow in well-drained soils. Habitats range from closed tropical forests to semideserts, the majority in tropical or subtropical climates in regions of predominantly summer rainfall. Cycads often occur on or are restricted to specialized and/or localized sites, such as nutritionally deficient sites, limestone or serpentinite outcrops, beach dune deposits or precipitously steep sites.

Contractile roots are present in all cycads (above), particularly in juvenile plants. These draw the sensitive growing apex of seedlings below the soil surface, affording protection against drought and the fires that are a frequent feature of many cycad habitats.

Coralloid roots host symbiotic cyanobacteria, which fix atmospheric nitrogen and contribute to the nutrient needs of the plant. This provides an advantage in the nutritionally deficient soils occurring in many cycad habitats.

Cycadaceae

The monogeneric Cycadaceae is apparently Laurasian in origin, and relatively recently dispersed into the Australasian region. This is supported by the fossil record, with *Cycas* fossils known only from the Eocene of China and Japan (Yokoyama, 1911; Liu *et al.*, 1991). Australia stands out as a major centre of speciation for *Cycas*, with some 27 of the *c.* 100 species.

Zamiaceae

Zamiaceae shows a distinct break into Laurasian and Gondwanan elements, possibly from an ancestral disjunction resulting from the breakup of Pangaea. Fossil evidence places the extant genera in Australia at least back into the Eocene (Cookson, 1953; Hill, 1978, 1980; Carpenter, 1991). *Macrozamia* has also speciated widely, with 38 species recognized in Australia. Many species are components of complexes with narrow geographic replacement patterns, suggesting that speciation is active and ongoing.

Bowenia and *Stangeria* were placed in Stangeriaceae, but more recent studies have failed to corroborate their sister relationship, and have indicated that both genera may be best included in Zamiaceae. Both are Gondwanan, with one genus in Australia and another in southern Africa. *Bowenia* occurs as understorey shrubs in moist eucalypt woodlands or forests, or in closed mesophyll forests. Fossil evidence places the genus *Bowenia* in southern Australia in the Early Tertiary (Hill, 1978).

Evolution and fossil record

While the extant cycads have been clearly shown to be a monophyletic group by both morphological and molecular studies (Stevenson, 1990; Chase *et al.*, 1993), ancestry and relationships of the group remain unclear. The group is acknowledged as extremely ancient, with a fossil record extending back to the Early Permian (Gao and Thomas, 1989). The Palaeozoic and Mesozoic cycads were, however, very different from those of the present day, and fossil evidence of the extant genera is known only from the Tertiary. The cycads have been proposed as the sister group to all other living seed plants (Nixon *et al.*, 1994), although other studies have suggested different relationships (Pryer *et al.*, 2001).

Relationships among the cycad genera are still not well understood. A detailed classification of taxa at and above generic level based on morphological and molecular data has been presented by Stevenson (1992), although more recent molecular studies indicate somewhat different relationships (Hill *et al.*, 2003; Rai *et al.*, 2003).

Cycads first appeared in the Late Carboniferous, and are thought to have arisen from the Medullosan seed ferns (Taylor and Taylor, 1993). Fossils from this period are somewhat problematical and unlike later or modern cycads. *Spermopteris* is characterized by two rows of ovules attached to the abaxial surface of *Taeniopteris*, a common simple leaf type from the Upper Paleozoic. This plant has been placed with the cycads on the basis of foliar features, mainly the haplocheilic stomata, in which the guard mother cell gives rise to only two guard cells. Venation patterns and a non-bifurcating leaf-base also represent derived characters shared between *Taeniopteris* and cycads. However, haplocheilic stomata also occur in several other types of seed plants, such as conifers, ginkgoes, *Ephedra*, glossopterids and *Cordaites*. Fossil foliage assigned to cycads is commonly pinnately compound (e.g. *Nilssonia*), but includes some simple, entire leaves (e.g. some *Nilssonia* and *Taeniopteris*). Cycad-like pinnate fossil foliage of the common Mesozoic group, Bennetitales, is separated by possession of syndetocheilic stomata, in which guard mother cells produce four guard cells, a feature shared with angiosperms.

During the Permian, fossil genera with greater resemblance to living cycads begin to appear, for example, *Crossozamia*. Among these are taxa that produce megasporophylls in a helical arrangement similar to the simple ovulate cones of extant *Zamia*, and taxa with foliaceous megasporophylls closely resembling extant *Cycas*.

The cycads reached their peak in species richness and ecological importance during the Mesozoic, and have been declining since that time (Harris, 1976). Many cycad genera from the Mesozoic have been reconstructed from fairly complete fossils, for example *Leptocycas*. Based on these reconstructions, some workers think that the Mesozoic cycads in general had relatively slender trunks with widely spaced leaves that abscised, and that modern cycads with the short, thick stems did not evolve until the Tertiary (Delevoryas, 1993).

Ginkgophyta

Present-day occurrence

The Ginkgophyta is represented today by a single species, *Ginkgo biloba*, which is restricted to central China in the wild, but extensively propagated as an ornamental tree. There have been suggestions that it has been extinct in the wild for centuries and maintained only in cultivation in temples. However, there have also been reports of wild occurrences in the Tianmu Mountains in Zhejiang province (Fu *et al.*, 1999).

Vegetative morphology

Ginkgophytes are large, long-lived, deciduous, dioecious trees. Roots are fibrous to woody and undifferentiated. Stems are differentiated into long and short shoots, most probably in a parallel development to that present in some conifers and angiosperms. In short shoots, internodes are very short, in contrast to long-internode long shoots. Secondary xylem in long shoots is pycnoxylic like that of conifers and Cordaites; wood in short shoots is rich in parenchyma, approaching the manoxylic condition of cycads. Leaves are exstipulate with multiple dichotomous venation and double vascular traces and develop on both long shoots and short shoots although the shape of leaves varies somewhat between long and short shoots. Leaves are simple although at times quite deeply lobed and triangular in shape with veins that bifurcate to fill space as the wedge of the leaf widens. *Ginkgo* has a leptocaul habit with well-developed, dense wood, and also has well-developed axillary buds and branching. Resin canals are absent.

Reproductive morphology

Ginkgo biloba is dioecious, a character in common between *Ginkgo* and the living cycads. Pollen is produced in paired microsporangia (also termed sporangiophores) on simple stalked microsporophylls, which are spirally aggregated into simple catkin-like strobili (lacking bracts), which are also carried on short shoots. The pollen grains are monosulcate, spherical and wind-dispersed (see also Foster and Gifford, 1989).

Ovules are erect and borne in pairs subtended by a collar of uncertain origin on axillary stalks on short shoots. The ovule has a single integument that becomes three-layered and develops in the seed into a fleshy outer sarcotesta, a stony inner sclerotesta and a thin endotesta. This is superficially similar to integument differentiation in the cycads and *Medullosa*, and is an adaptation to animal dispersal.

Development of the male gametophyte is very similar to that in the cycads except that there are two prothallial cells instead of one. The mature spermatozoid is similar to the cycad sperm, but it is smaller and has only 2.5 turns of the spiral, compared with 5 or 6 in the cycads. Numerous flagella (10,000–12,000 in cycads, uncounted in *Ginkgo*) are attached along the spiral.

Evolution and fossil record

The ginkgophytes first appeared in the fossil record in the Permian and became important components of Mesozoic ecosystems worldwide, apparently reaching maximum diversity in the Jurassic (Thomas and Spicer, 1986). Today this group is represented by a single species, *Ginkgo biloba*.

A range of Permian and Mesozoic ginkgophytes has been described based on leaves, wood and some reproductive structures (Thomas and Spicer, 1986). Some

ancient *Ginkgo* leaves closely resemble modern *Ginkgo*. Others are highly dissected and resemble the typical *Ginkgo* leaf but without tissue between the veins (Tralau, 1968). In living *Ginkgo*, leaf shape is highly variable, suggesting that more taxa may have been described on the basis of leaves only than might have actually existed.

Pinophyta (The Conifers)

The conifers are uniquely defined by the reduced (non-megaphyllous) leaves, the presence of resin canals, the compound female sporophylls, and the undifferentiated shoot apex of fertile axillary shoots. Molecular studies corroborate the monophyletic nature of the Pinophyta most distinctly in that all members show loss of the inverted repeat unit in the chloroplast genome (Raubeson and Jansen, 1992).

Present-day occurrence

Globally, there are about 650 species of conifers. These are placed in 68 genera in seven families (Farjon and Page, 2001).

Conifers occur on all continents except Antarctica, but their abundance is unevenly distributed in terms of both individuals and taxa (Table 3.1). Where the vast boreal conifer forests stretch across continents and contain billions of trees, they sustain no more than a handful of species. In contrast, more southerly latitudes in the northern hemisphere and all of the southern hemisphere have either scattered conifer forests,

Table 3.1. The extant conifer families: diversity, distribution and earliest appearance in the fossil record.

Family	Time range	Genera/species	Present distribution
Araucariaceae	Cretaceous to recent	3/37	Southern hemisphere
Cephalotaxaceae	Jurassic to recent	1/5	Northern hemisphere
Cupressaceae	Triassic to recent	30/157	Both hemispheres
Pinaceae	Cretaceous to recent	10/250	Northern hemisphere
Podocarpaceae	Triassic to recent	18/180	Southern hemisphere
Sciadopityaceae	Jurassic to recent	1/1	Northern hemisphere
Taxaceae	Jurassic to recent	5/17	Northern hemisphere

Note: distributions are given for extant taxa only.

or mixed conifer/hardwood forests in which conifers occur in low densities, dispersed among other trees or shrubs. Many species occupy very small areas, often as relict populations of once greater abundance. Some areas have a high diversity of species, but hardly any of these species are abundant enough to form forests of any appreciable size. A good example is New Caledonia in the south-west Pacific, an island with 43 species of conifers, all endemic, in an area about the size of Wales. Mexico has 42 species of pines (*Pinus*), compared with eight species in all of Canada and Alaska. About 200 species of conifers are restricted to the southern hemisphere, where vast conifer forests are unknown. It is this scattered diversity that is most threatened with extinction. Families and genera are unevenly distributed and show a number of relictual biogeographic patterns.

Despite the relatively low numbers in comparison with the flowering plants, the conifers are an economically important group of plants.

Vegetative morphology

All conifers are woody perennials with aerial stems. Roots are fibrous to woody and undifferentiated. Shoots are similar in all conifer families, with a leptocaul habit and pycnoxylic secondary xylem. An architecture with strongly differentiated plagiotropic and orthotropic shoots is present in a number of genera in different families, and several genera in the Pinaceae show a differentiation into long and short shoots. Most conifers are evergreen but a few (*Larix* and *Metasequoia*) are deciduous.

Conifer leaves are quite diverse in shape and form, although basic structure is rather uniform and no conifer has megaphyllous or compound leaves. Phyllotaxis may be opposite, verticillate or spiral, and all conifer leaves are simple and without stipules. A single vascular trace enters the leaf from the stele. Axillary buds are present, although frequently vestigial or reduced to small pads of undifferentiated meristematic tissue (Burrows, 1987). Cuticle is characteristically thick, and stomata sunken. Both epidermal and hypodermal cells are frequently lignified. Venation appears parallel and leaf vasculature does not show more than one discrete order, although highly variable, and venation is often characteristic at the family or genus level. When numerous vascular bundles are present, a clear midrib is absent and vasculature is often dichotomously branched. Leaf bases may or may not be persistent, reflecting presence or absence of an abscission layer.

Wood anatomy is relatively uniform across the conifers. Xylem is composed entirely of tracheids and wood is generally parenchyma poor and thus pycnoxylic (dense wood that contains little parenchyma). Tracheids of many conifers have a characteristic circular bordered pitting on element walls that has been used to identify fossil conifer wood to the family level. Both xylem and phloem tissues are characteristic at the family level, with slight differences between the families (Hardin *et al.*, 2001).

Reproductive morphology

Extant conifers are monoecious, with strongly dimorphic, unisexual, strobilate reproductive structures.

Pollen is borne in microsporangia carried on simple microsporophylls lacking subtending bracts, which are arranged in determinate strobili that may be solitary or clustered, and axillary or terminal. Sporophylls carry two or more abaxial microsporangia, which open by slits.

Pollen grains are monosulcate, with somewhat different morphologies according to family. Araucariaceae has spherical pollen with little ornament. Cupressaceae, Podocarpaceae and Pinaceae have saccate pollen with two distinct sacs or bladders that are thought to add buoyancy that assists with wind pollination.

Fertilization in all conifers is by wind. A pollen-drop mechanism is employed to capture pollen, and pollen tubes then grow haustorially to the archegonia in the apical section of the embryo sac. The male gametophyte has non-motile sperm cells.

Megasporangiate structures are somewhat more complex, with each fertile scale representing a modified short shoot with its subtending bract (Florin, 1951). In all extant conifers, the modified shoot is reduced to a single fertile scale-leaf, with one, two or more adaxial ovules (depending on family and genus). The fertile scale may be free from, or wholly or partially fused with, the subtending bract. Bract–scale complexes are borne in usually large, woody strobili, although these may be highly reduced, in some cases to single bract–scale structures that show very little resemblance to the original strobilate structure. Not all conifers produce woody strobili and, in some taxa, ovule-bearing structures are reduced to fleshy, berry-like forms (e.g. *Juniper*, *Taxus*, *Podocarpus*). These structures are autapomorphies derived in relation to seed dispersers. Ovules are unitegmic, and may be erect or inverted.

Endosperm is haploid, derived from the female gametophyte tissue. Seeds are medium to large, dry or fleshy, and winged or wingless. Dispersal is by wind (dry seeds) or by birds or small mammals (fleshy seeds). The fleshy component that functions as an attractant is variously derived, developing from an aril, epimatium or receptacle in different genera. The embryo is straight, with two or more cotyledons, and germination is cryptocotular or phanerocotular.

Modes of perennation range from resprouting to obligate reseeding. Population dynamics of individual species are generally reflected by the dynamics of the host communities: species occurring in rainforests where fires are infrequent are reseeders and show regular recruitment, whereas those in sclerophyllous habitats are either resprouters or reseeders with episodic recruitment patterns responsive to fire regimes. Mast seeding behaviour has also been reported in some subalpine taxa (Gibson *et al.*, 1995). Some episodic reseeders are dependent on fire for seed release (some *Callitris* species; Bowman and Harris, 1995), although most are not. Most display continuous recruitment in undisturbed habitats and episodic recruitment in disturbance-prone habitats.

Distribution and ecology

Many extant conifer taxa show relictual distributions and, at least at the familial and generic level, were far more diverse, widespread and abundant in the past than they are today (Hill, 1995). Distribution and ecology differ in the different families, and will be discussed below.

The conifer families

Pinaceae

Leaves are linear, acicular, spirally arranged, or dimorphic, bract-like and acicular, on specialized short shoots. Male cones are lateral, comprising numerous spirally arranged fertile scales, each scale with two abaxial microsporangia; pollen is winged, with two air sacs (except in *Larix* and *Pseudotsuga*). Female cones are terminal on specialized lateral shoots, comprising numerous spirally arranged imbricate woody or dry fertile scales; each scale has a free bract and two inverted adaxial ovules. Seeds have a single terminal wing. Germination is phanerocotular. Cotyledons 3–18.

A northern hemisphere family of ten genera; several genera include important forest community dominants, many of which are important economic timber sources. Many species are widely cultivated as ornamental plants.

Taxaceae

Pollen cones are axillary and solitary or clustered, with sporophylls bearing 2–16 microsporangia (pollen sacs); pollen is more or less spherical and not winged. Seed cones are reduced to one to two ovules subtended by inconspicuous, decussate bracts. Each 'cone' has one erect ovule, which produces one unwinged seed with a hard seed coat partially or wholly surrounded by a juicy, fleshy or leathery aril. Embryos have two cotyledons.

Five genera and 17 species, mainly northern hemisphere. Usually secondary components of the vegetation.

Cephalotaxaceae

Pollen cones are each subtended by one bract and are aggregated into axillary capitula of six to eight cones, each with 4–16 microsporophylls. Each microsporophyll has two to four (usually three) pollen sacs bearing subspherical, non-saccate pollen. Seed cones are borne from axils of terminal bud scales, with one to six or occasionally up to eight long pedunculate cones per bud. Each cone axis has several pairs of decussate bracts, each bearing two erect, axillary ovules. Seeds ripen in their second year and are drupelike and completely enclosed by a succulent aril. Embryos have two cotyledons, and germination is epigeal.

Cephalotaxaceae differs from Taxaceae in its seed cones, which have several two-ovulate bracts, instead of a single fertile, one-ovulate bract.

One genus and five species, northern hemisphere. Also usually secondary components of the vegetation.

Sciadopityaceae

Shoots are dimorphic (long or short) with leaves of two types, scale leaves on the stem, and photosynthetic leaves at the apex of both long and short shoots. Photosynthetic leaves variously interpreted as a pair of true leaves fused together, or as highly modified shoots (cladodes). Pollen cones are borne in dense terminal clusters. Seed cones are fragile, breaking up soon after seed release. Each cone has 15–40 thin fertile scales, each with five to nine flattened, narrowly winged seeds. Embryos have two cotyledons.

The family was formerly included as a genus within Taxodiaceae (now included in Cupressaceae), but recent genetic studies have shown that it is clearly not allied with that group (Brunsfeld et al., 1994).

A single genus with a single species, Sciadopitys verticillata, endemic in Japan.

Cupressaceae

The Cupressaceae first appear in the fossil record in the Late Triassic (Bock, 1969), although a number of earlier forms placed in the extinct Lebachiaceae have been suggested as possible ancestral Cupressaceae (Miller, 1982). The basal lineages formerly known as Taxodiaceae are well represented in the fossil record of the later Mesozoic and Tertiary; the lineages previously placed in the Cupressaceae in the narrower sense are much less so (Ohsawa, 1994). A number of extant genera of the latter have been recorded as fossils from the later Mesozoic and Tertiary, but these identifications are based on similarites rather than synapomorphies, and must be regarded as doubtful (Thomas and Spicer, 1986). Many fossils at first placed in the narrow Cupressaceae have since been proven to belong to the extinct conifer family Cheirolepidaceae (Miller, 1988).

Male cones are small and comprise opposite, whorled or spirally arranged scales with two to nine microsporangia on the abaxial surface. Pollen is not winged or saccate. Female cones comprise one or more (c. 20) fertile scales, each with a fully fused bract. Cone scales are alternate, opposite or whorled in the same phyllotaxis as the foliage leaves. Fertile scales are imbricate or valvate, persistent and usually woody at maturity (secondarily fleshy in Juniperus and Arceuthos). Each scale has 1–12 erect ovules on the adaxial surface. Seeds are winged or not, and embryos usually have two but rarely up to nine cotyledons.

A family of 30 genera (many monotypic) and about 155 species. Higher taxonomy of this group has been unstable, with the wide recognition of two families, Taxodiaceae and Cupressaceae. Recent studies indicate that these do not represent natural groups, and that the former concept of the Cupressaceae represents one of several lineages of common descent. The remaining lines were aggregated into the Taxodiaceae, creating a paraphyletic assemblage.

Podocarpaceae

Taxonomy of the Podocarpaceae has received a great deal of attention in recent years, and the formerly large and diverse genera Podocarpus and Dacrydium have been extensively split into smaller segregate genera (Quinn, 1970; de Laubenfels, 1978, 1987; Page, 1988; Molloy,

1995). *Phyllocladus* has at times been removed to its own family (Keng, 1978), but more recently has been shown to be nested within Podocarpaceae (Conran *et al.*, 2000).

Podocarpaceae is the first of the extant families to appear in the fossil record, with *Rissikia* from the Early Triassic (Miller, 1977). Many of the extant genera are known as fossils, but almost exclusively from the Tertiary (Hill, 1995). Strong evidence exists for a progressive decline in abundance and diversity of this family throughout the Tertiary, continuing until as recently as the Late Pleistocene (Hill, 1995).

Male cones are made up of numerous, spirally inserted sporophylls, each with two abaxial microsporangia. Pollen grains are saccate. Female cones consist of one to many fleshy or dry but not woody fertile scales with fully adnate bracts, each with one (or two) erect or inverted ovules. Cone scales are persistent or deciduous, and scales and axes are fleshy or dry, not woody at maturity. Germination is phanerocotular and embryos have two cotyledons.

A family of *c*. 18 genera and 180 species, largely southern hemisphere in its distribution, extending through much of Africa, from Japan through Malaysia to Australia and New Zealand, and through Central and South America.

Araucariaceae

Much of the present-day distribution of the family is relictual, although *Araucaria* and *Agathis* have radiated widely in New Caledonia. The family today is essentially Australasian in distribution, with a minor presence in South-east Asia and the western Pacific. Two species of *Araucaria* occur in South America, illustrating another Gondwanan link.

This family appears in the fossil record in the Early Cretaceous, and is abundant worldwide in deposits from the Late Mesozoic (Stockey, 1982). The family became restricted to the southern hemisphere (Gondwana) from the end of the Mesozoic. Araucariaceae is remarkable in that extant sections of the genus *Araucaria* can be recognized in the fossil record as far

back as the Middle Cretaceous. *Agathis*, on the other hand, is only known as fossils from the Tertiary, and from sites geographically close to present-day occurrences. *Wollemia* is today relictual, with an extremely reduced living population. Fossil pollen comparable to *Wollemia* indicates an appearance in the Middle Cretaceous, becoming more widespread and abundant in the Late Cretaceous and Early Tertiary, and persisting in some areas until the Pleistocene (MacPhail *et al.*, 1995).

Male cones are made up of numerous spirally arranged fertile scales, each with 4–9 pendulous abaxial microsporangia. Pollen is unwinged. Female cones are made up of numerous imbricate, spirally arranged, fertile scales, each with a fully or mostly adnate bract and a central, adaxial, inverted ovule. Seeds may be winged or not. Embryos have two or, less commonly, four cotyledons.

A family of three genera and *c*. 38 species, mostly Malaysian and western Pacific in distribution.

Evolution and fossil record

True conifers first appeared in the Carboniferous and increased their abundance, dominance and taxonomic richness during the Permian and into the Mesozoic. Palaeozoic conifers show little resemblance to modern or Mesozoic conifers (Florin, 1950; Rothwell, 1982), and modern families appeared only in the Mesozoic (Table 3.1). Fossil taxa from the Palaeozoic such as *Lebachia* and *Walchia* are regarded, however, as true conifers (Miller, 1982).

Conifers were ecosystem dominants through most of the Mesozoic and reached their pinnacle in species diversity during that time, declining with the diversification of the angiosperms in the Cretaceous and more so in the Tertiary. Many of the Mesozoic conifers in fact represent now-extinct families, for example, Cheirolepidaceae. Many extant taxa are relictual, although recent evolutionary radiations are evident in some genera such as *Pinus*, *Cupressus*, *Callitris* and *Podocarpus*. Many conifers are well represented as fossils and their distribution can clearly document the diminution of biogeographic range through the Tertiary.

Florin (1951, 1954) suggested that the ovuliferous cone scale with its subtending bract of the modern conifer cone was homologous to the reproductive short shoot and its subtending leaf in Cordaites. With this interpretation, the entire axis of ovuliferous *Cordaianthus* can be homologized with the conifer seed cone. Florin argued from this that Cordaites may be the ancestor to the conifers, although the two groups coexisted for a long period in the upper Carboniferous and Permian.

Cordaites (Cordaitales) was abundant in the Late Palaeozoic, and reconstructions suggest it was one of the largest trees of the Carboniferous. The leaves were long (up to 1 m) and strap-like, up to 12–15 cm wide, similar in appearance to some cycad leaflets. The reproductive organs were cones. Modified lateral shoots produced needle-like bracts subtending a short axis bearing overlapping, sterile bracts at the base, and a spiral of sporophylls above. Each microsporophyll usually had six terminal microsporangia exerted just beyond the apex of the cone at maturity. Ovules were produced in similar structures in place of microsporophylls, with one or sometimes more terminal ovules extending beyond the sterile bracts.

In conifer pollen cones, microsporangia are borne on a single bract without a trace of a second subtending structure. This is different from Cordaites in which the microsporangia were borne on a branch subtended by a bract on a branch system that was subtended by a second order of bract. Thus, the homologies between pollen-bearing structures in Cordaites and the conifers are not clear.

Saccate pollen is common among the conifers as well as many other Mesozoic seed plants. Cordaites, glossopterids, *Caytonia*, *Callistophyton* and corystosperms all have this type of pollen. Cycads, *Ginkgo*, peltasperms, gnetophytes and angiosperms lack saccate pollen. The saccate condition has been interpreted as primitive or plesiomorphic within the conifers.

Gnetophyta

The gnetophytes are highly diverse in gross morphology but are clearly shown to be a monophyletic group by numerous molecular studies (Chaw *et al.*, 2000; Donoghue and Doyle, 2000; Nickrent *et al.*, 2000; Pryer *et al.*, 2001).

Present-day occurrence

The group comprises three families, Gnetaceae, Ephedraceae and Welwitschiaceae, each with only a single extant genus (*Gnetum*, *Ephedra* and *Welwitschia*) and 30, 60 and 1 extant species, respectively.

Today, *Gnetum* is a tropical moist-forest plant of both the Old and New Worlds. In contrast, both *Ephedra* and *Welwitschia* are dry climate or desert plants. *Ephedra* contains about 30 species distributed on all continents except Australia and Antarctica, while *Welwitschia* is represented by a single species that is restricted to the Namib Desert in south-west Africa.

Vegetative morphology

Ephedra species are shrubs or occasionally clambering vines with jointed whorled or fascicled branches. Leaves are simple, scale-like, opposite and decussate or whorled, connate at the base forming a sheath, generally ephemeral and mostly not photosynthetic. Leaves are vascularized by a pair of traces that exit from the eustele and enter the leaf. This contrasts with conifers, where leaves are vascularized by a single trace.

Gnetum species are mostly woody climbers, rarely shrubs or trees. Leaves are opposite, simple, elliptic, petiolate, without stipules, pinnately veined with reticulate secondary venation and entire margins. Both *Ephedra* and *Gnetum* possess a leptocaul habit with axillary branching and pycnoxylic secondary xylem similar to the conifers and *Ginkgo*.

Welwitschia is a bizarre plant unlike anything else in the plant kingdom, with a short, woody, unbranched stem producing only two or rarely three strap-shaped leaves that grow from a basal meristem throughout the life of the plant. The leaves contain numerous subparallel veins that may anastomose or terminate blindly in the mesophyll (this character is unique in *Welwitschia* among the gymnosperms). The woody stem widens with age to become a concave disc up to a metre across. The branched reproductive shoots arise from near the leaf bases at the outside edge of the disc.

The gnetophytes show a number of vegetative features that were thought to ally them closely with the angiosperms. The shoot apex meristem is differentated into tunica and corpus. The leaves of *Gnetum* have reticulate venation that bears a striking resemblance to that of some dicot angiosperms, although this has not been shown to be a stricly homologous state. Also like the flowering plants, the gnetophytes possess vessels in their wood. However, gnetophyte vessels have a different developmental origin and thus may not be strictly homologous with those of angiosperms. Similarly, the sieve tube–companion cell associations are now thought to be parallel developments (Thomas and Spicer, 1986).

Reproductive morphology

All three genera are dioecious. Pollen and seeds are borne in complex strobilate structures, and cone, pollen and seed morphology varies among genera. All appear wind pollinated, although there are some suggestions that *Gnetum* may be insect pollinated. Spermatozoioids are non-motile. Ovules are erect and unitegmic or apparently bitegmic.

In *Ephedra*, microsporangiate strobili consist of several pairs of bracts. A shoot, bearing 'bracteoles' (a second level of bracts), arises in the axis of each larger bract. Each bracteole surrounds a stalked microsporophyll bearing two or more pollen sacs (microsporangia). The megasporangiate cone is similarly arranged with pairs of bracts, the top two of which subtend an ovule. The ovule appears to be surrounded by two integuments, although most inter-

pretations regard the inner layer as the true integument and the outer layer as a reduced bracteole similar to that surrounding the microsporangia. The two layers develop into a fleshy, leathery or corky outer layer and a woody inner layer as the seed matures, implying animal dispersal.

Strobili in *Gnetum* are arranged on axes with a conspicuous node–internode organization. Microsporangiate strobili have two fused bracts that form a cupule that surrounds fertile shoots. Each fertile shoot is composed of two fused bracteoles, which surround the microsporophyll. In the megasporangiate strobilus, the cupule subtends a whorl of ovules, which are wrapped in two layers of tissue outside of the integument. These external 'envelopes' of tissue may be sclerified and fused to the integument to form a seed-coat-like structure, which makes the *Gnetum* seed appear very angiosperm-like. The fleshy coat implies animal dispersal.

Pollen cones of *Welwitschia* are red and resemble those of *Ephedra*, appearing in groups of two to three terminally on each branch. Ovulate cones are also red and arise from branched reproductive shoots; each cone consists of a single unitegmic ovule and another layer derived from two confluent primordia (sometimes called a 'perianth') with two 'bracts'. Normally, only one seed develops within each cone; it is dispersed by wind with the 'perianth' as a wing.

Ephedra and *Gnetum* are also similar to angiosperms in displaying a form of double fertilization. In most angiosperms, two sperm cells are involved in fertilization. One unites with the egg cell to form a diploid zygote; the other fuses with the two polar nuclei to form a triploid endosperm. The endosperm serves as nutrition for the developing embryo. In *Ephedra*, two egg cells are produced, one of which is fertilized to become the embryo while the other begins but does not complete development into an embryo (Friedman, 1992). In *Ephedra*, this second embryo does not provide nutrition to the developing embryo as does the endosperm of angiosperms; however, the degenerate embryo is believed to be an intermediary step between the primitive seed plant condition (where the embryo is nourished by a larger gametophyte) and that of angiosperms (megagametophyte so reduced as to not be effective for nourishing the embryo). Another way to think of this is as a reduction in the polyembryony observed in many non-angiosperm seed plants. Polyembryonic cycad seeds have been reported, with several embryos actually capable of germination and growing to maturity. In some conifers, several embryos regularly form from fertilizations by several pollen grains, but only one embryo per seed survives. The multi-embryo condition appears to be plesiomorphic for the seed plants and has been reduced in both the gnetophyte and angiosperm lineages.

Evolution and fossil record

The gnetophytes have long been believed to be the closest living relatives of the flowering plants (Doyle and Donoghue, 1986, 1992; Donoghue and Doyle, 2000), although a number of recent molecular studies dispute this relationship (Winter *et al.*, 1999; Pryer *et al.*, 2001; Rydin *et al.*, 2002; Schmidt and Schneider-Poetsch, 2002; Soltis *et al.*, 2002). These studies do not, however, agree on the position of the *Gnetophyta* in the evolution of the seed plants, and this remains at present an unresolved question. The fossil record is meagre, with records of pollen grains that resemble *Ephedra* and *Welwitschia* from the Triassic and Cretaceous. *Welwitschia*-like fossil cones are known from the Late Triassic (Cornet, 1996). Fossil sporophyll structures associated with ephedroid pollen are known from the Jurassic (van Konijnenburg-van Cittert, 1992), and may represent gnetophyte progenitors. Leaf fossils with *Gnetum*-like venation patterns are known from the Early Cretaceous (Crane and Upchurch, 1987). This age is compatible with the placement of the Gnetophyta as sister to the angiosperms rather than in a monophyletic Gymnospermae.

References

Beck, C.B. (ed.) (1976) *Origin and Early Evolution of Angiosperms*. Cambridge University Press, Cambridge.

Bock, W. (1969) The American Triassic flora and global distribution. *Geological Center Research Series*, Volumes 3 and 4. Geological Center, North Wales, Pennsylvania.

Bowe, L.M., Coat, G. and dePamphilis, C.W. (2000) Phylogeny of seed plants based on all three genomic compartments: extant gymnosperms are monophyletic and Gnetales' closest relatives are conifers. *Proceedings of the National Academy of Sciences USA* 97, 4092–4097.

Bowman, D.M.J.S. and Harris, S. (1995) Conifers of Australia's dry forests and open woodlands. In: Enright, N.J. and Hill, R.S. (eds) *Ecology of the Southern Conifers*. Melbourne University Press, Melbourne, pp. 252–270.

Bremer, K., Humphries, C.J., Mishler, B.D. and Churchill, S.P. (1987) On cladistic relationships in green plants. *Taxon* 36, 339–349.

Brunsfeld, S.J., Soltis, P.E., Soltis, D.E., Gadek, P.A., Quinn, C.J., Strenge, D.D. and Ranker, T.A. (1994) Phylogenetic relationships among the genera of Taxodiaceae and Cupressaceae: evidence from *rbcL* sequences. *Systematic Botany* 19, 253–262.

Burbidge, A.H. and Whelan, R.J. (1982) Seed dispersal in a cycad, *Macrozamia riedlei*. *Australian Journal of Ecology* 7, 63–67.

Burrows, G.E. (1987) Leaf axil anatomy in the Araucariaceae. *Australian Journal of Botany* 35, 631–640.

Carpenter, R.J. (1991) Macrozamia from the early Tertiary of Tasmania and a study of the cuticles of extant species. *Australian Systematic Botany* 4, 433–444.

Chamberlain, C.J. (1935) *Gymnosperms. Structure and Evolution*. University of Chicago Press, Chicago, Illinois.

Chase, M.W., Soltis, D.E., Olmstead, R.G., Morgan, D., Les, D.H., Mishler, B.D., Duvall, M.R., Price, R.A., Hills, H.G., Qiu, Y.-L., Kron, K.A., Rettig, J.H., Conti, E., Palmer, J.D., Manhart, J.R., Sytsma, K.J., Michaels, H.J., Kress, W.J., Karol, K.G., Clark, W.D., Hendren, M., Gaut, B.S., Jansen, K.R., Kim, K.-J., Wimpee, C.F., Smith, J.F., Furnier, G.R., Strauss, S.H., Xiang, Q.-Y., Plunkett, G.M., Soltis, P.S., Swensen, S.M., Williams, S.E., Gadek, P.A., Quinn, C.J., Eguiarte, L.E., Golenburg, E., Learn, G.H., Graham, S.W., Barrett, S.C.H., Dayanandan, S. and Albert, V.A. (1993) Phylogenetics of seed plants: an analysis of nucleotide sequences from the plastid gene *rbcL*. *Annals of the Missouri Botanical Garden* 80, 528–580.

Chaw, S.-M., Zharkikh, A., Sung, H.-M., Lau, T.-C. and Li, W.-H. (1997) Molecular phylogeny of extant gymnosperms and seed plant evolution: analysis of nuclear 18S rRNA sequences. *Molecular Biology and Evolution* 4, 56–68.

Chaw, S.-M., Parkinson, C.L., Cheng, Y., Vincent, T.M. and Palmer, J.D. (2000) Seed plant phylogeny inferred from all three plant genomes: monophyly of extant gymnosperms and origin of Gnetales from conifers. *Proceedings of the National Academy of Sciences USA* 97, 4086–4086.

Cleal, C. (1989) Angiosperm phylogeny: evolution in hidden forests. *Nature* 339, 16.

Conran, J.G., Wood, G.A., Martin, P.G., Dowd, J.M., Quinn, C.J., Gadek, P.A. and Price, R.A. (2000) Generic relationships within and between the gymnosperm families Podocarpaceae and Phyllocladaceae based on an analysis of the CP gene *rbcL*. *Australian Systematic Botany* 48(6), 715–724.

Cookson, I.C. (1953) On *Macrozamia hopeites*: an early Tertiary cycad from eastern Australia. *Phytomorphology* 3, 306–312.

Cornet, B. (1996) A new gnetophyte from the late Carnian (Late Triassic) of Texas and its bearing on the origin of the angiosperm carpel and stamen. In: Taylor, D.W. and Hickey, L.J. (eds) *Angiosperm Origin, Evolution and Phylogeny*. Chapman & Hall, New York, pp. 32–67.

Crane, P.R. (1985a) Phylogenetic relationships in seed plants. *Cladistics* 1, 329–385.

Crane, P.R. (1985b) Phylogenetic analysis of seed plants and the origin of angiosperms. *Annals of the Missouri Botanical Garden* 72, 716–793.

Crane, P.R. (1988) Major clades and relationships in the higher gymnosperms. In: Beck, C.B. (ed.) *Origin and Evolution of the Gymnosperms*. Columbia University Press, New York, pp. 218–272.

Crane, P.R. and Upchurch, G.R. Jr (1987) *Drewria potomacensis* gen. et sp. nov., an Early Cretaceous member of Gnetales from the Potomac Group of Virginia. *American Journal of Botany* 74, 1723–1738.

de Laubenfels, D.J. (1978) The genus *Prumnopitys* (Podocarpaceae) in Malesia. *Blumea* 24, 189–190.

de Laubenfels, D.J. (1987) Revision of the genus *Nageia* (Podocarpaceae). *Blumea* 32, 209–211.

Dehgan, B. and Yuen, C.K.K.H. (1983) Seed morphology in relation to dispersal, evolution and propagation of *Cycas* L. *Botanical Gazette* 144, 412–418.

Delevoryas, T. (1993) Origin, evolution and growth patterns in cycads. In: Stevenson, D.W. (ed.) *Proceedings of CYCAD 90, the Second International Conference on Cycad Biology*. Palm and Cycad Societies of Australia Limited, Brisbane, pp. 236–245.

Donaldson, J.S., Nanni, I. and de Wet Bosenberg, J. (1995) Beetle pollination of *Encephalartos cycadifolius*. In: Vorster, P. (ed.) *Proceedings of the Third International Conference on Cycad Biology*. The Cycad Society of South Africa, Stellenbosch, pp. 423–434.

Donoghue, M.J. (1994) Progress and prospects in reconstructing plant phylogeny. *Annals of the Missouri Botanical Garden* 81, 405–418.

Donoghue, M.J. and Doyle, J.A. (2000) Seed plant phylogeny: demise of the anthophyte hypothesis? *Current Biology* 10, R106–R109.

Doyle, J.A. (1996) Seed plant phylogeny and the relationships of Gnetales. *International Journal of Plant Science* 157(6), Supplement S3–S39.

Doyle, J.A. (1998a) Molecules, morphology, fossils and the relationship of angiosperms and Gnetales. *Molecular Phylogenetics and Evolution* 9, 448–462.

Doyle, J.A. (1998b) Phylogeny of vascular plants. *Annual Review of Ecology and Systematics* 29, 567–599.

Doyle, J.A. and Donoghue, J.M. (1986) Seed plant phylogeny and the origin of the angiosperms: an experimental cladistic approach. *Botanical Review* 52, 331–429.

Doyle, J.A. and Donoghue, J.M. (1992) Fossils and plant phylogeny reanalyzed. *Brittonia* 44, 89–106.

Farjon, A. and Page, C.N. (2001) *Conifers Status Survey and Conservation Action Plan*. IUCN, Gland, Switzerland.

Florin, R. (1939) The morphology of the female fructifications in Cordaites and conifers of Palaeozoic age. *Botaniska Notiser* 36, 547–565.

Florin, R. (1950) Upper Carboniferous and Lower Permian conifers. *Botanical Review* 16, 258–282.

Florin, R. (1951) Evolution in Cordaites and Conifers. *Acta Horti Bergiani* 15, 285–388.

Florin, R. (1954) The female reproductive organs of conifers and taxads. *Biological Reviews* 29, 367–389.

Foster, A.S. and Gifford, E.M. Jr (1989) *Comparative Morphology of Vascular Plants*. W.H. Freeman and Company, San Francisco, California.

Friedman, W.E. (1992) Evidence of a pre-angiosperm origin of endosperm: implications for the evolution of flowering plants. *Science* 255, 336–339.

Friis, E.M., Chaloner, W.G. and Crane, P.R. (eds) (1987) *The Origins of Angiosperms and their Biological Consequences.* Cambridge University Press, Cambridge, UK.

Frohlich, M.W. and Meyerwitz, E.M. (1997) The search for flower homeotic gene homologs in basal angiosperms and Gnetales: a potential new source of data on the evolutionary origin of flowers. *International Journal of Plant Science* 158, S131–S142.

Fu, L.-G., Nan, L. and Mill, R.R. (1999) Ginkgoaceae. In: Wu, Z.-Y. and Raven, P.H. (eds) *Flora of China,* Vol. 4. Science Press, Beijing; Missouri Botanical Garden, St Louis, Missouri, p. 8.

Gao, Z.-F. and Thomas, B.A. (1989) A review of fossil cycad megasporophylls, with new evidence of Crossozamia Pomel and its associated leaves from the lower Permian of Taiyuan, China. *Review of Palaeobotany and Palynology* 60, 205–223.

Gibson, N., Barker, P.C.J., Cullen, P.J. and Shapcott, A. (1995) Conifers of southern Australia. In: Enright, N.J. and Hill, R.S. (eds) *Ecology of the Southern Conifers.* Melbourne University Press, Melbourne, pp. 252–270.

Goremykin, V.V., Bobrova, V.K., Pahnke, J., Troitsky, A.V., Antonov, A.S. and Martin, W. (1996) Noncoding sequences from the slowly evolving chloroplast inverted repeat in addition to *rbcL* data do not support gnetalean affinities of Angiosperms. *Molecular Biology and Evolution* 13, 383–396.

Gregory, T.J. and Chemnick, J. (2004) Hypotheses on the relationship between biogeography and speciation in Dioon (Zamiaceae). In: Walters, T. and Osborne, R. (eds) *Cycad Classification Concepts and Recommendations.* CAB International, Wallingford, UK, pp. 137–148.

Guppy, H.B. (1906) *Observations of a Naturalist in the Pacific.* Macmillan, London.

Hardin, J.W., Harding, J.W. and White, F.M. (2001) *Textbook of Dendrology.* McGraw-Hill, New York.

Harris, T.M. (1976) The Mesozoic gymnosperms. *Review of Palaeobotany and Palynology* 21, 119–134.

Hasebe, M., Kofuji, R., Ito, M., Kato, K., Iwatsuki, M. and Ueda, K. (1992) Phylogeny of gymnosperms inferred from *rbcL* gene sequences. *Botanical Magazine (Tokyo)* 105, 673–679.

Hill, C.R. and Crane, P.R. (1982) Evolutionary cladistics and the origin of angiosperms. In: Joysey, K.A. and Friday, E.A. (eds) *Problems of Phylogenetic Reconstruction.* Academic Press, London, pp. 269–361.

Hill, K.D. (1994) The *Cycas rumphii* complex (Cycadaceae) in New Guinea and the western Pacific. *Australian Systematic Botany* 7, 543–567.

Hill, K.D., Chase, M.W., Stevenson, D.W., Hills, H.G. and Schutzman, B. (2003) The families and genera of cycads: a molecular phylogenetic analysis of cycadophyta based on nuclear and plastid DNA sequences. *International Journal of Plant Science* 164(6), 933–948.

Hill, R.S. (1978) Two new species of *Bowenia* Hook. ex Hook. F. from the Eocene of eastern Australia. *Australian Journal of Botany* 26, 837–846.

Hill, R.S. (1980) Three new Eocene cycads from eastern Australia. *Australian Journal of Botany* 28, 105–122.

Hill, R.S. (1995) Conifer origin, evolution and diversification in the Southern Hemisphere. In: Enright, N.J. and Hill, R.S. (eds) *Ecology of the Southern Conifers.* Melbourne University Press, Melbourne, pp. 10–29.

Judd, W.S., Campbell, C.S., Kellogg, E.A., Stevens, P.F. and Donoghue, M.J. (2002) *Plant Systematics: a Phylogenetic Approach,* 2nd edn. Sinauer Associates, Sunderland, Massachusetts.

Keng, H. (1978) The genus *Phyllocladus* (Phyllocladaceae). *Journal of the Arnold Arboretum* 59, 249–273.

Lewin, R. (1988) A lopsided look at evolution. *Science* 241, 291.

Lindley, J. (1830) *An Introduction to the Natural System of Botany.* Longman & Co., London.

Liu, Y.S., Zhou, Z.Y. and Li, H.M. (1991) First discovery of *Cycas* fossil leaf in northeast China. *Science Bulletin* 22, 1758–1759.

Loconte, H. and Stevenson, D.W. (1990) Cladistics of the Spermatophyta. *Brittonia* 42, 197–211.

MacPhail, M., Hill, K., Partridge, A., Truswell, E. and Foster, C. (1995) 'Wollemi Pine' – old pollen records for a newly discovered genus of gymnosperm. *Geology Today* 11(2), 48–50.

Meyen, S. (1981) Ginkgo: a living pteridosperm. *IOP Newsletter* 15, 6–11.

Miller, C.N. (1977) Mesozoic conifers. *Botanical Review* 43, 218–280.

Miller, C.N. (1982) Current status of Palaeozoic and Mesozoic conifers. *Review of Palaeobotany and Palynology* 37, 99–114.

Miller, C.N. (1988) The origin of modern conifer families. In: Beck, C.B. (ed.) *Origin and Evolution of Gymnosperms.* Columbia University Press, New York, pp. 448–486.

Molloy, B.J.P. (1995) *Manoao* (Podocarpaceae), a new monotypic conifer genus endemic to New Zealand. *New Zealand Journal of Botany* 33, 183–201.

Nathanielsz, C.P. and Staff, I.A. (1975) On the occurrence of intra cellular blue-green algae in cortical cells of apogeotropic roots of *Macrozamia communis. Annals of Botany* 39, 363–368.

Nickrent, D.L., Parkinson, C.L., Palmer, J.D. and Duff, R.J. (2000) Multigene phylogeny of land plants with special reference to bryophytes and the earliest land plants. *Molecular Biology and Evolution* 17, 1885–1895.

Nixon, K.C., Crepet, W.L., Stevenson, D.W. and Friis, E.M. (1994) A reevaluation of seed plant phylogeny. *Annals of the Missouri Botanical Garden* 81, 484–533.

Norstog, K.J. and Nichols, T.J. (1997) *The Biology of the Cycads.* Cornell University Press, Ithaca, New York.

Norstog, K.J., Stevenson, D.W. and Niklas, K.J. (1986) The role of beetles in the pollination of *Zamia furfuracea* L. fil. (Zamiaceae). *Biotropica* 18, 300–306.

Ohsawa, T. (1994) Anatomy and relationships of petrified seed cones of the Cupressaceae, Taxodiaceae and Sciadopityaceae. *Journal of Plant Research* 107, 503–512.

Page, C.N. (1988) New and maintained genera in the conifer families Podocarpaceae and Pinaceae. *Notes from the Royal Botanic Gardens, Edinburgh* 45, 377–395.

Page, C.N. (1990) The families and genera of vascular plants, Vol. 1: Pteridophytes and gymnosperms. In: Kramer, K.U. and Green, P.S. (series eds) *The Families and Genera of Vascular Plants.* Springer-Verlag, Berlin.

Parenti, L.R. (1980) A phylogenetic analysis of land plants. *Journal of the Linnaean Society, Biology* 13, 225–242.

Pellmyr, O., Tang, W., Groth, I., Bergstrom, G. and Thien, L.B. (1991) Cycad cone and angiosperm floral volatiles: inferences for the evolution of insect pollination. *Biochemical Systematics and Ecology* 19, 623–627.

Pryer, K.M., Schneider, H., Smith, A.R., Cranfill, R., Wolf, P.G., Hunt, J.S. and Sipes, S.D. (2001) Horsetails and ferns are a monophyletic group and the closest living relatives to seed plants. *Nature* 409, 618–622.

Qiu, Y.-L. and Palmer, J.D. (1999) Phylogeny of early land plants: insights from genes and genomes. *Trends in Plant Science* 4, 26–30.

Qiu, Y.-L., Lee, J., Bernasconi-Quadroni, F., Soltis, D.E., Soltis, P.S., Zanis, M., Zimmer, E.A., Chen, Z., Savolainen, V. and Chase, M.W. (1999) The earliest angiosperms: evidence from mitochondrial, plastid and nuclear genomes. *Nature* 402, 404–407.

Quinn, C.J. (1970) Generic boundaries in the Podocarpaceae. *Proceedings of the Linnaean Society of New South Wales* 94, 166–172.

Rai, H.S., O'Brien, H.E., Reeves, P.A., Olmstead, R.G. and Graham, S.W. (2003) Inference of higher-order relationships in the cycads from a large chloroplast data set. *Molecular Phylogenetics and Evolution* 29(2), 350–359.

Raubeson, L.A. (1998) Chloroplast DNA structural similarities between conifers and gnetales: coincidence of common ancestry. *American Journal of Botany* (suppl.) 85, 149 [abstract].

Raubeson, L.A. and Jansen, R.K. (1992) A rare chloroplast-DNA mutation shared by all conifers. *Biochemical Systematics and Ecology* 20, 17–24.

Raven, P.H., Evert, R.F. and Eichhorn, S.E. (1992) *Biology of Plants,* 5th edn. Worth Publishers, New York.

Rothwell, G.W. (1982) New interpretations of the earliest conifers. *Review of Palaeobotany and Palynology* 37, 7–28.

Rothwell, G.W. and Serbet, R. (1994) Lignophyte phylogeny and the evolution of spermatophytes: a numerical cladistic analysis. *Systematic Botany* 19, 443–482.

Rothwell, G.W., Scheckler, S.E. and Gillespie, W.H. (1989) *Elkinsia* gen. nov., a late Devonian gymnosperm with cupulate ovules. *Botanical Gazette* 150, 170–189.

Rydin, C., Källersjö, M. and Friis, E.M. (2002) Seed plant relationships and the systematic position of gnetales based on nuclear and chloroplast DNA: conflicting data, rooting problems, and the monophyly of conifers. *International Journal of Plant Science* 163, 197–214.

Schmidt, M. and Schneider-Poetsch, H.A.W. (2002) The evolution of gymnosperms redrawn by phytochrome genes: the Gnetatae appear at the base of the gymnosperms. *Journal of Molecular Evolution* 54, 715–724.

Soltis, D.E., Soltis, P.A. and Zanis, M.J. (2002) Phylogeny of seed plants based on evidence from eight genes. *Journal of Botany* 89, 1670–1681.

Stefanovic, S., Jager, M., Deutsch, J., Broutin, J. and Masselot, M. (1998) Phylogenetic relationships of conifers inferred from partial 28S rRNA gene sequences. *American Journal of Botany* 85, 688–697.

Stevenson, D.W. (1980) Observations on root and stem contraction in cycads (Cycadales) with special reference to *Zamia pumila* L. *Botanical Journal of the Linnaean Society (London)* 81, 275–281.

Stevenson, D.W. (1981) Observations on ptyxis, phenology, and trichomes in the Cycadales and their systematic implications. *Journal of Botany* 68, 1104–1114.

Stevenson, D.W. (1990) Morphology and systematics of the Cycadales. *Memoirs of the New York Botanical Garden* 57, 8–55.

Stevenson, D.W. (1992) A formal classification of the extant cycads. *Brittonia* 44, 220–223.

Stevenson, D.W., Norstog, K.J. and Fawcet, P.K.S. (1998) Pollination biology of cycads. In: Owens, S.J. and Rudall, P.J. (eds) *Reproductive Biology*. Royal Botanic Gardens, Kew, UK, pp. 277–294.

Stewart, W.N. and Rothwell, G.W. (1993) *Paleobotany and the Evolution of Plants*. Cambridge University Press, New York.

Stockey, R.A. (1982) The Araucariaceae: an evolutionary perspective. *Review of Palaeobotany and Palynology* 37, 133–154.

Tang, W. (1987) Insect pollination in the cycad *Zamia pumila* (Zamiaceae). *American Journal of Botany* 74, 90–99.

Taylor, T.N. and Taylor, E.L. (1993) *The Biology and Evolution of Fossil Plants*. Prentice Hall, New Jersey.

Thomas, B. and Spicer, R.A. (1986) *Evolution & Palaeobiology of Land Plants*. Croom Helm, London.

Tralau, H. (1968) Evolutionary trends in the genus *Ginkgo*. *Lethaia* 1, 63–101.

van Konijnenburg-van Cittert, J.H.A. (1992) An enigmatic Liassic microsporophyll, yielding Ephedripites pollen. *Review of Palaeobotany and Palynology* 71, 239–254.

Winter, K.-U., Becker, A., Munster, T., Kim, J.T., Saedler, H. and Thiessen, G. (1999) The MADS-box genes reveal that gnetophytes are more closely related to conifers than to flowering plants. *Proceedings of the National Academy of Sciences USA* 96, 7342–7347.

Yokoyama, M. (1911) Some tertiary fossils from the Mippe coalfields. *Journal of the College of Sciences Imperial University of Tokyo* 20, 1–16.

4 Chloroplast genomes of plants

Linda A. Raubeson[1] and Robert K. Jansen[2]

[1]*Department of Biological Sciences, Central Washington University, Ellensburg, WA 98926-7537, USA;* [2]*Integrative Biology, University of Texas, Austin, TX 78712-0253, USA*

Introduction

The chloroplast is descended from a formerly free-living bacterium. Thus the chloroplast genome is eubacterial and as such is circular, attached to the inner organellar membrane, unassociated with proteins, and uses (at least in part) gene regulatory and replication machinery similar to that characterized in the model organism, *Escherichia coli*. Chloroplasts are thought to have descended from one primary endosymbiotic event, where the free-living bacterium was engulfed and enslaved by a host eukaryotic cell (Douglas, 1998; McFadden, 2001; Moreira and Phillipe, 2001; although see Palmer, 2003, and Delwiche and Palmer, 1997, for a consideration of controversies concerning the monophyly of plastids; and Stiller *et al*., 2003, for an opposing viewpoint). Over time a complex genetic symbiosis developed that involved the loss of genes from the chloroplast genome, with any essential genes being transferred to the nucleus. In order for this transfer to result in a protein that is functional in the organelle, the nuclear copy first must be expressed with the product then targeted back to the chloroplast. Apparently the original transition from an independent bacterium to an organelle involved such a complex series of events as

to occur only once in evolutionary history. Three major lineages of primary endosymbionts are extant: red, green and a small 'primitive' group, the glaucocystophytes. For information on algal chloroplast genomes the reader is referred to Palmer and Delwiche (1998) and Simpson and Stern (2002) as well as the references above. The focus of this chapter will be the chloroplast genome of the land plants, a derived group within the green lineage.

Sugiura (2003) has recently published a concise history of work on the chloroplast genome. To summarize here: through the 1950s and early 1960s scientists demonstrated that the chloroplast contained its own unique genome. Then attention turned to characterizing the chloroplast DNA (cpDNA) in various plants using restriction site mapping, electron microscopy, and other techniques of the times. The first completely sequenced chloroplast genomes (Table 4.1) were tobacco (*Nicotiana*, Fig. 4.1; Shinozaki *et al*., 1986) and liverwort (*Marchantia*; Ohyama *et al*., 1986). Since that time, many additional genomes have been completely sequenced (Table 4.1), techniques have advanced, and molecular biological research has progressed to functional genomics.

As molecular techniques have progressed and basic knowledge of the chloroplast

Table 4.1. Completely sequenced land plant chloroplast genomes (as of 1 November 2003).

Higher classification (informal)	Subclass (for angiosperms)	Family	Species (common name)	Accession number (GenBank)	Publication
Bryophytes		Marchantiaceae	*Marchantia polymorpha* (liverwort)	X04465	Ohyama et al. (1986)
		Anthocerotaceae	*Anthoceros formosae* (hornwort)	AB086179	Kugita et al. (2003a)
		Funariaceae	*Physcomitrella patens* (moss)	AP005672	Sugiura et al. (2003)
Vascular plants Pteridophytes Fern ally		Psilotaceae	*Psilotum nudum* (whisk fern)	AP004638	Wakasugi et al., unpublished
Fern		Adiantaceae	*Adiantum capillus-veneris* (maiden-hair fern)	AY178864	Wolf et al. (2003)
Seed plants Gymnosperms Conifers		Pinaceae	*Pinus thunbergii* (black pine)	D17510	Wakasugi et al. (1994)
		Pinaceae	*Pinus koraiensis* (Korean pine)	NC_004677	Noh et al., unpublished
Angiosperms Dicots	Magnoliidae	Calycanthaceae	*Calycanthus fertilis*	AJ428413	Goremykin et al. (2003b)
	Magnoliidae	Amborellaceae	*Amborella trichopoda*	AJ506156	Goremykin et al. (2003a)
	Caryophyllidae	Chenopodiaceae	*Spinacia oleracea* (spinach)	AJ400848	Schmitz-Linneweber et al. (2001)
	Dilleniidae	Brassicaceae	*Arabidopsis thaliana*	AP000423	Sato et al. (1999)
	Rosidae	Onagraceae	*Oenothera elata* (evening primrose)	AJ271079	Hupfer et al. (2000)
	Rosidae	Fabaceae	*Lotus corniculatus* (lotus)	AP002983	Kato et al. (2000)
	Rosidae	Fabaceae	*Medicago truncatula* (lucerne)	AC093544	Lin et al., unpublished

	Asteridae	Scrophulariaceae	*Epifagus virginiana* (beech drops)	M81884	Wolfe *et al.* (1992)
	Asteridae	Solanaceae	*Nicotiana tabacum* (tobacco)	Z00044	Shinozaki *et al.* (1986)
	Asteridae	Solanaceae	*Atropa belladonna* (belladonna)	AJ316582	Schmitz-Linneweber *et al.* (2002)
Monocots	Commelinidae	Poaceae	*Triticum aestivum* (wheat)	AB042240	Ogihara *et al.* (2002)
	Commelinidae	Poaceae	*Zea mays* (maize)	X86563	Maier *et al.* (1995)
	Commelinidae	Poaceae	*Oryza sativa* (rice)	X15901	Hiratsuka *et al.* (1989)

Updated list of completely sequenced chloroplast genomes (for land plants and algae) can be found at http://megasun.bch.umontreal.ca/ogmp/projects/other/cp_list.html

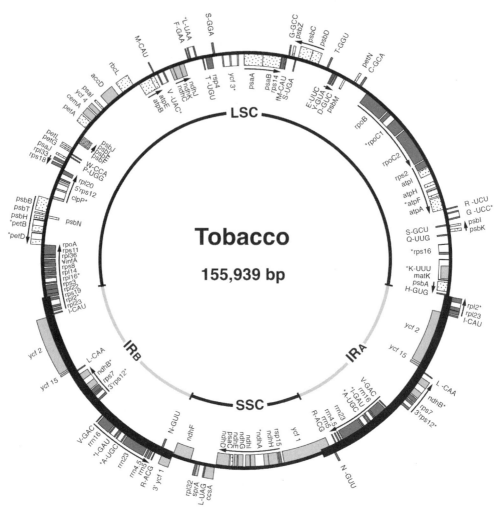

Fig. 4.1. Gene map of tobacco chloroplast genome (adapted from Wakasugi *et al.*, 1998). The inner circle shows the four major regions of the genome – the two copies of the inverted repeat (IRA and IRB) and the large and small single-copy regions (LSC and SSC). The outer circle represents the tobacco genome with the transcribed regions shown as boxes proportional to gene size. Genes shown on the inside of the circle are transcribed in a clockwise direction, whereas genes on the outside of the circle are transcribed anti-clockwise. The IR extent is also shown by the increased width of the circle representing the tobacco genome. Genes with introns are marked with asterisks (*). Arrows between the gene boxes and gene names show those operons known to occur in tobacco cpDNA. Other operons could be present. Genes coding for products that function in protein synthesis are darker grey; genes coding for products that function in photosynthesis are stippled; genes coding for products with various other functions are lighter grey.

genome has expanded, evolutionary biologists have been able to extend the use of cpDNA in comparative studies. Such studies have contributed to the understanding of mutational processes operating in chloroplast genomes as well as providing data for phylogenetic purposes. The chloroplast genome has been utilized more than any other plant genome as a marker for investigating plant evolution and diversity due to its many advantages. Because of the genome's small size (generally 120–160 kilobase pairs, kbp) and high copy number (as many as 1000 per cell), it is relatively straightforward to isolate and characterize cpDNA. Plus, the conservative nature of the genome allows for the use of DNA probes from even distantly related species and the design of 'universal' primers. In addition, the genome is a good phylogenetic marker because rates of nucleotide change (while overall being slower than in the nuclear genome) show a range of rates making different parts of the genome appropriate for different levels of comparison, rare changes in gene order can be informative even at deep phylogenetic levels, and the usual pattern of uniparental inheritance and lack of recombination simplify analysis.

The use of cpDNA in phylogenetic studies dates back to the early 1980s when a few plant biologists used the genome to address species relationships in several groups of crop plants by comparing fragment patterns of purified cpDNA digested with restriction enzymes (Palmer and Zamir, 1982; Bowman et al., 1983; Clegg et al., 1984; Hosaka et al., 1984). Later studies compared restriction site changes via filter hybridization at higher taxonomic levels (e.g. Sytsma and Gottlieb, 1986; Jansen and Palmer, 1988). Some studies also mapped gene order and used rearrangements to address evolutionary relationships (e.g. Jansen and Palmer, 1987a; Raubeson and Jansen, 1992a,b). More recently the vast majority of phylogenetic and systematic studies have employed sequence data for cpDNA-based phylogenetic comparisons. The first sequencing studies (Doebley et al., 1990; Soltis et al., 1990) utilized the large subunit of ribulose 1,5-bisphosphate carboxylase/oxygenase (rbcL). This has been the most widely sequenced chloroplast gene and the emphasis on this gene culminated in a multi-authored study involving 499 species of seed plants (Chase et al., 1993). Many other individual chloroplast genes and intergenic regions have now been utilized (reviewed in Soltis and Soltis, 1998). Several recent studies have used ten or more protein-coding genes from partially or completely sequenced chloroplast genomes to estimate phylogenetic relationships of plants (e.g. Graham and Olmstead, 2000; Lemieux et al., 2000; Martin et al., 2002).

For the remainder of this chapter, we will focus on two aspects of the plant chloroplast genome: (i) its organization and evolution; and (ii) the phylogenetic utility of different approaches to cpDNA characterization.

Organization and Evolution of Land Plant Chloroplast Genomes

Chloroplast genome organization is highly conserved within land plants. Most land plant genomes have a quadripartite structure with two copies of a large inverted repeat (IR) separating two single copy regions (refer to the inner circle of Fig. 4.1). As the two regions of unique genes are of unequal size, these regions are referred to as the large and small single copy regions (LSC and SSC, respectively). Land plant cpDNAs usually contain 110–130 different genes. The majority of these genes (about 80; see Table 4.1) code for proteins, mostly involved in photosynthesis or gene expression; the remainder are transfer RNA (about 30) or ribosomal RNA (4) genes. Most chloroplast genes are part of polycistronic transcription units (Fig. 4.1; Palmer, 1991; Mullet, 1993); that is, they occur in operons where the genes within each operon are under the control of the same promoter. Often operons contain multiple promoters that allow transcription of a subset of genes within the operon (e.g. Miyagi et al., 1998; Kuroda and Maliga, 2002). Some polycistronic transcripts are subject to cis- (within the same transcript) or trans- (between different transcripts) splicing or both. For example (Hubschmann et al., 1996), the rps12 gene exists as three exons. The 5′ exon occurs as part of the clpP operon. The remaining two exons occur (in a quite distant location) together in an operon with ndhB. In the construction of the mature rps12 mRNA, the intron between exons 2 and 3 is removed and the exons joined (cis-splicing) and exon 1 (from the separate operon) is joined to exons 2 and 3 (trans-splicing). Both group I and group II types of self-splicing introns are found in land plant cpDNAs; the majority are group II (Palmer, 1991). Although intron content is quite variable in algal genomes, it is conserved in land plants. Nineteen of the 20 introns found in the tobacco genome (Wakasugi *et al.*, 1998) occur also in the hornwort, *Anthoceros*, genome (Kugita *et al.*, 2003a); the only intron not shared between the two genomes occurs in *rps16*, a gene that is absent from the hornwort genome (and other non-seed plant cpDNAs).

Although the gene content and organization of the chloroplast genome is evolutionarily conservative, changes do occur. In the remainder of this section we will consider in more detail three classes of gene order changes in the land plant chloroplast genome: gene and intron loss; inverted repeat changes; and inversions. These types of mutations are often referred to as structural changes or rearrangements in this literature.

Gene and intron loss

As mentioned above, the loss of genetic information from the chloroplast genome has been a general pattern over evolutionary time since the original free-living prokaryote was first engulfed. Most genes were lost early in the process of endosymbiosis (Martin *et al.*, 1998); however, some loss events have continued to occur during land plant diversification. Most losses within land plants are restricted to individual lineages; thus, gene content is mostly shared among chloroplast genomes. For example,

78 of 81 protein-coding genes found in the tobacco genome also occur in the genome of liverwort (this and additional comparisons are shown in Table 4.2).

Loss events can occur if genes are not essential to organismal function or if a functional copy of the gene can be transferred to the nucleus. The entire gene could be lost simply by deletion in a single mutational event but more likely point mutations make the gene non-functional and then the pseudogene gradually decays until it is no longer recognizable in the genome. For instance, in the conifer *Pinus thunbergii* chloroplast (Wakasugi *et al.*, 1994), the 11 *ndh* genes are lost (four) or present only as pseudogenes (seven). The *ndh* genes are homologues of the mitochondrial NADH dehydrogenase genes. In organisms with functional chloroplast copies, the gene products are active within the chloroplast; for example, expression increases when plants are under oxidative stress (Casano *et al.*, 2001). In a second example of the loss of multiple related genes, none of the genes involved in photosynthesis is present in the highly reduced chloroplast genome of the non-photosynthetic plant *Epifagus* (Wolfe *et al.*, 1992). In a third example, angiosperms *Welwitschia* and *Psilotum* have all lost (presumably independently) copies of *chlL*, *chlN* and *chlB* from their chloroplasts (Burke *et al.*, 1993). These three genes encode the three subunits of a protein that allows chlorophyll to mature in the absence of light. A separate nuclear-encoded protein

Table 4.2. Distribution of protein-coding genes in a bryophyte (*Marchantia*), a conifer (*Pinus*), a dicot (*Nicotiana*) and a monocot (*Zea*). Calculated from Table 5 of Martin *et al.* (2002). Bold values on the diagonal indicate the number of genes in each genome. Above the diagonal, the number of genes shared between genomes is shown. Below the diagonal (and following the slash on the diagonal) are the number of unique genes found in a single genome or found in only one pair of genomes.

	Marchantia	*Pinus*	*Nicotiana*	*Zea*
Marchantia	**86**/4	72	78	75
Pinus	5	**73**/0	67	64
Nicotiana	0	0	**81**/0	78
Zea	0	1	3	**79**/0

matures chlorophyll but only in the presence of light. Because angiosperms are unable to green (i.e. mature chlorophyll) in the dark we can infer a simple loss of the *chl* genes from the chloroplast without a functional transfer to the nucleus. In a final example, an exceptional case involving a single gene, *infA* (coding for a translation initiation factor), has been lost multiple times (about 24) within angiosperms with an estimated four independent functional transfers of the gene to the nucleus (Millen *et al.*, 2001).

Introns are also lost from within genes in the chloroplast. Recombination of the processed mRNA with the genomic DNA is probably the mechanism responsible (Palmer, 1991). As long as the intron is precisely removed this mutation would be selectively neutral, unless regulatory or other functional elements are contained within the intron sequence. Some instances of intron loss have occurred repeatedly. For example, the *rpoC1* intron has been lost independently in grasses (Katayama and Ogihara, 1996) and in a subfamily of Cactaceae (Wallace and Cota, 1996), as well as a minimum of four additional times within dicots (Downie *et al.*, 1996). These gene and intron loss events, which can occur multiple times independently, might be locally useful characters for phylogenetic inference; however, they must be treated cautiously in broader comparisons.

Inverted repeat

As mentioned above, the vast majority of land plant chloroplast genomes contain a large duplicated region, the inverted repeat, where the two copies are reverse complements of each other. The genes that form the core of the repeat encode the ribosomal RNAs (23S, 16S, 5S and 4.5S). This rDNA-containing IR appears to be an ancestral genomic feature as it is found in charophytes, basal green algae and some red algae (Turmel *et al.*, 1999). Gene content other than the rDNA varies. Genes, formerly single copy at the boundaries of the single copy regions, can be duplicated and incor-

porated into the IR. Within land plants, the length of the IR has 'grown' from 10 kbp in liverwort to 25 kbp in tobacco (Palmer and Stein, 1986). Most, but not all, examples of gene duplication within the chloroplast genome occur through IR expansion.

The underlying mechanism of IR expansion is not well understood. Gene conversion is thought to be involved since the existing DNA sequence is used as the template for forming the new copy. Evidence supports gene conversion or copy correction acting on the chloroplast genome. Rates of nucleotide substitution are reduced in the IR relative to single-copy regions (Palmer, 1991) and in the same genes when in the IR versus when single copy (Perry and Wolfe, 2002). Where examined the two copies of the repeat are identical (Palmer, 1991). Presumably, these patterns (of rate and identity) occur because large amounts of homologous recombination (and copy correction) take place between the two copies of the repeat. So much recombination occurs in fact that two different versions of the genome are present (with opposite orientations of the small and large single copy regions relative to one another) in equamolar quantities (Palmer, 1983; Stein *et al.*, 1986). Minor changes, of about 100 base pairs (bp) or less, in the endpoints of the IR are probably relatively common and gene conversion alone is an adequate explanation (Goulding *et al.*, 1996). However, this mechanism does not account for the fact that, in some cases, no existing material is lost. Thus major changes (those incorporating one or multiple genes into the IR) must be explained by additional mechanisms such as double reciprocal recombination (Palmer *et al.*, 1985; Yamada, 1991) or double-stranded break repair (Goulding *et al.*, 1996) combined with gene conversion.

An increase in the length of the IR is much more common than its decrease, although decrease is easier to explain mechanistically by deletion. Only two accounts of significant decrease in IR gene content are published: in the Apiaceae where a series of sequential deletion forms has been characterized (Plunkett and Downie, 2000) and in *Cuscuta*, where a probable 7 kbp contraction has occurred (Bömmer *et al.*, 1993). Increase, in contrast, has been exemplified on many occasions: the growth into the LSC within land plants mentioned previously (Palmer and Stein, 1986; Raubeson, 1991); a phylogenetically informative addition of about 11.5 kbp of the LSC within Berberidaceae (Kim and Jansen, 1998); incorporation of SSC genes into the IR shared among families within the Campanulales (Knox and Palmer, 1999); and a spectacular expansion of the IR to 75 kbp in *Pelargonium* (Geraniaceae; Palmer *et al.*, 1987b) among others. Where the IR has grown extensively beyond the boundaries seen in tobacco, such as in *Pelargonium* (Price *et al.*, 1990) and also in the Campanulaceae (Cosner *et al.*, 1997), the growth is associated with multiple additional changes in genome organization.

It has been suggested that the presence of the IR promotes stability (i.e. reduces gene order changes) in the remainder of the molecule (Palmer *et al.*, 1987a). Reasons why the IR may facilitate gene order conservation include: (i) that enzymes mediating recombination are active at the IR, leaving few copies of these enzymes available to modify other parts of the molecule; or (ii) that the interactions between the two IR copies physically hold the SSC and LSC in a more open orientation, diminishing the likelihood that portions interact and recombine. (For a more detailed discussion and additional possible reasons, see Palmer, 1991.) The correlation is not perfect, but most genomes without the IR or with a greatly enlarged IR have unusually high numbers of changes in gene order. Perhaps the same mechanism that promotes gene order changes also promotes changes in extent of the IR, or perhaps there is some stability provided to the molecule by the 'normal' IR.

In a few lineages the IR has been lost; one copy has been eliminated leaving each gene only in the retained copy. A complete loss of the IR is known or suspected from the chloroplast genomes of some members of the legume family, two members of the Geraniaceae (Price *et al.*, 1990), *Conophilis* (a non-photosynthetic plant in the Orobanchaceae; Downie and Palmer, 1992) and *Striga* (Scrophulariaceae; Palmer, 1991),

whereas an almost-complete loss is known from conifers (Tsudzuki *et al.*, 1992). Thus, on a minimum of five independent occasions the IR has been lost from land plant chloroplast genomes. At times the legume loss and the conifer loss have been equated and thus seen as an example of homoplasy. However, the two cases differ in the extent of the loss, in the gene content of the IR prior to loss, and in the copy of the IR that is lost in the event (Fig. 4.2). The loss of the IR defines six tribes of legumes (Fabaceae; Palmer *et al.*, 1987b; Lavin *et al.*, 1990), whereas the conifer loss event supports conifer monophyly (Raubeson and Jansen, 1992a). In both of these instances, other gene order changes usually co-occur with the loss of the IR.

Inversions

So far we have discussed deletion of genetic information in the context of gene loss, intron loss and IR loss or contraction as well as the addition of information in the context of IR expansion. To conclude this section on genome organization we will discuss changes in gene order and orientation within the genome. The most common mechanism leading to gene order change in the chloroplast genome is inversion, where a section of the genome is reversed in order and orientation relative to the remainder of the genome (Palmer, 1991). Inversions can occur via homologous recombination between small inverted repeats or through double-stranded break repair (Palmer, 1991). In discussing the nature and utility of inversion characters, it is important to distinguish scale. Inversions commonly occur within non-coding regions of cpDNA sequence where small repeats associated with hairpin or stem–loop structures provide foci for inversions (Kelchner, 2000). These small-scale (*c.* 2–200 bp) inversions may be very prone to homoplasy and complicate interpretation of non-coding sequence in phylogenetic studies (Kelchner, 2000). However, large-scale changes where inversions reverse the order and orientation of multiple genes have different characteristics that will be discussed below.

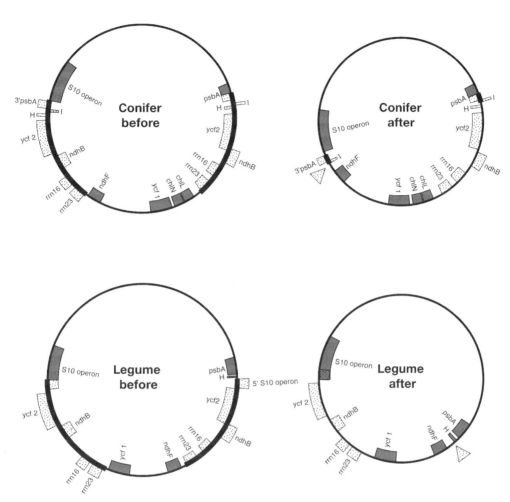

Fig. 4.2. Inverted repeat (IR) loss in conifer and legume chloroplast genomes. Genes in the IR prior to the loss events are shown stippled in the 'before' circles. Genes in the remaining copy are still stippled in the 'after' circles and the site of the lost copy is shown as a stippled triangle. Only selected genes are shown and distance between the genes is not to scale. Extent of the IR is shown as a bar along the genome circle. Note that the extent of the IR differs between 'conifer before' and 'legume before'; *trnH, trnI* and 3' *psbA* are duplicated in the conifer but not the legume, whereas in the legume a portion of the S10 operon is in the IR but is single copy in the conifer. A comparison of the two 'after' circles will reveal a difference in the extent of the loss; the conifer loss is partial (*trnI* and 3' *psbA* remain as a small remnant IR), whereas in the legume the loss is complete.

The majority of chloroplast genomes that have been characterized and compared lack any changes in gene order (Palmer, 1991; Downie and Palmer, 1992). It appears that, in most lineages, gene order remains unchanged over vast periods of evolutionary time. For example, *Amborella* (in several phylogenetic studies, the most basal extant angiosperm) and tobacco (a derived asterid dicot) cpDNAs have an identical gene order. However, in some lineages, changes do occur (last reviewed in Downie and Palmer, 1992). Most inversions occur with the endpoints in non-coding regions so that no genes are disrupted, and only on rare occasions are operons split (Palmer, 1991). Additionally, both endpoints of inversions usually occur within the LSC, perhaps simply because this region is largest and therefore has the most regions that could serve as endpoints that would not disrupt genes or operons. In a few instances, inversions have occurred that have one endpoint in the LSC and the other in the IR. Most such inversions are probably lethal because they result in direct repeats that would promote deletion of genes within the inversion. The few junction-spanning inversions that occur (e.g. in the leptosporangiate ferns (Stein *et al.*, 1992; Raubeson and Stein, 1995), buckwheat (Ali *et al.*, 1997) and adzuki bean (Perry *et al.*, 2002)) are associated with expansion of the IR. Incorporation of the inversion into the IR would eliminate the potentially disruptive direct repeats.

Even a small number of inversions, depending on their distribution, can serve as powerful phylogenetic markers. For example, a single inversion identified the basal members of the Asteraceae (as will be discussed in more detail later in this chapter). In a second case, a 30-kbp inversion (first recognized as a difference between liverwort and tobacco) was found to occur in all vascular plant cpDNAs except those of lycopsids (Fig. 4.3), marking lycopsids as basal lineage of vascular plants (Raubeson and Jansen, 1992b). Although somewhat controversial at the time, sequence-based studies since (e.g. Nickrent *et al.*, 2000; Pryer *et al.*, 2001) have supported this basal position of the lycopsids. Additionally, two inversions and an expansion of the IR clarify basal nodes in leptosporangiate ferns (Stein *et al.*, 1992; Raubeson and Stein, 1995) and informative inversions have been characterized in legumes (Lavin *et al.*, 1990). Of three inversions shared throughout the Poaceae, one is restricted to the family, one is shared with Joinvilleaceae and one is shared with Joinvilleaceae and Restoniaceae, thus clarifying the sister groups of the grasses (Doyle *et al.*, 1992).

Few cases have been published comparing genomes of taxa among which numerous inversions co-occur. In one of the earliest such studies, Hoot and Palmer (1994) generated restriction site and mapping data for members of the Ranunculaceae. The distribution of two of the inversion characters was incongruent with the most parsimonious trees based on the restriction site characters. Hoot and Palmer suggested that the conflicting inversions occurred in parallel. It is possible for identical inversions to occur independently if they form due to homologous recombination across the same repeat structure. However, repetitive sequences are uncommon in land plant chloroplast genomes (Palmer, 1991). Also, it is unclear whether, in general, inversions occur because of repeats or whether repeats occur because of inversions. For example, during double-stranded break repair it is common for small duplicated segments of DNA (filler DNA) to be inserted at the site of the break (Gorbunova and Levy, 1999). This may explain why repeats, including the duplication of transfer RNA genes or portions of larger genes, are associated with the endpoints of inversions (or other rearrangements). Even where repeats occur in genomes and many inversions have occurred, independent occurrence of identical inversions is unusual. In a study of the Campanulaceae, over 40 inversions occur in a data set of 18 taxa with very little homoplasy evident in the data (Cosner, 1993).

Thus, as rare and complex genomic changes, inversions are especially useful phylogenetic markers (Rokas and Holland, 2000). Of course no characters are perfect and these, as any other, should be carefully investigated and interpreted in the light of all other evidence. In the remainder of the chapter we will more explicitly compare the utility of three types of cpDNA data in phylogenetic studies: restriction site polymorphisms, rearrangement characters and nucleotide sequence data.

Phylogenetic Utility of cpDNA Data

Chloroplast genomes have been characterized for studies of plant diversity using three different approaches: restriction fragment/site comparisons, structural rearrangements, and sequencing of genes or non-coding regions. The utility of these various approaches has been reviewed in detail in several papers (Palmer *et al.*, 1988; Downie and Palmer,

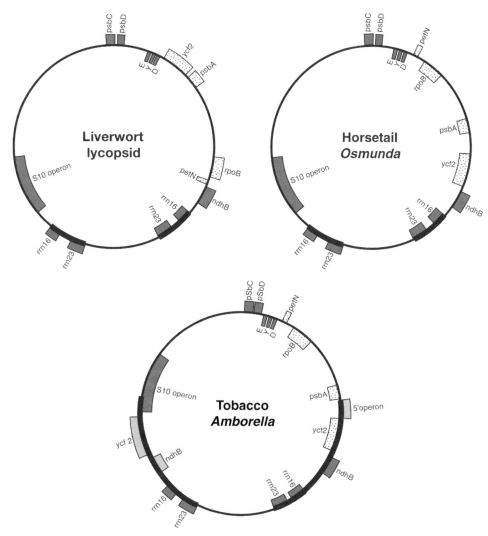

Fig. 4.3. Inversion distribution in land plants. Bryophytes (e.g. liverwort) and lycopsids have the ancestral gene order for land plants, whereas horsetails and the fern *Osmunda* differ only in the orientation of the 30-kbp region, shown stippled. Angiosperm (represented by tobacco, *Amborella*) cpDNAs have the 30-kbp inversion plus the further modification of additional genes incorporated into the IR (shown in lighter grey). Only selected genes are shown as landmarks. Distance between the genes is not to scale. Extent of the IR is shown as a bar along the genome circle.

1992; Doyle, 1993; Olmstead and Palmer, 1994; Jansen *et al.*, 1998; Soltis and Soltis, 1998; Graham *et al.*, 2000; Rokas and Holland, 2000). Below we briefly review each of the three primary methods for using cpDNA and provide examples of their use. We end by discussing the relative utility of these three approaches for reconstructing the phylogenetic history of plants.

Restriction fragment/site comparisons

Until about 1998, restriction enzyme approaches were the most widely used techniques for estimating phylogenetic relationships among plants (see Jansen *et al.*, 1998). The earliest applications of this approach used highly purified cpDNA (e.g. Palmer and Zamir, 1982). By the late 1980s, researchers utilized total genomic DNA from which the cpDNA could be visualized by Southern hybridization (reviewed in Palmer *et al.*, 1988). The latter approach overcame the difficulties of isolating sufficient quantities of pure cpDNA, allowed for a better assessment of the homology of restriction fragments and has been used in numerous studies from a wide diversity of plants at a wide range of taxonomic levels (reviewed in Jansen *et al.*, 1998). At lower taxonomic levels where cpDNA variation is generally quite low (< 1%) it has been possible to estimate fragment homology with a great deal of confidence by simple inspection of fragment patterns. At higher levels of sequence divergence, the more labour-intensive mapping of restriction sites was essential to accurately assess character homology. Two advantages (Givnish and Sytsma, 1997; Jansen *et al.*, 1998) of the restriction site mapping approach over DNA sequencing of individual genes are that: (i) by using a large number of enzymes that recognize sequences scattered throughout the entire chloroplast genome it is possible to gather a very large number of phylogenetically informative characters; and (ii) comparisons of cpDNA sequences of individual genes and whole chloroplast genome restriction site studies suggest that the restriction site data exhibit less homoplasy than DNA sequences.

To illustrate the power of this approach, we will describe an early study of *Heterogaura* (Sytsma and Gottlieb, 1986) because of its historical importance and the surprising results that it uncovered. *Heterogaura* is a monotypic genus in the evening primrose family (Onagraceae) that has been considered distinct since 1866. It is closely related to *Clarkia*, a genus that has served as a model system for studies of speciation in plants, but differs from *Clarkia* in several features, especially floral morphology. Sytsma and Gottlieb (1986) mapped sites for 29 restriction enzymes for eight species of *Heterogaura* and *Clarkia* using the Southern hybridization approach. They surveyed 605 restriction sites and found 119 variable sites, 55 of which were shared by two or more taxa (i.e. were parsimony informative). Phylogenetic analyses of these data generated a single most parsimonious tree with a consistency index of 0.95 (Fig. 4.4). Surprisingly, the genus *Heterogaura* was nested within *Clarkia*, sister to *Clarkia dudleyana*. These results clearly indicated that the morphologically distinct *Heterogaura* should be merged with *Clarkia* and that previous morphological comparisons were misleading with regard to the relationships in this group.

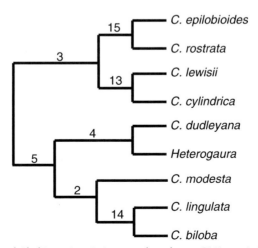

Fig. 4.4. Phylogenetic tree of *Clarkia* section *Peripetasma* based on cpDNA restriction site data. Numbers above nodes indicate number of restriction site changes. Adapted from Systma and Gottieb (1986).

The results of the *Heterogaura* study are particularly noteworthy for several reasons. First, the Onagraceae was viewed as one of the best-studied angiosperm families, so it was shocking to learn that a genus that was recognized as distinct for 120 years was nested within the well-studied genus *Clarkia*. Second, *Clarkia* was considered a model system for studying speciation in plants so it was surprising that such a novel result could have gone undetected by previous workers. Third, this was one of the earliest studies that used this approach in plant systematics, and, in combination with several other early studies (e.g. Sytsma and Schaal, 1985; Coates and Cullis, 1987; Jansen and Palmer, 1988), it set the stage for a rapid surge in the use of cpDNA restriction site comparisons for examining plant diversity.

Structural rearrangements

The second approach for using chloroplast genomes for reconstructing phylogenies of plants involves major structural rearrangements. As stated earlier, the overall structure of the chloroplast genome is highly conserved among land plants and major structural changes, including inversions, deletions of genes and introns, expansion/contraction of the inverted repeat, and loss of the inverted repeat, are relatively uncommon events. In most cases, cpDNA structural rearrangements have little or no homoplasy making them excellent characters for phylogenetic analysis (Palmer *et al.*, 1988). Some types of changes, such as gene and intron losses and expansion and contraction of the inverted repeat, have occurred multiple times (as discussed above) but others, especially inversions, have virtually no homoplasy (Soltis and Soltis, 1998). Here we will focus on inversions, which, because of their rare occurrence and low levels of homoplasy, make especially robust phylogenetic indicators, especially for deep nodes in the phylogeny of plants. Early approaches for examining cpDNA structure involved the very labour-intensive method of constructing restriction site and gene maps. In some studies, other faster methods were developed to

survey for the distribution of structural rearrangements that were initially identified by gene mapping. This included the hybridization of cloned cpDNA fragments that spanned the rearrangement endpoints (e.g. Jansen and Palmer, 1987a) or polymerase chain reaction (PCR) using primers that closely flank rearrangement endpoints (e.g. Doyle *et al.*, 1996). Hundreds of chloroplast genomes were mapped using the Southern hybridization approach (see Table 1 in Downie and Palmer, 1992) and, in a number of the groups investigated, structural changes of various types were detected. In most cases, only one or a few structural rearrangements occurred (e.g. in the angiosperm families Asteraceae, Fabaceae, Poaceae; see Table 2 in Downie and Palmer, 1992). However, there were several plant groups in which the chloroplast genomes were highly rearranged (i.e. conifers and the angiosperm families Campanulaceae, Geraniaceae and Lobeliaceae).

One of the most notable examples, demonstrating the powerful utility of cpDNA rearrangements for phylogenetic studies in plants, comes from the angiosperm family Asteraceae. We describe this example for three reasons: (i) it was the first, extensive study to demonstrate the power of cpDNA rearrangements for assessing relationships among deep nodes; (ii) it resolved a long-standing controversy regarding the identification of the basal lineage of this large, extensively studied family; and (iii) the surprising result obtained from the cpDNA rearrangement generated considerable controversy among angiosperm systematists but was later confirmed by multiple lines of evidence.

The Asteraceae (composite or daisy family) is one of the largest flowering-plant families, with approximately 1535 genera and 25,000 species (Bremer, 1994). Although the family has been the focus of numerous studies, considerable controversy existed about the identity of the basal lineage. Five of the 16 recognized tribes of Asteraceae had been suggested as being ancestral based on morphological, biogeographical and chemical evidence. Comparative restriction site and gene mapping studies by Jansen and Palmer

(1987b) identified a 22-kbp inversion (Fig. 4.5a) in the large single-copy region of the chloroplast genome of lettuce (*Lactuca sativa*, tribe Lactuceae), although the inversion was absent from the chloroplast genome of *Barnadesia caryophylla* (tribe Mutisieae). Eighty species representing all tribes of Asteraceae and ten related families were examined for the distribution of this inversion using cloned cpDNA fragments that spanned the inversion endpoints (Jansen and Palmer, 1987a). The results showed that all related families and members of the subtribe Barnadesiinae of the tribe Mutisieae lack the inversion, whereas all other members of the Asteraceae have this structural change (Fig. 4.5b). This suggested that the Barnadesiinae were basal in the Asteraceae and that the cpDNA inversion marks an ancient evolutionary split in the family. The relationships inferred from the inversion distribution were later confirmed by morphological (Bremer, 1987) and DNA sequence data (Kim *et al.*, 1992; Kim and Jansen, 1995; Jansen and Kim, 1996). The implications of this finding were very significant in altering the classification of the Asteraceae (Barnadesiinae were elevated to subfamilial status) and improving our understanding of the biogeography and character evolution in the family.

More recently, complete chloroplast genome sequences have been generated for 31 taxa, including 19 land plants (Table 4.1). These data are facilitating the exploration of

(a)

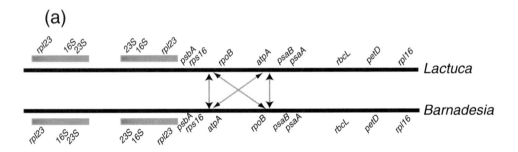

(b)

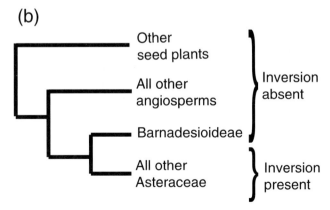

Fig. 4.5. (a) Comparison of chloroplast genome organization between two members of the Asteraceae (*Barnadesia* and *Lactuca*) that differ by a 22-kbp inversion. Grey bars indicate the extent of the inverted repeats. Arrows show the region in which inversion occurs. (b) Phylogenetic tree showing the taxonomic distribution of the 22-kbp inversion. Adapted from Jansen and Palmer (1987b).

structural rearrangements in phylogeny reconstruction at the deepest nodes of the plant evolutionary tree. Ongoing genomic sequencing projects by our group and others will greatly expand this sampling during the next several years, with a special emphasis on sequencing genomes from all of the major lineages of green plants. The accumulation of more chloroplast genome sequences from a wider diversity of taxa will make it necessary to develop better computational methods to deal with gene order characters for phylogeny reconstruction. Considerable work has already been done in this area (Cosner *et al.*, 2000; Moret *et al.*, 2001; Wang *et al.*, 2002; Bourque and Pevzner, 2002), but additional research is needed in order to analyse more highly rearranged genomes than are sequenced currently.

DNA sequencing

The third approach for comparing chloroplast genomes in studies of plant phylogenies is DNA sequencing. Early sequence-based phylogenetic studies were hampered by several factors, including the need to clone genes being sequenced, the lack of universal primers and the labour-intensive nature and high expense of manual DNA sequencing. Thus, many early studies suffered from limited taxon sampling and the use of inappropriate genes (i.e. ones with inadequate levels of variation). The advent of PCR technology and automated DNA sequencing has made it possible to sequence many more taxa and to explore the utility of additional chloroplast genes. Once these new methods increased the capacity for comparative chloroplast DNA sequencing projects, genes could be selected based on criteria of appropriateness rather than simple logistics. The gene sequenced should exhibit the appropriate amount of variation for the taxonomic level and group being studied.

In general, nucleotide substitution rates in cpDNA are slower than those of the nuclear genome and faster than those of the mitochondrial DNA (Wolfe *et al.*, 1987). Substitution rates are lower in the IR where

they equal rates of mitochondrial single-copy genes (Gaut, 1998). Whereas some studies have found the rates of nucleotide substitution in chloroplast-coding regions more conservative than those in non-coding regions (see references in Kelchner, 2000), other studies have suggested that some non-coding regions may evolve no faster than coding regions in the chloroplast genome (e.g. Manen and Natali, 1995). Lineage effects in nucleotide substitution rates have been detected; for example, grasses have an elevated rate relative to tobacco or pine (Muse and Gaut, 1997). Locus-dependent differences in rates of non-synonymous substitution also occur in some genes (Gaut *et al.*, 1997; Muse and Gaut, 1997; Matsuoka *et al.*, 2002). RNA editing has been detected in land plant cpDNAs (Miyamoto *et al.*, 2002; Sabater *et al.*, 2002; Kugita *et al.*, 2003b) and rate of DNA change is usually accelerated in genes with editing of transcripts (Shields and Wolfe, 1997; Bock, 2000).

When comparing DNA sequence in a phylogenetic study, too little variation results in trees that are highly unresolved and there are numerous examples of this in the literature. Too much variation can result in excessively high levels of homoplasy leading to suspect relationships, a problem encountered (as just one example) in a phylogenetic analysis of *rbcL* among all land plants and their green algal relatives (Manhart, 1994). Too much variation can also lead to difficulties in alignment, caused by high rates of nucleotide substitution or the presence of many deletions and insertions (indels). In general, coding regions tend to be more easily aligned because indels must be in multiples of three to maintain functionality of the genes. The increased use of many more variable intergenic regions and introns has caused considerable difficulty in alignment of these chloroplast sequences (see Kelchner, 2000, for a review).

There are numerous examples of the application of cpDNA sequences for estimating phylogenetic studies of plants. One notable example is the analysis of 499 *rbcL* sequences from seed plants (Chase *et al.*, 1993). At the time of its publication this study represented the largest data set of DNA sequences for any group of organisms. Forty-two scientists contributed to the analysis, making the study one of the most amazing examples of cooperation among the systematics community. The resulting phylogenies defined many major clades of flowering plants, which formed a set of hypotheses of relationships that could be tested with other data sets (Fig. 4.6). And finally, the computational phylogenetics community has utilized this data set extensively for evaluating many issues, including the effects of taxon sampling on the accuracy of phylogeny reconstruction (Hillis, 1998) and the development of faster parsimony methods to handle such large data sets (Rice *et al.*, 1997; Nixon, 1999).

The 499-taxon *rbcL* data set generated a well-resolved tree with some very important implications for resolving the major clades of seed plants (Fig. 4.6; readers should refer to Chase *et al.* (1993) for the details of relationships within these major clades of seed plants). Some notable results included the placement of the Gnetales as the sister group of the flowering plants, the position of the aquatic genus *Ceratophyllum* in a basal position in angiosperms, the division of angiosperms into two major groups corresponding to those taxa with uniaperturate and triaperturate pollen, the occurrence of the Magnoliidae as a polyphyletic group at the base of the angiosperms, and the

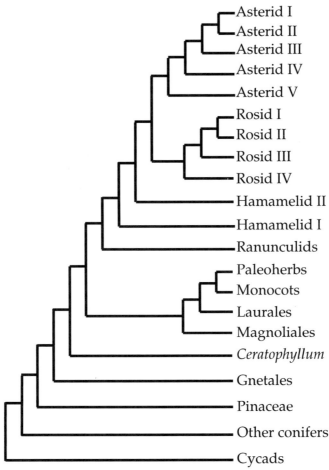

Fig. 4.6. Phylogenetic tree of angiosperms based on *rbcL* sequences. This tree illustrates the major clades that were part of the analysis including 499 species. Adapted from Chase *et al.* (1993).

presence of several large clades that correspond with the major recognized subclasses of angiosperms. Although more recent molecular phylogenies based on other genes and/or different phylogenetic methods have suggested alternative relationships for some of these groups, many of the major clades, especially within angiosperms, have been confirmed (Soltis *et al.*, 2000).

The number of chloroplast genes available for sequence-based studies has grown rapidly with the widespread use of PCR and automated sequencing (Table 4.3). The most widely sequenced coding regions include *rbcL*, *atpB*, *matK* and *ndhF*. Large data sets have been examined for all angiosperms, or for large clades within angiosperms, for each of the genes or combinations of these genes. These analyses have provided many new insights into phylogenetic relationships in a wide diversity of plants from the earliest land plants to the most derived clades of angiosperms. There are too many results from phylogenetic studies of plants using these genes to summarize here so we refer the reader to some of the recent papers using these four genes at various taxonomic levels (Olmstead and Palmer, 1994; Soltis and Soltis, 1998; Graham and Olmstead, 2000). Two of these genes, *atpB* and *rbcL*, are highly conserved in nucleotide sequence and have been most useful for assessing relationships among the major lineages of plants. The other two, *matK* and *ndhF*, provide two to four times more phylogenetically informative characters and, therefore, have been used to examine relationships among more recently diverged taxa in the families Acanthaceae, Asteraceae, Brassicaceae, Orchidaceae, Poaceae, Polemoniaceae, Saxifragaceae, Scrophulariaceae and Solanaceae (reviewed in Soltis and Soltis, 1998). At lower taxonomic levels many systematists have utilized sequences from introns and intergenic regions to reconstruct phylogenies (Table 4.3). Although this approach has been successful in many instances, there are a number of problems associated with using these markers. A number of molecular mechanisms (indels, secondary structure, slipped-strand mispairing and localized intramolecular recombination) can make it difficult to align these sequences (see Kelchner, 2000, for a detailed review).

Comparative utility of these approaches

All three of the approaches for using chloroplast DNA data for studies of plant diversification have been shown to be very valuable for resolving phylogenetic relationships. The first approach, restriction site/fragment comparisons, is not widely used anymore, primarily because it is now much easier to generate DNA sequence data than it was in the past. The primary advantage of this approach was the capacity to examine many restriction sites (scattered sequences of 4–8 bp) from throughout the genome. However, it is now possible to easily sequence numerous genes (or regions) or even to undertake relatively automated sequencing of entire chloroplast genomes. The other two types of characters, structural rearrangements and DNA sequences will continue to be widely used in the future. Structural characters are extremely valuable for assessing relationships at the deepest nodes as they exhibit less homoplasy than DNA sequence data. The rapid increase in the availability of complete genome sequences will provide more of these types of characters in the future. The only limitation of structural changes is that there are many fewer characters available; however, this should not discourage plant systematists from utilizing these characters where they are present. Chloroplast DNA sequence data certainly will continue to be widely used for reconstructing the phylogeny of plants. The systematics community is moving away from relying on one or a few chloroplast genes or regions as evidenced by the increase in the number of papers using multiple sets of genes or whole genomes (Turmel *et al.*, 1999; Graham and Olmstead, 2000; Lemieux *et al.*, 2000; Martin *et al.*, 2002; Maul *et al.*, 2002; Rai *et al.*, 2003).

Summary

We have reviewed aspects of chloroplast genome diversity and evolution in land plants, especially with regard to their phylogenetic utility. In general, the molecule is evolutionarily conservative in both struc-

Table 4.3. Chloroplast genes and regions used for phylogenetic studies of plants arranged by the number of sequences in GenBank (>100) as of 28 July 2003. LSC = large single copy region; SSC = small single copy region; IR = inverted repeat (gene position, if variable, given for tobacco); NA = not applicable. Gene and intergenic region length is in base pairs using the tobacco genome as a reference.

Gene or region	Location	Length	Number in GenBank	Protein product
rbcL	LSC	1,434	17,656	Rubisco – large subunit
trnL-trnF spacer	LSC	358	10,938	NA
trnL intron	LSC	503	9,951	NA
matK	LSC	1,530	8,572	Maturase within trnK intron
atpB	LSC	1,497	5,475	ATP synthase beta subunit
ndhF	SSC	2,133	4,422	NADH dehydrogenase ND5
rps16 intron	LSC	860	2,324	NA
atpB-rbcL spacer	LSC	711	2,302	NA
psbA	LSC	1,062	1,515	Photosystem II 32 kDa protein
rpoB	LSC	3,213	1,494	RNA polymerase beta subunit
trnT-trnL spacer	LSC	711	1,441	NA
atpA	LSC	1,524	1,290	ATP synthase alpha subunit
rpl16 intron	LSC	1,020	1,269	NA
rpoC1	LSC	2,046	870	RNA polymerase beta' subunit
psbA-trnH spacer	LSC	454	768	NA
psaA	LSC	2,253	520	Photosystem I P700 apoprotein A1
rpoA	LSC	1,014	447	RNA polymerase alpha subunit
psaB	LSC	2,205	399	Photosystem I P700 apoprotein A2
rps7	IR	468	387	Ribosomal protein S7
psbB	LSC	1,527	379	Photosystem II 47 kDa protein
rpl2	IR	1,491	318	Ribosomal protein L2
petB	LSC	1,401	303	Cytochrome b6/f apoprotein
ndhI	SSC	504	302	NADH dehydrogenase subunit I
psbC	LSC	1,386	283	Photosystem II 44 kDa protein
psbE	LSC	252	283	Photosystem II 8 kDa subunit
petD	LSC	1,225	266	Cytochrome b/f complex subunit IV
rpoC2	LSC	4,179	260	RNA polymerase beta' subunit
psbT	LSC	105	255	Photosystem II T-protein
psbF	LSC	120	248	Photosystem II 4 kDa protein
psbL	LSC	117	237	Photosystem II L-protein
psbJ	LSC	123	231	Photosystem II J-protein
psbN	LSC	132	212	Photosystem II N-protein
psbH	LSC	222	205	Photosystem II 10 kDa phosphoprotein
ndhD	SSC	1,503	198	NADH dehydrogenase subunit D
ndhG	SSC	531	177	NADH dehydrogenase subunit G
psbD	LSC	1,062	162	Photosystem II D2 protein
ndhB	IR	2,212	160	NADH dehydrogenase subunit B

ture (gene order and organization) and rates of nucleotide substitution. However, structural rearrangements (e.g. gene loss, IR loss or expansion, inversions) do occur in some lineages. These changes can be used to infer relationships of plant lineages and suggest types and patterns of mutational processes affecting gene organization in the chloroplast genome. In addition, changes at the nucleotide level can also be used to reconstruct phylogenies of plant groups. The vast majority of phylogenetic studies utilize cpDNA characteristics as markers. Most evolutionary studies compare nucleotide sequence data, although historically comparing restriction digest patterns was important. Where available, structural changes have provided some important insights into the evolution of land plants. Nuclear and mitochondrial markers are likely to be better utilized by the plant systematics community in the future; however, the chloroplast genome will continue to be important due to its logistical advantages and its evolutionary characteristics.

Acknowledgements

We thank Steve Wagner and our students (Rhiannon Peery, Nichole Fine, Melissa Phillips, Tim Chumley, Andy Alverson, Stacia Wyman, Josh McDill, Aneke Padolina, Ruth Timme, Elizabeth Ruck, Mike Moore and Mary Guisinger) for reading and commenting on an earlier version of our manuscript. NSF has provided continued funding for our work on chloroplast genomes (currently, RUI/DEB0075700 to LAR and DEB0120709 to RKJ and LAR). Figures 4.1, 4.2 and 4.3 were prepared by Gwen Gage.

References

Ali, J., Kishima, Y., Mikami, T. and Adachi, T. (1997) Expansion of the IR in the chloroplast genomes of buckwheat species is due to incorporation of an SSC sequence that could be mediated by an inversion. *Current Genetics* 31, 276–279.

Bock, R. (2000) Sense from nonsense: how the genetic information of chloroplasts is altered by RNA editing. *Biochimie* 82, 549–557.

Bömmer, D., Haberhausen, G. and Zetsche, K. (1993) A large deletion in the plastid DNA of the holoparasitic flowering plant *Cuscuta relexa* concerning two ribosomal proteins (*rpl2, rpl23*), on transfer RNA (*trnI*) and an ORF2280 homologue. *Current Genetics* 24, 171–176.

Bourque, G. and Pevzner, P.A. (2002) Genome-scale evolution: reconstructing gene order in the ancestral species. *Genome Research* 12, 26–36.

Bowman, C.M., Bonnard, G. and Dyer, K. (1983) Chloroplast DNA variation between species of *Triticum* and *Aegilops*. Location of variation on the chloroplast genome and its relevance to the inheritance and classification of the cytoplasm. *Theoretical and Applied Genetics* 65, 247–262.

Bremer, K. (1987) Tribal interrelationships of the Asteraceae. *Cladistics* 3, 210–253.

Bremer, K. (1994) *Asteraceae: Cladistics and Classification*. Timber Press, Portland, Oregon.

Burke, D.H., Raubeson, L.A., Alberti, M., Hearst, J.E., Jordan, E.T., Kirch, S.A., Valinski, A.E.C., Conant, D.S. and Stein, D.B. (1993) The *chlL* (*frxC*) gene: phylogenetic distribution in vascular plants and DNA sequence from *Polystichum acrostichoides* (Pteridophyta) and *Synechococcus* sp. 7002 (Cyanobacteria). *Plant Systematics and Evolution* 187, 89–102.

Casano, L.M., Martin, M. and Sabater, B. (2001) Hydrogen peroxide mediates the induction of chloroplastic *ndh* complex under photooxidative stress in barley. *Plant Physiology* 125, 1450–1458.

Chase, M., Soltis, D., Olmstead, R., Morgan, D., Les, D., Mishler, B., Duvall, M., Price, R., Hills, H., Qui, Y.-L., Kron, K., Rettig, J., Conti, E., Palmer, J., Manhart, J., Sytsma, K., Michaels, H., Kress, J., Karol, K., Clark, D., Hedren, M., Gaut, B., Jansen, R., Kim, K.-J., Wimpee, C., Smith, J., Furnier, G., Straus, S., Xiang, Q.-Y., Plunkett, G., Soltis, P., Swensen, S., Williams, S., Gadek, P., Quinn, C., Equiarte, L., Golenberg, E., Learn, G. Jr, Graham, S., Barrett, S., Dayanandan, S. and Albert, V. (1993) Phylogenetics of seed plants: an analysis of nucleotide sequences from the plastid gene *rbcL*. *Annals of the Missouri Botanical Garden* 80, 528–580.

Clegg, M.T., Brown, A.D.H. and Whitfield, P.R. (1984) Chloroplast DNA diversity in wild and cultivated barley: implications for genetic conservation. *Genetic Research* 43, 339–343.

Coates, D. and Cullis, C.A. (1987) Chloroplast DNA variability among *Linum* species. *American Journal of Botany* 74, 260–268.

Cosner, M.E. (1993) Phylogenetic and molecular evolutionary studies of chloroplast DNA variation in the Campanulaceae. PhD dissertation, Ohio State University, Columbus, Ohio.

Cosner, M.E., Jansen, R.K., Palmer, J.D. and Downie, S.R. (1997) The highly rearranged chloroplast genome of *Trachelium caeruleum* (Campanulaceae): multiple inversions, inverted repeat expansion and contraction, transposition, insertions/deletions, and several repeat families. *Current Genetics* 31, 419–429.

Cosner, M.E., Jansen, R.K., Moret, B.M.E., Raubeson, L.A., Wang, L.-S., Warnow, T. and Wyman, S. (2000) A new fast heuristic for computing the breakpoint phylogeny and experimental analyses of real and synthetic data. In: Bourne, P., Gribskov, M., Altman, R., Jensen, N., Hope, D., Lengauer, T., Mitchell, J., Scheeff, E., Smith, C., Strande, S. and Weissig, H. (eds) *Proceedings of the 8th International Conference on Intelligent Systems for Molecular Biology*. La Jolla, California, pp. 104–115.

Delwiche, C.F. and Palmer, J.D. (1997) The origin of plastids and their spread via secondary symbiosis. *Plant Systematics and Evolution* 11 (suppl.), 53–86.

Doebley, J.F., Durbin, M., Golenberg, E.M., Clegg, M.T. and Ma, D.P. (1990) Evolutionary analysis of the large subunit of carboxylase (*rbcL*) nucleotide sequence among the grasses (Gramineae). *Evolution* 44, 1097–1108.

Douglas, S.E. (1998) Plastid evolution: origins, diversity, trends. *Current Opinions in Genetics and Development* 8, 655–661.

Downie, S.R. and Palmer, J.D. (1992) Use of chloroplast DNA rearrangements in reconstructing plant phylogeny. In: Soltis, P.S., Soltis, D.E. and Doyle, J.J. (eds) *Molecular Systematics of Plants*. Chapman & Hall, New York, pp. 14–35.

Downie, S.R., Llanas, E. and Katz-Downie, D.S. (1996) Multiple independent losses of the rpoC1 intron in angiosperm chloroplast DNAs. *Sytematic Botany* 21, 135–151.

Doyle, J.J. (1993) DNA, phylogeny, and the flowering of systematics. *Bioscience* 43, 380–389.

Doyle, J.J., Davis, J.I., Soreng, R.J., Garvin, D. and Anderson, M.J. (1992) Chloroplast DNA inversions and the origin of the grass family (Poaceae). *Proceedings of the National Academy of Sciences USA* 89, 7723–7726.

Doyle, J.J., Doyle, J.L. and Palmer, J.D. (1996) The distribution and phylogenetic significance of a 50-kbp chloroplast DNA inversion in the flowering plant family Leguminosae. *Molecular Phylogenetics and Evolution* 5, 429–438.

Gaut, B.S. (1998) Molecular clocks and nucleotide substitution rates in higher plants. In: Hecht, M.K., MacIntyre, R.J. and Clegg, M.T. (eds) *Evolutionary Biology*, Vol. 30. Plenum Press, New York, pp. 93–120.

Gaut, B.S., Clark, L.G., Wendel, J.F. and Muse, S.V. (1997) Comparisons of the molecular evolutionary process at *rbcL* and *ndhF* in the grass family (Poaceae). *Molecular Biology and Evolution* 14, 769–777.

Givnish, T.J and Sytsma, K.J. (1997) Homoplasy in molecular vs. morphological data: the likelihood of correct phylogenetic inference. In: Givnish, T.J. and Sytsma, K.J. (eds) *Molecular Evolution and Adaptive Radiation*. Cambridge University Press, Cambridge, pp. 55–101.

Gorbunova, V. and Levy, A.A. (1999) How plants make ends meet: DNA double-strand break repair. *Trends in Plant Science* 4, 263–269.

Goremykin, V.V., Hirsch-Ernst, K.I., Wolfl, S. and Hellwig, F.H. (2003a) Analysis of the *Amborella trichopoda* chloroplast genome sequence suggests that *Amborella* is not a basal angiosperm. *Molecular Biology and Evolution* 20, 1499–1505.

Goremykin, V.V., Hirsch-Ernst, K.I., Wolfl, S. and Hellwig, F.H. (2003b) The chloroplast genome of the 'basal' angiosperm *Calycanthus fertilis* – structural and phylogenetic analyses. *Plant Systematics and Evolution* 242, 119–135.

Goulding, S.E., Olmstead, R.G., Morden, C.W. and Wolfe, K.H. (1996) Ebb and flow of the chloroplast inverted repeat. *Molecular General Genetics* 252, 195–206.

Graham, S.W. and Olmstead, R.W. (2000) Utility of 17 chloroplast genes from inferring the phylogeny of the basal angiosperms. *American Journal of Botany* 87, 1712–1730.

Graham, S.W., Reeves, P.A., Burns, A.C.E. and Olmstead, R.G. (2000) Microstructural changes in noncoding chloroplast DNA: interpretation, evolution, and utility of indels and inversions in basal angiosperm phylogenetic inference. *International Journal of Plant Science* 161, S83–S96.

Hillis, D.M. (1998) Taxon sampling, phylogenetic accuracy, and investigator bias. *Systematic Biology* 47, 3–8.

Hiratsuka, J., Shimada, H., Whittier, R., Ishibashi, T., Sakamoto, M., Mori, M., Kondo, C., Honji, Y., Sun, C.R., Meng, B.Y., Li, Y.Q., Kanno, A., Nishizawa, Y., Hirai, A., Shinozaki, K. and Sugiura, M. (1989) The complete sequence of the rice (*Oryza sativa*) chloroplast genome: intermolecular recombination between distinct tRNA genes accounts for a major plastid DNA inversion during the evolution of the cereals. *Molecular and General Genetics* 217, 185–194.

Hoot, S.B. and Palmer, J.D. (1994) Structural rearrangements, including parallel inversions, within the chloroplast genome of *Anemone* and related genera. *Journal of Molecular Evolution* 38, 274.

Hosaka, K., Ogihara, Y., Matsubayaski, M. and Tsunewaki, K. (1984) Phylogenetic relationships between the tuberous *Solanum* species as revealed by restriction endonuclease analysis of chloroplast DNA. *Japanese Journal of Genetics* 59, 349–369.

Hubschmann, T., Hess, W.R. and Borner, T. (1996) Impaired splicing of the *rps12* transcript in ribosome-deficient plastids. *Plant Molecular Biology* 30, 109–123.

Hupfer, H., Swiatek, M., Hornung, S., Herrmann, R.G., Maier, R.M., Chiu, W.L. and Sears, B. (2000) Complete nucleotide sequence of the *Oenothera elata* plastid chromosome, representing plastome I of the five distinguishable euoenothera plastomes. *Molecular and General Genetics* 263, 581–585.

Jansen, R.K. and Kim, K.-J. (1996) Implications of chloroplast DNA data for the classification and phylogeny of the Asteraceae. In: Hind, N., Beentje, H. and Pope, G. (eds) *Proceedings of Compositae Symposium*. Royal Botanic Gardens, Kew, UK, pp. 317–339.

Jansen, R.K. and Palmer, J.D. (1987a) A chloroplast DNA inversion marks an ancient evolutionary split in the sunflower family (Asteraceae). *Proceedings of the National Academy of Sciences USA* 84, 5818–5822.

Jansen, R.K. and Palmer, J.D. (1987b) Chloroplast DNA from lettuce and *Barnadesia* (Asteraceae): structure, gene localization, and characterization of a large inversion. *Current Genetics* 11, 553–564.

Jansen, R.K. and Palmer, J.D. (1988) Phylogenetic implications of chloroplast DNA restriction site variation in the Mutisieae (Asteraceae). *American Journal of Botany* 75, 751–764.

Jansen, R.K., Wee, J.L. and Millie, D. (1998) Comparative utility of restriction site and DNA sequence data for phylogenetic studies in plants. In: Soltis, D.E., Soltis, P.S. and Doyle, J.J. (eds) *Molecular Systematics of Plants* II: *DNA Sequencing*. Chapman & Hall, New York, pp. 87–100.

Katayama, H. and Ogihara, Y. (1996) Phylogenetic affinities of the grasses to other monocots as revealed by molecular analysis of chloroplast DNA. *Current Genetics* 29, 572–581.

Kato, T., Kaneko, T., Sato, S., Nakamura, Y. and Tabata, S. (2000) Complete structure of the chloroplast genome of a legume, *Lotus japonicus*. *DNA Research* 7, 323–330.

Kelchner, S.A. (2000) The evolution of non-coding chloroplast DNA and its application in plant systematics. *Annals of the Missouri Botanical Garden* 87, 482–498.

Kim, K.-J. and Jansen, R.K. (1995) *ndh*F sequence evolution and the major clades in the sunflower family. *Proceedings of the National Academy of Sciences USA* 92, 10379–10383.

Kim, Y.-D. and Jansen, R.K. (1998) Chloroplast DNA restriction site variation and phylogeny of the Berberidaceae. *American Journal of Botany* 85, 1766–1778.

Kim, K.-J., Jansen, R.K., Wallace, R.S., Michaels, H.J. and Palmer, J.D. (1992) Phylogenetic implications of *rbc*L sequence variation in the Asteraceae. *Annals of the Missouri Botanical Garden* 79, 428–445.

Knox, E.B. and Palmer, J.D. (1999) The chloroplast genome arrangement of *Lobelia thuliana* (Lobeliaceae): expansion of the inverted repeat in an ancestor or the Campanulales. *Plant Systematics and Evolution* 214, 49–64.

Kugita, M., Kaneko, A., Yamamoto, Y., Takeya, Y., Matsumoto, T. and Yoshinaga, K. (2003a) The complete nucleotide sequence of the hornwort (*Anthoceros formosae*) chloroplast genome: insight into the earliest land plants. *Nucleic Acids Research* 31, 716–721.

Kugita, M., Yamamoto, Y., Fukikawa, T. and Matsumoto, T. (2003b) RNA editing in hornwort chloroplasts makes more than half the genes functional. *Nucleic Acids Research* 31, 2417–2423.

Kuroda, H. and Maliga, P. (2002) Overexpression of the *clpP* 5′-untranslated region in a chimeric context causes a mutant phenotype, suggesting competition for a *clpP*-specific RNA maturation factor in tobacco chloroplasts. *Plant Physiology* 129, 1600–1606.

Lavin, M., Doyle, J.J. and Palmer, J.D. (1990) Systematic and evolutionary significance of the loss of the large chloroplast DNA inverted repeat in the family Leguminosae. *Evolution* 44, 390–402.

Lemieux, C., Otis, C. and Turmel, M. (2000) Ancestral chloroplast genome in *Mesostigma viride* reveals an early branch of green plant evolution. *Nature* 403, 649–652.

Maier, R.M., Neckermann, K., Igloi, G.L. and Kossel, H. (1995) Complete sequence of the maize chloroplast genome: gene content, hotspots of divergence and fine tuning of genetic information by transcript editing. *Journal of Molecular Biology* 251, 614–628.

Manen, J-F. and Natali, A. (1995) Comparison of the evolution of ribulose-1,5-bisphosphate carboxylase (*rbcL*) and *atpB-rbcL* non-coding spacer sequences in a recent plant group in the tribe Rubieae (Rubiaceae). *Journal of Molecular Evolution* 41, 920–927.

Manhart, J.R. (1994) Phylogenetic analysis of green plant *rbcL* sequences. *Molecular Phylogenetics and Evolution* 3, 114–127.

Martin, M., Rujan, T., Richly, T., Hansen, A., Cornelsen, S., Lins, T., Leister, D., Stoebe, B., Hasegawa, M. and Penny, D. (2002) Evolutionary analysis of *Arabidopsis*, cyanobacterial, and chloroplast genomes reveals plastid phylogeny and thousands of cyanobacterial genes in the nucleus. *Proceedings of the National Academy of Sciences USA* 99, 12246–12251.

Martin, W., Stoebe, B., Goremykin, V., Hansmann, S., Hasegawa, M. and Kowallik, K.V. (1998) Gene transfer to the nucleus and the evolution of chloroplasts. *Nature* 393, 162–165.

Matsuoka, Y., Yamazaki, Y., Ogihara, Y. and Tsunewaki, K. (2002) Whole chloroplast genome comparison of rice, maize, and wheat: implications for chloroplast gene diversification and phylogeny of cereals. *Molecular Biology and Evolution* 19, 2084–2091.

Maul, J.E., Lilly, J.W., Cui, L., dePamphilis, C.W., Miller, W., Harris, E.H. and Stern, D.B. (2002) The *Chlamydomonas reinhardtii* plastid chromosome: islands of genes in a sea of repeats. *The Plant Cell* 14, 1–22.

McFadden, G.I. (2001) Primary and secondary endosymbiosis and the origin of plastids. *Journal of Phycology* 37, 951–959.

Millen, R.S., Olmstead, R.G., Adams, K.L., Palmer, J.D., Lao, N.T., Heggie, L., Kavanagh, T.A., Hibberd, J.M., Gray, J.C., Morden, C.W., Calie, P.J., Jermlin, L.S. and Wolfe, K.H. (2001) Many parallel losses of *infA* from chloroplast DNA during angiosperm evolution with multiple independent transfers to the nucleus. *The Plant Cell* 13, 645–658.

Miyagi, T., Kapoor, S., Sugita, M. and Sugiura, M. (1998) Transcript analysis of the tobacco plastid operon *rps2/atpI/H/F/A* reveals the existence of a non-consensus type (NCII) promoter upstream of the *atpI* coding sequence. *Molecular and General Genetics* 257, 299–307.

Miyamoto, T., Obokata, J. and Sugiura, M. (2002) Recognition of RNA editing sites is directed by unique proteins in chloroplasts: biochemical identification of *cis*-acting elements and *trans*-acting factors involved in RNA editing in tobacco and pea chloroplasts. *Molecular and Cellular Biology* 22, 6726–6734.

Moreira, D. and Philippe, H. (2001) Sure facts and open questions about the origin and evolution of photo-synthetic plastids. *Research in Microbiology* 152, 771–780.

Moret, B., Wang, L., Warnow, T. and Wyman, S. (2001) New approaches for reconstructing phylogenies from gene order data. In: *Proceedings of the International Conference on Intelligent Systems in Molecular Biology (ISMB 2001), 21–25 July 2001, Copenhagen, Denmark*, pp. 165–173.

Mullet, J.E. (1993) Dynamic regulation of chloroplast transcription. *Plant Physiology* 103, 309–313.

Muse, S.V. and Gaut, B.S. (1997) Comparing patterns of nucleotide substitution rates among chloroplast loci using relative ratio test. *Genetics* 146, 393–399.

Nickrent, D.L., Parkinson, C.L., Palmer, J.D. and Duff, R.J. (2000) Multigene phylogeny of land plants with special reference to bryophytes and the earliest land plants. *Molecular Biology and Evolution* 17, 1885–1895.

Nixon, K. (1999) The parsimony ratchet, a new method for rapid parsimony analysis. *Cladistics* 15, 407–414.

Ogihara, Y., Isono, K., Kojima, T., Endo, A., Hanaoka, M., Shiina, T., Terachi, T., Utsugi, S., Murata, M., Mori, N., Takumi, S., Ikeo, K., Gojobori, T., Murai, R., Murai, K., Matsuoka, Y., Ohnishi, Y., Tajiri, H. and Tsunewaki, K. (2002) Structural features of a wheat plastome as revealed by complete sequencing of chloroplast DNA. *Molecular Genetics and Genomics* 266, 740–746.

Ohyama, K., Fukuzawa, H., Kohchi, T., Shirai, H., Sano, T., Sano, S., Umesono, K., Shiki, Y., Takeuchi, M., Chang, Z., Aota, S., Inokuchi, H. and Ozeki, H. (1986) Chloroplast gene organization deduced from complete sequence of liverwort *Marchantia polymorpha* chloroplast DNA. *Nature* 322, 572–574.

Olmstead, R.G. and Palmer, J.D. (1994) Chloroplast DNA systematics: a review of methods and data analysis. *American Journal of Botany* 81, 1205–1224.

Palmer, J.D. (1983) Chloroplast DNA exists in two orientations. *Nature* 301, 92–93.

Palmer, J.D. (1991) Plastid chromosomes: structure and evolution. In: Hermann, R.G. (ed.) *The Molecular Biology of Plastids. Cell Culture and Somatic Cell Genetics of Plants*, Vol. 7A. Springer-Verlag, Vienna, pp. 5–53.

Palmer, J.D. (2003) The symbiotic birth and spread of plastids: how many times and whodunit? *Journal of Phycology* 39, 4–11.

Palmer, J.D. and Delwiche, C.F. (1998) Origin and evolution of plastids and their genomes. In: Soltis, D.E., Soltis, P.S. and Doyle, J.J. (eds) *Molecular Systematics of Plants* II: *DNA Sequencing*. Chapman & Hall, New York, pp. 375–409.

Palmer, J.D. and Stein, D.B. (1986) Conservation of chloroplast genome structure among vascular plants. *Current Genetics* 10, 823–833.

Palmer, J.D. and Zamir, D. (1982) Chloroplast DNA evolution and phylogenetic relationships in *Lycopersicon*. *Proceedings of the National Academy of Sciences USA* 81, 8014–8018.

Palmer, J.D., Boynton, J.E., Gillham, N.W. and Harris, E.H. (1985) Evolution and recombination of the large inverted repeat in *Chlamydomonas* chloroplast DNA. In: Steinbeck, K.E., Bonitz, S., Arntzen, C.J. and Bogorad, L. (eds) *Molecular Biology of the Photosynthetic Apparatus*. Cold Spring Harbor Laboratory Press, New York, pp. 269–278.

Palmer, J.D., Aldrich, J. and Thompson, W.F. (1987a) Chloroplast DNA evolution among legumes: loss of a large inverted repeat occurred prior to other sequence rearrangements. *Current Genetics* 11, 275–286.

Palmer, J.D., Nugent, J.M. and Herbon, L.A. (1987b) Unusual structure of geranium chloroplast DNA: a triple-sized inverted repeat, extensive gene duplications, multiple inversions and two repeat families. *Proceedings of the National Academy of Sciences USA* 84, 769–773.

Palmer, J.D., Jansen, R.K., Michaels, H., Manhart, J. and Chase, M. (1988) Chloroplast DNA variation and plant phylogeny. *Annals of the Missouri Botanical Garden* 75, 1180–1206.

Perry, A.S. and Wolfe, K.H. (2002) Nucleotide substitution rates in legume chloroplast DNA depend on the presence of the inverted repeat. *Journal of Molecular Evolution* 55, 501–508.

Perry, A.S., Brennan, S., Murphy, D.J., Kavanagh, T.A. and Wolfe, K.H. (2002) Evolutionary re-organisation of a large operon in adzuki bean chloroplast DNA caused by inverted repeat movement. *DNA Research* 9, 157–162.

Plunkett, G.M. and Downie, S.R. (2000) Expansion and contraction of the chloroplast inverted repeat in Apiaceae subfamily Apiodeae. *Systematic Botany* 25, 648–667.

Price, R.A., Caile, P.J., Downie, S.R., Logsdon, J.M. Jr and Palmer, J.D. (1990) Chloroplast DNA variation in the Geraniaceae: a preliminary report. In: Vorster, P. (ed.) *Proceedings of the International Geraniaceae Symposium, 24–26 September 1990, Stellenbosch, South Africa*, pp. 1–8.

Pryer, K.M., Schneider, H., Smith, A.R., Cranfill, R., Wolf, P.G., Hunt, J.S. and Sipes, S.D. (2001) Horsetails and ferns are a monophyletic group and the closest living relatives to seed plants. *Nature* 409, 618–621.

Rai, H.S., O'Brien, H.E., Reeves, P.A., Olmstead, R.G. and Graham, S.W. (2003) Inference of higher-order relationships in the cycads from a large chloroplast data set. *Molecular Phylogenetics and Evolution* 29, 350–359.

Raubeson, L.A. (1991) Structural variation in the chloroplast genome of vascular land plants. PhD dissertation, Yale University, New Haven, Connecticut.

Raubeson, L.A. and Jansen, R.K. (1992a) A rare chloroplast-DNA structural mutation is shared by all conifers. *Biochemical Systematics and Ecology* 20, 17–24.

Raubeson, L.A. and Jansen, R.K. (1992b) Chloroplast DNA evidence on the ancient evolutionary split in vascular land plants. *Science* 255, 1697–1699.

Raubeson, L.A. and Stein, D.B. (1995) Insights into fern evolution from mapping chloroplast genomes. *American Fern Journal* 85, 193–204.

Rice, K.A., Donoghue, M.J. and Olmstead, R.G. (1997) Analyzing large data sets: *rbcL* 500 revisited. *Systematic Biology* 46, 554–563.

Rokas, A. and Holland, P.W.H. (2000) Rare genomic changes as tools for phylogenetics. *Trends in Ecology and Evolution* 15, 454–459.

Sabater, B., Martin, M., Schmitz-Linneweber, C. and Maier, R.M. (2002) Is clustering of plastid RNA editing sites a consequence of transitory loss of gene function? Implications for past environmental and evolutionary events in plants. *Perspectives in Plant Ecology, Evolution and Systematics* 5, 81–89.

Sato, S., Nakamura, Y., Kaneko, T., Asamizu, E. and Tabata, S. (1999) Complete structure of the chloroplast genome of *Arabidopsis thaliana*. *DNA Research* 6, 283–290.

Schmitz-Linneweber, C., Maier, R.M., Alcaraz, J.P., Cottet, A., Herrmann, R.G. and Mache, R. (2001) The plastid chromosome of spinach (*Spinacia oleracea*): complete nucleotide sequence and gene organization. *Plant Molecular Biology* 45, 307–315.

Schmitz-Linneweber, C., Regel, R., Du, T.G., Hupfer, H., Herrmann, R.G. and Maier, R.M. (2002) The plastid chromosome of *Atropa belladonna* and its comparison with that of *Nicotiana tabacum*: the role of RNA editing in generating divergence in the process of plant speciation. *Molecular Biology and Evolution* 19, 1602–1612.

Shields, D.C. and Wolfe, K.H. (1997) Accelerated rates of evolution of sites undergoing RNA editing in plant mitochondria and chloroplasts. *Molecular Biology and Evolution* 14, 344–349.

Shinozaki, K., Ohme, M., Tanaka, M., Wakasugi, T., Hayashida, N., Matsubayashi, T., Zaita, N., Chunwongse, J., Obokata, J., Yamaguchi-Shinozaki, K., Ohto, C., Torazawa, K., Meng, B.Y., Sugita, M., Deno, H., Kamogashira, T., Yamada, K., Kusuda, J., Takaiwa, F., Kato, A., Tohdoh, N., Shimada, H. and Sugiura, M. (1986) The complete nucleotide sequence of the tobacco chloroplast genome: its gene organization and expression. *EMBO Journal* 5, 2043–2049.

Simpson, C.L. and Stern, D.B. (2002) The treasure trove of algal chloroplast genomes. Surprises in architecture and gene content, and their functional implications. *Plant Physiology* 129, 957–966.

Soltis, D.E. and Soltis, P.S. (1998) Choosing an approach and an appropriate gene for phylogenetic analysis. In: Soltis, D.E., Soltis, P.S. and Doyle, J.J. (eds) *Molecular Systematics of Plants* II: *DNA Sequencing*. Chapman & Hall, New York, pp. 1–42.

Soltis, D.E., Soltis, P.S., Clegg, M.T. and Durbin, M. (1990) *rbcL* sequence divergence and phylogenetic relationships among Saxifragaceae sensu lato. *Proceedings of the National Academy of Sciences USA* 87,
4640–4644.

Soltis, D.E., Soltis, P.S., Chase, M.W., Mort, M.W., Albach, D.C., Zanis, M., Savolainen, V., Hahn, W.H.,
Hoot, S.B., Fay, M.F., Axtell, M., Swensen, S.M., Prince, L.M., Kress, W.J., Nixon, K.J. and Farris, J.S.
(2000) Angiosperm phylogeny inferred from 18S rDNA, *rbcL*, and *atpB* sequences. *Botanical Journal of
the Linnaean Society* 133, 381–461.

Stein, D.B., Palmer, J.D. and Thompson, W.F. (1986) Structural evolution and flip-flop recombination of
chloroplast DNA in the fern genus *Osmunda*. *Current Genetics* 10, 835–841.

Stein, D.B., Conant, D.S., Ahearn, M.E., Jordan, E.T., Kirch, S.A., Hasebe, M., Iwatsuki, K., Tan, M.K. and
Thomson, J.A. (1992) Structural rearrangements of the chloroplast genome provide an important phylogenetic link in ferns. *Proceedings of the National Academy of Sciences USA* 89, 1856–1860.

Stiller, J.W., Reel, D.C. and Johnson, J.C. (2003) A single origin of plastids revisited: convergent evolution in
organellar gene content. *Journal of Phycology* 39, 95–105.

Sugiura, C., Kobayashi, Y., Aoki, S., Sugita, C. and Sugita, M. (2003) Complete chloroplast DNA sequence of
the moss *Physcomitrella patens*: evidence for the loss and relocation of *rpoA* from the chloroplast to
the nucleus. *Nucleic Acids Research* 31, 5324–5331.

Sugiura, M. (2003) History of chloroplast genomics. *Photosynthesis Research* 76, 371–377.

Sytsma, K.J. and Gottlieb, L.D. (1986) Chloroplast DNA evidence for the origin of the genus *Heterogaura*
from a species of *Clarkia* (Onagraceae). *Proceedings of the National Academy of Sciences USA* 83,
5554–5557.

Sytsma, K.J. and Schaal, B.A. (1985) Phylogenetics of the *Lisianthius skinneri* (Gentianaceae) species complex in Panama utilizing DNA restriction fragment analysis. *Evolution* 39, 594–608.

Tsudzuki, J., Nakashima, K., Tsudzuki, T., Hiratsuka, J., Shibata, M., Wakasugi, T. and Sugiura, M. (1992)
Chloroplast DNA of black pine retains a residual inverted repeat lacking rRNA genes: nucleotide
sequences of *trnQ, trnK, psbA, trnI* and *trnH* and the absence of *rps16*. *Molecular and General
Genetics* 232, 206–214.

Turmel, M., Otis, C. and Lemieux, C. (1999) The complete chloroplast DNA sequence of the green alga
Nephroselmis olivacea: insights into architecture of ancestral chloroplast genomes. *Proceedings of the
National Academy of Sciences USA* 96, 10248–10253.

Wakasugi, T., Tsudzuki, J., Ito, S., Nakashima, K., Tsudzuki, T. and Sugiura, M. (1994) Loss of all *ndh* genes
as determined by sequencing the entire chloroplast genome of the black pine *Pinus thunbergii*.
Proceedings of the National Academy of Sciences USA 91, 9794–9798.

Wakasugi, T., Sugita, M., Tsudzuki, T. and Sugiura, M. (1998) Updated gene map of tobacco chloroplast
DNA. *Plant Molecular Biology Reporter* 16, 231–241.

Wallace, R.S. and Cota, J.H. (1996) An intron loss in the chloroplast gene *rpoC1* supports a monophyletic
origin for the subfamily Cactoideae of the Cactaceae. *Current Genetics* 29, 275–281.

Wang, L.-S., Jansen, R.K., Moret, B.M.E., Raubeson, L.A. and Warnow, T. (2002) Fast phylogenetic methods
for analysis of genome rearrangement data: an empirical study. In: *Proceedings of the 7th Pacific
Symposium Biocomputing PSB 2002, 3–7 January 2002, Lihue, Hawaii*, pp. 524–535.

Wolf, P.G., Rowe, C.A., Sinclair, R.B. and Hasebe, M. (2003) Complete nucleotide sequence of the chloroplast genome from a leptosporangiate fern, *Adiantum capillus-veneris* L. *DNA Research* 10, 59–65.

Wolfe, K.H., Li, W.-H. and Sharp, P.M. (1987) Rates of nucleotide substitution vary greatly among plant
mitochondria, chloroplast, and nuclear DNAs. *Proceedings of the National Academy of Sciences USA*
84, 9054–9058.

Wolfe, K.H., Morden, C.W. and Palmer, J.D. (1992) Function and evolution of a minimal plastid genome
from a nonphotosynthetic parasitic plant. *Proceedings of the National Academy of Sciences USA* 89,
10648–10652.

Yamada, T. (1991) Repetitive sequence-mediated rearrangements in *Chlorella ellipsoidea* chloroplast DNA:
completion of nucleotide sequence of the large inverted repeat. *Current Genetics* 19, 139–147.

5   The mitochondrial genome of higher plants: a target for natural adaptation

Sally A. Mackenzie

Plant Science Initiative, N305 Beadle Center for Genetics Research, University of Nebraska, Lincoln, NE 68588-0660, USA

Introduction

Some of the most intriguing examples of adaptation in eukaryotes arise within the plant kingdom, many in response to a plant's immotility and consequent inability to escape environmental stresses (Hawkesford and Buchner, 2001). These unique attributes occur in various forms to produce wonders of plant architecture, specialized physiology and reproductive strategies. At a cellular level, several unique features of plant metabolism and organellar genome maintenance are evident in plants (Mackenzie and McIntosh, 1999). Some of these cellular attributes are thought to be the outcome of the endosymbiotic process that has led to the present-day plastid and consequent mitochondrial–plastid co-evolution (Allen, 1993; Adams *et al.*, 2002; Elo *et al.*, 2003).

As in the case of most animal systems, organellar genomes generally show strict maternal inheritance in plants. However, there are exceptions to this pattern. In some cases, paternal inheritance is observed, though varying degrees of biparental inheritance are also seen (Reboud and Zeyl, 1994; Zhang *et al.*, 2003). In nearly all such exceptions, the plastid has been more likely to show variation from strict maternal inheritance than the mitochondrion. Why the relaxation of strict maternal inheritance pat-terns might be tolerated more in plant sys-tems than in animal, and the mechanisms underlying these selective organelle trans-mission patterns, are not yet well under-stood. Mechanisms for selective exclusion of paternal organelles vary. Whereas some sys-tems appear to target paternally derived organellar DNA for selective destruction or suppressed replication (Nagata *et al.*, 1999; Sodmergen *et al.*, 2002; Moriyama and Kawano, 2003), some animal systems are postulated to exclude or destroy the pater-nal organelles themselves (Sutovsky, 2003).

When considering organelle inheritance and segregation processes, one must keep in mind the distinct dynamics of organelle behaviour. In contrast to nuclear genetic information, which undergoes replication at a precise point within a tightly regulated cell cycle, segregating to daughter cells in equal and unchanging copy number each cellular generation, organellar genomes obey very different rules. Organellar DNA replication does not maintain tight synchrony with the cell cycle (Birky, 1983), and the numbers of genomes per organelle and organelles per cell vary dramatically depending on tissue type. In general, mitochondrial biogenesis is highest in meristematic and reproductive tissues, where numbers of genomes per mitochondrion and mitochondria per cell generally range in the hundreds, while mitochondrial numbers decrease markedly in vegetative tissues (Lamppa and Bendich, 1984; Fujie *et al.*, 1993; Robertson *et al.*, 1995). The 'segregation' of organelles, existing as multi-genomic populations within cells, occurs by a process of cytoplasmic sorting throughout development. Although cytoplasmic sorting is generally considered a stochastic process, nuclear gene influence on cytoplasmic segregation is evident.

This review will describe the unusual nature of plant mitochondrial genomes, contrasting their features and behaviour with what is known of mammalian and fungal systems. One anticipates that the considerable divergence observed in plants derives, at least in part, from the unusual plant cellular context of mitochondrial–chloroplast co-evolution. With availability of complete plant genome sequence information, considerable evidence has accumulated recently in support of this assumption. However, it is also important to note that the vast majority of genes essential for mitochondrial processes are nuclear encoded; the mitochondrial genome, though essential, encodes less than 5% of the information required for its varied functions. Therefore, it is impossible to consider plant mitochondrial genome evolution and adaptation without addressing the critical role of the nucleus in these ongoing processes.

Nuclear Regulation of Mitochondrial DNA and RNA Metabolism

The availability of complete genome sequences for *Arabidopsis* and rice has allowed the identification of several candidate genes predicted to function in organellar DNA and RNA metabolism functions. Two striking features of nuclear genes that appear to direct organellar genome maintenance are evident. The first surprising property of these genes is their organization within the nuclear genome. Nuclear genes encoding organellar DNA and RNA metabolism loci appear to be largely clustered in a few regions of the plant genome (Elo *et al.*, 2003). Moreover, this genomic arrangement may not be limited to plants alone (Lefai *et al.*, 2000).

The endosymbiotic processes believed to have given rise to present day organelles are generally thought to have involved the transfer of large amounts of genetic information from mitochondrial and plastid progenitor genomes to the nucleus. Very early on, these transfers might have involved large genomic segments that encompassed many genes simultaneously. More recent gene transfers, following the advent of RNA editing processes, have probably occurred as singular gene events that involve an RNA intermediate in the process (Nugent and Palmer, 1991; Covello and Gray, 1992). With the massive nuclear genomic rearrangements that have occurred in plants subsequent to the endosymbiotic events (Blanc *et al.*, 2000), it is difficult to envisage how genetic linkage has been maintained for large numbers of transferred genes without selection. One possibility is that maintenance of related genes in a linkage might facilitate coordinate gene regulation during key points in development (Boutanaev *et al.*, 2002). For example, at the point immediately following pollination, a maternally derived cytoplasm must immediately establish compatibility with the newly introduced paternal nuclear contribution. Possibly epigenetic regulation would be crucial for re-establishing necessary intergenomic compatibility.

A second intriguing observation regarding the nuclear genes that participate in organellar genome maintenance is the large number predicted to encode proteins functional in both mitochondria and plastids (Hedtke *et al.*, 1997; Beardslee *et al.*, 2002; Elo *et al.*, 2003). This assumption is supported by individual genes that encode dual-targeted proteins, as well as genes that have undergone duplication, with duplicate members targeting distinct organelle types. The prevalence of proteins that apparently function in both mitochondria and plastids suggests that a substantial component of the DNA and RNA metabolism apparatus is overlapping in the two organelle types (Small *et al.*, 1998; Peeters and Small, 2001). At one time this would have seemed an incongruous idea, given the numerous differences that exist in mitochondrial and plastid functions, genome structure and gene organization. These distinctions are likely to be remnant features of their progenitors, as the number of reports of shared, probably acquired, features increases.

The Mitochondrial Genome of Higher Plants

Plant mitochondrial genomes are distinguished by their extreme variation in size, ranging from about 200 to 2400 kbp, in comparison to the 16-kbp genome of human mitochondria and intermediate but less variably sized genomes of fungi. With the recent availability of complete mitochondrial genome sequences for at least four plant species, we now find that the dramatic differences in genome size are not accounted for by vast discrepancies in coding capacity. In fact, plant mitochondrial genomes encode somewhere between 55 and 70 genes (Oda *et al.*, 1992; Unseld *et al.*, 1997; Notsu *et al.*, 2002; Handa, 2003), less than twice the number of genes found in the human mitochondrion. Considerable sequence redundancy, integration of non-mitochondrial DNA, and ectopic recombination have contributed to the observed variation.

The mitochondrial genome of plants consists of a heterogeneous population of both circular and linear DNA molecules, many existing in highly branched configurations (Backert *et al.*, 1996; Bendich, 1996; Oldenburg and Bendich, 1996; Backert and Borner, 2000). To date, an origin of replication has not been defined in plants, and evidence suggests that replication may occur, at least in part, by a rolling circle mechanism. In fact, it has been suggested that replication may initiate by a strand invasion process perhaps resembling that of T4 phage (Backert and Borner, 2000).

In contrast to mammalian mitochondria, plant mitochondrial genomes are unusually dynamic in their structure, in part a consequence of prolific intra- and intergenomic recombination activity. Within most plant mitochondrial genomes are dispersed several repeated sequences, present in both inverted and direct orientations. High-frequency DNA exchange at these sites produces a complex assemblage of large inversions and subgenomic DNA molecules, each containing a portion of the genome. Whether this recombination activity is continuous throughout plant development, or restricted to a particular cell type, is not clear. The technical difficulties that have complicated these studies arise from the entwined nature of replicative and recombinational processes.

In addition to high-frequency homologous recombination, plant mitochondrial genomes commonly undergo low-frequency recombinations at non-homologous, often intragenic, sites. This ectopic recombination activity gives rise to chimeric gene configurations, often expressive, within a wide array of plant species. In many cases, these unusual gene chimeras are discovered by their causative association with cytoplasmic male sterility (Schnable and Wise, 1998). However, not all ectopic recombinations necessarily produce a detectable phenotype (Marienfeld *et al.*, 1997).

Cytoplasmic male sterility (CMS) is a condition in which the plant is unable to produce or shed viable pollen as a consequence of mitochondrial mutation. In nearly all cases investigated, the associated mitochondrial mutations are dominant, stemming from expression of unique sequence chimeras that form aberrant open reading frames. To date, all cases of CMS appear to be associated with ectopic recombination within the genome, implying an adaptive advantage to this activity. No two CMS mutations have been identical, although striking similarities have been observed in some cases (Tang *et al.*, 1996). Interestingly, the frequency of non-homologous recombination, or the relative copy number of the derived recombinant forms, appears to be controlled by nuclear genes.

Nuclear Regulation of Mitochondrial Genome Structure

Whereas evolutionary pressures appear to have selected for a highly conservative, stable and compact mitochondrial genome configuration within animals, adaptive selection within the plant kingdom has produced a system that appears to benefit from a high degree of variability (Marienfeld *et al.*, 1999). A fascinating example of the dynamic nature of the plant mitochondrial genome is the widespread phenomenon termed stoichiometric shifting (Small *et al.*, 1987).

Certain subgenomic mitochondrial DNA configurations change dramatically in relative copy number during the development of the plant. When present substoichiometrically, these components of the mitochondrial genome have been estimated at one copy per every 100–200 cells of the plant (Arrieta-Montiel *et al.*, 2001), representing a heteroplasmic (heterogeneous cytoplasmic) condition. However, stoichiometric shifting can result in preferential amplification of these molecules to levels equimolar with the principal mitochondrial genome. The selective amplification or suppression of particular portions of the mitochondrial genome is influenced by nuclear genotype.

Stoichiometric shifting has been reported in a wide array of plant species (Mackenzie and McIntosh, 1999), and shifting events are apparently induced under conditions of cell culture or cybridization (Kanazawa *et al.*, 1994; Gutierres *et al.*, 1997; Bellaoui *et al.*, 1998), alloplasmy (Kaul, 1988), spontaneous CMS reversion to pollen fertility (Mackenzie *et al.*, 1988; Smith and Chowdhury, 1991), and the introduction or mutation of specific nuclear genes.

In the CMS system of common bean, the mitochondrial sequence associated with pollen sterility, designated *pvs-orf239*, resides on a mitochondrial molecule that is shifted to substoichiometric levels in response to the introduction of the dominant nuclear gene *Fr* (Mackenzie and Chase, 1990). This is an appealing system for study because introduction of *Fr*, by standard crossing, results in Mendelian segregation for a particular, reproducible mitochondrial rearrangement. The *pvs-orf239* sequence, when present in high copy number, is expressed and the plant is male sterile. When *Fr* is introduced, and the mitochondrial *pvs-orf239* sequence is reduced to substoichiometric levels, the

plant is male fertile. Interestingly, the condition of male fertility is not reversed by the segregation of *Fr*, suggesting that the product of *Fr* acts unidirectionally or in a limited context, and the reversal of *Fr* action might require additional nuclear components.

In *Arabidopsis*, mutation at the nuclear locus *CHM* results in the copy number amplification of a mitochondrial chimeric DNA configuration and the appearance of green–white variegation (Martinez-Zapater *et al.*, 1992; Sakamoto *et al.*, 1996). Several mutant alleles of *CHM* are available in *Arabidopsis*, presenting an opportunity to clone the gene.

Recently, the product of the *CHM* locus was shown to resemble the MUTS protein of *Escherichia coli*, and the gene has now been designated *MSH1* (MutS Homologue 1; Abdelnoor *et al.*, 2003). *MutS* is a component of the DNA mismatch repair apparatus, and several of its homologues have been identified to function within the nuclear genome of higher eukaryotes. One other nuclear gene encoding a mitochondrial *MutS* homologue was reported several years ago in yeast (Reenan and Kolodner, 1992). In that case, the mitochondrial protein was suggested to function in mismatch repair (Chi and Kolodner, 1994).

Mismatch repair components appear to serve two important functions within the eukaryotic genome: to bind and repair nucleotide mismatches and to suppress non-homologous recombination activity (Modrich and Lahue, 1996; Harfe and Jinks-Robertson, 2000). Enhanced ectopic recombination appears to be the effect within the plant mitochondrial genome in response to *MSH1* mutation (Abdelnoor *et al.*, 2003).

Interestingly, allelic variation for *MutS* in microbial populations appears to provide certain adaptive advantage under severe selection conditions by enhancing mutation frequency (LeClerc *et al.*, 1996; LeClerc and Cebula, 1997; Bjedov *et al.*, 2003; Chopra *et al.*, 2003). This 'mutator' phenomenon, arising from *MutS* variation, has been reported in a number of organisms including humans, where such variation contributes to cancer incidence (Li, 1999). A phenomenon resembling mitochondrial stoichiometric shifting has also been described in *Drosophila*, although the nuclear effector is not yet identified (Le Goff *et al.*, 2002). It now appears likely that evolutionary advantage might have been realized by permitting a degree of genetic variation within the mitochondrial *MSH1* gene of higher plants. A possible adaptive role of *MSH1* variation in plant populations will be discussed in a later section.

Nuclear Regulation of Mitochondrial Transcript Processing and Fertility Restoration

A striking example of a derived plant feature shared by both mitochondria and plastids is their dependence on RNA processing, editing and stabilizing functions for organellar gene expression (Hoffmann *et al.*, 2001; Binder and Brennicke, 2003). RNA processing, which involves the cleavage of RNA at precise sites as part of RNA maturation, and RNA editing, which generally involves specific C to U conversions within a transcript, are both found to occur in a wide array of plastid and mitochondrial RNAs. The expansion of these processes has apparently been accompanied, or driven by, a concomitant expansion in the number of nuclear genes associated with these functions.

The pentatricopeptide repeat (PPR) family of proteins in *Arabidopsis* numbers over 500 members, with over two-thirds encoding proteins predicted to target mitochondria or plastids (Small and Peeters, 2000). The PPR proteins share almost no detectable sequence homology. Rather, they are linked by their unusual structural similarities. Although highly divergent at their amino termini, each PPR protein contains a series of 35-amino-acid repeat structures, present in variable numbers. These repeats are predicted to confer a helical structure to the protein that is postulated to interact directly with RNA or proteins. It has been suggested that this family of nuclear proteins may provide the RNA recognition specificity necessary for RNA processing activities.

Interestingly, molecular studies of fertility restoration mechanisms in several CMS systems reveal a role of most restorer genes in RNA processing (Schnable and Wise, 1998). Over the past few years, five nuclear genes that restore pollen fertility to CMS mutants have been cloned. Of these, four have been shown to encode PPR proteins. These include the restorers of fertility in petunia (Bentolila *et al.*, 2002), Kosena radish (Koizuka *et al.*, 2003), Ogura radish (Brown *et al.*, 2003; Desloire *et al.*, 2003) and rice (Kazama and Toriyama, 2003). It would not be surprising to find PPR proteins implicated in the fertility restoration of several other CMS plant species in the near future. The RNA specificity that is postulated by PPR protein:RNA binding, combined with the intragenic recombination origins of most CMS mutations, appears an ideal system for nuclear control of CMS-associated aberrant gene expression.

Evolutionary Implications of Mitochondrial Genome Dynamics in Higher Plants

The mitochondrial mutations that confer cytoplasmic male sterility have been of great interest for their value to the hybrid seed industry (Fig. 5.1). In a broad range of plant species, the phenomenon of heterosis, or hybrid vigour, is well documented (Tsaftaris and Kafka, 1998; Rieseberg *et al.*, 2000). The genomic condition produced by hybridization, probably associated with higher levels of gene heterozygosity and perhaps epigenetic processes, provides markedly enhanced reproductive capacity and plant vigour. Although of obvious agricultural benefit, the heterotic state is also clearly advantageous in natural populations (Rieseberg *et al.*, 2000).

Most domesticated crop species are categorized as predominantly outcrossing or self-pollinating. However, in natural populations exist the more heterogeneous and dynamic conditions of gynodioecy. Gynodioecious populations are composed of both female and hermaphroditic individuals, permitting population expansion within geographically isolated environments

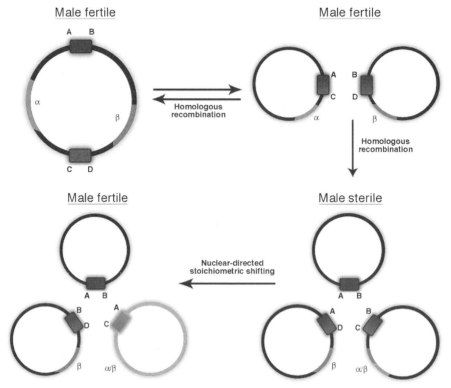

Fig. 5.1. Patterns of alteration distinguishing the plant mitochondrial genome. Homologous intra- and intermolecular recombination occurs at repeated sequences (boxes) that, when in direct orientation, can produce an equilibrium of non-recombinant and recombinant subgenomic molecules. Intragenic (α–β) ectopic recombination can occur to produce sequence chimeras. What is shown is the simplest scenario; often these chimeric sequences are derived from multiple recombination and/or insertion events (Schnable and Wise, 1998). Note that the final population of molecules does not necessarily include all parental and recombinant forms. Stoichiometric shifting represents a nuclear-directed process that can modulate the relative copy number of particular recombinant molecules within the genome, often reducing them to one copy per 100–200 cells of the plant. The shifting mechanism is not yet understood. Mitochondrial DNA molecules are shown in a circular form, as they map, for convenience. *In vivo* structures are predominantly linear and often multimeric.

together with enhanced genomic recombination. Gynodioecy comprises self- and cross-pollination activity in dynamic equilibrium (Couvet *et al.*, 1990). Whereas many of our domesticated crop species are predominantly self-pollinating, it is possible that several of these species originated from gynodioecious natural populations.

Most detailed studies of gynodioecy have been conducted in non-crop species such as *Silene vulgaris* (Olson and McCauley, 2002), *Thymus vulgaris* (Manicacci *et al.*, 1997) and *Plantago lanceolata* (Van Damme, 1983). In the various natural populations studied,

gynodioecy involves the interaction of one to multiple mitochondrial mutations that condition CMS together with maintainer (non-fertility restoring) and fertility restorer nuclear genotypes. The CMS cytoplasm, in combination with a fertility-restoring nucleus, constitutes a hermaphrodite capable of self-pollination, while a CMS cytoplasm combined with a maintainer genotype constitutes a female, outcrossing form. Extensive cross-pollination activity eventually introduces the restorer genotype to the female, shifting the localized frequency of hermaphrodites.

Several research groups have investigated the environmental factors that influence frequencies of fertility restoring, non-restoring and sterility-determining components of this process within naturally occurring and widely dispersed populations (Bailey *et al.*, 2003; Jacobs and Wade, 2003). However, these models generally do not account for an additional genetic component that is probably essential to understanding the maintenance of females in the population. This missing factor is the spontaneous inter-convertibility of the male-sterile and male-fertile condition. The nuclear–mitochondrial interactions that effect mitochondrial stoichiometric shifting permit the spontaneous inter-conversion of females and hermaphrodites. Female plants located in isolation are dependent on the process of low frequency spontaneous reversion to fertility to facilitate reproduction late in the plant's cycle. Spontaneous reversion of a CMS plant to male fertility has been described in several plant systems (Laughnan and Gabay-Laughnan, 1983; Smith and Chowdhury, 1991; Bellaoui *et al.*, 1998; Janska *et al.*, 1998; Andersson, 1999), where the frequency of stoichiometric shifting depends on nuclear background.

Although not yet documented in nature, it appears likely that genetic variability at the *MSH1* locus accounts for some portion of the inter-conversion of females and hermaphrodites. Whether *MSH1* displays higher-than-average natural mutability in plants has not yet been investigated, but its unusually complex gene structure, comprising 21 introns in at least five plant species studied to date (Abdelnoor *et al.*, 2003; Abdelnoor and Mackenzie, unpublished), raises the possibility of alternative splicing (Black, 2003).

Recent literature suggests that several of our present-day crop plants probably derive from gynodioecious natural populations. These include beet (*Beta vulgaris*; Cuguen *et al.*, 1994), pearl millet (*Pennisetum americanum*; Delorme *et al.*, 1997), sunflower (*Helianthus* spp.; Jan, 2000) and common bean (*Phaseolus vulgaris*; Mackenzie, 1991). In the common bean, domestication occurred from multiple centres of origin extending from Mexico to Ecuador (Gepts,

1998). Surveys of core germplasm collections, encompassing all available genetic diversity from natural populations, reveal the presence of the CMS-associated *pvs-orf239* mitochondrial sequence, highly conserved, in 100% of the lines, although the sequence is substoichiometric in over 90% of them (Arrieta-Montiel *et al.*, 2001). The *pvs-orf239* sequence is also present in *Phaseolus coccineus*, *Phaseolus polyanthus* and *Phaseolus acutafolius* (Hervieu *et al.*, 1993), implying that its evolution predates *Phaseolus* speciation.

In bean, hermaphrodites arise by two distinct mechanisms: suppression of the male sterility phenotype can be conditioned by the prevalent *Fr2* locus (Mackenzie, 1991), while substoichiometric shifting of *pvs-orf239* is conditioned by the more rare *Fr* locus (Mackenzie and Chase, 1990). Spontaneous reversion to fertility ranges in frequency depending on nuclear background (Mackenzie *et al.*, 1988), but even the most effective maintainer genotypes often result in a small number of seed pods produced just prior to plant senescence if the plant has not been artificially pollinated (Mackenzie, unpublished data).

These observations suggest the existence of a naturally occurring, highly refined genetic system to facilitate cross-pollination in a species that, upon domestication, has become largely inbreeding. This fascinating genetic system apparently integrates nuclear suppressors of mitochondrial gene expression with nuclear mechanisms that influence heteroplasmic sorting to reversibly modulate female:hermaphrodite ratios. A system integrated in this manner would permit a more dynamic response to environmental changes than is generally predicted in most models.

Evidence suggests that nuclear–mitochondrial genetic interactions, and the dynamic features of the plant mitochondrial genome, are highly amenable to adaptation for reproductive fitness in natural populations. Thus, it is not surprising that the most common problems arising in the deployment of cytoplasmic male sterility for commercial hybrid seed production are cytoplasmic instability and incomplete fertility restoration. These features, undesirable within the highly controlled environmental conditions of today's agriculture, are likely to have served plant populations well for reproductive success in their ever-changing natural environments.

References

Abdelnoor, R.V., Yule, R., Elo, A., Christensen, A., Meyer-Gauen, G. and Mackenzie, S. (2003) Substoichiometric shifting in the plant mitochondrial genome is influenced by a gene homologous to *MutS*. *Proceedings of the National Academy of Sciences USA* 100, 5968–5973.

Adams, K.L., Dalen, D.O., Whelan, J. and Palmer, J.D. (2002) Genes for two mitochondrial ribosomal proteins in flowering plants are derived from their chloroplast or cytosolic counterparts. *Plant Cell* 14, 931–943.

Allen, J.F. (1993) Control of gene expression by redox potential and the requirement of chloroplast and mitochondrial genomes. *Journal of Theoretical Biology* 165, 609–631.

Andersson, H. (1999) Female and hermaphrodite flowers on a chimeric gynomonoecious *Silene vulgaris* plant produce offspring with different genders: a case of heteroplasmic sex determination. *Journal of Heredity* 90, 563–565.

Arrieta-Montiel, M., Lyznik, A., Woloszynska, M., Janska, H., Tohme, J. and Mackenzie, S. (2001) Tracing evolutionary and developmental implications of mitochondrial genomic shifting in the common bean. *Genetics* 158, 851–864.

Backert, S. and Borner, T. (2000) Phage T4-like intermediates of DNA replication and recombination in the mitochondria of the higher plant *Chenopodium album* (L.). *Current Genetics* 37, 304–314.

Backert, S., Lurz, R. and Borner, T. (1996) Electron microscopic investigation of mitochondrial DNA from *Chenopodium album* (L.). *Current Genetics* 29, 427–436.

Bailey, M.F., Delph, L.F. and Lively, C.A. (2003) Modeling gynodioecy: novel scenarios for maintaining polymorphism. *American Naturalist* 161, 762–776.

Beardslee, T.A., Roy-Chowdhury, S., Jaiswal, P., Buhot, L., Lerbs-Mache, S., Stern, D.B. and Allison, L.A. (2002) A nucleus-encoded maize protein with sigma factor activity accumulates in mitochondria and chloroplasts. *Plant Journal* 31, 199–209.

Bellaoui, M., Martin-Canadell, A., Pelletier, G. and Budar, F. (1998) Low-copy-number molecules are produced by recombination, actively maintained and can be amplified in the mitochondrial genome of Brassicaceae: relationship to reversion of the male sterile phenotype in some cybrids. *Molecular and General Genetics* 257, 177–185.

Bendich, A.J. (1996) Structural analysis of mitochondrial DNA molecules from fungi and plants using moving pictures and pulsed-field gel electrophoresis. *Journal of Molecular Biology* 255, 564–588.

Bentolila, S., Alfonso, A.A. and Hanson, M.R. (2002) A pentatricopeptide repeat-containing gene restores fertility to cytoplasmic male-sterile plants. *Proceedings of the National Academy of Sciences USA* 99, 10887–10892.

Binder, S. and Brennicke, A. (2003) Gene expression in plant mitochondria: transcriptional and post-transcriptional control. *Philosophical Transactions of the Royal Society of London Series B: Biological Sciences* 358, 181–188.

Birky, C.W. Jr (1983) Relaxed cellular controls and organelle heredity. *Science* 222, 468–475.

Bjedov, I., Tenaillon, O., Gerard, B., Souza, V., Denamur, E., Radman, M., Taddei, F. and Matic, I. (2003) Stress-induced mutagenesis in bacteria. *Science* 300, 1404–1409.

Black, D.L. (2003) Mechanisms of alternative pre-messenger RNA splicing. *Annual Review of Biochemistry* 72, 291–336.

Blanc, G., Barakat, A., Guyot, R., Cooke, R. and Delseny, M. (2000) Extensive duplication and reshuffling in the *Arabidopsis* genome. *Plant Cell* 12, 1093–1101.

Boutanaev, A.M., Kalmykova, A.I., Sheveloyov, Y.Y. and Nuriminsky, D.I. (2002) Large clusters of co-expressed genes in the *Drosophila* genome. *Nature* 420, 666–669.

Brown, G.G., Formanova, N., Jin, H., Wargachuk, R., Dendy, C., Patil, P., Laforest, M., Zhang, J.F., Cheung, W.Y. and Landry, B.S. (2003) The radish *Rfo* restorer gene of Ogura cytoplasmic male sterility encodes a protein with multiple pentatricopeptide repeats. *Plant Journal* 35, 262–272.

Chi, N.W. and Kolodner, R.D. (1994) Purification and characterization of MSH1, a yeast mitochondrial protein that binds to DNA mismatches. *Journal of Biological Chemistry* 269, 29984–29992.

Chopra, I., O'Neill, A.J. and Miller, K. (2003) The role of mutators in the emergence of antibiotic-resistant bacteria. *Drug Resistance Updates* 6, 137–145.

Couvet, D., Atlan, A., Belhassen, E., Gliddon, C.J., Gouyon, P.H. and Kjellberg, F. (1990) Co-evolution between two symbionts: the case of cytoplasmic male-sterility in higher plants. In: Futuyama, D. and Antonovics, J. (eds) *Oxford Survey in Evolutionary Biology*, Vol. 7. Oxford University Press, Oxford, pp. 225–249.

Covello, P.S. and Gray, M.W. (1992) Silent mitochondrial and active nuclear genes for subunit 2 of cytochrome c oxidase (Cox2) in soybean: evidence for RNA-mediated gene transfer. *EMBO Journal* 11, 3815–3820.

Cuguen, J., Wattier, R., Saumitou-Laprade, P., Forcioli, D., Morchen, M., Van Dijk, H. and Vernet, P. (1994) Gynodioecy and mitochondrial DNA polymorphism in natural populations of *Beta vulgaris* ssp. *maritima*. *Genetics Selection Evolution* 26, 87s–101s.

Delorme, V., Keen, C.L., Rai, K.N. and Leaver, C.J. (1997) Cytoplasmic-nuclear male sterility in pearl millet: comparative RFLP and transcript analyses of isonuclear male-sterile lines. *Theoretical and Applied Genetics* 95, 961–968.

Desloire, S., Gherbi, H., Laloui, W., Marhadour, S., Clouet, V., Cattolico, L., Falentin, C., Giancola, S., Renard, M., Budar, F., Small, I., Caboche, M., Delourme, R. and Bendahmane, A. (2003) Identification of the fertility restoration locus, *Rfo*, in radish, as a member of the pentatricopeptide-repeat protein family. *EMBO Reports* 4, 588–594.

Elo, A., Lyznik, A., Gonzalez, D.O., Kachman, S.D. and Mackenzie, S. (2003) Nuclear genes encoding mitochondrial proteins for DNA and RNA metabolism are clustered in the Arabidopsis genome. *Plant Cell* 15, 1619–1631.

Fujie, M., Kuroiwa, H., Kawano, S. and Kuroiwa, T. (1993) Studies on the behavior of organelles and their nucleoids in the root apical meristem of *Arabidopsis thaliana* (L.) Col. *Planta* 189, 443–452.

Gepts, P. (1998) Origin and evolution of common bean: past events and recent trends. *HortScience* 33, 1124–1130.

Gutierres, S., Sabar, M., Lelandais, G., Chetrit, P., Diolez, P., Degand, H., Boutry, M., Vedel, F., de Kouchkovsky, Y. and DePaepe, R. (1997) Lack of mitochondrial and nuclear-encoded subunits of complex I and alteration of the respiratory chain in *Nicotiana sylvestris* mitochondrial deletion mutants. *Proceedings of the National Academy of Sciences USA* 94, 3436–3441.

Handa, H. (2003) The complete nucleotide sequence and RNA editing content of the mitochondrial genome of rapeseed (*Brassica napus* L.): comparative analysis of the mitochondrial genomes of rapeseed and *Arabidopsis thaliana*. *Nucleic Acids Research* 31, 5907–5916.

Harfe, B.D. and Jinks-Robertson, S. (2000) DNA mismatch repair and genetic instability. *Annual Review of Genetics* 34, 359–399.

Hawkesford, M.J. and Buchner, P. (eds) (2001) *Molecular Analysis of Plant Adaptation to the Environment*, 1st edn. Kluwer Academic Publishers, Dordrecht, The Netherlands.

Hedtke, B., Borner, T. and Weihe, A. (1997) Mitochondrial and chloroplast phage-type RNA polymerases in Arabidopsis. *Science* 277, 809–811.

Hervieu, F., Charbonnier, L., Bannerot, H. and Pelletier, G. (1993) The cytoplasmic male sterility (CMS) determinant of common bean is widespread in *Phaseolus coccineus* L. and *Phaseolus vulgaris*. L. *Current Genetics* 24, 149–155.

Hoffmann, M., Kuhn, J., Daschner, K. and Binder, S. (2001) The RNA world of plant mitochondria. *Nucleic Acid Research & Molecular Biology* 70, 119–154.

Jacobs, M.S. and Wade, M.J. (2003) A synthetic review of the theory of gynodioecy. *American Naturalist* 161, 837–851.

Jan, C.C. (2000) Cytoplasmic male sterility in two wild *Helianthus annuus* L. accessions and their fertility restoration. *Crop Science* 40, 1535–1538.

Janska, H., Sarria, R., Woloszynska, M., Arrieta-Montiel, M. and Mackenzie, S. (1998) Stoichiometric shifts in the common bean mitochondrial genome leading to male sterility and spontaneous reversion to fertility. *Plant Cell* 10, 1163–1180.

Kanazawa, A., Tsutsumi, N. and Hirai, A. (1994) Reversible changes in the composition of the population of mtDNAs during dedifferentiation and regeneration in tobacco. *Genetics* 138, 865–870.

Kaul, M.L.H. (1988) *Male Sterility in Higher Plants*. Springer, Berlin.

Kazama, T. and Toriyama, K. (2003) A pentatricopeptide repeat-containing gene that promotes the processing of aberrant atp6 RNA of cytoplasmic male-sterile rice. *FEBS Letters* 544, 99–102.

Koizuka, N., Imai, R., Fujimoto, H., Hayakawa, T., Kumura, Y., Kohno-Murase, J., Sakai, T., Kawasaki, S. and Imamura, J. (2003) Genetic characterization of a pentatricopeptide repeat protein gene, *orf687*, that restores fertility in the cytoplasmic male-sterile Kosena radish. *Plant Journal* 34, 407–415.

Lamppa, G.K. and Bendich, A.J. (1984) Changes in mitochondrial DNA levels during development of pea (*Pisum sativum* L.). *Planta* 162, 463–468.

Laughnan, J.R. and Gabay-Laughnan, S. (1983) Cytoplasmic male sterility in maize. *Annual Review of Genetics* 17, 27–48.

LeClerc, J.E. and Cebula, T.A. (1997) Hypermutability and homeologous recombination: ingredients for rapid evolution. *Bulletin Institute Pasteur* 95, 97–106.

LeClerc, J.E., Li, B., Payne, W.L. and Cebula, T.A. (1996) High mutation frequencies among *Escherichia coli* and *Salmonella* pathogens. *Science* 274, 1208–1211.

Lefai, E., Fernandez-Moreno, M.A., Kaguni, L.S. and Garesse, R. (2000) The highly compact structure of the mitochondrial DNA polymerase genomic region of *Drosophila melanogaster*: functional and evolutionary implications. *Insect Molecular Biology* 9, 315–322.

Le Goff, S., Lachaume, P., Touraille, S. and Alziari, S. (2002) The nuclear genome of a *Drosophila* mutant strain increases the frequency of rearranged mitochondrial DNA molecules. *Current Genetics* 40, 345–354.

Li, G.M. (1999) The role of mismatch repair in DNA damage-induced apoptosis. *Oncology Research* 11, 393–400.

Mackenzie, S. (1991) Identification of a sterility-inducing cytoplasm in a fertile accession line of *Phaseolus vulgaris* L. *Genetics* 127, 411–416.

Mackenzie, S. and Chase, C. (1990) Fertility restoration is associated with loss of a portion of the mitochondrial genome in cytoplasmic male sterile common bean. *Plant Cell* 2, 905–912.

Mackenzie, S. and McIntosh, L. (1999) Higher plant mitochondria. *Plant Cell* 11, 571–585.

Mackenzie, S., Pring, D., Bassett, M. and Chase, C. (1988) Mitochondrial DNA rearrangement associated with fertility restoration and reversion to fertility in cytoplasmic male sterile *Phaseolus vulgaris* L. *Proceedings of the National Academy of Sciences USA* 85, 2714–2717.

Manicacci, D., Atlan, A. and Couvet, D. (1997) Spatial structure of nuclear factors involved in sex determination in the gynodioecious *Thymus vulgaris* L. *Journal of Evolutionary Biology* 10, 889–907.

Marienfeld, J.R., Unseld, M., Brandt, P. and Brennicke, A. (1997) Mosaic open reading frames in the *Arabidopsis thaliana* mitochondrial genome. *Biological Chemistry* 378, 859–862.

Marienfeld, J., Unseld, M. and Brennicke, A. (1999) The mitochondrial genome of *Arabidopsis* is composed of both native and immigrant information. *Trends in Plant Science* 4, 495–502.

Martinez-Zapater, J., Gil, P., Capel, J. and Somerville, C. (1992) Mutations at the *Arabidopsis CHM3* locus promote rearrangements of the mitochondrial genome. *Plant Cell* 4, 889–899.

Modrich, P. and Lahue, R. (1996) Mismatch repair in replication fidelity, genetic recombination, and cancer biology. *Annual Review of Biochemistry* 65, 101–133.

Moriyama, Y. and Kawano, S. (2003) Rapid, selective digestion of mitochondrial DNA in accordance with the matA hierarchy of multiallelic mating types in the mitochondrial inheritance of *Physarum polycephalum*. *Genetics* 164, 963–975.

Nagata, N., Saito, C., Sakai, A., Kuroiwa, H. and Kuroiwa, T. (1999) The selective increase or decrease of organellar DNA in generative cells just after pollen mitosis one controls cytoplasmic inheritance. *Planta* 209, 53–65.

Notsu, Y., Masood, S., Nishikawa, T., Kubo, N., Akiduki, G., Nakazono, M., Hirai, A. and Kadowaki, K. (2002) The complete sequence of the rice (*Oryza sativa* L.) mitochondrial genome: frequent DNA sequence acquisition and loss during the evolution of flowering plants. *Molecular Genetics and Genomics* 268, 434–445.

Nugent, J.M. and Palmer, J.D. (1991) RNA-mediated transfer of the gene *coxII* from the mitochondrion to the nucleus during flowering plant evolution. *Cell* 66, 473–481.

Oda, K., Yamato, K., Ohta, E., Nakamura, Y., Takemura, M., Nozato, N., Akashi, K., Kanegae, T., Ogura, Y., Kohchi, T. and Ohyama, K. (1992) Gene organization deduced from the complete sequence of liverwort *Marchantia polymorpha* mitochondrial DNA: a primitive form of plant mitochondrial genome. *Journal of Molecular Biology* 223, 1–7.

Oldenburg, D.J. and Bendich, A.J. (1996) Size and structure of replicating mitochondrial DNA in cultured tobacco cells. *Plant Cell* 8, 447–461.

Olson, M.S. and McCauley, D.E. (2002) Mitochondrial DNA diversity, population structure, and gender association in the gynodioecious plant *Silene vulgaris*. *Evolution* 56, 253–262.

Peeters, N. and Small, I. (2001) Dual targeting to mitochondria and chloroplasts. *Biochimica et Biophysica Acta* 1541, 54–63.

Reboud, X. and Zeyl, C. (1994) Organelle inheritance in plants. *Heredity* 72, 132–140.

Reenan, R.A.G. and Kolodner, R.D. (1992) Isolation and characterization of two *Saccharomyces cerevisiae* genes encoding homologs of the bacterial HexA and MutS mismatch repair proteins. *Genetics* 132, 963–973.

Rieseberg, L.H., Baird, S.J.E. and Gardner, K.A. (2000) Hybridization, introgression, and linkage evolution. *Plant Molecular Biology* 42, 205–224.

Robertson, E.J., Williams, M., Harwood, J.L., Lindsay, J.G., Leaver, C.J. and Leech, R.M. (1995) Mitochondria increase three-fold and mitochondrial proteins and lipid change dramatically in postmeristematic cells in young wheat leaves grown in elevated CO_2. *Plant Physiology* 108, 469–474.

Sakamoto, W., Kondo, H., Murata, M. and Motoyoshi, F. (1996) Altered mitochondrial genome expression in a maternal distorted leaf mutant of Arabidopsis induced by chloroplast mutator. *Plant Cell* 8, 1377–1390.

Schnable, P.S. and Wise, R.P. (1998) The molecular basis of cytoplasmic male sterility and fertility restoration. *Trends in Plant Science* 3, 175–180.

Small, I. and Peeters, N. (2000) The PPR motif, a TPR-related motif prevalent in plant organellar proteins. *Trends in Biological Sciences* 25, 46–47.

Small, I.D., Isaac, P.G. and Leaver, C.J. (1987) Stoichiometric differences in DNA molecules containing the *atpA* gene suggest mechanisms for the generation of mitochondrial genome diversity in maize. *EMBO Journal* 6, 865–869.

Small, I., Wintz, H., Akashi, K. and Mireau, H. (1998) Two birds with one stone: genes that encode products targeted to two or more compartments. *Plant Molecular Biology* 38, 265–277.

Smith, R.L. and Chowdhury, M.K.U. (1991) Characterization of pearl millet mitochondrial DNA fragments rearranged by reversion from cytoplasmic male sterility to fertility. *Theoretical and Applied Genetics* 81, 793–799.

Sodmergen, S., Zhang, Q.A., Zhang, Y.T., Sakamoto, W. and Kuroiwa, T. (2002) Reduction in amounts of mitochondrial DNA in the sperm cells as a mechanism for maternal inheritance in *Hordeum vulgare*. *Planta* 216, 235–244.

Sutovsky, P. (2003) Ubiquitin-dependent proteolysis in mammalian spermatogenesis, fertilization, and sperm quality control: killing three birds with one stone. *Microscopy Research and Technique* 61, 88–102.

Tang, H.V., Pring, D.R., Shaw, L.C., Salazar, R.A., Muza, F.R., Yan, B. and Schertz, K.F. (1996) Transcript processing internal to the mitochondrial open reading frame is correlated with fertility restoration in male-sterile sorghum. *Plant Journal* 10, 123–133.

Tsaftaris, A.S. and Kafka, M. (1998) Mechanisms of heterosis in crop plants. *Journal of Crop Production* 1, 95–111.

Unseld, M., Marienfeld, J.R., Brandt, P. and Brennicke, A. (1997) The mitochondrial genome of *Arabidopsis thaliana* contains 57 genes in 366,924 nucleotides. *Nature Genetics* 15, 57–61.

Van Damme, J.M.M. (1983) Gynodioecy in *Plantago lanceolata* L. II. Inheritance of three male sterility types. *Heredity* 50, 253–273.

Zhang, Q., Liu, Y. and Sodmergen, S. (2003) Examination of the cytoplasmic DNA in male reproductive cells to determine the potential for cytoplasmic inheritance in 295 angiosperm species. *Plant and Cell Physiology* 44, 941–951.

6   Reticulate evolution in higher plants

Gay McKinnon

School of Plant Science, University of Tasmania, Private Bag 55, Hobart, Tasmania 7001, Australia

Introduction

Many evolutionary analyses focus exclusively on divergence. Evolution is seen as a branching tree, from which new lineages are continually splitting or budding. Most commonly used methods for calculating the evolutionary history of a group are designed to estimate the order in which different genera or species diverged. Yet divergence is only part of the evolutionary pattern. Particularly in the plant kingdom, populations and species may undergo repeated episodes of divergence, followed by episodes of recombination. Through hybridization between taxa, new genetic combinations are formed. This process, whereby branches of the evolutionary tree exchange genes or are grafted together, is called reticulate (net-like) evolution.

The study of reticulate evolution is a challenging and dynamic area of research. Methods for reconstructing reticulate evolutionary pathways are only now being developed (e.g. Lapointe, 2000; Xu, 2000). These methods should find wide application, as new molecular studies confirm that reticulation is both widespread and important to evolution and diversity in higher plants. In some cases reticulation is creative, leading to enhanced diversity through the establishment of new hybrid species or the expansion of a species' gene pool. In others, reticulation may be destructive of diversity, causing the merging or assimilation of species. This chapter reviews evidence for reticulation in higher plants and discusses its significance as an evolutionary mechanism.

Natural Hybridization in Plants

In the plant kingdom, periods of genetic divergence followed by recombination take place at various taxonomic levels, including the population and subspecies levels. However, the term 'reticulate evolution' is normally restricted to the description of gene flow between taxonomically distinct species. If two species are able to cross successfully to form a first-generation (F_1) hybrid (Fig. 6.1a), a range of different consequences may ensue:

1. The F_1 hybrid fails to reproduce, and forms an evolutionary dead-end.
2. The F_1 hybrid self-pollinates, or mates with other F_1 hybrids, to give second-generation (F_2) hybrid progeny. High genotypic and phenotypic variability is expected in the F_2 generation, because genes from the parental species can segregate into numerous different combinations (Fig. 6.1b). Some F_2 progeny may closely resemble the F_1 generation or the original parental species.

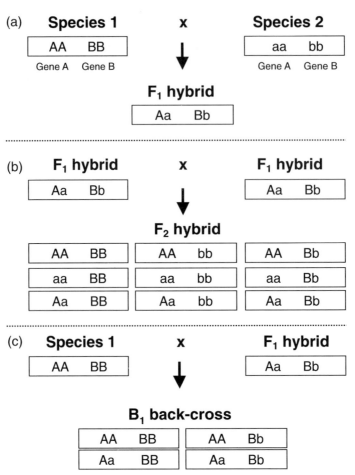

Fig. 6.1. Possible genetic recombinations at two segregating nuclear loci (A and B) in the diploid progeny of two hybridizing diploid species (1 and 2). (a) First-generation hybrids; (b) second-generation hybrids; (c) back-crosses to species 1.

3. The F_1 hybrid mates with members of either or both parental species to produce back-crossed (B_1) progeny. Given that members of the parental species are often both more numerous in the population and more fertile than F_1 hybrids, this is a likely scenario. Back-crossed progeny also have a range of different genotypes and phenotypes, which may resemble the F_1 generation or the original parental species (Fig. 6.1c). In some cases back-crossing occurs with one parental species preferentially, biasing the genomic composition of subsequent generations.

If numerous mating combinations are possible, a complex hybrid zone may be established, containing a mixture of F_1 progeny, advanced generation hybrids, back-crosses and pure parental species. In some cases, such hybrid zones are both local and ephemeral, and may have little evolutionary significance. In others, hybridization can have far-reaching consequences including the formation of new varieties, subspecies, species or polyploid complexes.

The frequency of natural hybridization

Natural hybridization is quite common in higher plants, but its frequency is not evenly distributed among taxa. Certain genera and families, as well as certain regions, show high hybridization frequencies relative to others. Collated evidence on the distribution of spontaneous plant hybrids from five floras (the British Isles, Scandinavia, the Hawaiian Islands, and the Great Plains and Intermountain West of North America) shows that within each flora, only 16 to 34% of families have any reported hybrids (Ellstrand *et al.*, 1996). Most of the hybrids are concentrated within a few genera, which share certain characteristics that may promote hybrid formation and persistence: outcrossing, perennial habit and mechanisms for clonal reproduction (e.g. vegetative spread). Families that show high hybrid frequencies in more than one flora are the Poaceae, Cyperaceae, Scrophulariaceae, Salicaceae, Rosaceae and Asteraceae. Hybridization is also common in the Betulaceae, Onagraceae, Orchidaceae and Pinaceae (Stace, 1975).

The failure to observe hybridization within some plant families today does not mean that the evolutionary history of those families has been free of reticulation. According to one estimate, about 70% of angiosperms have polyploidy in their history (Masterson, 1994). Since many polyploids have apparently arisen following hybridization, a major evolutionary role for hybridization can be inferred for higher plants.

Barriers to hybridization

What factors control the extent of hybridization among plant taxa? Successful hybridization requires: (i) transfer of pollen between species; (ii) successful ovule fertilization and seed maturation; and (iii) ability of hybrid progeny to survive and attain reproductive maturity.

Barriers to one or more of these requirements commonly operate to prevent gene flow between plant species. Pre-pollination barriers include differences in flowering time, adaptation to different pollinators, and ecological differences that prevent species growing in sympatry. If pollen is transferred between species, successful fertilization requires pollen germination, penetration of the stigma, growth of the pollen tube through the style, discharge of gametes with fertilization of the ovule, and the production of viable seed. Incompatibility between species can affect any of these stages, resulting in pollen tube arrest or embryo abortion. For example, hybridization between *Rhododendron* species with differing flower structures is restricted by mismatches between style length and pollen tube length. Pollen tubes from species with much shorter or longer styles than the recipient species either fail to penetrate the ovary, or grow too far and fail to release their gametes (Williams *et al.*, 1986; Williams and Rouse, 1988). A complex combination of barriers to fertilization can coexist in a given species pair. In *Penstemon*, hybridization between the naturally sympatric species *Penstemon spectabilis* and *Penstemon centranthifolius* is limited by reduced pollen tube growth and seed set when *P. spectabilis* is the ovule parent, and by poor pollen grain germination and fruit set when *P. centranthifolius* is the ovule parent (Chari and Wilson, 2001). In addition, partial pollinator specificity helps to maintain isolation between these species (Chari and Wilson, 2001).

Even when two species are reproductively compatible, heterospecific (other-species) pollen must generally overcome competition with conspecific (same-species) pollen to fertilize the ovules. Such competition is usually shown to favour conspecific pollen as the seed sire, for example in *Hibiscus* (Klips, 1999), *Piriqueta* (Wang and Cruzan, 1998) and *Helianthus* (Rieseberg *et al.*, 1995a). The success of pollen competition may be dependent on the direction of the cross, leading to asymmetric hybridization. In crosses between *Mimulus nasutus* and *Mimulus guttatus*, pollen tubes from *M. nasutus* grow much more slowly than those of *M. guttatus* in styles of *M. guttatus*, whereas pollen tubes from either species grow equally fast in styles of *M. nasutus*. As predicted, mixed pollen loads produce far more hybrid seed in *M. nasutus* than in *M. guttatus* (Diaz and Macnair, 1999).

Hybrid fitness

The fitness of hybrids, once formed, is variable. The new genetic combinations generated by hybridization range from advantageous to deleterious. First-generation hybrids often show superior growth or size, termed hybrid vigour or heterosis, compared with parental species (particularly where the parents are inbred). However, their fertility may be low. This can result from chromosomal differences between the parents that prevent successful pairing during meiosis, or from unfavourable genic interactions, including cytonuclear interactions (reviewed in Burke and Arnold, 2001). If hybrids succeed in reproducing, fitness may be reduced in later generations, through the loss of favourably interacting gene complexes or the generation of lethal gene combinations. This is referred to as hybrid breakdown.

Nevertheless, some hybrid combinations equal or excel their parent species in fecundity or tolerance to environmental stresses. New, invasive lineages can evolve following hybridization, as documented in 12 different plant families by Ellstrand and Schierenbeck (2000). Their success may be due to fixed heterosis and/or to evolutionary novelty. Later-generation hybrids frequently demonstrate transgressive segregation: the production of phenotypes that are novel and extreme, rather than intermediate between their parents. Quantitative genetic studies show that this is chiefly due to the action of complementary genes (Rieseberg et al., 1999a). These novel phenotypes may contribute to evolutionary success. The invasive cordgrass, *Spartina anglica*, a hybrid derivative of *Spartina maritima* and *Spartina alterniflora*, is cited as a possible example of success through both fixed heterosis and evolutionary novelty (Ellstrand and Schierenbeck, 2000).

The role of environment

Environmental conditions are important in determining the frequency and results of hybridization. A large body of evidence shows that hybrids are particularly common in disrupted habitats. For example, scarlet oak (*Quercus coccinea*) and black oak (*Quercus velutina*), both of which occur naturally in the eastern USA, are normally reproductively isolated by their preferences for moist low areas and dry well-drained areas, respectively. Hybrids between the two species are common only when the natural environment has been disturbed by human activity such as cutting and burning. In *Rorippa*, different types of habitat disturbance promote different patterns of hybridization (Bleeker and Hurka, 2001). In the River Elbe, which shows a natural dynamic of erosion leading to periodic habitat disturbance, hybridization with bi-directional introgression occurs between *Rorippa amphibia* and *Rorippa sylvestris*; in artificial drainage ditches, hybridization occurs instead between *R. amphibia* and *Rorippa palustris*, with unidirectional introgression of genetic markers into *R. amphibia*.

There are several ways in which habitat disturbance is likely to promote interspecific hybridization. The first is by altering species distributions, thereby creating new mating opportunities. Upheavals caused by floods, fires, landslides, farming and road building create corridors for dispersal that increase contact between species. The second is by the provision of novel or open environments in which hybrids are able to establish themselves. While new hybrids are unlikely to be as well adapted as their parents to the habitats in which those parents evolved, they may be successful in habitats that have been cleared of competitors and/or that contain new ecological niches. Anderson (1948, 1949) attributed the presence of numerous phenotypically variable *Iris* hybrids in farmland on the Mississippi Delta to the rich variety of novel habitats created by land use. Later molecular analyses of *Iris* hybrid zones showed that certain hybrid genotypes were indeed associated with novel habitats (Cruzan and Arnold, 1993). A third, less commonly reported consequence of habitat disturbance is a change in flowering synchrony between species. Lamont et al. (2003) showed that hybrids between *Banksia hookeriana* and

Banksia prionotes occurred in disturbed vegetation as a result of a shift in flowering phenology. Affected populations showed increased fecundity, earlier flowering of *B. hookeriana* and prolonged flowering of *B. prionotes*, so that interspecific pollination became possible. Undisturbed interspersed populations rarely coflowered and showed no hybridization.

Climate change and hybridization

Climate change is a form of natural disturbance that causes both species migrations and habitat transformations. The Quaternary Ice Ages led to cyclic expansion and contraction of many species' ranges in response to changes in aridity and temperature (Hewitt, 1999, 2000). Alpine plants descended and ascended mountains, while lowland species underwent major geographical redistributions. Worldwide, the lowering of sea levels linked landmasses periodically, allowing contact between different floras. In some cases, these changes apparently led to hybridization. For instance, chloroplast (cp) DNA diversity in northern populations of *Packera pseudaurea* in Alberta suggests former hybridization with other *Packera* species which migrated southwards during periods of glaciation (Yates *et al.*, 1999; Golden and Bain, 2000). In south-east Spain, species of *Armeria*, which now occur at different altitudes, may have hybridized when in temporary sympatry (Larena *et al.*, 2002). On the island of Tasmania, cpDNA patterns in *Eucalyptus* are consistent with hybridization between local endemics and species which migrated into Tasmania from mainland Australia (McKinnon *et al.*, 2004).

One of the best-studied examples of a species complex that was redistributed during the Ice Ages is that of the oaks. Pollen evidence and molecular markers show that oaks have recolonized Europe from refugia in Iberia, Italy and the Balkans since the Last Glacial Maximum. A systematic sharing of local cpDNA markers, revealing widespread hybridization, has been found in seven species throughout Europe (Dumolin-Lapègue *et al.*, 1997). Initially, it was thought that hybridization between pedunculate oak (*Quercus robur*) and sessile oak (*Quercus petraea*) probably occurred at a time of low population size when the two species were confined to a glacial refuge (Ferris *et al.*, 1993). Later post-glacial expansion could then have led to the two species sharing cpDNA haplotypes in recolonized terrain. However, a fine-scale analysis (Petit *et al.*, 1997) showed that hybridization could have played an important role in the recolonization process itself. *Q. robur* may have acted as the pioneer species recolonizing new territories through seed dispersal, while *Q. petraea* followed by pollinating established populations of *Q. robur* and ultimately replacing it through back-crossing.

Forms of Reticulate Evolution

The term 'reticulate evolution' embraces a range of evolutionary outcomes which may follow hybridization between species. These depend on many factors such as the degree of genetic compatibility between the parental species, ecological preferences of the parents and hybrids, and the fitness and fertility of hybrids. Although intuitively it would seem that low levels of hybrid formation are unlikely to attain evolutionary significance, this is not always the case. Many species which have quite low interfertility nevertheless form stable and persistent hybrid zones, or give rise to hybrid lineages. For example, the formation of F_1 hybrid seed between the North American milkweeds *Asclepias exaltata* and *Asclepias syriaca* is a rare event, yet natural hybrids occur throughout areas of sympatry, and act as bridges for interspecific gene flow through back-crossing (Broyles, 2002). Only 5.6% of pollen is viable in F_1 hybrids between *Helianthus annuus* and *Helianthus petiolaris*, yet over 90% fertility is regained after only four generations of sib-mating or back-crossing (Ungerer *et al.*, 1998). In nature, these two species commonly form hybrid zones and are believed to be the progenitors of three different hybrid species: *Helianthus anomalus*, *Helianthus deserticola* and *Helianthus paradoxus* (Rieseberg, 1991).

The three major forms of reticulate evolution, which will be considered in more detail below, are:

1. The formation of new species, termed hybrid speciation.
2. The transfer of genes between species, termed introgression.
3. Merging of species or extinction by assimilation.

Hybrid speciation

There is little doubt that hybrid speciation is of major evolutionary importance in the plant kingdom. The potential role of hybridization in plant speciation was recognized by early botanists such as Linnaeus and Mendel, and present-day understanding of hybrid speciation is based on over two centuries' worth of investigation (reviewed in Rieseberg, 1997). The current view is that some recognized 'species' have arisen more than once through separate hybridization events. Furthermore, in some genera, hybridization between the same parental taxa has given rise to different hybrid species. Certain hybrid taxa are derived from more than two parental species; for example, *Iris nelsonii* is apparently a derivative of three species: *Iris hexagona*, *Iris fulva* and *Iris breviligulata* (Arnold, 1993). These findings illustrate the complexity of species relationships in plants and the inadequacy of simple bifurcating phylogenies to describe such relationships.

The formation of a new species through hybridization requires the development of a reproductive barrier between the newly formed hybrid lineage and its parents. Without such a barrier, the new lineage will be swamped by gene flow with one or both parental species, particularly during the early stages when the parents and hybrids are sympatric. For this reason, hybrid speciation often involves a change in ploidy, or some other form of chromosomal or genic incompatibility between a hybrid and its parents. The ability to reproduce clonally, for instance through apomixis, may help to stabilize a newly formed hybrid lineage.

Ecological divergence between the hybrid species and its parents is also likely to reinforce genetic isolation of the hybrid.

Studying hybrid speciation

Studies of hybrid speciation fall into two broad categories. The first is the investigation of naturally occurring hybrid species. Modern molecular techniques make it possible to confirm or disprove hypotheses of hybrid speciation, since hybrid species are expected to carry a combination of genetic markers from their putative parents. In some cases, hybrid speciation is discovered during phylogenetic analysis of a species complex. For instance, the sunflower *Helianthus anomalus* was found to combine the chloroplast and nuclear ribosomal markers of two species, *H. annuus* and *H. petiolaris*, suggesting a hybrid origin (Rieseberg, 1991). Matching evidence from multiple, unlinked genetic markers provides the strongest support for a hypothesis of hybrid speciation. Many studies now employ combined evidence from cytological analysis, nuclear markers (e.g. allozymes, nuclear ribosomal DNA, low-copy nuclear genes, random amplified polymorphic DNA) and/or chloroplast markers.

The second category is the manipulation of experimental hybrid lineages to determine the mechanisms governing hybrid speciation. Mapped molecular markers have been used to study gene segregation in synthetic hybrids, with remarkable results. Rieseberg *et al.* (1996) used 197 markers covering the sunflower genome to investigate three experimentally created hybrid lineages between the sunflowers *H. annuus* and *H. petiolaris*. If chance governed the segregation of markers in later generation hybrids, independent hybrid lineages would have widely differing genotypes. Astonishingly, however, all three lineages were very similar in genomic composition to each other and to the naturally occurring hybrid species *H. anomalus*. The fact that almost identical lineages can be reproduced by independent hybridization events shows that selection, rather than chance, governs the genomic composition of hybrid species.

This research lends support to the contention that many hybrid species have arisen repeatedly.

Hybrid speciation with an increase in ploidy

Polyploidy is a common mode of speciation in plants, and in many cases arises following hybridization between genetically differentiated species whose chromosomes are too dissimilar to pair correctly during meiosis. Meiotic failure in the F_1 hybrid leads to the production of unreduced ($2n$) gametes. The union of two unreduced $2n$ gametes gives rise to a $4n$ (tetraploid) zygote, termed an allotetraploid. Alternatively, union of a $2n$ gamete with a normal haploid gamete gives rise to a triploid ($3n$) offspring, which may produce triploid gametes; these can then unite with haploid gametes to produce tetraploid progeny. The latter mechanism was demonstrated by Müntzing in crosses of mint (*Galeopsis pubescens* × *Galeopsis speciosa*) as long ago as 1930. Higher level allopolyploids, which combine three or more genomes, may also arise following hybridization.

Speciation by polyploidy is a form of instantaneous, sympatric speciation. The new allopolyploid is often fully fertile, and fully or partially reproductively isolated from its nearest relatives. Polyploid speciation may give rise to polyploid series (with multiples of the basic chromosome number, as for instance in *Chrysanthemum*) or complexes in which species with different basic chromosome numbers have hybridized and become polyploid (for instance *Clarkia*; Lewis and Lewis, 1955). Current research indicates that many polyploid species have arisen recurrently, contradicting the principle that biological species have a unique, monophyletic origin (Soltis and Soltis, 1999). In fact, multiple origins for polyploid species may be the rule rather than the exception. Molecular studies on the *Tragopogon* tetraploids, *Tragopogon miscellus* and *Tragopogon mirus*, indicate that spread of each species is occurring not through dispersal from a single origin but through repeated instances of recreation (Soltis *et al.*, 1995; Cook *et al.*, 1998). These two species may have formed as often as 20 and 12

times, respectively, in 70 years. Polyploid species of *Draba* and *Saxifraga* may also have had multiple origins from diploid progenitors (Brochmann and Elven, 1992; Brochmann *et al.*, 1998).

Since most parental species will exhibit some genetic variation across their natural geographic ranges, polyploid species arising from separate hybridization events in different localities will constitute a series of genetically differentiated populations. Gene flow may then occur between the different polyploid populations, creating even greater genetic variability. This variability will be further enhanced by different chromosomal rearrangements arising in different populations following polyploidy. Recent evidence shows that allopolyploids undergo extensive and rapid genomic reshuffling after formation (Soltis and Soltis, 1999). In *Brassica*, extensive genetic and phenotypic diversity developed after only a few generations in experimentally created allopolyploids (Song *et al.*, 1995). Thus, polyploid hybrid speciation represents a particularly dynamic form of reticulate evolution.

Homoploid hybrid speciation

Hybridization can also give rise to new species with the same ploidy as the parental species. Molecular studies have confirmed the natural occurrence of homoploid hybrid speciation in *Stephanomeria* (Gallez and Gottlieb, 1982), *Helianthus* (Rieseberg, 1991), *Iris* (Arnold, 1993), *Pinus* (Wang and Szmidt, 1994; Wang *et al.*, 2001), *Paeonia* (Sang *et al.*, 1995) and *Penstemon* (Wolfe *et al.*, 1998). Evidence from *Paeonia* (Sang *et al.*, 1995) suggests that homoploid hybrid species in some cases may go on to found speciose lineages. Homoploid hybrid speciation is often reported for diploids, but may also occur naturally between allotetraploids without an intermediate stage of genome diploidization or a further increase in ploidy (Ferguson and Sang, 2001).

How does a hybrid develop reproductive isolation from its parental species and found a new lineage without a change in ploidy? Stebbins (1950) proposed a mechanism, later termed 'recombinational speciation'

(Grant, 1971), involving parental species differing by two or more separable chromosomal rearrangements. In theory, two such species could found a new hybrid lineage through the following sequence of events:

1. The F_1 hybrid between the two parental species is formed but has low fertility, since most of its gametes carry unbalanced chromosome complements.
2. Some offspring of the F_1 hybrid, through chance segregation and recombination, are homozygous for balanced chromosome complements.
3. These offspring are fertile, but reproductively isolated from other such homozygotes and from the parental species, enabling the establishment of a new lineage.

The mechanism of recombinational speciation has been verified experimentally in a number of genera, including *Nicotiana* (Smith and Daly, 1959) and *Gilia* (Grant, 1966). More general models for homoploid hybrid speciation have since been proposed (e.g. Templeton, 1981). The modern view is that a variety of mechanisms including both chromosomal and genic incompatibility, and ecological divergence and selection for hybrids, can promote this form of speciation.

Recent studies have shown that, like allopolyploids, diploid hybrid species may arise recurrently. Schwarzbach and Rieseberg (2002) deduced from chloroplast DNA and crossability data that the diploid hybrid sunflower species *H. anomalus* probably arose on three occasions independently from crosses between its parental species, *H. annuus* and *H. petiolaris*. Recurrent diploid speciation has also been suggested for *Pinus densata* (Wang *et al.*, 2001) and *Argyranthemum sundingii* (Brochmann *et al.*, 2000). In addition, different diploid hybrid species can arise naturally from the same cross. For example, *Argyranthemum lemsiii* and *A. sundingii* share the parental species *Argyranthemum frutescens* and *Argyranthemum broussonetii*, but are derived by crosses in opposite directions and show different chromosomal rearrangements (Borgen *et al.*, 2003). The sunflowers *H. deserticola* and *H. paradoxus* have the same parental species as

H. anomalus (Rieseberg, 1991), but occur in different habitats, suggesting a role for ecological selection.

Introgression

Another common consequence of hybridization between species is introgression, the infiltration of genes from one species into the gene pool of the other. F_1 hybrids, once formed, act as a bridge to gene flow through back-crossing to either or both of the parental species. Introgression may be unidirectional, with genes flowing into only one of the species involved, or bidirectional. Its nature depends on a complex combination of factors, including mating patterns and chromosomal and genic incompatibilities. A distinction is drawn between localized introgression, which refers to the exchange of genetic markers within an obvious hybrid zone, and dispersed introgression, which refers to the flow of genes from one species into another at a distance from the hybrid zone. Dispersed introgression may be due to: (i) flow of introgressed genes across a population through pollen dispersal; (ii) seed dispersal of progeny carrying introgressed genes; or (iii) the movement or disappearance over time of a hybrid zone, leaving behind introgressed individuals.

Introgressive hybridization has been proposed as an important mechanism leading to race formation in plant groups including *Pinus*, *Abies*, *Quercus*, *Purshia*, *Cistus*, *Coprosma*, *Dracophyllum*, *Helianthus*, *Gilia* and *Tradescantia* (Stebbins, 1950; Grant, 1971). Molecular evidence confirms that certain intraspecific taxa such as the sunflower subspecies, *H. annuus* ssp. *texanus*, and the groundsel variant, *Senecio vulgaris* var. *hibernicus*, have arisen by introgression (reviewed in Abbott, 1992). In theory, introgression should increase the genetic diversity of a species and allow it to occupy new habitats through the capture or development of useful adaptations. Such increased genetic diversity has been demonstrated in species of *Cypripedium* (Klier *et al.*, 1991) and *Aesculus* (dePamphilis and Wyatt, 1990). However, the role of introgression as a means for transferring or creating beneficial adaptations remains difficult to prove through direct evidence. A possible example is that of *Rhododendron ponticum* in the British Isles. Combined molecular and biogeographic evidence suggest that this species may have acquired enhanced cold tolerance through introgressive hybridization with *Rhododendron catawbiense* (Milne and Abbott, 2000).

Studying introgression

Introgressive hybridization is sometimes difficult to distinguish from other processes which give rise to similar patterns of morphological variation. Individuals that are morphologically intermediate between two recognized species might have arisen by hybridization. Alternatively, they might be remnants of an ancestral population from which the two species arose, or members of different species converging in morphology under natural selection. In addition, advanced generation hybrids sometimes resemble their parental species so strongly that their hybrid nature goes undetected. For this reason molecular markers are widely applied to the study of introgression. Evidence from cytoplasmic (chloroplast, cp, or mitochondrial, mt) DNA markers is commonly used in combination with evidence from nuclear genomic markers. The latter include allozymes, random amplified polymorphic DNA (RAPDs), ribosomal DNA (rDNA) sequences, microsatellites and restriction fragment length polymorphisms (RFLPs).

Recent studies have shown that introgression can be remarkably selective. Typically, cytoplasmic markers such as cpDNA are exchanged far more readily than nuclear markers (Rieseberg and Soltis, 1991). This is due partly to their uniparental mode of inheritance (Fig. 6.2). In most (although not all) flowering plants, both mitochondria and chloroplasts are inherited from the maternal parent. An F_1 hybrid between two species, A (male) and B (female) therefore inherits mtDNA and cpDNA only from B. If the hybrid is pollinated by A, its progeny will carry nuclear markers characteristic of A,

combined with the cytoplasmic markers of B. Repeated generations of back-crossing in the same direction will dilute the nuclear markers of B until they are difficult to detect, but cannot remove the cytoplasmic markers of B. Another possible reason for higher levels of cytoplasmic marker introgression is selection against alien nuclear genes, but not alien cytoplasmic genes.

The most exciting recent research on introgression uses molecular markers which have been mapped to different chromosomal regions. These mapped markers allow tracking of the movement of chromosomal segments between species. A detailed study by Martinsen *et al.* (2001) of contemporary

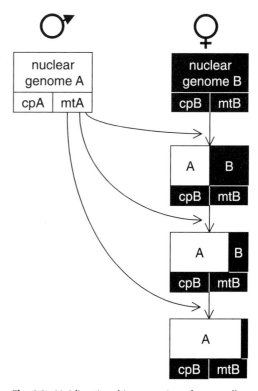

Fig. 6.2. Unidirectional introgression of maternally inherited cp and mtDNA following hybridization and back-crossing between two species, A and B. Species A acts as the pollen parent in all crosses. First-generation hybrids carry the combined nuclear genes of A and B, with the cp and mtDNA of B. Successive back-crosses to pollen parent A dilute the nuclear genomic contribution of B, but can never erase the cp and mtDNA of B.

unidirectional introgression between Fremont cottonwood (*Populus fremontii*) and the higher-altitude species, narrowleaf cottonwood (*Populus angustifolia*), along the Weber River in Utah, used 35 genetically mapped RFLP markers. These markers showed that the majority of the nuclear genome was not exchanged between species. However, a small percentage (21%) of nuclear markers from Fremont cottonwood was able to introgress into 'pure' narrowleaf cottonwood. Different markers showed different levels of introgression. Some were found in narrowleaf cottonwood only short distances from the 13 km long hybrid zone, but others showed dispersed introgression up to 100 km from the hybrid zone. This pattern may have arisen when a hybrid zone between the two species moved downhill gradually in response to climate change. The authors suggested that hybrids could act as evolutionary filters that allow introgression of beneficial genes between species, while preventing the transfer of deleterious genes.

In *Helianthus*, the genetic architecture of barriers to introgression of nuclear genes is now being uncovered. The species *H. annuus* and *H. petiolaris*, described above as the progenitors of three different hybrid species, also demonstrate localized introgression. Both species are diploid (*n* = 17) but only seven of their chromosomes are co-linear; the remaining ten differ by a number of translocations and inversions. Rieseberg *et al.* (1995b) used mapped RAPD markers to study the introgression of nuclear genomic segments from *H. petiolaris* into *H. annuus* in experimental hybrids. They found that chromosomal rearrangements acted as a strong barrier to introgression. Only 2.4% of the rearranged portion of the genome introgressed, whereas 40% of the co-linear portion was able to introgress. For both portions of the genome, marker introgression was significantly less than would be predicted by chance, although a few markers introgressed at higher than expected frequencies. Thus, both chromosomal rearrangements and selection against certain *H. petiolaris* genes appeared to be limiting introgression.

A follow-up study (Rieseberg *et al.*, 1999b) used three sympatric populations of *H. petiolaris* and *H. annuus* to investigate introgression in natural hybrid zones. Patterns of introgression were similar to those seen in experimental hybrids, although with greater recombination and introgression across rearranged parts of the genome. They were also remarkably consistent across the three different hybrid zones, suggesting that chromosomal segments were under similar selective regimes in different populations. Many of the chromosomal segments that failed to introgress were associated with reduced pollen fertility, which would create a selective disadvantage in hybrids. These studies show that, like hybrid speciation, introgression is a non-random process that can produce similar patterns of genetic variation in separate locations.

Introgression and phylogenetic incongruence

The ability of some species to capture genes from others by introgressive hybridization has important consequences for molecular phylogenetic analysis in plants. Until quite recently, evolutionary relationships among plant species were often estimated by phylogenetic analysis of cpDNA sequences, using one or a few individuals to represent each species. More extensive sampling has now shown that many plant species carry multiple cpDNA lineages, and that these lineages are not always species-specific. In some cases, this is due to a phenomenon called incomplete lineage sorting (Fig. 6.3a). Under lineage sorting, the ancestor to a group of species carries multiple divergent cpDNA lineages (A, B, C, D), all or some of which are passed on to each descendant species. These lineages are subject to random drift in the daughter species, so that by chance some species will retain only A, some retain A and D, others retain B and C, and so on. As a result, the cpDNA phylogeny may not accurately reflect the species relationships. Introgression of cpDNA following hybridization between species (Fig. 6.3b) can also create a pattern of shared lineages that obscures the real species relationships. The same principles apply to nuclear gene phylogenies. As a result, phylogenies generated

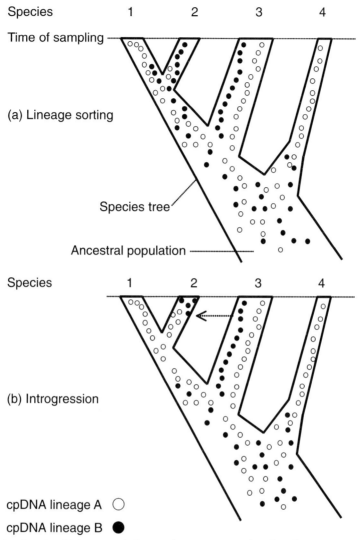

Species 1 2 3 4

Time of sampling

(a) Lineage sorting

Species tree

Ancestral population

Species 1 2 3 4

(b) Introgression

cpDNA lineage A ○

cpDNA lineage B ●

Fig. 6.3. Two situations in which a gene phylogeny does not accurately reflect the species phylogeny. Both situations may occur together. (a) Lineage sorting. The ancestral population to species 1–4 carries two different cpDNA lineages, A and B. By chance, B is eliminated from species 1 and 4, and A is eliminated from species 2. From the cpDNA relationships, species 1 appears more closely related to species 4 than to species 2. (b) Introgression. Species 2 has acquired lineage B through hybridization with species 3. At the time of sampling, species 2 appears more closely related to species 3 than to species 1.

using cp and nuclear DNA sequences may be in conflict with one another, or with phylogenies based on morphological characters.

How can phylogenetic incongruence caused by introgression be distinguished from incongruence due to lineage sorting? A number of criteria are helpful in separating the two processes:

1. Under lineage sorting, enough time may have elapsed since speciation for further sequence divergence within lineages A to D. Species which inherited B from their common ancestor will therefore carry somewhat divergent copies of B. By contrast, recent introgression will result in different species carrying identical copies of

B. The more ancient the introgression, the more difficult it will be to distinguish it from lineage sorting by this criterion. Separation of the two phenomena is complicated by the fact that hybridization within some genera is a continuous process. According to fossil analysis, hybridization between species of *Populus* has been occurring for at least 12 million years (Eckenwalder, 1984). This is likely to generate a complex pattern resulting from both recent and ancient gene flow between species.

2. Assuming that the marker in question is not under selection, lineage sorting is unlikely to give a matching geographical pattern of genetic variation across species. A pattern of shared markers between two species only in regions of sympatry is therefore more likely to result from introgression. This is particularly the case if multiple markers are shared in the region of sympatry.

Phylogenetic incongruence has led to the discovery of unsuspected, historical introgression in higher plant genera. Wendel *et al.* (1995) 'stumbled across' a case of ancient, cryptic introgression in *Gossypium* during phylogenetic analysis of the nuclear rDNA. Sequence data for the rDNA of American *Gossypium gossypioides* placed it in the same clade as African species of *Gossypium*, conflicting with evidence from fertility relationships, cytogenetics, morphology, allozymes and cpDNA. They concluded that an ancient hybridization event must have taken place between species that now occupy different hemispheres. Comes and Abbott (1999) found that, for two species of *Senecio*, both rDNA and cpDNA evidence conflicted with morphological classification. The most likely explanation was historical capture of both rDNA and cpDNA, following introgressive hybridization among species. In one of the two species, this former capture was quite undetectable by RAPD profile or morphology. An increasing number of such discoveries shows that extensive sampling of species and markers is wise when analysing evolutionary histories.

Merging of species and genetic assimilation

An extreme consequence of introgressive hybridization is the merging of species at one or more sympatric populations. In plants with short generation times, this can happen rapidly. Carney *et al.* (2000a) studied the change in genetic and morphological composition of a hybridizing population of the sunflowers *Helianthus bolanderi* and *H. annuus* after 50 years. They found that few genetically pure parental plants remained in the population. The average phenotype had shifted in bias from *H. bolanderi* to *H. annuus* in this time. The trend was towards assimilation of *H. bolanderi* in this population, and potentially others throughout its range.

Genetic assimilation is presently receiving attention because of its implications for the conservation of rare species. Hybrids formed between a rare species and a more abundant congener may contribute to the demise of the rare species by replacing its conspecific (pure) progeny with increasing numbers of hybrid and back-crossed progeny in each generation. One well-known example is that of the Catalina Island mahogany, *Cercocarpus traskiae*, whose population size has shrunk to about six 'pure' individuals (Rieseberg and Gerber, 1995). This species has declined in number over the last century through overgrazing, but is also apparently under threat through hybridization with its more abundant congener *Cercocarpus betuloides* var. *blancheae*. RAPD data show that in addition to the six pure *Cercocarpus traskiae* individuals, five adult hybrid trees and at least five seedlings of hybrid origin are present in the surviving population.

Island plants appear to be particularly susceptible to genetic assimilation through hybridization. This is due to their small population size, the likelihood of invasion of their habitat by related congeners from other landmasses, and sometimes a lack of strong reproductive barriers between species because of unspecialized pollinators and/or relatively recent divergence (for example, following adaptive radiations). In the Canary Islands, the rare endemic *Argyranthemum coronopifolium* is undergoing assimilation by hybridization with a widespread weed, *A. frutescens* (Levin *et al.*, 1996). Following an encounter between the two species, a population of *A. coronopifolium* was gradually swamped by hybrids and *A. frutescens* over a period of 30 years. Numerous other examples of hybridization between rare species and abundant congeners have been documented in the British Isles (e.g. *Saxifraga, Salix, Sorbus*) and the Hawaiian Islands (e.g. *Cyrtandra, Dubautia*; reviewed in Levin *et al.*, 1996, and Carney *et al.*, 2000b).

Factors such as absolute and relative population sizes, geographical proximity, rates of hybridization and the fitness of hybrids are all likely to influence the rate of genetic assimilation. The typical scenario is that of a rare species undergoing assimilation by a more abundant invader. However, studies of *Spartina* (Anttila *et al.*, 1998) show that even an abundant species may be threatened by serial hybridization with a small population of an invader that produces large quantities of superior pollen. The evolutionary consequences for the assimilating species are rarely considered, but must include the acquisition of new genetic variability. In cases where a species has been completely assimilated, this variability may appear to have arisen within the assimilating species, when in fact it has been acquired through reticulation. Harlan and deWet (1963) proposed the term 'compilospecies' to describe an aggressive species that plunders the gene pools of congeners, thereby increasing its own ecological tolerance and geographic range.

Paradoxically, it has been suggested that hybridization might be one way to conserve genes from extremely rare or threatened species. When only a few individuals of a species remain, inbreeding is likely to become catastrophic. Hybridization with a congener may produce healthy progeny that can propagate the genes of the endangered species. This method has been used to preserve the genes of the St Helena redwood (*Trochetiopsis erythroxylon*) and the St Helena ebony (*Trochetiopsis ebenus*). The two species are almost extinct, but have been crossed to produce vigorous hybrids (Cronk, 1995). In theory, natural hybridization could therefore enrich an endangered species by contributing genes that raise its fitness. Hybrid populations could act as a genetic reservoir for reconstituting the parental genotypes under favourable conditions (Anderson, 1949).

Summary

1. Reticulation (hybridization between divergent taxa) contributes to biodiversity in higher plants through the creation of new hybrid lineages and the transfer of genes between species. The frequency of hybridization varies greatly between plant genera, and some large plant families have few reported hybrids. Nevertheless, it is estimated that most flowering plants have episodes of reticulation in their evolutionary history.

2. Natural hybridization between species is controlled by numerous barriers to gene flow. The frequency and success of hybridization depend on the compatibility of the parent species and on environmental conditions. Natural or anthropogenic disturbances of the environment can promote hybridization by bringing allopatric species together, creating new habitats that favour hybrids, and/or changing flowering synchrony.

3. Even low levels of hybrid formation can have a significant evolutionary outcome. Some species with low inter-fertility nevertheless form persistent hybrid zones, and are known to have founded new hybrid lineages.

4. Hybrid speciation is of major importance in the plant kingdom and can occur with or without a change in ploidy. Recent research shows that many hybrid species have multiple origins. The genomic composition of hybrid species is not governed by chance segregation, but by selection for certain gene combinations.

5. Introgressive hybridization contributes to genetic variability, and probably adaptability, within established species. The exchange of genes between species by introgression is remarkably selective. Cytoplasmic genes are exchanged more readily than nuclear genes, and some nuclear genes are exchanged more readily than others. Like hybrid speciation, introgression is apparently directed by selection.

6. Hybridization can lead to extinction by assimilation in rare species.

7. The study of reticulate evolution is demanding, and requires data of high quality and quantity. Reticulation is problematic for phylogenetic reconstruction. Most methods for phylogenetic analysis assume that species have evolved through a simple, branching evolutionary pathway. Introgression, hybrid speciation and multiple origins for hybrid species do not fit within this assumption. The importance of reticulation is becoming increasingly apparent, and future phylogenetic analyses are likely to place more emphasis on this important feature of plant evolution.

References

Abbott, R.J. (1992) Plant invasions, interspecific hybridization and the evolution of new plant taxa. *Trends in Ecology and Evolution* 7, 401–405.

Anderson, E. (1948) Hybridization of the habitat. *Evolution* 2, 1–9.

Anderson, E. (1949) *Introgressive Hybridization.* John Wiley & Sons, New York.

Anttila, C.K., Daehler, C.C., Rank, N.E. and Strong, D.R. (1998) Greater male fitness of a rare invader (*Spartina alterniflora*, Poaceae) threatens a common native (*Spartina foliosa*) with hybridization. *American Journal of Botany* 85, 1597–1601.

Arnold, M.L. (1993) *Iris nelsonii* (Iridaceae); origin and genetic composition of a homoploid hybrid species. *American Journal of Botany* 80, 577–583.

Bleeker, W. and Hurka, H. (2001) Introgressive hybridization in *Rorippa* (Brassicaceae): gene flow and its consequences in natural and anthropogenic habitats. *Molecular Ecology* 10, 2013–2022.

Borgen, L., Leitch, I. and Santos-Guerra, A.T.I. (2003) Genome organization in diploid hybrid species of *Argyranthemum* (Asteraceae) in the Canary Islands. *Botanical Journal of the Linnaean Society* 141, 491–501.

Brochmann, C. and Elven, R. (1992) Ecological and genetic consequences of polyploidy in arctic *Draba* (Brassicaceae). *Evolutionary Trends in Plants* 6, 111–124.

Brochmann, C., Xiang, Q.Y., Brunsfeld, S.J., Soltis, D.E. and Soltis, P.S. (1998) Molecular evidence for polyploid origins in *Saxifraga* (Saxifragaceae): the narrow arctic endemic *S. svalbardensis* and its widespread allies. *American Journal of Botany* 85, 135–143.

Brochmann, C., Borgen, L. and Stabbetorp, O.E. (2000) Multiple diploid hybrid speciation of the Canary Island endemic *Argyranthemum sundingii* (Asteraceae). *Plant Systematics and Evolution* 220, 77–92.

Broyles, S.B. (2002) Hybrid bridges to gene flow: a case study in milkweeds (*Asclepias*). *Evolution* 56, 1943–1953.

Burke, J.M. and Arnold, M.L. (2001) Genetics and the fitness of hybrids. *Annual Review of Genetics* 35, 31–52.

Carney, S.E., Gardner, K.A. and Rieseberg, L.H. (2000a) Evolutionary changes over the fifty-year history of a hybrid population of sunflowers (*Helianthus*). *Evolution* 54, 462–474.

Carney, S.E., Wolf, D.E. and Rieseberg, L.H. (2000b) Hybridization and forest conservation. In: Young, A., Boshier, D. and Boyle, T. (eds) *Forest Conservation Genetics: Principles and Practice.* CSIRO Publishing, Collingwood, Australia, pp.167–182.

Chari, J. and Wilson, P. (2001) Factors limiting hybridization between *Penstemon spectabilis* and *Penstemon centranthifolius*. *Canadian Journal of Botany – Revue Cananadienne de Botanique* 79, 1439–1448.

Comes, H.P. and Abbott, R.J. (1999) Reticulate evolution in the Mediterranean species complex of *Senecio* sect. *Senecio*: uniting phylogenetic and population-level approaches. In: Hollingsworth, P.M., Bateman, R.M. and Gornall, R.J. (eds) *Molecular Systematics and Plant Evolution.* Taylor & Francis, London, pp. 171–198.

Cook, L.M., Soltis, P.S., Brunsfeld, S.J. and Soltis, D.E. (1998) Multiple independent formations of *Tragopogon* tetraploids (Asteraceae): evidence from RAPD markers. *Molecular Ecology* 7, 1293–1302.

Cronk, Q.C.B. (1995) A new species and hybrid in the St. Helena endemic genus *Trochetiopsis*. *Edinburgh Journal of Botany* 52, 205–213.

Cruzan, M.B. and Arnold, M.L. (1993) Ecological and genetic associations in an *Iris* hybrid zone. *Evolution* 47, 1432–1445.

dePamphilis, C.W. and Wyatt, R. (1990) Electrophoretic confirmation of interspecific hybridization in *Aesculus* (Hippocastanaceae) and the genetic structure of a broad hybrid zone. *Evolution* 44, 1295–1317.

Diaz, A. and Macnair, M.R. (1999) Pollen tube competition as a mechanism of prezygotic reproductive isolation between *Mimulus nasutus* and its presumed progenitor *M. guttatus*. *New Phytologist* 144, 471–478.

Dumolin-Lapègue, S., Demesure, B., Fineschi, S., Le Corre, V. and Petit, R.J. (1997) Phylogeographic structure of white oaks throughout the European continent. *Genetics* 146, 1475–1487.

Eckenwalder, J.E. (1984) Natural intersectional hybridization between North American species of *Populus* (Salicaceae) in sections *Aiegeiros* and *Tacamahaca*. III. Paleobotany and evolution. *Canadian Journal of Botany* 62, 336–342.

Ellstrand, N.C. and Schierenbeck, K.A. (2000) Hybridization as a stimulus for the evolution of invasiveness in plants? *Proceedings of the National Academy of Sciences USA* 97, 7043–7050.

Ellstrand, N.C., Whitkus, R. and Rieseberg, L.H. (1996) Distribution of spontaneous plant hybrids. *Proceedings of the National Academy of Sciences USA* 93, 5090–5093.

Ferguson, D. and Sang, T.T.I. (2001) Speciation through homoploid hybridization between allotetraploids in peonies (*Paeonia*). *Proceedings of the National Academy of Sciences USA* 98, 3915–3919.

Ferris, C., Oliver, R.P., Davy, A.J. and Hewitt, G.M. (1993) Native oak chloroplasts reveal an ancient divide across Europe. *Molecular Ecology* 2, 337–344.

Gallez, G.P. and Gottlieb, L.D. (1982) Genetic evidence for the hybrid origin of the diploid plant *Stephanomeria diegensis*. *Evolution* 36, 1158–1167.

Golden, J.L. and Bain, J.F. (2000) Phylogeographic patterns and high levels of chloroplast DNA diversity in four *Packera* (Asteraceae) species in southwestern Alberta. *Evolution* 54, 1566–1579.

Grant, V. (1966) The origin of a new species of *Gilia* in a hybridization experiment. *Genetics* 54, 1189–1199.

Grant, V. (1971) *Plant Speciation*. Columbia University Press, New York.

Harlan, J.R. and deWet, J.M.J. (1963) The compilospecies concept. *Evolution* 17, 497–501.

Hewitt, G.M. (1999) Post-glacial recolonization of European biota. *Biological Journal of the Linnaean Society* 68, 87–112.

Hewitt, G. (2000) The genetic legacy of the Quaternary ice ages. *Nature* 405, 907–913.

Klier, K., Leoschke, M.J. and Wendel, J.F. (1991) Hybridization and introgression in white and yellow lady-slipper orchids (*Cypripedium candidum* and *C. pubescens*). *Journal of Heredity* 82, 305–319.

Klips, R.A. (1999) Pollen competition as a reproductive isolating mechanism between two sympatric *Hibiscus* species (Malvaceae). *American Journal of Botany* 86, 269–272.

Lamont, B.B., He, T., Enright, N.J., Krauss, S.L. and Miller, B.P. (2003) Anthropogenic disturbance promotes hybridization between *Banksia* species by altering their biology. *Journal of Evolutionary Biology* 16, 551–557.

Lapointe, F.-J. (2000) How to account for reticulation events in phylogenetic analysis: a comparison of distance-based methods. *Journal of Classification* 17, 175–184.

Larena, B.G., Aguilar, J.F. and Feliner, G.N. (2002) Glacial-induced altitudinal migrations in *Armeria* (Plumbaginaceae) inferred from patterns of chloroplast DNA haplotype sharing. *Molecular Ecology* 11, 1965–1974.

Levin, D.A., Francisco-Ortega, J.K. and Jansen, R.K. (1996) Hybridization and the extinction of rare plant species. *Conservation Biology* 10, 10–16.

Lewis, H. and Lewis, M. (1955) The genus *Clarkia*. *University of California Publications in Botany* 20, 241–392.

Martinsen, G.D., Whitham, T.G., Turek, R.J. and Keim, P. (2001) Hybrid populations selectively filter gene introgression between species. *Evolution* 55, 1325–1335.

Masterson, J. (1994) Stomatal size in fossil plants: evidence for polyploidy in majority of angiosperms. *Science* 264, 421–423.

McKinnon, G.E., Jordan, G.J., Vaillancourt, R.E., Steane, D.A. and Potts, B.M. (2004) Glacial refugia and reticulate evolution: the case of the Tasmanian eucalypts. *Philosophical Transactions of the Royal Society of London Series B: Biological Sciences* 359, 275–284.

Milne, R.I. and Abbott, R.J. (2000) Origin and evolution of invasive naturalized material of *Rhododendron ponticum* L. in the British Isles. *Molecular Ecology* 9, 541–556.

Petit, R.J., Pineau, E., Demesure, B., Bacilieri, R., Ducousso, A. and Kremer, A. (1997) Chloroplast DNA footprints of postglacial recolonization by oaks. *Proceedings of the National Academy of Sciences USA* 94, 9996–10001.

Rieseberg, L.H. (1991) Homoploid reticulate evolution in *Helianthus* (Asteraceae): evidence from ribosomal genes. *American Journal of Botany* 78, 1218–1237.

Rieseberg, L.H. (1997) Hybrid origins of plant species. *Annual Review of Ecology and Systematics* 28, 359–389.

Rieseberg, L.H. and Gerber, D. (1995) Hybridization in the Catalina Island Mountain Mahogany (*Cercocarpus traskiae*): RAPD evidence. *Conservation Biology* 9, 199–203.

Rieseberg, L.H. and Soltis, D.E. (1991) Phylogenetic consequences of cytoplasmic gene flow in plants. *Evolutionary Trends in Plants* 5, 65–84.

Rieseberg, L.H., DesRochers, A.M. and Youn, S.J. (1995a) Interspecific pollen competition as a reproductive barrier between sympatric species of *Helianthus* (Asteraceae). *American Journal of Botany* 82, 515–519.

Rieseberg, L.H., Linder, C.R. and Seiler, G.J. (1995b) Chromosomal and genic barriers to introgression in *Helianthus*. *Genetics* 141, 1163–1171.

Rieseberg, L.H., Sinervo, B., Linder, C.R., Ungerer, M.C. and Arias, D.M. (1996) Role of gene interactions in hybrid speciation: evidence from ancient and experimental hybrids. *Science* 272, 741–745.

Rieseberg, L.H., Archer, M.A. and Wayne, R.K. (1999a) Transgressive segregation, adaptation and speciation. *Heredity* 83, 363–372.

Rieseberg, L.H., Whitton, J. and Gardner, K. (1999b) Hybrid zones and the genetic architecture of a barrier to gene flow between two sunflower species. *Genetics* 152, 713–727.

Sang, T., Crawford, D.J. and Stuessy, T.F. (1995) Documentation of reticulate evolution in peonies (*Paeonia*) using ITS sequences of nrDNA: implications for biogeography and concerted evolution. *Proceedings of the National Academy of Sciences USA* 92, 6813–6817.

Schwarzbach, A.E. and Rieseberg, L.H. (2002) Likely multiple origins of a diploid hybrid sunflower species. *Molecular Ecology* 11, 1703–1715.

Smith, H.H. and Daly, K. (1959) Discrete populations derived by interspecific hybridization and selection. *Evolution* 13, 476–487.

Soltis, D.E. and Soltis, P.S. (1999) Polyploidy: recurrent formation and genome evolution. *Trends in Ecology and Evolution* 14, 348–352.

Soltis, P.S., Plunkett, G.M., Novak, S.J. and Soltis, D.E. (1995) Genetic variation in *Tragopogon* species: additional origins of the allotetraploids *T. mirus* and *T. miscellus* (Compositae). *American Journal of Botany* 82, 1329–1341.

Song, K., Lu, P., Tang, K. and Osborn, T.C. (1995) Rapid genome change in synthetic polyploids of *Brassica* and its implications for polyploid evolution. *Proceedings of the National Academy of Sciences USA* 92, 7719–7723.

Stace, C.A. (1975) Introduction. In: Stace, C.A. (ed.) *Hybridization and the Flora of the British Isles.* Academic Press, London, pp. 1–90.

Stebbins, G.L. Jr (1950) *Variation and Evolution in Plants.* Columbia University Press, New York.

Templeton, A.R. (1981) Mechanisms of speciation: a population genetic approach. *Annual Review of Ecology and Systematics* 12, 23–48.

Ungerer, M.C., Baird, S.J.E., Pan, J. and Rieseberg, L.H. (1998) Rapid hybrid speciation in wild sunflowers. *Proceedings of the National Academy of Sciences USA* 95, 11757–11762.

Wang, J.Q. and Cruzan, M.B. (1998) Interspecific mating in the *Piriqueta caroliniana* (Turneracea) complex: effects of pollen load size and competition. *American Journal of Botany* 85, 1172–1179.

Wang, X.R. and Szmidt, A.E. (1994) Hybridization and chloroplast DNA variation in a *Pinus* complex from Asia. *Evolution* 48, 1020–1031.

Wang, X.R., Szmidt, A.E. and Savolainen, O.T.I. (2001) Genetic composition and diploid hybrid speciation of a high mountain pine, *Pinus densata*, native to the Tibetan plateau. *Genetics* 159, 337–346.

Wendel, J.F., Schnabel, A. and Seelanan, T. (1995) An unusual ribosomal DNA sequence from *Gossypium gossypioides* reveals ancient, cryptic, intergenomic introgression. *Molecular Phylogenetics and Evolution* 4, 298–313.

Williams, E.G. and Rouse, J.L. (1988) Disparate style lengths contribute to isolation of species in *Rhododendron*. *Australian Journal of Botany* 36, 183–191.

Williams, E.G., Kaul, V., Rouse, J.L. and Palser, B.F. (1986) Overgrowth of pollen tubes in embryo sacs of *Rhododendron* following interspecific pollination. *Australian Journal of Botany* 34, 413–423.

Wolfe, A.D., Xiange, Q.Y. and Kephart, S.R. (1998) Diploid hybrid speciation in *Penstemon* (Scrophulariaceae). *Proceedings of the National Academy of Sciences USA* 95, 5112–5115.

Xu, S. (2000) Phylogenetic analysis under reticulate evolution. *Molecular Biology and Evolution* 17, 897–907.

Yates, J.S., Golden, J.L. and Bain, J.F. (1999) A preliminary phytogeographical analysis of inter- and intra-populational chloroplast DNA variation in *Packera pseudaurea* (Asteraceae: Senecioneae) from south-western Alberta and adjacent Montana. *Canadian Journal of Botany* 77, 305–311.

7 Polyploidy and evolution in plants

Jonathan Wendel[1] and Jeff Doyle[2]

[1]Department of Ecology, Evolution and Organismal Biology, Iowa State University, Ames, IA 50011, USA; [2]Department of Plant Biology, 228 Plant Science Building, Cornell University, Ithaca, NY 14853-4301, USA

One of the central goals of evolutionary biology is to understand the origin of new lineages and species. Accordingly, there is an abiding interest in the processes by which biodiversity arises, and in elucidating the full spectrum of intrinsic mechanisms and extrinsic forces that shape the speciation process. In plants, one of the more prominent mechanisms of speciation involves genome doubling, or polyploidy. This phenomenon exemplifies the complex interplay between 'intrinsic mechanisms' and 'extrinsic forces', as it entails a suite of internal genetic, genomic and physiological processes as well as external population-level and ecological forces. Thus, a chapter devoted to the subject of polyploidy is interesting not only because of its importance to plant speciation, but also because of how much the subject enriches our understanding of the evolutionary process. In this chapter we summarize some of the salient features of polyploidy in plants, including a brief description of its prevalence and modes of formation. We also introduce several model systems for the study of polyploids and provide example case studies, hoping to illuminate more richly how the 'internal' and 'external' processes associated with polyploidy contribute to evolutionary success and to the generation of biodiversity.

Prevalence of Polyploidy in Plants

Although genome sequencing and comparative mapping studies have demonstrated that all eukaryotes have experienced one or more rounds of genome doubling at some point in their evolutionary history (Wolfe and Shields, 1997; Pébusque et al., 1998; Hughes et al., 2000; Gu et al., 2002), the phenomenon appears to have been especially prevalent in higher plants. In fact, it is difficult to overstate the importance of polyploidy in the evolutionary history of flowering plants. Based on the distribution of chromosome numbers among extant species (Stebbins, 1950, 1971; Lewis, 1980b; Grant, 1981), or by comparisons of stomatal size in living and fossil taxa (Masterson, 1994), it has been estimated that perhaps three-quarters of angiosperms have experienced one or more episodes of ancient chromosome doubling. Although polyploidy is uncommon in gymnosperms and liverworts, it is common in mosses (Kuta and Przywara, 1997) and exceptionally so in ferns: perhaps 95–100% of pteridophytes have experienced at least one round of polyploidization in their past (Masterson, 1994; Leitch and Bennett, 1997; Otto and Whitton, 2000). Thus, the notion that 'polyploidy has contributed little to progressive evolution' (Stebbins, 1971) has been replaced by a consensus view that polyploidy is a prominent force in plant evolution (Leitch and Bennett, 1997; Otto and Whitton, 2000; Soltis and Soltis, 2000; Wendel, 2000; Soltis *et al.*, 2004).

It is helpful to provide some additional perspective on the use of the term 'polyploidy'. Because genome doubling has been continuing since angiosperms first appeared in the Cretaceous and because this remains an active, ongoing process (Otto and Whitton, 2000), many angiosperm genomes have experienced several cycles of polyploidization at various times in the past. Thus, most angiosperms are appropriately considered to have 'paleopolyploid' genomes resulting from one or more rounds of genome doubling. The more ancient of these past events may be difficult to discern because of potentially rapid evolutionary restoration of diploid-like chromosomal behaviour and/or other evolutionary changes following polyploidization. Moreover, relatively recent episodes of genome doubling may become superimposed on earlier cycles of polyploidization. Consequently, the polyploid nature of many plant genomes was not evident until the advent of comparative genomics and genome sequencing projects, which commonly reveal duplicate (or higher multiples) genomic regions or chromosomes that are most readily explained by polyploidy (Gaut and Doebley, 1997; Wilson *et al.*, 1999; Devos and Gale, 2000; Paterson *et al.*, 2000; Vision *et al.*, 2000; Wendel, 2000; Simillion *et al.*, 2002; Blanc *et al.*, 2003; Paterson *et al.*, 2003). Prominent examples include many of our most important crop species (Paterson *et al.*, 2003; Arnold *et al.*, 2004), as well as the model plant *Arabidopsis* (Vision *et al.*, 2000; Simillion *et al.*, 2002; Blanc *et al.*, 2003), which was once considered a quintessential diploid because of its small genome and low chromosome number. Given these and many other examples, it is probably safe to state that *no* higher plant has escaped the influence of polyploidy.

Frequency of Polyploidy

If polyploid formation was only a rare and ancient phenomenon, it would not deserve so much attention. In plants, though, it is an active, ongoing process in many lineages (Stebbins, 1950; Lewis, 1980b; Grant, 1981). Many plant genera contain a high percentage of polyploid species as well as diploids, showing that polyploid formation occurred repeatedly in each genus since its origin. Also, polyploid series are commonly observed within angiosperm and pteridophyte genera; these comprise species that differ in multiples of a single base chromosome number, for example, 7, 14, 28... (for many examples, see Stebbins, 1950; Grant, 1981). Thus, genome multiplication occurs beyond the single round giving rise to a tetraploid, generating higher ploidy levels. Extraordinary examples abound of polyploid series that attain very high ploidy levels, including *Potentilla* (up to 16-ploid), *Chrysanthemum* (up to 22-ploid) and *Poa* (up to 38-ploid). According to Grant (1981), *Kalanchoe* ranks among the leaders in this category, with chromosome multiples approaching 60-ploid, whereas even higher ploidy levels have been suggested for *Sedum* and *Ophioglossum* (see Otto and Whitton, 2000). These widespread observations provide cytogenetic evidence that polyploidy occurs repeatedly on the evolutionary timescale of individual genera and that it is widely dispersed among angiosperms and other plant groups. Additional evidence on the frequency of polyploid speciation events has come from studies of the distribution of haploid chromosome numbers (Otto and Whitton, 2000). Because genome doubling will immediately create even haploid numbers, there should be an excess of even as opposed to odd haploid numbers if polyploidy is common. Otto and Whitton (2000) demonstrated that this is the case for both ferns and angiosperms, and provided minimal estimates of the percentage of speciation events that result from polyploidization events. The conclusion suggested by these and other studies is that polyploidy represents the most common mode of sympatric speciation in plants, and hence is extraordinarily important to a discussion of plant diversity.

Types of Polyploids and Modes of Formation

So far we have not addressed the manner in which genome doubling takes place or the modes by which polyploids form. Traditionally, polyploidy has been thought to result from either duplication of a single genome (autopolyploidy) or from the combination of two or more differentiated genomes (allopolyploidy) (Stebbins, 1947, 1950; Grant, 1981). However, polyploids form in many different ways, running the full gamut from a single genetically uniform diploid plant doubling its chromosome complement to hybridization between individuals from highly divergent species. For systematists, the primary distinction is taxonomic: an autopolyploid arises within a species, and an allopolyploid involves hybridization between two or more species (Lewis, 1980b; Grant, 1981; Ramsey and Schemske, 1998; Soltis *et al.*, 2004). Although these definitions work well in many taxa, the terms would be more useful if species themselves were both objectively more definable and taxonomically more stable. Taxonomic definitions are hampered by the realities of natural variation: some named species harbour tremendous amounts of genetic and chromosomal variation, and may in fact comprise several cryptic species, whereas other species may be genetically nearly identical to their closest relatives. Geneticists are more interested in the behaviour of genes and chromosomes following genome doubling than in taxonomic definitions; here the important distinction is whether or not chromosomes from the different (homoeologous) complements are capable of pairing with one another at meiosis. If so, the plant is autopolyploid, if not, and chromosome complements instead are maintained as two separate sets that generally do not interact, the plant is an allopolyploid.

Clearly, there is expected to be broad overlap between the taxonomic and genetic definitions of polyploids, and in fact much of the richness of the complexity of polyploid formation is lost when it is pigeon-holed into the terms 'autopolyploidy' and 'allopolyploidy.' In actuality, these two modes of formation represent endpoints in a taxonomic–genetic continuum. In general, individuals within species will have genomes that are less diverged from one another than will individuals from different species, and, to the extent that pairing of homoeologous chromosomes becomes more difficult as genomes diverge, taxonomic allopolyploids will more often be genetically allopolyploid than will taxonomic autopolyploids.

An important consequence of the mode of formation is that the two endpoints of autopolyploidy and allopolyploidy often make different predictions with respect to chromosome behaviour and genetic segregation. These data, in turn, may offer insight into the mode of formation of any particular polyploid. In strict autopolyploids, there are four homologous chromosomes capable of associating with one another at meiosis, resulting in random bivalents and, often, in multivalents. In contrast, in a genetic allopolyploid the homologous chromosomes contributed by the two diploid progenitor species are by definition unable to pair at meiosis, and as a consequence only bivalents are formed at meiosis in the allotetraploid. Because of these differences in chromosome behaviour, autopolyploids and allopolyploids display different patterns of genetic segregation. The case with allopolyploids is straightforward; although each gene is doubled, the two homoeologous genomes behave independently and genetic segregation at each is disomic, as in its diploid progenitors. That is, the genetic behaviour of any single locus in an allopolyploid is expected to be very much like that of a diploid (interesting and mysterious exceptions to this expectation will be discussed below). Random pairing and multivalent formation in autopolyploids, however, leads to quasi-random patterns of segregation among the multiple chromosome copies, and hence patterns of genetic segregation that differ from simple, disomic, Mendelian inheritance. In an autotetraploid, for example, tetrasomic ratios are observed, either chromosomal or chromatidal, depending on whether or not the locus in question resides close to the centromere (for a lucid description, see pp. 240ff. in Grant, 1975). In practice, the segregation ratio observed in any situation depends on many variables, including the mode of polyploid formation, amount of pairing among doubled chromosomes, time since polyploid formation and chromosomal location of the gene in question.

The genetic behaviour of duplicated loci has important consequences. Genetic and evolutionary advantages accrue to both classes of polyploids, but in very different ways: allopolyploids offer the presumed advantage of fixed heterozygosity at all loci, whereas for autopolyploids the smaller proportion of homozygotes due to tetrasomic inheritance buffers them against the loss of genetic variation. The coalescent times of loci under disomic and tetrasomic inheritance can differ dramatically. Disomically inherited loci typical of allopolyploids trace their origin to the diploid progenitors, and thus should coalesce more deeply than tetrasomically inherited loci, where segregational loss of alleles continues after polyploid formation. The difference in coalescent times has been used to infer a complex history for the maize genome; Gaut and Doebley (1997) found a bimodal distribution of coalescent times among low copy nuclear genes, and suggested that maize is a segmental allopolyploid, a class of polyploid intermediate between strict auto- and allopolyploidy. Presumably many polyploids show some features of both auto- and allopolyploids, and the full spectrum of evolutionary possibilities is not adequately captured with only two or three terms.

Both autopolyploids and allopolyploids are known to be formed by several different mechanisms (Harlan and DeWet, 1975; Ramsey and Schemske, 1998). One key feature of these various mechanisms is that meiosis is an imperfect process. Specifically, failure in chromosome segregation may lead to the formation of 'unreduced' or '$2n$' gametes, which have a somatic complement of chromosomes. A union of two unreduced gametes may subsequently lead to the formation of a tetraploid embryo, for example, whereas the union of a normal, reduced gamete ($1n$) with an unreduced gamete will generate a triploid.

Autopolyploids in nature are known to arise either from such mechanisms as spontaneous somatic doubling, merger of two unreduced gametes, or by a triploid 'bridge', the latter having arisen as described above. Triploid individuals, although usually highly sterile, typically produce an appreciable percentage of viable gametes, thereby facilitating autotetraploid formation either by self-pollination or crossing with diploid individuals that also produce a low percentage of unreduced gametes. Allotetraploids also arise by several means, including 'one-step' and 'two-step' pathways (Harlan and DeWet, 1975; Ramsey and Schemske, 1998), the former from the merger of two unreduced gametes from two different species, and the latter either via a triploid bridge or from spontaneous somatic doubling of a sterile, interspecific diploid. Unreduced gamete formation is nearly 50-fold greater in hybrids than in non-hybrids (Ramsey and Schemske, 1998), increasing the likelihood of allopolyploid formation by this route, after an interspecific hybrid has been formed.

Allopolyploidy is probably more common than autopolyploidy in nature (Ramsey and Schemske, 1998; Soltis et al., 2004), although the latter is far more prevalent than was once thought (Lewis, 1980a). Autopolyploidy has historically been considered maladaptive or at least uncommon relative to allopolyploidy, in part because of fertility reductions associated with multivalent formation (and the attendant production of gametes with an unbalanced chromosome complement), but also because of the fitness advantages presumed to accompany the 'fixed heterozygosity' of allopolyploids (Stebbins, 1950, 1971; Grant, 1981). More recent empirical studies have drawn attention to many examples of successful autopolyploidy in plants (Soltis and Soltis, 1993, 1999; Soltis et al., 2004). In addition, Ramsey and Schemske (1998) point out that the relative frequencies of autopolyploidy and allopolyploidy are strongly dependent on the frequency of interspecific F_1 hybrid formation as well as mating system, because of the aforementioned boost in unreduced gamete formation in hybrid relative to non-hybrid individuals. Although quantitative data on the frequency of interspecific hybrid formation in plants are lacking, hybridization is known to be common (Grant, 1981; Rieseberg and Wendel, 1993; Rieseberg, 1995; Arnold, 1997). Ramsey and Schemske, however, conclude that interspecific hybridization rates are probably too low in many situations to generate an abundance of allopolyploids, and hence that the rate of formation of autopolyploids may often exceed that of allopolyploids. More recently, Ramsey and Schemske (2002) have challenged the primary assumption that autopolyploidy is associated with high fertility cost caused by chromosomally unbalanced gametes, because, as mentioned above, natural selection acts quickly to restore fertility. Taken together, the evidence suggests that both allopolyploidy and autopolyploidy are common evolutionary outcomes; much remains to be learned about overall frequencies, relative rates in specific genera, or the long-term evolutionary fates of these alternative products.

Processes in Polyploids that Contribute to Biological Diversity

The pervasiveness of polyploidy in the plant kingdom, as discussed above, offers the most obvious measure of the importance of the phenomenon with respect to the genesis of biodiversity. Yet the full significance of polyploidy requires an understanding of the ecological context in which polyploids form and the full suite of interactions with their animal pollinators and herbivores. Recent work has underscored this important point, showing how sympatric diploid and autotetraploid *Heuchera grossulariifolia* plants differ with respect to the pollinators that they attract and in their phytophagous insects (Nuismer and Thompson, 2001; Thompson *et al.*, 2004). Thompson *et al.* (2004) review the relevance of polyploidy to biodiversity in animals, and conclude that polyploidy in plants represents a significant diversifying force in animals by virtue of the many ecological interactions with herbivores and pollinators.

Embedded within the biology of polyploids are other processes that play subtle roles in fostering genetic and phenotypic variation within and among plant popula-

tions (Figs 7.1–7.6). These include: (i) the surprising and relatively recent realization that many polyploid taxa have experienced multiple origins from genotypically similar antecedents, which is expected not only to increase genetic diversity within and among polyploid populations but also to provide an opportunity for (ii) novel recombinational products at the polyploid level. The possibilities created by these twin processes lead directly to (iii) speciation mechanisms not available at the diploid level. Each of these is discussed briefly in turn below. To illustrate these and related points more clearly, examples are provided from some of the more thoroughly studied angiosperm genera.

Multiple origins and genetic variation in polyploids

Of all the accomplishments of molecular phylogenetics in the area of polyploidy, perhaps none has received as much attention as the rigorous documentation that polyploids can form recurrently. This is not entirely a

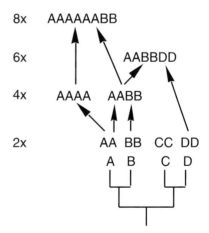

Fig. 7.1. Four diploid (2x) species (A–D), along with their phylogenetic relationships. Two types of tetraploids (4x) are formed from these species: AAAA is an autotetraploid, and AABB is an allopolyploid. Additional levels of complexity occur when the AABB allopolyploid hybridizes with species D (DD diploid) to form a hexaploid (6x) AABBDD, and when the two tetraploid species form the autoalloocotoploid (8x) AAAAAABB.

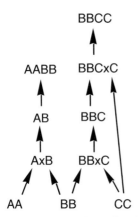

Fig. 7.2. Modes of origin of two allopolyploids are shown. Diploid species AA and BB each produce haploid gametes (A and B) that unite (AxB) to form the diploid hybrid AB, which will usually be sterile because of poor chromosome pairing. If this plant doubles its chromosome number, a fixed hybrid allotetraploid (AABB) is formed, in which case homologous pairing can occur and fertility is restored. A comparable allopolyploid (BBCC) is formed by a 'triploid bridge' (BBC). Diploid CC produces a reduced gamete (C), but diploid BB produces an unreduced, diploid gamete (BB); these unite to form triploid BBC. Meiotic aberrations in the triploid increase the frequency of unreduced gamete production; in the example shown, a triploid gamete (BBC) unites with a haploid gamete from the CC diploid to form the BBCC tetraploid.

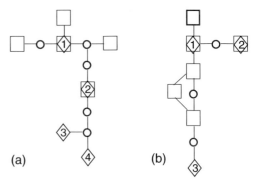

Fig. 7.3. Allele networks for a pair of homoeologous single-copy nuclear gene loci in allopolyploid AABB and its diploid progenitors (AA and BB). Alleles sampled from diploids are shown as squares, alleles sampled from tetraploid individuals are shown as diamonds; lines connecting alleles represent one mutational step; small circles represent unsampled alleles. (a) Alleles from diploid AA and the A-homoeologue alleles from the polyploid. In this network, two alleles from the polyploid are identical to alleles from the diploid progenitor; each is inferred to be derived from a separate origin of the polyploid. Alleles 3 and 4 from the polyploid are differentiated from any diploid alleles. They could represent either additional independent origins or could simply reflect a lack of sampling of the diploid or extinction of allele lineages in the diploid. Allele network (b) represents the BB diploid and the B-homoeologues of the polyploid. The network illustrates recombination among alleles from the diploid (the loop in the network).

new concept, and in fact was implicit if not directly stated early in the biosystematics era (Soltis *et al.*, 2004). Nevertheless, the ability to assess variation at the level of DNA sequences, and, in particular, to identify with great precision particular nuclear gene alleles and chloroplast haplotypes shared by a polyploid and its diploid progenitors eliminated any notion that most polyploids are products of single events (Soltis and Soltis, 1999; Wendel, 2000). Although in some cases the number of origins may be overestimated, when, for example, the possibility of heterozygous unreduced gametes is ignored (Watanabe *et al.*, 1991; Vogel *et al.*, 1999), more commonly the number of independent origins will be underestimated, because of the joint processes of lineage extinction in diploids and polyploids and allelic divergence over time.

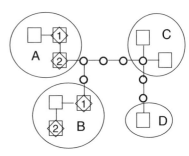

Fig. 7.4. Chloroplast haplotype network for species A–D. Polyploid AABB has haplotypes from both the A and B diploid species, illustrating bidirectional origins if chloroplast DNA is maternally inherited.

The phenomenon of multiple, independent origins of polyploids is central to our understanding of polyploid evolution and the generation of biodiversity. Most obvious

| | Locus | | | |
Plant	A	B	cp	Genotype
1a	1	3	C1	P
1b	1	3	C1	P
1c	4	2	B2	Q
1d	4	2	B2	Q
2a	1	3	C1	P
2b	4	2	B2	Q
2c	4	3	B2	R
2d	1	2	C1	S

Fig. 7.5. Lineage recombination. The genotypes of four AABB polyploid plants (a–d) are shown, based on the networks shown in Figs 7.3 and 7.4. Two cases are shown. In the first (1a–1d), there are two classes of plants, each with the same alleles at the two homoeologous loci and having the same chloroplast haplotype; these are classified as genotypes P and Q. In the second example (plants 2a–2d), these same two genotypes are also found, but lineage recombination has occurred to produce two new genotypes, R and S. In Case 1 there is evidence for two origins of the AABB polyploid. In Case 2 there is also evidence of more than one origin, but the number of different genotypes can be explained either by four separate origins or by lineage recombination following gene exchange between plants from a smaller number of origins.

is the inference that polyploids have the potential to sample extensively from the pool of genetic variation found in their progenitors. This is true spatially, through the contemporaneous formation of polyploids from different diploid populations with different genotypes. It is also true temporally, if polyploid formation continues

over a long enough time to permit sampling of progenitors which themselves are experiencing allelic divergence from their antecedents. In such cases, the polyploid gene pool would receive periodic infusions of new allelic variation, the importance of which is not fully understood but which clearly could provide fodder for evolutionary diversification. There is as yet no documentation of the phenomenon of temporally recurrent polyploid formation, and it may be experimentally challenging to gather the quality of evidence necessary to rule out alternative explanations for an observed pattern of diversity within a given polyploid lineage. For example, one phylogenetic signature of this scenario might be that a polyploid would share some alleles with a diploid progenitor and others that are unique to the polyploid. This pattern, however, may also be produced by incomplete sampling of, or lineage extinction within, the diploid. Notwithstanding the absence of compelling empirical examples, there is no reason to suspect that polyploids do not form recurrently over time, at least in those cases where long-term sympatry among progenitors may be expected and where barriers to hybridization are relatively weak.

Multiple origins are well illustrated by *Tragopogon*, a genus of Asteraceae that provides the best-studied example of recent allopolyploid origin. Early work by Ownbey (1950) showed that two polyploid species, *Tragopogon miscellus* and *Tragopogon mirus*, originated through allopolyploidy in the western United States around the turn of the

| | Locus | | | | | Locus | |
Plant	A	B	cp		1a x 1c	A	B
1a	1	(3)	C1		F1-1	1	(3)
1b	1	(3)	C1	→	F1-2	1	2
1c	(4)	2	B2		F1-3	(4)	2
1d	(4)	2	B2		F1-4	(4)	(3)

Fig. 7.6. Reciprocal silencing. Plants from the two separate origins of the AABB polyploid are shown, as in Fig. 7.5. Brackets around alleles from the two homoeologous loci indicate an allele that is constitutively silenced and may be a pseudogene. All plants are viable, because in each there is one functional homoeologous locus. However, when plants from the two different origins are crossed, some F$_1$ individuals (represented by F1–4) have non-functional alleles at both homoeologous loci; such plants are inviable. When this occurs at many unlinked loci, the result will be a barrier to gene flow that can lead to speciation.

20th century. This novel creation of entirely new species turned out to be a consequence of the introduction to the USA of three Eurasion diploid species followed by range expansion and subsequent sympatry among pairs of these three species. Occasionally, these sympatric species pairs underwent hybridization followed by allopolyploid formation. This now classic story is referred to as the *Tragopogon* triangle.

The evolutionary history of these nascent allopolyploids has been extensively studied by Douglas and Pamela Soltis and their colleagues (Soltis *et al.*, 2004). Among the more interesting aspects of this complex is its dynamic nature over a relatively limited range, with new local populations having originated multiple times, and with polyploids replacing diploids as prevalent weeds. Not only have the polyploids originated multiple times, but in some cases in both directions; namely, *T. miscellus* has formed by crosses involving each of its diploid progenitors as maternal parent, resulting in reciprocal morphological differences between *T. miscellus* individuals depending on maternal parentage (Soltis *et al.*, 2004). Another interesting observation is that the *Tragopogon* allopolyploids have formed only in North America, and not in the native European ranges of the diploid species.

Among the more enduring lessons to be learned from *Tragopogon* is that new polyploid species may originate in a telescoped timeframe that is apparent within individual human lifetimes. This lesson has been retaught on several other occasions in other plant genera, most notably in the fascinating stories of *Spartina anglica* (Ainouche *et al.*, 2004) and *Senecio cambrensis* (Abbott and Comes, 2004). The latter species has arisen more than once, whereas *Spartina anglica* appears to have only a single origin.

Recombination among allopolyploid lineages

In addition to the potential evolutionary significance of infusion of new variation, recurrent polyploid formation may be important in facilitating novel recombinant genotypes to be produced at the polyploid level by gene exchange between two or more recurrently formed entities. This, in fact, may be critical to the survival and diversification of polyploid lineages because, in the absence of recurrent formation, many nascent polyploids are expected to be relatively depauperate genetically. With gene flow between polyploids ('lineage recombination', *sensu* Doyle *et al.*, 1999), a genetically diverse and coherent polyploid species is formed, which in principle may be enriched by periodic and ongoing infusions of genetic variation from diploid progenitors.

The best-documented example of this is in the soybean genus (*Glycine*), which includes a large allopolyploid complex involving at least eight Australian diploid species and eight allopolyploids that combine diploid genomes in various ways (Doyle *et al.*, 2004a,b). In some of these polyploids, all alleles are identical or nearly identical to alleles in the diploids, suggesting an origin within the last 50,000 years, coincident with climatic changes in Australia that may have been anthropogenic (Doyle *et al.*, 1999, 2004a,b).

The history of the complex genetic and taxonomic system comprising *Glycine* diploids and allopolyploids has been investigated in detail using molecular markers. This work has demonstrated that different, closely related polyploids vary in their manner and frequency of formation, ecology, life history, geographical range and patterns of molecular evolution (Doyle *et al.*, 2004a,b). Nearly all polyploids have arisen more than once, but some show exclusively unidirectional maternal origins whereas in others both diploid progenitors have contributed chloroplast genomes. Some show evidence of extensive interbreeding among polyploid populations with separate origins, resulting in extensive lineage recombination, whereas in others multilocus genotypes formed by different combinations of the same diploid genomes have persisted intact. Several polyploids have extensive ranges outside of Australia, and appear to be far better colonizers than their diploid progenitors, whereas other polyploids have highly restricted ranges. This diversity in life histories and population histories is paralleled by the range of molecular evolutionary responses; some polyploids have retained nearly equal amounts of homoeologous nuclear ribosomal DNA repeats, whereas in others one homoeologue predominates (Rauscher *et al.*, 2004), apparently as a consequence of repeat loss and concerted evolution (Joly *et al.*, 2004).

Novel mechanisms of speciation

In general, polyploids and their diploid progenitors are assumed to be reproductively isolated from one another due to cytological incompatibilities and triploid sterility. This makes polyploidy a pervasive mode of sympatric speciation. However, the issue of isolation between cytotypes has received relatively little study (but see e.g. Trinti and Scali, 1996; Husband and Schemske, 2000; Husband *et al.*, 2002). More generally, the role of the cytoplasm in polyploid evolution is not well understood (Wendel, 2000; Levin, 2003). One aspect of particular relevance to speciation at the polyploid level is the possibility of cytonuclear interactions that create barriers to gene flow between polyploid lineages that may share nuclear genomes but have different cytoplasms. As discussed above, bidirectional formation of allopolyploids has been demonstrated in at least some groups. In principle, reproductive isolation may arise following either multiple origins or recombination among different allopolyploid lineages, thereby promoting diversification.

In addition to speciation promoted by cytonuclear differentiation, other mechanisms may drive polyploid diversification. Dramatic and potentially rapid molecular evolution of polyploids (discussed below) might lead to speciation. For example, Barrier *et al.* (2001) have suggested that rapid evolution of floral transcription factors is in part responsible for the dramatic radiation of polyploid Hawaiian silverswords. As discussed in more detail below, polyploidy may be associated with a high level of latent epigenetic variation, some of which may have phenotypic effects that are potentially visible to selection. An interesting example is the epigenetic control of flowering time variation in synthetic *Brassica* allopolyploids (Schranz and Osborn, 2000); because flowering phenology is so obviously important to reproduction, one can imagine that epigenetic divergence can lead to reproductive isolation, even in the face of little or no genetic differentiation.

Following polyploid formation, gene silencing is one of the several possible fates of the genome-wide duplication of all genes (reviewed in Wendel, 2000; Lawton-Rauh, 2003). This process occurs with the onset of polyploid formation but increases with time, such that one of the two duplicate genes is expected to be silenced at many different loci. Werth and Windham (1991) hypothesized that what they termed 'reciprocal silencing' – the loss of expression of different duplicated gene copies in allopatric polyploid populations – could lead to reproductive isolation among polyploids, and thus to speciation. The basic idea is that hybrids between individuals bearing reciprocally silenced duplicated genes would segregate progeny that are silenced at both gene copies for one to many loci. These double mutants would presumably be inviable or deleterious due to their negative phenotypic or physiological effects. This idea has recently been elaborated under the term 'divergent resolution' by Lynch and Force (2000), and is suggested as a possible factor in the evolution of teleost fishes (Taylor *et al.*, 2001).

Reasons for the Evolutionary Success of Polyploids

The abundance of polyploid plant species in nature suggests that polyploidy confers a selective advantage over diploidy in some settings. Traditional explanations for the success of polyploids have included a diversity of proposals that, broadly speaking, may be divided into 'ecological' and 'genetic'. The former include fitness advantages inferred from the greater ecological breadth and amplitude, or colonization of new habitats, which is observed more frequently in polyploids than in their diploid antecedents

(Stebbins, 1950; Grant, 1981). Genetic explanations typically invoke a form of heterosis (hybrid vigour) in polyploids, particularly in allopolyploids, resulting from the merger into a single nucleus of complementary alleles. In actuality, ecological and genetic perspectives are not exclusive but instead may be complementary, the latter representing an explanation for the former. For example, the early molecular study of *Tragopogon* by Roose and Gottlieb (1976) documented 'fixed hybridity' at isozyme loci in polyploids (as did many subsequent studies), from which they speculated that this genetic diversity *enabled* the polyploids to become successful weeds and extend their ranges. Other genetic explanations for the adaptive superiority of polyploids invoke genetic 'buffering' conferred by heterozygosity (Grant, 1981) or the perceived favourable consequences of gene redundancy and its attendant release from functional constraint for one gene copy, and/or functional divergence (Harland, 1936; Ohno, 1970; Force *et al.*, 1999; Lynch and Conery, 2000; Wendel, 2000; Lynch, 2002; Lawton-Rauh, 2003). More recent genetic and genomic proposals include adaptive genome-wide allelic and/or non-allelic interactions (Pikaard, 2002), and altered regulatory interactions that generate novel variation that is visible to selection (Osborn *et al.*, 2003; Riddle and Birchler, 2003).

Otholog(ue)s, Paralog(ue)s and Hom(o)eolog(ue)s: Some Essential Terminology

With the discovery that most plant genes belong to gene families, molecular biologists and systematists alike have come to embrace the terminology for duplicate genes developed many years ago by Fitch (1970). This issue is more than simply a nomenclatural exercise, in that its understanding is central to an appreciation of a diversity of biological phenomena, such as functional diversification of duplicated genes, as well as the appropriate use of gene sequence data for phylogeny reconstruction. Fitch defined orthologous genes ('orthologues' or

'orthologs'; see Koonin *et al.*, 2004) as being those that originated by speciation, in contrast to paralogous genes, which are formed by gene duplication. Discriminating paralogous genes is theoretically a simple task: if two non-allelic homologues (= homologs) are present in a plant genome, then by definition they must have arisen by duplication, and not by speciation, and they are paralogues (= paralogs). Orthology is more difficult to establish, requiring a phylogenetic test in which the gene tree must be identical to a known species phylogeny.

Where do homoeologues (= homoeologs) fit in? In the genome of an allopolyploid, loci that were orthologous in the two diploid progenitors become homoeologous. The term 'homoeologous' predates the orthology/paralogy terminology by nearly 40 years and is defined in the *Glossary of Genetics and Cytogenetics* (Rieger *et al.*, 1976) as 'the residual homology of originally completely homologous chromosomes'. 'Homology' is used here in the cytogenetic sense of the term: 'chromosomes or chromosome segments ... identical with respect to their constituent genetic loci (the same loci in the same order) and their visible structure'; definition 1 in Rieger *et al.* (1976). Definition 2 is the more familiar evolutionary definition of homology: similarity due to common origin.

So, are homoeologues paralogues or orthologues? The answer is yes! They have characteristics of both, but they are more like paralogues than like orthologues. Because homologous chromosomes become homoeologous due to divergence following speciation, the orthologous genes of two diploid sister species could be said to be homoeologous. However, once these homoeologues are united in the compound genome of an allopolyploid, they meet the criterion for paralogues: they are formed by gene duplication, like any other paralogue, the only difference being that they are formed by whole genome duplication. Perhaps more importantly from a practical perspective, today's obvious homoeologues are tomorrow's paralogues, and their origin as orthologues may be obscured as evidence for polyploidy is lost owing to diploidization.

As discussed above, the episodic process of polyploidization that we now know is typical of plant genomes generates repeated cycles of homoeology followed by gene divergence which, over vast amounts of evolutionary time, become multigene families consisting of paralogues with varying degrees of sequence similarity.

Gene and Genome Evolution in Polyploids

Among the many aspects of polyploidy that have been studied recently, one of the more intriguing has been the question of how two divergent genomes coordinately adjust and evolve to guide growth and development once they become united in a common nucleus. The pervasiveness of polyploidy constitutes prima facie evidence that such adjustments occur and that some fraction of them have positive fitness consequences. Hence it is of interest to ask about the genetic, genomic and adaptive consequences of genome doubling.

A useful device for conceptualizing gene and genome evolution in polyploids is offered by Fig. 7.7. The most immediate and important genomic consequence of polyploidization is the simultaneous duplication of all nuclear genes, a phenomenon long thought to be central to the evolutionary success of polyploids (Stebbins, 1950; Stephens, 1951a,b; Ohno, 1970; Lewis, 1980b; Levin, 1983). As explained above, genes duplicated by polyploidy are termed homoeologues. As modelled in Fig. 7.7, at the time of polyploid formation each gene in the genome will become duplicated, such that two homoeologues ('At' and 'Bt') will exist for each locus, with each homoeologue being phylogenetically sister to its counterpart in the progenitor diploid (A with At, B with Bt). One possible evolutionary outcome is the long-term preservation of both homoeologues, as well as retention of ancestral functions by both copies. This scenario provides a useful null model against which other possibilities may be evaluated. One long-recognized possibility is relaxation of selection, allowing divergence between the

duplicated genes and the acquisition of new function (Ohno, 1970; Force et al., 1999; Lynch and Conery, 2000; Lynch and Force, 2000; Lynch et al., 2001; Altschmied et al., 2002; Lynch, 2002; Lawton-Rauh, 2003). This process is widely perceived to provide the raw material for adaptive diversification. An alternative and common outcome of gene doubling is that one member of the duplicated gene pair will become silenced and ultimately degenerate as a pseudogene (reviewed in Wendel, 2000; Lawton-Rauh, 2003). Several other possibilities also exist, as modelled in Fig. 7.7, and these have only recently come to light as a consequence of molecular genetic investigations conducted in several plant systems, including *Brassica* (Song et al., 1995), wheat (Feldman et al., 1997; Liu et al., 1997, 1998a,b; Ozkan et al., 2001; Shaked et al., 2001; Kashkush et al., 2002), *Arabidopsis* (Comai et al., 2000; Lee and Chen, 2001; Madlung et al., 2002) and cotton (Wendel et al., 1995; Hanson et al., 1998, 1999; Jiang et al., 1998; Zhao et al., 1998).

Thus, there has been a growing awareness of a diversity of phenomena associated with polyploidy that were previously unknown or unsuspected. The notion that has emerged is that polyploid genomes are 'dynamic' (Soltis and Soltis, 1995) at the molecular level, generating an array of novel genomic instabilities or changes during the initial stages of polyploid formation or over longer time spans. Some of these alterations are not readily explained by Mendelian principles, but may none the less have contributed to the evolutionary success of polyploids (Soltis and Soltis, 1995; Wendel, 2000; Finnegan, 2001a; Rieseberg, 2001; Liu and Wendel, 2002; Pikaard, 2002; Osborn et al., 2003). Examples of recent insights into the genetic and genomic behaviour of polyploids include: (i) rapid genomic changes and intergenomic interactions that become possible as a consequence of the merger of two genomes into a single nucleus; and (ii) epigenetic alterations that may accompany new polyploids. We summarize briefly some of the important observations and explore the possible biological significance of the various phenomena.

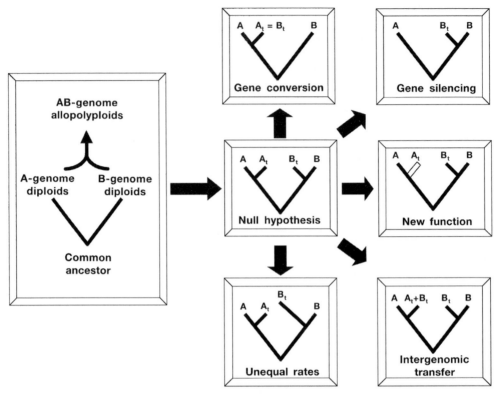

Fig. 7.7. Gene duplication and evolution in polyploids. Shown on the left is a hypothetical set of organisms comprising two diploids and their allopolyploid derivative. With the onset of allopolyploid formation all genes become duplicated. The two homoeologues (At, Bt) are not expected to be phylogenetically sister to each other, but are instead expected to be sister to their counterparts from each respective diploid (A, B). In addition, all else being equal, evolutionary rates are expected to be equivalent. These expectations provide convenient null hypotheses (middle), which may be falsified by a number of processes, including gene conversion (top centre), unequal rates (bottom centre), gene silencing (top right), the evolution of new function (centre right) and intergenomic transfer (bottom right). Recent work has demonstrated all of these possibilities in one or more natural and/or synthetic allopolyploids.

Genome change and intergenomic interactions resulting from polyploidy

Recent explorations of polyploidy have led to the discovery of a number of somewhat mysterious phenomena associated with polyploid formation. Though we still know relatively little about most details of genomic merger, the initial stages evidently are moulded by an array of molecular genetic mechanisms and processes that collectively lead to polyploid stabilization (Song *et al.*, 1995; Soltis and Soltis, 1999; Comai *et al.*, 2000; Wendel, 2000; Lee and Chen, 2001; Kashkush *et al.*, 2002; Liu and Wendel, 2002; Soltis *et al.*, 2004). An

important early paper by Song *et al.* (1995) demonstrated the novel appearance as well as disappearance of different restriction fragments in synthetic *Brassica* polyploids and their progeny. This work was soon followed by similar observations in tetraploid and hexaploid wheats, with the added twist that some of the changes observed in newly synthesized wheat allopolyploids mimicked those observed in natural wheats with the same genomic composition (Feldman *et al.*, 1997; Liu *et al.*, 1998a,b; Shaked *et al.*, 2001). The latter observation implied that the non-Mendelian response to allopolyploid formation was to a certain extent 'directed' by the specificities of the genomes involved. These and other studies showed that polyploid genomes are neither static nor need be strictly additive with respect to the genomes of progenitor diploids. Instead, the merger of two different genomes in a common nucleus may be accompanied by genomic 'reorganization' of unknown aetiology (Wendel, 2000; Liu and Wendel, 2002). Proposed mechanisms to account for this non-Mendelian behaviour include homologous and non-homologous recombination, methylation alterations and other epigenetic changes (see below) and perhaps deletional processes that are not well understood.

A different but related phenomenology associated with polyploidy concerns transposable elements. Genome merger in an allopolyploid creates the potential for the spread of transposable elements between two formerly isolated genomes. Transposable elements are ubiquitous in plant genomes (Kumar and Bennetzen, 1999; Bennetzen, 2000), where they contribute to genome evolution and genetic diversity by transposition and the attendant effects on gene expression (Kidwell and Lisch, 2001; Wessler, 2001). Most transposable elements are inactive under normal conditions, but they may become activated under stress (McClintock, 1984; Hirochika et al., 1996; Wessler, 1996; Grandbastien, 1998; Beguiristain et al., 2001). In diploid hybrids, enhanced transposable element activity is likely to be maladaptive because insertions may disrupt essential gene functions. In polyploids, however, the harmful effects of transposable element activity may be buffered by genomic redundancy, and hence insertions would be more likely to be tolerated (Matzke et al., 1999; Wendel, 2000). Hence, it is noteworthy that in newly generated allotetraploid wheat plants (Shaked et al., 2001) and in Orzya × Zizania hybrids (Liu and Wendel, 2000), transposable elements have been shown to be activated.

Although few natural plant hybrids and allopolyploids have been experimentally evaluated for transposable element activity, studies to date suggest that wide hybridization and allopolyploidy may trigger activation of dormant transposable elements. The extent and tempo at which these events will occur undoubtedly varies among plant species and genome combinations. In general, however, the inherently higher level of tolerance to insertions makes it more likely that transposable elements have played a role in genome evolution in polyploid than in diploid species. Evidence in support of this supposition is circumstantial, consisting primarily of the observation that transposable elements and other repeated sequences have spread among genomes in tetraploid cotton (Hanson et al., 1998, 1999) and wheat (Belyayev et al., 2000). In Nicotiana, there apparently has been a massive proliferation of pararetroviruses following polyploid formation (Matzke et al., 2004). In all cases, the adaptively relevant effects, if any, of these inter-genomic, intra-nuclear colonizations are not known.

In addition to the spread of transposable elements among genomes, polyploidy creates the opportunity for various types of interactions between homoeologous genes or repeated sequences. Tandemly repeated sequences, such as ribosomal genes, have been demonstrated to experience interlocus homogenization or concerted evolution, whereby sequences from one genome overwrite the homoeologous sequences from the other genome. First convincingly demonstrated in allopolyploid Gossypium (Wendel et al., 1995), the phenomenon has now been described in a number of genera (Wendel, 2000; Joly et al., 2004). New twists on this phenomenon were recently reported in synthetic Nicotiana allopolyploids (Skalicka et al., 2003; Kovarik et al., 2004), where the rapid evolution of rDNA types was observed within a few generations, as was the apparent birth of a new rDNA locus. This latter observation may well represent the real-time capture of the well-known but unexplained phenomenon of birth and death of ribosomal arrays (Dubcovsky and Dvorák, 1995).

Epigenetics and polyploid evolution

Epigenetics refers to heritable alterations in gene expression that do not entail changes in nucleotide sequence, but which nevertheless may have phenotypic and hence evolutionary consequences. Epigenetic effects can be accomplished by several interrelated covalent modifications of DNA and/or chromosomal proteins, such as DNA methylation and histone modifications (Nathan *et al.*, 2003), and by chromatin remodelling, such as repositioning of nucleosomes. These heritable modifications are collectively termed 'epigenetic codes' (reviewed in Richards and Elgin, 2002). Programmed epigenetic control of gene expression is essential during normal growth and development (Wolffe and Matzke, 1999; Finnegan, 2001b) and, because of this, the epigenetic arena is a vibrant field in current biological research. Given this fundamental significance, it is of interest to discuss the possible connections between epigenetics and polyploidy.

An integral component of the developmental control of gene expression is programmed cytosine methylation (Richards, 1997; Finnegan *et al.*, 2000). Hypermethylation is usually a hallmark of heterochromatin and is characteristic of euchromatic gene silencing, whereas hypomethylation is often associated with active gene expression (Martienssen and Colot, 2001; Grewal and Moazed, 2003). In plants, cytosine methylation patterns are usually stably maintained through meiosis and over generations. Experimental disruption of cytosine methylation patterns may lead to aberrant plant morphology (Finnegan *et al.*, 1998, 2000; Finnegan, 2001b). As a potential genome defence system (Yoder *et al.*, 1997), the cytosine methylation machinery may respond to environmental or genomic challenges by causing alterations in methylation that are thought to mediate physiologically meaningful responses. Polyploidy, by uniting divergent genomes into one nucleus, may constitute such a challenge, or 'genomic shock' (McClintock, 1984; Comai *et al.*, 2003).

This suggestion has garnered recent experimental support. In synthetic *Brassica* (Song *et al.*, 1995) and wheat allopolyploids (Liu *et al.*, 1998a; Shaked *et al.*, 2001), DNA methylation changes, including both hypo- and hypermethylations, were shown to occur at anonymous genomic loci and in cDNAs. Using a genome-wide fingerprinting approach, Shaked *et al.* (2001) showed that, in wheat, cytosine methylation alterations are genomically widespread but may significantly differ in frequency between the two constituent genomes of an allopolyploid; in first generation allotetraploid wheat, ten of the 11 bands that showed heritable methylation changes were from one of the two parental genomes. Genome-wide and non-random changes in DNA methylation patterns are also observed in synthetic allotetraploid *Arabidopsis* and *Arabidopsis arenosa* (Madlung *et al.*, 2002). Given the importance of DNA methylation to gene expression, the foregoing examples indicate that polyploid formation could have genome-wide epigenetic consequences of relevance to gene expression and hence polyploid evolution.

This suggestion of epigenetic effects on gene expression may be related to the general observation that polyploids are often associated with variation and instability in phenotypes that cannot be accounted for by conventional Mendelian transmission genetics or chromosomal aberrations (Comai, 2000; Comai *et al.*, 2000). The affected traits are diverse, including timing of flowering, overall plant habit, leaf morphology and homeotic transformations in floral morphology. These allopolyploidy-associated changes in phenotypes may arise from altered gene expression due to variation in dosage-regulated gene expression, altered regulatory interactions, and rapid genetic and epigenetic changes (reviewed in Osborn *et al.*, 2003).

One of the surprising recent findings concerning polyploids has been the degree to which gene expression may be altered by genome doubling (Liu and Wendel, 2002; Adams *et al.*, 2003; Comai *et al.*, 2003; Osborn *et al.*, 2003; Riddle and Birchler, 2003). Studies of gene expression in natural and synthetic polyploids have shown that genes may be silenced immediately upon or shortly following polyploidization. For example, ribosomal RNA arrays from one parent may be silenced in some organs of *Brassica napus*, although both parental rRNA sets are expressed in floral organs (Chen and Pikaard, 1997). This is the well-known phenomenon of nucleolar dominance, where in hybrids or allopolyploids nucleoli form in association with ribosomal RNA genes on chromosomes inherited from only one of the two parents (Pikaard, 1999, 2000a,b). These demonstrations for ribosomal genes suggested the possibility that hybridization and polyploidization might similarly induce epigenetic modifications of protein-coding genes. This suspicion has now been confirmed in several model plant systems (Comai, 2000; Comai et al., 2000; Lee and Chen, 2001; Kashkush et al., 2002; Madlung et al., 2002), including Arabidopsis, wheat and cotton polyploids, where silencing of numerous protein-coding genes has been demonstrated (Comai et al., 2000; Lee and Chen, 2001; Kashkush et al., 2002; He et al., 2003).

These studies indicate that allopolyploid formation in Arabidopsis is accompanied by epigenetic gene silencing, and that this silencing may affect a variety of genes with diverse biological functions. The silencing events may occur rapidly (F$_2$ generation of synthetic allopolyploid) or over a longer evolutionary time span, but their reversibility may be retained in natural allopolyploid species for thousands to perhaps millions of years. Of particular significance is the remarkable similarity or concordance in the silencing patterns between synthetic and natural allopolyploids, which suggests that allopolyploidy not only induces epigenetic changes but also that the changes may be visible to natural selection and, judging from their persistence, adaptive.

The scale of the phenomenon, and hence its potential level of evolutionary importance, is illustrated by recent work involving synthetic and natural allopolyploid wheat (Kashkush et al., 2002) and cotton (Gossypium) (Adams et al., 2003). In newly generated tetraploid wheat, an appreciable frequency (1% to 5%) of the genes surveyed experienced silencing within the first generation, and novel transcripts were occasionally observed. Interestingly, the novel transcripts activated by polyploidy that could be assigned a function are retrotransposons, suggesting release from epigenetic control or suppression (see below).

In cotton, natural and artificially generated allotetraploids were studied by Adams et al. (2003) using an electrophoretic approach to separate the transcripts of 40 different duplicate gene pairs. A remarkably high level of transcription bias was observed, with respect to the duplicated copies, in that about 25% of the genes studied exhibited altered expression in one or more plant organs. The most surprising result was the observation of organ-specific gene silencing that in some cases was reciprocal, meaning that one duplicate was expressed in one organ (e.g. stamens), while its counterpart was expressed in a different organ (e.g. carpels). Moreover, this organ-specific partitioning of duplicate expression was also evident in newly generated allotetraploids. Remarkably, the silencing patterns observed in natural cotton allopolyploids, estimated to be approximately 1.5 million years old (Senchina et al., 2003; Wendel and Cronn, 2003), were rather similar in some cases to those observed in the newly generated tetraploids. This observation implicates either long-term evolutionary maintenance of epigenetically induced expression states, or subsequent fixation of expression states in the natural tetraploids through normal mutational processes during the 1–2 million years since polyploid formation in the genus.

Collectively, recent studies in several model plant systems have revealed that polyploid formation may be accompanied by epigenetic alterations in gene expression throughout the genome. These epigenetic changes may occur with the onset of polyploidy or accrue more slowly on an evolutionary timeframe. In at least some cases, rapid epigenetic modifications that arise with the onset of allopolyploidy may be preserved on an evolutionary timescale through multiple speciation events. A more intriguing suggestion is that genome doubling or merger creates a massive and sudden pulse of novel epigenetic variation, which may be released and become visible to natural selection over periods of time ranging up to millions of years.

The studies discussed in this section illustrate a number of important phenomena that may bear directly on the evolution of polyploids. This includes non-random genomic changes observed in natural allopolyploid species and their synthetic counterparts, gene silencing, novel gene expression, the possibility of organ-specific partitioning of duplicate gene function, and transposable element activation. How each of these processes translates into phenotypic or physiological variation that may be visible to natural selection is not yet known in most cases, but the scale and scope of epigenetic alterations accompanying polyploidy and the importance of epigenetics to growth and development suggest that these connections have significance to our understanding of polyploid evolution. In this respect a particularly relevant example may be epigenetically controlled flowering time variation in synthetic *Brassica* polyploids (Schranz and Osborn, 2000), because flowering phenology is so obviously important to reproduction. A second example of an epigenetic modification that is evolutionarily consequential is the natural flower symmetry mutant (from wild-type bilateral to radial) in *Linaria vulgaris*, originally described by Linnaeus more than 250 years ago. The molecular basis of this mutation is hypermethylation and silencing of a gene controlling flower form (Cubas *et al.*, 1999). When one extrapolates these examples of epigenetic regulation of single genes to the entire genome, it becomes evident that allopolyploid lineages may harbour a nearly infinite and latent reservoir of epigenetic/genetic combinations for later release and evaluation by natural selection, perhaps after millions of years (Adams *et al.*, 2003). This intriguing possibility requires further study, but it may be important to our understanding of the evolution of polyploids.

Conclusion

As summarized in this review, genome doubling has been a pervasive phenomenon in plant evolution and remains a prominent process by which biodiversity is generated today. We have illustrated some of the intrinsic features of polyploids, generated by genome doubling and/or merger, which provide novel opportunities for creating phenotypic variation, and have highlighted some of the extrinsic factors that guide polyploid formation and subsequent evolution. Much remains to be learned regarding the functional consequences of gene and genome doubling and the array of molecular genetic mechanisms that they both engender and are subject to, as well as the interplay between these internal forces and external ecological and population-level phenomena. Many insights are likely in the near future, however, as molecular genetic and genomic approaches are increasingly brought to bear on natural and artificial model polyploid systems.

References

Abbott, R.J. and Comes, H.P. (2004) Evolution in the Arctic: a phylogeographic analysis of the circumarctic plant, *Saxifraga oppositifolia* (Purple saxifrage). *New Phytologist* 161, 211–224.

Adams, K.L., Cronn, R., Percifield, R. and Wendel, J.F. (2003) Genes duplicated by polyploidy show unequal contributions to the transcriptome and organ-specific reciprocal silencing. *Proceedings of the National Academy of Sciences USA* 100, 4649–4654.

Ainouche, M.L., Baumel, A., Salmon, A. and Yannic, G. (2004) Hybridization, polyploidy and speciation in *Spartina* (Poaceae). *New Phytologist* 161, 165–172.

Altschmied, J., Delfgaauw, J., Wilde, B., Duschl, J., Bouneau, L., Volff, J.-N. and Schartl, M. (2002) Subfunctionalization of duplicate *mitf* genes associated with differential degeneration of alternative exons in fish. *Genetics* 161, 259–267.

Arnold, M.L. (1997) *Natural Hybridization and Evolution*. Oxford University Press, New York.

Arnold, M.L., Bouck, A.C. and Scott Cornman, R. (2004) Verne Grant and Louisiana Irises: is there anything new under the sun? *New Phytologist* 161, 143–149.

Barrier, M., Robichaux, R.H. and Purugganan, M.D. (2001) Accelerated regulatory gene evolution in an adaptive radiation. *Proceedings of the National Academy of Sciences USA* 98, 10208–10213.

Beguiristain, T., Grandbastien, M.-A., Puigdomenech, P. and Casacuberta, J.M. (2001) Three Tnt1 subfamilies show different stress-associated patterns of expression in tobacco. Consequences for retrotransposon control and evolution in plants. *Plant Physiology* 127, 212–221.

Belyayev, A., Raskina, O., Korol, A. and Nevo, E. (2000) Coevolution of A and B genomes in allotetraploid *Triticum dicoccoides*. *Genome* 43, 1021–1026.

Bennetzen, J.L. (2000) Transposable element contributions to plant gene and genome evolution. *Plant Molecular Biology* 42, 251–269.

Blanc, G., Hokamp, K. and Wolfe, K.H. (2003) A recent polyploidy superimposed on older large-scale duplications in the *Arabidopsis* genome. *Genome Research* 13, 137–144.

Chen, Z.J. and Pikaard, C.S. (1997) Transcriptional analysis of nucleolar dominance in polyploid plants: biased expression/silencing of progenitor rRNA genes is developmentally regulated in *Brassica*. *Proceedings of the National Academy of Sciences USA* 94, 3442–3447.

Comai, L. (2000) Genetic and epigenetic interactions in allopolyploid plants. *Plant Molecular Biology* 43, 387–399.

Comai, L., Tyagi, A.P., Winter, K., Holmes-Davis, R., Reynolds, S.H., Stevens, Y. and Byers, B. (2000) Phenotypic instability and rapid gene silencing in newly formed *Arabidopsis* allotetraploids. *The Plant Cell* 12, 1551–1567.

Comai, L., Madlung, A., Joseffson, C. and Tyagi, A. (2003) Do the different parental 'heteromes' cause genomic shock in newly formed allopolyploids? *Philosophical Transactions of the Royal Society of London B* 358, 1149–1155.

Cubas, P., Vincent, C. and Coen, E. (1999) An epigenetic mutation responsible for natural variation in floral symmetry. *Nature* 401, 157–161.

Devos, K.M. and Gale, M.D. (2000) Genome relationships: the grass model in current research. *The Plant Cell* 12, 637–646.

Doyle, J.J., Doyle, J.L. and Brown, A.H.D. (1999) Origins, colonization, and lineage recombination in a widespread perennial soybean polyploid complex. *Proceedings of the National Academy of Sciences USA* 96, 10741–10745.

Doyle, J.J., Doyle, J.L., Rauscher, J.T. and Brown, A.H.D. (2004a) Diploid and polyploid reticulate evolution throughout the history of the perennial soybeans (*Glycine* subgenus *Glycine*). *New Phytologist* 161, 121–132.

Doyle, J.J., Doyle, J.L., Rauscher, J.T. and Brown, A.H.D. (2004b) Evolution of the perennial soybean polyploid complex (*Glycine* subgenus *Glycine*): a study of contrasts. *Biological Journal of the Linnean Society* 82, 583–597.

Dubcovsky, J. and Dvorák, J. (1995) Ribosomal RNA multigene loci: nomads of the Triticeae genomes. *Genetics* 140, 1367–1377.

Feldman, M., Liu, B., Segal, G., Abbo, S., Levy, A.A. and Vega, J.M. (1997) Rapid elimination of low-copy DNA sequences in polyploid wheat: a possible mechanism for differentiation of homoeologous chromosomes. *Genetics* 147, 1381–1387.

Finnegan, E.J. (2001a) Epialleles: a source of random variation in times of stress. *Current Opinion in Plant Biology* 5, 101–106.

Finnegan, E.J. (2001b) Is plant gene expression regulated globally. *Trends in Genetics* 17, 361–365.

Finnegan, E.J., Genger, R.K., Peacock, W.J. and Dennis, E.S. (1998) DNA methylation in plants. *Annual Review of Plant Physiology and Plant Molecular Biology* 49, 223–247.

Finnegan, E.J., Peacock, W.J. and Dennis, E.S. (2000) DNA methylation, a key regulator of plant development and other processes. *Current Opinion in Genetics and Development* 10, 217–223.

Fitch, W.M. (1970) Distinguishing homologous from analogous proteins. *Systematic Zoology* 19, 99–113.

Force, A., Lynch, M., Pickett, F.B., Amores, A., Yan, Y.-L. and Postlethwait, J. (1999) Preservation of duplicate genes by complementary, degenerative mutations. *Genetics* 151, 1531–1545.

Gaut, B.S. and Doebley, J.F. (1997) DNA sequence evidence for the segmental allotetraploid origin of maize. *Proceedings of the National Academy of Sciences USA* 94, 6808–6814.

Grandbastien, M.-A. (1998) Activation of plant retrotransposons under stress conditions. *Trends in Plant Science* 3, 181–187.

Grant, V. (1975) *Genetics of Flowering Plants*. Columbia University Press, New York.

Grant, V. (1981) *Plant Speciation*. Columbia University Press, New York.

Grewal, S.I.S. and Moazed, D. (2003) Heterochromatin and epigenetic control of gene expression. *Science* 301, 798–802.

Gu, X., Wang, Y. and Gu, J. (2002) Age distribution of human gene families shows significant roles of both large- and small-scale duplications in vertebrate evolution. *Nature Genetics* 31, 205–209.

Hanson, R.E., Zhao, X.-P., Islam-Faridi, M.N., Paterson, A.H., Zwick, M.S., Crane, C.F., McKnight, T.D., Stelly, D.M. and Price, H.J. (1998) Evolution of interspersed repetitive elements in *Gossypium* (Malvaceae). *American Journal of Botany* 85, 1364–1368.

Hanson, R.E., Islam-Faridi, M.N., Crane, C.F., Zwick, M.S., Czeschin, D.G., Wendel, J.F., McKnight, T.D., Price, H.J. and Stelly, D.M. (1999) Ty1-*copia*-retrotransposon behavior in a polyploid cotton. *Chromosome Research* 8, 73–76.

Harlan, J.R. and DeWet, J.M.J. (1975) On Ö. Winge and a prayer: the origins of polyploidy. *Botanical Review* 41, 361–390.

Harland, S.C. (1936) The genetical conception of the species. *Cambridge Philosophical Society Biological Reviews* 11, 83–112.

He, P., Friebe, B., Gill, B. and Zhou, J.-M. (2003) Allopolyploidy alters gene expression in the highly stable hexaploid wheat. *Plant Molecular Biology* 52, 401–414.

Hirochika, H., Sugimoto, K., Otsuki, Y. and Kanda, M. (1996) Retrotransposons of rice involved in mutations induced by tissue culture. *Proceedings of the National Academy of Sciences USA* 93, 7783–7788.

Hughes, A.L., Green, J.A., Garbayo, J.M. and Roberts, R.M. (2000) Adaptive diversification within a large family of recently duplicated, placentally expressed genes. *Proceedings of the National Academy of Sciences USA* 97, 3319–3323.

Husband, B.C. and Schemske, D.W. (2000) Ecological mechanisms of reproductive isolation between diploid and tetraploid *Chamerion angustifolium*. *Journal of Ecology* 88, 689–701.

Husband, B.C., Schemske, D.W., Burton, T.L. and Goodwillie, C. (2002) Pollen competition as a unilateral reproductive barrier between sympatric *Chamerion angustifolium*. *Proceedings of the Royal Society of London Series B: Biological Sciences* 269, 2565–2571.

Jiang, C., Wright, R., El-Zik, K. and Paterson, A. (1998) Polyploid formation created unique avenues for response to selection in *Gossypium* (cotton). *Proceedings of the National Academy of Sciences USA* 95, 4419–4424.

Joly, S., Rauscher, J.T., Sherman-Broyles, S.L., Brown, A.H.D. and Doyle, J.J. (2004) Evolution of the 18S-5.8S-26S nuclear ribosomal gene family and its expression in natural and artificial *Glycine* allopolyloids. *Molecular Biology and Evolution* 21, 1409–1421.

Kashkush, K., Feldman, M. and Levy, A.A. (2002) Gene loss, silencing, and activation in a newly synthesized wheat allotetraploid. *Genetics* 160, 1651–1659.

Kidwell, M.G. and Lisch, D.R. (2001) Perspective: transposable elements, parasitic DNA, and genome evolution. *Evolution* 55, 1–24.

Koonin, E., Fedorova, N., Jackson, J., Jacobs, A., Krylov, D., Makarova, K., Mazumder, R., Mekhedov, S., Nikolskaya, A., Rao, B., Rogozin, I., Smirnov, S., Sorokin, A., Sverdlov, A., Vasudevan, S., Wolf, Y., Yin, J. and Natale, D. (2004) A comprehensive evolutionary classification of proteins encoded in complete eukaryotic genomes. *Genome Biology* 5, R7.

Kovarik, A., Matyasek, R., Lim, K.Y., Skalicka, K., Koukalova, B., Knapp, S., Chase, M.W. and Leitch, A.R. (2004) Concerted evolution of 18-5.8-26S rDNA repeats in Nicotiana allotetraploids. *Biological Journal of the Linnean Society* 82, 615–625.

Kumar, A. and Bennetzen, J.L. (1999) Plant retrotransposons. *Annual Review of Genetics* 33, 479–532.

Kuta, E. and Przywara, L. (1997) Polyploidy in mosses. *Acta Biologica Cracoviensia* 39, 17–26.

Lawton-Rauh, A. (2003) Evolutionary dynamics of duplicated genes in plants. *Molecular Phylogenetics and Evolution* 29, 396–409.

Lee, H.-S. and Chen, Z.J. (2001) Protein-coding genes are epigenetically regulated in *Arabidopsis* polyploids. *Proceedings of the National Academy of Sciences USA* 98, 6753–6758.

Leitch, I.J. and Bennett, M.D. (1997) Polyploidy in angiosperms. *Trends in Plant Science* 2, 470–476.

Levin, D.A. (1983) Polyploidy and novelty in flowering plants. *American Naturalist* 122, 1–25.

Levin, D.A. (2003) The cytoplasmic factor in plant speciation. *Systematic Botany* 28, 5–11.

Lewis, W.H. (1980a) Polyploidy in species populations. In: Lewis, W.H. (ed.) *Polyploidy: Biological Relevance*. Plenum, New York, pp. 103–144.

Lewis, W.H. (1980b) *Polyploidy: Biological Relevance*. Plenum, New York.

Liu, B. and Wendel, J.F. (2000) Retroelement activation followed by rapid repression in interspecific hybrid plants. *Genome* 43, 874–880.

Liu, B. and Wendel, J.F. (2002) Non-Mendelian phenomena in allopolyploid genome evolution. *Current Genomics* 3, 489–506.

Liu, B., Segal, G., Vega, J.M., Feldman, M. and Abbo, S. (1997) Isolation and characterization of chromosome-specific sequences from a chromosome arm genomic library of common wheat. *Plant Journal* 11, 959–965.

Liu, B., Vega, J.M. and Feldman, M. (1998a) Rapid genomic changes in newly synthesized amphiploids of *Triticum* and *Aegilops*. II. Changes in low-copy coding DNA sequences. *Genome* 41, 535–542.

Liu, B., Vega, J.M., Segal, G., Abbo, S., Rodova, M. and Feldman, M. (1998b) Rapid genomic changes in newly synthesized amphiploids of *Triticum* and *Aegilops*. I. Changes in low-copy non-coding DNA sequences. *Genome* 41, 272–277.

Lynch, M. (2002) Genomics: gene duplication and evolution. *Science* 297, 945–947.

Lynch, M. and Conery, J.S. (2000) The evolutionary fate and consequences of duplicate genes. *Science* 290, 1151–1155.

Lynch, M. and Force, A. (2000) The probability of duplicate gene preservation by subfunctionalization. *Genetics* 154, 459–473.

Lynch, M., O'Hely, M., Walsh, B. and Force, A. (2001) The probability of preservation of a newly arisen gene duplicate. *Genetics* 159, 1789–1804.

Madlung, A., Masuelli, R.W., Watson, B., Reynolds, S.H., Davison, J. and Comai, L. (2002) Remodeling of DNA methylation and phenotypic and transcriptional changes in synthetic *Arabidopsis* allotetraploids. *Plant Physiology* 129, 733–746.

Martienssen, R.A. and Colot, V. (2001) DNA methylation and epigenetic inheritance in plants and filamentous fungi. *Science* 293, 1070–1074.

Masterson, J. (1994) Stomatal size in fossil plants: evidence for polyploidy in majority of angiosperms. *Science* 264, 421–424.

Matzke, M.A., Scheid, O.M. and Matzke, A.J.M. (1999) Rapid structural and epigenetic changes in polyploid and aneuploid genomes. *Bioessays* 21, 761–767.

Matzke, M.A., Gregor, W., Mette, M.F., Aufsatz, W., Kanno, T., Jakowitsch, J. and Matzke, J.M. (2004) Endogenous pararetroviruses of polyploid tobacco (*Nicotiana tabacum*) and its diploid progenitors, *N. sylvestris* and *N. tomentosiformis*. *Biological Journal of the Linnean Society* 82, 627–638.

McClintock, B. (1984) The significance of responses of the genome to challenge. *Science* 226, 792–801.

Nathan, D., Sterner, D.E. and Berger, S.L. (2003) Histone modifications: now summoning sumoylation. *Proceedings of the National Academy of Sciences USA* 100, 13118–13120.

Nuismer, S.L. and Thompson, J.N. (2001) Plant polyploidy and non-uniform effects on insect herbivores. *Proceedings of the Royal Society of London Series B: Biological Sciences* 268, 1937–1940.

Ohno, S. (1970) *Evolution by Gene Duplication*. Springer-Verlag, New York.

Osborn, T.C., Pires, J.C., Birchler, J.A., Auger, D.L., Chen, Z.J., Lee, H.-S., Comai, L., Madlung, A., Doerge, R.W., Colot, V. and Martienssen, R.A. (2003) Understanding mechanisms of novel gene expression in polyploids. *Trends in Genetics* 19, 141–147.

Otto, S.P. and Whitton, J. (2000) Polyploid incidence and evolution. *Annual Review of Genetics* 34, 401–437.

Ownbey, M. (1950) Natural hybridization and amphiploidy in the genus *Tragopogon*. *American Journal of Botany* 37, 487–499.

Ozkan, H., Levy, A.A. and Feldman, M. (2001) Allopolyploidy-induced rapid genome evolution in the wheat (*Aegilops-Triticum*) group. *The Plant Cell* 13, 1735–1747.

Paterson, A.H., Bowers, J.E., Burow, M.D., Draye, X., Elsik, C.G., Jiang, C.-X., Katsar, C.S., Lan, Y.-R., Ming, R. and Wright, R.J. (2000) Comparative genomics of plant chromosomes. *The Plant Cell* 12, 1523–1539.

Paterson, A.H., Bowers, J.E., Peterson, D.G., Estill, J.C. and Chapman, B.A. (2003) Structure and evolution of cereal genomes. *Current Opinion in Genetics and Development* 13, 644–650.

Pébusque, M.-J., Coulier, F., Birnbaum, D. and Pontarotti, P. (1998) Ancient large-scale genome duplications: phylogenetic and linkage analyses shed light on chordate genome evolution. *Molecular Biology and Evolution* 15, 1145–1159.

Pikaard, C.S. (1999) Nucleolar dominance and silencing of transcription. *Trends in Plant Science* 4, 478–483.

Pikaard, C.S. (2000a) The epigenetics of nucleolar dominance. *Trends in Genetics* 16, 495–500.

Pikaard, C.S. (2000b) Nucleolar dominance: uniparental gene silencing on a multi-megabase scale in genetic hybrids. *Plant Molecular Biology* 43, 163–177.

Pikaard, C.S. (2002) Genomic change and gene silencing in polyploids. *Trends in Genetics* 17, 675–677.

Ramsey, J. and Schemske, D.W. (1998) Pathways, mechanisms, and rates of polyploid formation in flowering plants. *Annual Review of Ecology and Systematics* 29, 467–501.

Ramsey, J. and Schemske, D.W. (2002) Neopolyploidy in flowering plants. *Annual Review of Ecology and Systematics* 33, 589–639.

Rauscher, J.T., Doyle, J.J. and Brown, A.H.D. (2004) Multiple origins and nrDNA ITS homoeologue evolution in the *Glycine tomentella* (Leguminosae) allopolyploid complex. *Genetics* 166, 987–998.

Richards, E. (1997) DNA methylation and plant development. *Trends in Genetics* 13, 319–323.

Richards, E.J. and Elgin, S.C.R. (2002) Epigenetic codes for heterochromatin formation and silencing: rounding up the usual suspects. *Cell* 108, 489–500.

Riddle, N.C. and Birchler, J.A. (2003) Effects of reunited diverged regulatory hierarchies in allopolyploids and species hybrids. *Trends in Genetics* 19, 597–600.

Rieger, R., Michaelis, A. and Green, M.M. (1976) *Glossary of Genetics and Cytogenetics*. Springer-Verlag, New York.

Rieseberg, L.H. (1995) The role of hybridization in evolution: old wine in new skins. *American Journal of Botany* 82, 944–953.

Rieseberg, L.H. (2001) Polyploid evolution: keeping the peace at genomic reunions. *Current Biology* 13, R925–R928.

Rieseberg, L.H. and Wendel, J.F. (1993) Introgression and its consequences in plants. In: Harrison, R. (ed.) *Hybrid Zones and the Evolutionary Process*. Oxford University Press, New York, pp. 70–109.

Roose, M.L. and Gottlieb, L.D. (1976) Genetic and biochemical consequences of polyploidy in *Tragopogon*. *Evolution* 30, 818–830.

Schranz, M.E. and Osborn, T.C. (2000) Novel flowering time variation in the resynthesized polyploid *Brassica napus*. *Journal of Heredity* 91, 242–246.

Senchina, D., Alvarez, I., Cronn, R., Liu, B., Rong, J., Noyes, R., Paterson, A.H., Wing, R.A., Wilkins, T.A. and Wendel, J.F. (2003) Rate variation among nuclear genes and the age of polyploidy in *Gossypium*. *Molecular Biology and Evolution* 20, 633–643.

Shaked, H., Kashkush, K., Ozkan, H., Feldman, M. and Levy, A.A. (2001) Sequence elimination and cytosine methylation are rapid and reproducible responses of the genome to wide hybridization and allopolyploidy in wheat. *The Plant Cell* 13, 1749–1759.

Simillion, C., Vandepoele, K., Van Montagu, M.C.E., Zabeau, M. and Van de Peer, Y. (2002) The hidden duplication past of *Arabidopsis thaliana*. *Proceedings of the National Academy of Sciences USA* 99, 13627–13632.

Skalicka, K., Lim, K.Y., Matyasek, R., Koukalova, B., Leitch, A.R. and Kovarik, A. (2003) Rapid evolution of parental rDNA in a synthetic tobacco allotetraploid line. *American Journal of Botany* 90, 988–996.

Soltis, D.E. and Soltis, P.S. (1993) Molecular data and the dynamic nature of polyploidy. *Critical Reviews in Plant Science* 12, 243–273.

Soltis, D.E. and Soltis, P.S. (1995) The dynamic nature of polyploid genomes. *Proceedings of the National Academy of Sciences USA*, 92, 8089–8091.

Soltis, D.E. and Soltis, P.S. (1999) Polyploidy: origins of species and genome evolution. *Trends in Ecology and Evolution* 9, 348–352.

Soltis, P.S. and Soltis, D.E. (2000) The role of genetic and genomic attributes in the success of polyploids. *Proceedings of the National Academy of Sciences USA* 97, 7051–7057.

Soltis, D.E., Soltis, P.S. and Tate, J.A. (2004) Advances in the study of polyploidy since plant speciation. *New Phytologist* 161, 173–191.

Song, K., Lu, P., Tang, K. and Osborn, T.C. (1995) Rapid genome change in synthetic polyploids of *Brassica* and its implications for polyploid evolution. *Proceedings of the National Academy of Sciences USA* 92, 7719–7723.

Stebbins, G.L. (1947) Types of polyploids: their classification and significance. *Advances in Genetics* 1, 403–429.

Stebbins, G.L. (1950) *Variation and Evolution in Plants*. Columbia University Press, New York.

Stebbins, G.L. (1971) *Chromosomal Evolution in Higher Plants*. Edward Arnold, London.

Stephens, S.G. (1951a) Evolution of the gene: 'homologous' genetic loci in *Gossypium*. *Cold Spring Harbor Symposium on Quantitative Biology* 16, 131–140.

Stephens, S.G. (1951b) Possible significance of duplication in evolution. *Advances in Genetics* 4, 247–265.

Taylor, J.S., Van de Peer, Y., Braasch, I. and Meyer, A. (2001) Comparative genomics provides evidence for an ancient genome duplication event in fish. *Proceedings of the Royal Society of London Series B: Biological Sciences* 356, 1661–1679.

Thompson, J.N., Nuismer, S.L. and Merg, K. (2004) Plant polyploidy and the evolutionary ecology of plant/animal interactions. *Biological Journal of the Linnean Society* 82, 511–519.

Trinti, F. and Scali, V. (1996) Androgenetics and triploids from an interacting parthenogenetic hybrid and its ancestors in stick insects. *Evolution* 50, 1251–1258.

Vision, T.J., Brown, D.G. and Tanksley, S.D. (2000) The origins of genomic duplications in *Arabidopsis*. *Science* 290, 2114–2117.

Vogel, J.C., Barrett, J.A., Rumsey, F.J. and Gibby, M. (1999) Identifying multiple origins in polyploid homo-sporous pteridophytes. In: Hollingsworth, P.M., Bateman, R.M. and Gornall, R.J. (eds) *Molecular Systematics and Plant Evolution.* Taylor & Francis, London, pp. 101–117.

Watanabe, K., Peloquin, S.J. and Endo, M. (1991) Genetic significance of mode of polyploidization: somatic doubling or 2n gametes. *Genome* 34, 28–34.

Wendel, J.F. (2000) Genome evolution in polyploids. *Plant Molecular Biology* 42, 225–249.

Wendel, J.F. and Cronn, R.C. (2003) Polyploidy and the evolutionary history of cotton. *Advances in Agronomy* 78, 139–186.

Wendel, J.F., Schnabel, A. and Seelanan, T. (1995) Bidirectional interlocus concerted evolution following allopolyploid speciation in cotton (*Gossypium*). *Proceedings of the National Academy of Sciences USA* 92, 280–284.

Werth, C.R. and Windham, M.D. (1991) A model for divergent, allopatric speciation of polyploid pterido-phytes resulting from silencing of duplicate-gene expression. *American Naturalist* 137, 515–526.

Wessler, S.R. (1996) Plant retrotransposons: turned on by stress. *Current Biology* 6, 959–961.

Wessler, S.R. (2001) Plant transposable elements: a hard act to follow. *Plant Physiology* 125, 149–151.

Wilson, W.A., Harrington, S.E., Woodman, W.L., Lee, M., Sorrells, M.E. and McCouch, S.R. (1999) Inferences on the genome structure of progenitor maize through comparative analysis of rice, maize and the domesticated Panicoids. *Genetics* 153, 453–473.

Wolfe, K.H. and Shields, D.C. (1997) Molecular evidence for an ancient duplication of the entire yeast genome. *Nature* 387, 708–713.

Wolffe, A.P. and Matzke, M.A. (1999) Epigenetics: regulation through repression. *Science* 286, 481–486.

Yoder, J.A., Walsh, C.P. and Bestor, T.H. (1997) Cytosine methylation and the ecology of intragenomic para-sites. *Trends in Genetics* 13, 335–340.

Zhao, X.-P., Si, Y., Hanson, R.E., Crane, C.F., Price, H.J., Stelly, D.M., Wendel, J.F. and Paterson, A.H. (1998) Dispersed repetitive DNA has spread to new genomes since polyploid formation in cotton. *Genome Research* 8, 479–492.

8 Crucifer evolution in the post-genomic era

Thomas Mitchell-Olds,[1] Ihsan A. Al-Shehbaz,[2] Marcus A. Koch[3] and Tim F. Sharbel[4]

[1]Department of Genetics and Evolution, Max Planck Institute of Chemical Ecology, Hans-Knoll Strasse 8, 07745, Jena, Germany; [2]Missouri Botanical Gardens, PO Box 299, St Louis, MO 63166-0299, USA; [3]Heidelberg Institute of Plant Sciences, Biodiversity and Plant Systematics, Im Neuenheimer Feld 345, D-69129 Heidelberg, Germany; [4]Laboratoire IFREMER de Genetique et Pathologie, 17390 La Tremblade, France

Introduction

The worldwide genomics initiative in *Arabidopsis thaliana* has facilitated a renaissance in evolutionary studies of the Brassicaceae. More than 20,000 published papers examine aspects of *Arabidopsis* biology, providing detailed understanding of molecular biology, genetics, biochemistry, physiology and development of this model plant. Increasingly, comparative analyses (Hall *et al.*, 2002a; Mitchell-Olds and Clauss, 2002) build upon *Arabidopsis* genomics to elucidate biology of crucifer species, and the evolutionary processes which influence adaptation and diversification in the Brassicaceae. Here we review aspects of systematics, speciation and functional variation in this important plant family.

Systematics and Taxonomy

Family characteristics and importance

The mustard family (Brassicaceae or Cruciferae) includes some 340 genera and about 3350 species distributed worldwide, especially in the temperate and alpine regions of the northern hemisphere (Al-Shehbaz, 1984; Appel and Al-Shehbaz, 2003). It includes important crop plants cultivated worldwide as vegetables and ornamentals and as sources of cooking and industrial oils, condiments and forage (Koch *et al.*, 2003a). One species, *Arabidopsis thaliana* (thale cress), is the model flowering plant in nearly every field of experimental biology, and its entire genome has recently been sequenced (The Arabidopsis Genome Initiative, 2000). The Brassicaceae are easily recognized by having flowers with four petals forming a cross (sometimes reduced or lacking), six stamens (the outer two being shorter than the inner four, although sometimes only two or four stamens are present), and often a two-valved capsule with a septum dividing it into two chambers.

Family limits and relatives

The Brassicaceae were previously thought to have evolved in the New World, either through the putatively basal tribe Thelypodieae (Stanleyeae) from the Capparaceae subfamily Cleomoideae (Janchen, 1942; Al-Shehbaz, 1973; Takhtajan, 1997), or to share a common ancestry with that subfamily (Al-Shehbaz, 1985). Dvořák (1973) suggested an alternative origin in the Old World and considered the tribe Hesperideae as the link between the Brassicaceae and Cleomoideae. Hauser and Crovello (1982) tested these hypotheses and favoured a New World origin from the Cleomoideae.

Based on a small sample (two genera of Brassicaceae and four of Capparaceae) and a set of only 16 characters, Judd *et al.* (1994) concluded in a cladistic morphological study that the Brassicaceae is nested within the Capparaceae and that the two families should be united in one, Brassicaceae. Although this merger was followed (Angiosperm Phylogeny Group, 1998) or recommended (Appel and Al-Shehbaz, 2003), thorough molecular and morphological data (Rodman *et al.*, 1996, 1998) suggested a closer relationship of the Brassicaceae to *Cleome* than to the rest of Capparaceae. Hall *et al.* (2002b), who conducted detailed molecular studies using the chloroplast regions *trn*L-*trn*F and *ndh*F, advocated that three, well-supported monophyletic families (Brassicaceae, Cleomaceae, Capparaceae) should be recognized, with the Brassicaceae and Cleomaceae as sister families sharing a common ancestor.

The remarkable morphological similarities between Thelypodieae and Cleomaceae (exserted stamens equal in length, sessile stigmas, linear fruits, dense racemes, long gynophores, linear anthers coiled at dehiscence, to name some) all appear to be the result of convergence rather than synapomorphy. Using internal transcribed spacer of nuclear ribosomal DNA (ITS) data, Warwick *et al.* (2002) have clearly demonstrated that the Thelypodieae and many New World genera form an unresolved, rather advanced, terminal polytomy. These findings agree with those of Galloway *et al.* (1998).

All broad-based molecular studies of Brassicaceae (Zunk *et al.*, 1996, 1999; Galloway *et al.*, 1998; Koch *et al.*, 2001a, 2003a) demonstrated that *Aethionema* occupies the most basal position in the family. *Aethionema*, a highly polymorphic genus of 50–60 species, is distributed primarily in Turkey and the Middle East (Appel and Al-Shehbaz, 2003). It has angustiseptate fruits

(flattened at a right angle to the septum), a feature not found among the Old World Cleomaceae. Although published molecular data are based on two, highly variable species, *Aethionema grandiflorum* and *Aethionema saxatile*, phylogenetic studies in progress (Menke, personal communication) should resolve its monophyly, determine its nearest relatives and shed light on what makes it basal in the Brassicaceae.

ITS and other molecular markers at the tribal level

All major classification systems (Prantl, 1891; Hayek, 1911; Schulz, 1936; Janchen, 1942), which divide the family into 4–19 tribes, are based on a limited number of morphological characters and do not recognize convergence as a factor in the evolution of Brassicaceae. Molecular studies (Price *et al.*, 1994; Zunk *et al.*, 1996, 1999; Koch *et al.*, 1999a, 2000, 2001a, 2003a; Bailey *et al.*, 2002; Koch, 2003; O'Kane and Al-Shehbaz, 2003) have demonstrated the polyphyly and artificiality of almost all tribes recognized in these systems. For example, *Capsella* and *Arabidopsis*, treated by Schulz (1936) in unrelated tribes, have been shown (Koch *et al.*, 2001a; O'Kane and Al-Shehbaz, 2003) to be very closely related. Numerous other examples can be cited, and the interested reader should consult Koch *et al.* (2003a).

Extensive molecular data, summarized by Warwick and Black (1997a,b), show that the Brassiceae, characterized by segmented fruits and/or conduplicate cotyledons (Appel and Al-Shehbaz, 2003), is the only monophyletic group among Schulz's (1936) 19 tribes. However, generic limits, as traditionally recognized (Gómez-Campo, 1999), remain problematic, and a major revision of the boundaries of most genera is needed in light of molecular data (Koch *et al.*, 2003a).

Based on ITS sequence data (Kropf *et al.*, 2002; Koch *et al.*, 2003a; O'Kane and Al-Shehbaz, 2003) and other markers (Galloway *et al.*, 1998; Koch *et al.*, 2000, 2001a, 2003a), several monophyletic clades are readily recognized (Fig. 8.1). One such group, the *Brassica* alliance (*c.* 550 species),

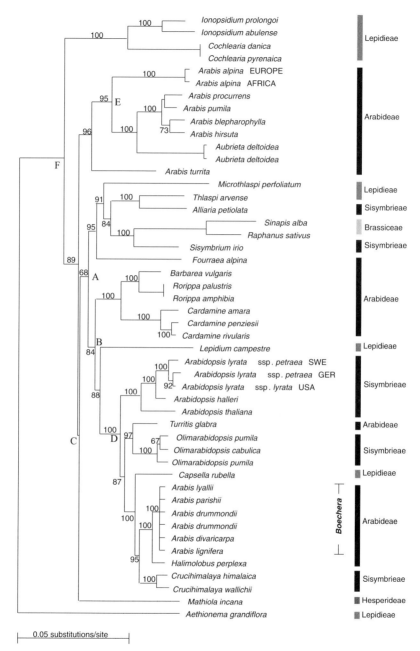

Fig. 8.1. Neighbour-joining distance tree based on *matK* and *Chs* sequences (Koch *et al.*, 2001a). Percentage bootstrap values from 1000 replicates are shown on each branch. Of the six monophyletic clades discussed in the text, four are represented here: *Brassica* alliance (*Sinapis, Raphanus, Alliaria, Thlaspi, Microthlaspi*), *Arabidopsis* alliance (*Arabidopsis, Boechera, Halimolobos, Capsella, Olimarabidopsis, Crucihimalaya, Turritis*), *Arabis* alliance (*Arabis* and *Aubrieta*) and Cardaminine alliance (*Cardamine, Barbarea, Rorippa*). Tribal assignments are given on the right. Approximate divergence dates (million years before present) for nodes A–F are: A, 16–21; B, 13–19; C, 19–25; D, 10–14; E, 15–17; and F, 26–32. Several taxa mentioned in the text are not shown in this figure: *Brassica* is near *Raphanus* and *Sinapis*, and *Leavenworthia* is related to *Barbarea*. Reproduced, with permission, from Koch *et al.* (2001a).

includes all members of the tribe Brassiceae, the expanded New World Thelypodieae and *Sisymbrium sensu* Warwick *et al.* (2002), and *Thlaspi* and its segregates (Mummenhoff *et al.*, 1997a,b; Koch and Mummenhoff, 2001). The vast majority of genera in this clade have species either glabrous or with simple trichomes, though branched trichomes apparently evolved independently a few times, especially in some South American Thelypodieae and southern African *Sisymbrium*. ITS sequence data provide little resolution within the Thelypodieae. This suggests a relatively recent evolution of the group and insufficient time for ITS sequences to diverge, in agreement with the absence of morphological differentiation and the difficulty in recognizing individual genera in the group (Warwick *et al.*, 2002).

The second clade (*c.* 300 species), designated herein as the *Arabidopsis* alliance, includes *Boechera sensu* Al-Shehbaz (2003b), the halimolobine clade (*Halimolobos, Mancoa, Pennellia, Sphaerocardamum*) *sensu* Bailey *et al.* (2002), *Arabidopsis* and its recent segregates (Al-Shehbaz *et al.*, 1999; O'Kane and Al-Shehbaz, 2003), *Camelina, Capsella, Cusickiella, Neslia*, the polycolpate clade (*Dimorphocarpa, Dithyrea, Lyrocarpa, Nerisyrenia, Paysonia, Physaria* including *Lesquerella, Synthlipsis*) *sensu* O'Kane and Al-Shehbaz (2003), *Pachycladon sensu* Heenan *et al.* (2002), *Erysimum* and *Transberingia* (*Beringia sensu* Price *et al.*, 2001). This clade is characterized by the preponderance of forked and/or stellate trichomes.

A third clade, the *Arabis* alliance (*c.* 450 species), involves *Arabis* (but excludes *Boechera, Turritis* and *Fourraea*), *Draba* (including *Erophila* and *Drabopsis*) and *Aubrieta*. This group has branched trichomes, accumbent cotyledons, and often latiseptate fruits (flattened parallel to the septum). Although the same combination of characters is found in *Boechera*, the similarities are superficial and result from convergence rather than synapomorphy (Koch *et al.*, 1999a; Koch and Al-Shehbaz, 2002).

A fourth monophyletic clade (*c.* 250 species) contains *Lepidium* including *Cardaria, Coronopus* and *Stroganowia*. This clade is well supported by molecular

(Mummenhoff, 1995; Bruggemann, 2000; Mummenhoff *et al.*, 2001) and morphological (Al-Shehbaz *et al.*, 2002) evidence. The genera *Acanthocardamum, Delpinophytum, Winklera, Stubendorffia, Megacarpaea* and *Biscutella* should be studied in connection with this clade, and it is likely that some, if not all, are allied to *Lepidium*. The clade has angustiseptate fruits, two ovules per ovary and simple or no trichomes.

Another clade is the Cardaminine alliance (*c.* 350 species), which includes *Cardamine* (including *Dentaria* and *Iti*), *Armoracia, Barbarea, Iodanthus, Leavenworthia, Nasturtium, Planodes, Rorippa, Selenia*, and perhaps several Himalayan genera. Members of this clade have accumbent cotyledons, latiseptate fruits, dissected or compound leaves, and simple or no trichomes, and they frequently occupy aquatic, wet or mesic habitats (Franzke *et al.*, 1998; Mitchell and Heenan, 2000; Sweeney and Price, 2000).

Another alliance much in need of further studies includes *Alyssum* and related genera (*c.* 220 species). The remaining 1200 species of Brassicaceae perhaps fall within these six major clades.

Molecular data and generic delimitation

Differences in fruit morphology and embryo position have been used extensively in the delimitation of genera (Al-Shehbaz, 1984; Rollins, 1993; Appel and Al-Shehbaz, 2003). However, such differences are often overemphasized and vegetative and floral characters are largely neglected. Sequence comparisons of relatively rapidly evolving regions such as ITS and *ndh*F suggest that fruit morphology and embryo type are subject to frequent convergence. The high degree of sequence similarity among taxa with different fruits (Fig. 8.2) emphasizes the rapid rate at which major changes in fruits and embryo morphology can occur. It is plausible that the number of genes responsible for changes in fruit shape and embryo position may be relatively small, thus allowing rapid bursts of evolution uncoupled from other aspects of morphology or molecular markers.

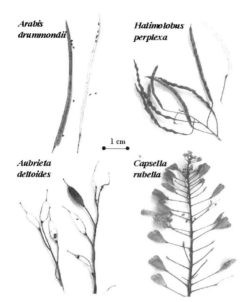

Fig. 8.2. Some examples of fruit morphologies among relatives of *Arabidopsis*. Three of these species are members of the *Arabidopsis* alliance (*Halimolobos perplexa*, *Capsella rubella* and *Arabis drummondii* [syn. *Boechera stricta*]). Reproduced, with permission, from Koch *et al.* (1999a).

More than 225 genera (60%) of Brassicaceae include four species or less (Appel and Al-Shehbaz, 2003). It is likely that this number will be reduced significantly with detailed molecular studies, as illustrated by two examples. First, based on chloroplast DNA restriction site variation (Warwick and Black, 1994) and ITS sequence data (Crespo *et al.*, 2000), *Euzumodendron* and *Boleum* are well nested within and hardly differ from *Vella*. Although these genera have very different fruit morphology, all include shrubs with *x*=17 and united paired stamens, a character combination not found elsewhere in the tribe Brassiceae. Molecular data, coupled with unique morphology and cytology, led Warwick and Al-Shehbaz (1998) to unite these genera in one, *Vella*, a position now widely accepted.

Another classic example involves the Chilean *Agallis* and Californian *Twisselmannia* and *Tropidocarpum*. These genera also have dramatically different fruit morphology but are indistinguishable in floral and vegetative characters. All three are basically identical in

ITS and *ndh*F sequence data, and they differ only in one base substitution for each marker (Price, personal communication). They are now recognized as one genus, *Tropidocarpum* (Al-Shehbaz and Price, 2001; Al-Shehbaz, 2003a).

The other extreme involves taxa that are very difficult to distinguish based on fruit morphology but their ITS and other sequence data show enormous divergence well supported by high bootstrap values. Three classic examples are given. First, ITS (Koch *et al.*, 1999a; O'Kane and Al-Shehbaz, 2003) and chalcone synthase and alcohol dehydrogenase data (Koch *et al.*, 2000) clearly demonstrate that the North American *Arabis sensu* Rollins (1993) is polyphyletic and that all except 15 of the 80 species should be assigned to *Boechera* (Al-Shehbaz, 2003b). Although the fruits of *Boechera* and *Arabis* are quite similar, significant morphological differences have been found, and the two genera are clearly unrelated (Al-Shehbaz, 2003b). Second, ITS data (Koch *et al.*, 1999a; O'Kane and Al-Shehbaz, 2003) strongly support splitting the c. 60 species of *Arabidopsis* into several genera differing significantly in characters other than fruit morphology (Al-Shehbaz *et al.*, 1999; Al-Shehbaz and O'Kane, 2002). Finally, molecular data on *Thlaspi sensu* lato (Mummenhoff and Koch, 1994; Mummenhoff *et al.*, 1997a,b; Koch and Mummenhoff, 2001) provide ample support for most of Meyer's (1973, 1979) segregates of the genus based on seed-coat anatomy.

In conclusion, extreme care should be taken in evaluating generic limits based solely on fruit and/or embryo morphology because both structures are highly homoplastic. Monotypic genera are always suspect, and they should not be erected without thorough molecular studies using both nuclear and chloroplast markers.

Speciation and Differentiation

Hybridization and polyploidization

Since at least Stebbins's (1940) work it has been clear that hybridization and polyploidization have played an important role in the evolution of angiosperms (Ehrendorfer, 1980; Soltis and Soltis, 1999). This is also true for the Brassicaceae. The majority of taxa are recent or ancient polyploids with largely duplicated genomes (Appel and Al-Shehbaz, 2003; Koch *et al.*, 2003a), and chromosome numbers varying from $n = 4$ in some species of the North American *Physaria* and Australian *Stenopetalum* to $n = 128$ in *Cardamine concatenata* (as *Cardamine laciniata*, see Al-Shehbaz, 1984). Polyploid complexes with reticulate evolutionary patterns are found frequently and date back to different time periods. A late Pleistocene history of speciation has been characterized in genera such as *Cochlearia* (Koch *et al.*, 1996, 1999b; Koch, 2002) and Central European *Cardamine* (Franzke and Hurka, 2000; Marhold *et al.*, 2002) or closely related *Nasturtium* and *Rorippa* (Bleeker *et al.*, 1999; Bleeker, 2003). Other genera such as *Yinshania* (Koch and Al-Shehbaz, 2000) and *Draba* (Koch and Al-Shehbaz, 2002) have been identified as polyploid complexes with speciation processes dating back to the Tertiary (mostly Pliocene). Pleistocene differentiation, reported for North American *Boechera* (Dobes *et al.*, 2004), was also greatly affected by various glacial and interglacial cycles.

Within the tribe Brassiceae (an apparently monophyletic lineage; see above) the genome underwent extensive duplications of large genomic regions (Kowalski *et al.*, 1994; Lagercrantz and Lydiate, 1996; Lagercrantz, 1998), which led to the conclusion that 'diploid' species such as *Brassica oleracea* and *Brassica rapa* ($n = 9$ and $n = 10$, respectively) are ancestral hexaploids. The impact of hybridization on several traits has been demonstrated within *Lepidium* for flower morphology (Bowman *et al.*, 1999; Lee *et al.*, 2002) or life history traits within *Capsella* (Hurka and Neuffer, 1997). Detailed studies show hybridization between different parental taxa with identical chromosome numbers accompanied by subsequent polyploidization (e.g. *Capsella*: Mummenhoff and Hurka, 1990; *Draba*: Brochmann *et al.*, 1992; *Cardamine*: Urbanska *et al.*, 1997; *Microthlaspi*: Koch and Hurka, 1999). However, there are also detailed descriptions of hybridization between taxa with dif-

ferent base chromosome numbers, such as among *Brassica* species (U, 1935; Erickson *et al.*, 1983; Palmer *et al.*, 1983) or *Arabidopsis* (Mummenhoff and Hurka, 1994, 1995; O'Kane and Al-Shehbaz, 1997; O'Kane *et al.*, 1997; Sall *et al.*, 2003). In the latter case artificial hybrids have been obtained between *A. thaliana* ($n = 5$) and other *Arabidopsis* species with $n = 8$ (Comai *et al.*, 2000, 2003; Nasrallah *et al.*, 2000), which can be analysed genetically and physiologically.

Species divergence

Evolutionary studies of species differences employ two main approaches: (i) genetic mapping of trait differences (Schemske and Bradshaw, 1999; Rieseberg *et al.*, 2003); and (ii) divergence population genetics (Kliman *et al.*, 2000). These approaches provide complementary insights into the evolutionary processes and functional changes during speciation. Several outstanding studies of quantitative genetic differences between sister species have been reported (e.g. Bradshaw and Schemske, 2003; Rieseberg *et al.*, 2003). Many studies have examined quantitative trait loci (QTL) among cultivars or subspecies in *Brassica* (e.g. Lan and Paterson, 2000; Schranz *et al.*, 2002). To date, QTL mapping from undisturbed natural populations of crucifers has not yet been reported, although a linkage map of molecular markers is now available for *Arabidopsis lyrata* (Kuittinen *et al.*, 2004).

Divergence population genetics compares allele genealogies for multiple loci among several species (Kliman *et al.*, 2000). Early in speciation, alleles will be shared between sister species. In contrast, species-specific polymorphisms predominate later in speciation. Between these two extremes, polymorphisms are especially informative at an intermediate stage of speciation, when some genes have shared alleles, while other loci have private alleles. These differences among loci may be attributable to natural selection on functional differences, random variation across the genome, or occasional gene flow between species.

Ramos-Onsins *et al.* (2004) examined nucleotide polymorphism at eight unlinked loci in species-wide samples of four taxa in *Arabidopsis*, comparing the highly-inbreeding *A. thaliana* with the closely related out-crossing *Arabidopsis halleri*, *A. lyrata* ssp. *petraea* and *A. lyrata* ssp. *lyrata*. Average levels of nucleotide polymorphism were highest in ssp. *petraea* and lowest in ssp. *lyrata*, presumably reflecting differences in effective population size between subspecies. This relatively low nucleotide polymorphism in ssp. *lyrata* may reflect a population bottleneck during Pleistocene colonization of North America (Wright *et al.*, 2003; Ramos-Onsins *et al.*, 2004). Population genetic analysis suggests that introgression has occurred between *A. halleri* and *A. lyrata* ssp. *petraea* subsequent to speciation (Ramos-Onsins *et al.*, 2004).

Gamete recognition genes play an important role in speciation (Howard, 1999), and they experience rapid adaptive evolution in a number of animal systems (Swanson and Vacquier, 2002a,b). Similar rapidly evolving mate recognition genes have also been suggested in plants. Binding of pollen to *A. thaliana* stigmas is controlled by an oleosin-domain protein belonging to a small gene family (Mayfield *et al.*, 2001). Genomic comparisons from *A. thaliana*, *Boechera stricta* (formerly *Arabis drummondii*) and *Brassica oleracea* show rapid evolution due to gene duplication and deletion, accelerated amino acid substitution, and insertions and deletions within the coding region (Mayfield *et al.*, 2001; Schein *et al.*, 2004). These results are consistent with a hypothesized function in species recognition. Further functional and population genetic analyses are required to elucidate causes of rapid evolution in this gene family.

Phylogeography

Since its development by Avise *et al.* (1987), phylogeography has become an increasingly important field of research within biogeography. Originally the aim was to describe the distribution of genetic variation in space and time. More recently, understanding of his-

torical and population processes has emerged as a central focus. For these purposes molecular markers have been utilized to elaborate geographic distribution patterns of presumably neutral loci. Maternally inherited DNA markers (e.g. the plastome of most angiosperms, see Harris and Ingram, 1991; Reboud and Zeyl, 1994) can be used to trace maternal lineages. A variety of nuclear markers are available (Sunnucks, 2000), including co-dominant isozymes and microsatellites, dominant markers such as RAPDs (random amplified polymorphic DNAs) or AFLPs (amplified fragment length polymorphisms), nuclear DNA sequences such as ITS1 and ITS2, single-copy nuclear genes (Savolainen *et al.*, 2000; Ramos-Onsins *et al.*, 2004), and SNPs (single nucleotide polymorphisms, e.g. Brumfield *et al.*, 2003; Schmid *et al.*, 2003).

Among crucifers, phylogeographic studies are available from all regions of their distribution and on very different geographic scales. Worldwide phylogeographies were conducted on *Lepidium* (Mummenhoff *et al.*, 2001) and *Arabidopsis* (Ramos-Onsins *et al.*, 2004), and central European phylogeographies were focused on *Biscutella* (Tremetsberger *et al.*, 2002), *Arabidopsis* (Sharbel *et al.*, 2000), *Cochlearia* (Koch, 2002; Koch *et al.*, 2003b), *Microthlaspi* (Koch *et al.*, 1998; Koch and Bernhardt, 2004), *Hornungia* (as *Pritzelago*) (Kropf *et al.*, 2003) and alpine *Draba* (Widmer and Baltisberger, 1999). These studies elucidated colonization routes from refugial areas into formerly glaciated areas of north and central Europe, and they also identified Pleistocene refugial areas in the Iberian Peninsula, northern Italy, or the Balkans (Comes and Kadereit, 1998).

A complex speciation and migration scenario in *Draba* was elaborated for North, Central, and South America (Koch and Al-Shehbaz, 2002). This genus also shows strong affinities to high alpine regions (e.g. in the Alps, Scandinavia, the Himalayas, Rocky Mountains and the Andes) and therefore its evolution was influenced by glacial and interglacial periods throughout the Pleistocene. Phylogeographic disjunction between North and South America was demonstrated in

Noccaea (Koch and Al-Shehbaz, 2004) and *Halimolobus/Sphaerocardamon* (Bailey *et al.*, 2002). Studies on the Chinese *Yinshania* (Koch and Al-Shehbaz, 2000) and several Australian and New Zealand mustards (Bleeker *et al.*, 2002; Heenan and Mitchell, 2003) demonstrated rapid range expansion and evolutionary radiation. Detailed molecular studies in *Boechera* demonstrated the impact of glacial and interglacial cycles on reticulate evolution and radiation, as well as migration and extinction (Sharbel and Mitchell-Olds, 2001; Koch *et al.*, 2003a; Dobes *et al.*, 2004). This genus is greatly affected by asexual reproduction (see below), which has important impacts on patterns of genetic and phenotypic diversity.

Differentiation and wild populations

Many studies have focused on speciation processes and phylogenetic relationships at or below the species level. Several phenotypic traits and characters have been investigated, either to learn more about their evolutionary significance or to elucidate plasticity and ecological relevance. These traits include local adaptation across climatic gradients in *Arabis fecunda* (McKay *et al.*, 2001) and *Capsella* (Neuffer and Hurka, 1999), survivorship in *Boechera* (as *Arabis*) *laevigata* (Bloom *et al.*, 2001), glucosinolate accumulation during plant/insect interaction in *A. thaliana* (Kliebenstein *et al.*, 2001), herbivore resistance (Agrawal *et al.*, 2002; Kroymann *et al.*, 2003; Weinig *et al.*, 2003b), pollination (Strauss *et al.*, 1999), as well as leaf morphology, flowering and maternal effects in *Capsella* (Neuffer, 1989, 1990; Neuffer and Koch, 1996). Detailed analyses of host–pathogen interaction have been presented in *Boechera* (*Arabis*) (Roy, 2001) and *A. thaliana* (Stahl *et al.*, 1999; Tian *et al.*, 2002, 2003). Several studies have focused on the evolutionary significance of phenotypic plasticity and reaction norms in *A. thaliana* (Pigliucci and Byrd, 1998; Pigliucci *et al.*, 1999; Pigliucci and Marlow, 2001; Pollard *et al.*, 2001; Weinig *et al.*, 2002, 2003a; Ungerer *et al.*, 2003).

Most population studies neglected the genetic diversity stored in the soil seed bank.

However, depending on the type of seed bank and reproductive biology of the species, the seed bank of a particular population may be essential for recruitment and establishment of new cohorts. Depending on the spatial genetic structure of the subpopulations (surface and aboveground populations stored as seeds or other diaspores), major changes in the genetic constitution of plant populations may occur during their history (Levin, 1990). Of less than ten studies focusing on seed-bank genetics, two are on members of the Brassicaceae (Cabin, 1996; Koch *et al.*, 2003c).

Comparative physiology and development

Arabidopsis genomics has revolutionized our understanding of plant biology and enabled functional analyses of many scientifically and economically important traits. Nevertheless, some mechanistic and evolutionary hypotheses are better addressed using wild relatives of *A. thaliana*, which are not confined to well-watered, temperate, ephemeral environments that follow agricultural disturbance. In particular, understanding of resistance to drought, heavy metals and a broad range of microbial pathogens can benefit from comparative genomics using the wild relatives of *A. thaliana*.

Water availability is fundamental to almost all aspects of plant physiology (Bray, 1997), and plant distribution and abundance in agricultural and natural ecosystems are largely determined by water availability. Although *Arabidopsis* genomics has enabled much progress in understanding responses to drought (Abe *et al.*, 2003; Boyce *et al.*, 2003; Cheong *et al.*, 2003; McKay *et al.*, 2003; Oono *et al.*, 2003), *A. thaliana* is confined to mesic habitats, and therefore provides an incomplete view of adaptive changes in water relations. In contrast, other crucifers are adapted to desert, mesic and aquatic habitats (Rollins, 1993; Bressan *et al.*, 2001; Mitchell-Olds, 2001; Xiong and Zhu, 2002) and display a broad range of adaptive differences in water use efficiency (McKay *et al.*, 2001) and salt tolerance, which can be elucidated by comparative genomics.

Resistance to deleterious effects of heavy metals occurs among several mustards, including *Thlaspi* (Persans *et al.*, 2001; Lombi *et al.*, 2002; Pineros and Kochian, 2003), *A. halleri* (Sarret *et al.*, 2002; Bert *et al.*, 2003) and *Alyssum* (Kerkeb and Kramer, 2003). Both natural variation (Sarret *et al.*, 2002; Bert *et al.*, 2003) and genomic methods (Persans *et al.*, 2001; Sarret *et al.*, 2002; Pineros and Kochian, 2003) have been employed to elucidate the physiological and molecular basis of heavy metal tolerance.

Crucifers are attacked by a wide variety of pests, offering potential for functional and evolutionary genomic studies of biotic interactions. In addition to the extensive literature on microbial pathogens of *A. thaliana* (e.g. Kunkel and Brooks, 2002; Nurnberger and Brunner, 2002; Farmer *et al.*, 2003; Shah, 2003), comparative genomics in the Brassicaceae will allow studies of pathogens whose host range does not include *Arabidopsis*, including *Puccinia* (Basidiomycetes; Roy, 1993), *Pyrenopeziza brassicae* (Discomycetes; Singh *et al.*, 1999) and *Leptosphaeria maculans* (Loculoascomycetes; Mitchell-Olds *et al.*, 1995). Furthermore, although earlier studies concluded that arbuscular mycorrhizal fungi do not colonize Brassicaceae, recent research has found intraradical hyphae, vesicles, coils and arbuscules formed by mycorrhizal fungi in the roots of several *Thlaspi* species (Regvar *et al.*, 2003).

Some crucifer species have patterns of flower and inflorescence development that differ substantially from *A. thaliana*. For example, most species of *Lepidium* have reduced petals and/or stamen number (Lee *et al.*, 2002). Hybridization and polyploidy have played a major role in *Lepidium* floral evolution, apparently because of dominant mutations causing loss of lateral stamens. In addition, Shu *et al.* (2000) compared the elongated inflorescence of *A. thaliana* with the rosette-flowering crucifer *Ionopsidium acaule*, where flowers are borne singly in the axils of rosette leaves. *In situ* hybridization suggested that orthologues of *LEAFY* may control evolutionary changes in inflorescence architecture.

Comparative genomics can identify conserved regulatory elements by comparing genomic sequences between related species (Cooper and Sidow, 2003). Phylogenetic footprinting considers a small number of distant evolutionary comparisons, whereas phylogenetic shadowing examines a set of closely related species (Boffelli *et al.*, 2003). Koch *et al.* (2001b) applied this approach to promoters of chalcone synthase (*Chs*) in 22 crucifer species at increasing evolutionary distances from *Arabidopsis*. They identified conserved regions of the *Chs* promoter and verified their functional importance by expressing promoter fragments from six crucifer species in *Arabidopsis* protoplasts. Hong *et al.* (2003) examined *cis*-regulatory sequences of the *Agamous* locus in 29 Brassicaceae species. Although they identified motifs conserved among taxa, some previously identified, functionally important *LFY* and *WUS* binding sites were not highly conserved. Functional significance of several conserved motifs was verified by reporter gene analyses. Phylogenetic footprinting can identify important regulatory regions in many species, and is now being applied to large-scale analyses in grasses (Inada *et al.*, 2003) and many animals (Cooper and Sidow, 2003).

Breeding systems

Self-incompatibility is ancestral in the Brassicaceae (Bateman, 1955; Kachroo *et al.*, 2002). Multiple independent evolutionary origins of self-compatibility have occurred in diploid crucifers, including *Leavenworthia uniflora* and *Leavenworthia crassa* (Liu *et al.*, 1998), *Capsella rubella*, *A. thaliana*, *Arabidopsis cebennensis*, *Arabidopsis croatica* and *Boechera* species (Mitchell-Olds, unpublished). Inbreeding is especially common in weedy crucifers (Appel and Al-Shehbaz, 2003). Theory predicts that self-fertilization should reduce genetic variation by 50% in comparison with complete outcrossing (Pollack, 1987; Charlesworth, 2003). Further reductions of genetic variability in inbreeding species may also occur because of effects of background selection and genetic hitchhiking (Maynard Smith and Haigh, 1974; Charlesworth *et al.*, 1993; Nordborg *et al.*, 1996). Comparisons of nucleotide polymorphism in inbreeding and outbreeding species of *Leavenworthia* in North America, as well as *Arabidopsis* native to Europe (Clauss and Mitchell-Olds, 2003; Wright *et al.*, 2003; Ramos-Onsins *et al.*, 2004) support these predictions. Other breeding systems, including wind pollination, monoecy, dioecy and gynodioecy are occasionally found in the Brassicaceae (Appel and Al-Shehbaz, 2003).

Asexual reproduction, or apomixis (Koltunow and Grossniklaus, 2003; Richards, 2003; Spielman *et al.*, 2003), occurs in *Draba verna*, *Smelowskia calycina*, *Draba oligosperma* and several species of *Boechera* (Böcher, 1951; Roy, 1995; Sharbel and Mitchell-Olds, 2001; Appel and Al-Shehbaz, 2003). The base chromosome number in *Boechera* is $n = 7$, and its asexual taxa also exhibit polyploidy (predominantly $3n$) and aneuploidy (Böcher, 1951; Roy, 1995; Sharbel and Mitchell-Olds, 2001). Studies of polyploidy and apomixis have concentrated on the *Boechera holboellii* group, a complex composed of *B. holboellii*, *B. stricta* (syn. *Arabis drummondi*) and their hybrid *Boechera divaricarpa* (Dobes *et al.*, 2003, 2004). Interestingly, both polyploidy and aneuploidy have evolved repeatedly within the *B. holboellii* complex (Sharbel and Mitchell-Olds, 2001). In addition, Böcher (1951) provided evidence for occasional diploid apomicts, an extremely rare condition among asexual plants.

Species of *Boechera* reproduce via diplosporous apomixis (Böcher, 1951). Böcher (1951) demonstrated that in *B. holboellii* originating from Greenland and Alaska gametogenesis and embryo formation can be extremely variable. Normal diploid meiosis, followed by typical dyad and pollen formation, occurs in sexual *B. stricta* (= *Arabis drummondii*), a predominantly selfing species that occasionally outcrosses (Roy, 1995). Furthermore, synaptic, partially synaptic and asynaptic microsporogenesis (pollen formation) in diploid and triploid individuals are also possible (Böcher, 1951). Variability in chromosome structure and unequal sister chromatid exchange are proposed as mechanisms leading to the lack of synapsis in some lineages.

Emasculation of apomictic plants results in sterility, hence apomictic individuals are also pseudogamous (i.e. fertilization is required for embryo development, although no genetic contribution is made by the male). Böcher (1951) showed that apomictic diploid and triploid individuals produced tetraploid and hexaploid endosperms, respectively, and thus pseudogamy can lead to autonomous endosperm formation. There is evidence that endosperm ploidy can be variable (Matzk *et al.*, 2000; Sharbel, unpublished), and it is likely that endosperm fertilization in some apomictic lineages is also possible.

In addition to the chromosomal variability in apomicts, aneuploid chromosome fragments in diploid and triploid individuals of *Boechera* were found (Böcher, 1951). Subsequent work using flow cytometry and chloroplast DNA sequencing has demonstrated that the aneuploid chromosome is widely distributed among different geographical locations and can be found in diverse haplotype backgrounds (Sharbel and Mitchell-Olds, 2001). Studies using microsatellite markers have shown that a similar chromosome fragment is involved with aneuploidy in genetically diverse apomictic lineages (Sharbel *et al.*, unpublished), and karyological and DNA sequencing work have demonstrated that the aneuploid chromosome is a non-recombining B chromosome that may undergo both structural and sequence degeneration.

The potential development of apomixis technology for crop plant research could have substantial impacts on agriculture (Hoisington *et al.*, 1999; van Dijk and van Damme, 2000), as it represents a method through which genetic heterozygosity and hybrid vigour could be fixed and faithfully propagated. Consequently, there is considerable interest in deciphering the molecular genetic and/or physiological mechanisms behind apomixis expression in the *B. holboellii* complex. The genetic diversity and geographic distribution of apomictic individuals within the *B. holboellii* complex imply that this group may have some predisposition to expressing this form of reproduction, and that apomixis has been stable and successful during the post-Pleistocene history of North America. If apomixis expression is associated with polyploid gene expression, then repeated evolution of polyploidy within this group may also be correlated with repeated expression of this reproductive mode from sexual ancestors.

References

Abe, H., Urao, T., Ito, T., Seki, M., Shinozaki, K. and Yamaguchi-Shinozaki, K. (2003) *Arabidopsis AtMYC2* (bHLH) and *AtMYB2* (MYB) function as transcriptional activators in abscisic acid signaling. *Plant Cell* 15, 63–78.

Agrawal, A., Conner, J., Johnson, M. and Wallsgrove, R. (2002) Ecological genetics of an induced plant defense against herbivores: additive genetic variance and costs of phenotypic plasticity. *Evolution* 56, 2206–2213.

Al-Shehbaz, I.A. (1973) The biosystematics of the genus *Thelypodium* Cruciferae. *Contributions From the Gray Herbarium of Harvard University*, 3–148.

Al-Shehbaz, I.A. (1984) The tribes of Cruciferae (Brassicaceae) in the southeastern United States. *Journal of the Arnold Arboretum* 65, 343–373.

Al-Shehbaz, I.A. (1985) The genera of Thelypodieae (Cruciferae; Brassicaceae) in the southeastern United States. *Journal of the Arnold Arboretum* 66, 95–111.

Al-Shehbaz, I.A. (2003a) A synopsis of *Tropidocarpum* (Brassicaceae). *Novon* 13, 392–395.

Al-Shehbaz, I.A. (2003b) Transfer of most North American species of *Arabis* to *Boechera* (Brassicaceae). *Novon* 13, 381–391.

Al-Shehbaz, I.A. and O'Kane, S.J. Jr (2002) Taxonomy and phylogeny of *Arabidopsis* (Brassicaceae). In: Somerville, C.R. and Meyerowitz, E.M. (eds) *The Arabidopsis Book*. American Society of Plant Biologists, Rockville, Maryland, doi/10.1199/tab.0001. Available at: www.aspb.org/publications/arabidopsis/

Al-Shehbaz, I.A. and Price, R.A. (2001) The Chilean *Agallis* and Californian *Tropidocarpum* (Brassicaceae) are congeneric. *Novon* 11, 292–293.

Al-Shehbaz, I.A., O'Kane, S.L. Jr and Price, R.A. (1999) Generic placement of species excluded from *Arabidopsis* (Brassicaceae). *Novon* 9, 296–307.

Al-Shehbaz, I.A., Mummenhoff, K. and Appel, O. (2002) *Cardaria, Coronopus*, and *Stroganowia* are united with *Lepidium* (Brassicaceae). *Novon* 12, 5–11.

Angiosperm Phylogeny Group (1998) An ordinal classification for the families of flowering plants. *Annals of the Missouri Botanical Garden* 85, 531–553.

Appel, O. and Al-Shehbaz, I.A. (2003) Cruciferae. In: Kubitzki, K. and Bayer, C. (eds) *The Families and Genera of Vascular Plants*. Springer-Verlag, Berlin, pp. 75–174.

Avise, J.C., Arnold, J., Ball, R.M., Bermingham, E., Lamb, T., Neigel, J.E., Reeb, C.A. and Saunders, N.C. (1987) Intraspecific phylogeography: the mitochondrial-DNA bridge between population genetics and systematics. *Annual Review of Ecology and Systematics* 18, 489–522.

Bailey, C.D., Price, R.A. and Doyle, J.J. (2002) Systematics of the halimolobine Brassicaceae: evidence from three loci and morphology. *Systematic Botany* 27, 318–332.

Bateman, A.J. (1955) Self-incompatibility systems in angiosperms. 3. Cruciferae. *Heredity* 9, 53–68.

Bert, V., Meerts, P., Saumitou-Laprade, P., Salis, P., Gruber, W. and Verbruggen, N. (2003) Genetic basis of Cd tolerance and hyperaccumulation in *Arabidopsis halleri*. *Plant and Soil* 249, 9–18.

Bleeker, W. (2003) Hybridization and *Rorippa austriaca* (Brassicaceae) invasion in Germany. *Molecular Ecology* 12, 1831–1841.

Bleeker, W., Huthmann, M. and Hurka, H. (1999) Evolution of hybrid taxa in *Nasturtium* R.Br. (Brassicaceae). *Folia Geobotanica and Phytotaxonomica* 34, 421–433.

Bleeker, W., Franzke, A., Pollmann, K., Brown, A.H.D. and Hurka, H. (2002) Phylogeny and biogeography of southern hemisphere high-mountain *Cardamine* species (Brassicaceae). *Australian Systematic Botany* 15, 575–581.

Bloom, T.C., Baskin, J.M. and Baskin, C.C. (2001) Ecological life history of the facultative woodland biennial *Arabis laevigata* variety *laevigata* (Brassicaceae): survivorship. *Journal of the Torrey Botanical Society* 128, 93–108.

Böcher, T.W. (1951) Cytological and embryological studies in the amphi-apomictic *Arabis holboellii* complex. *Det Kongelige Danske Videnskabernes Selskab* 6, 1–59.

Boffelli, D., McAuliffe, J., Ovcharenko, D., Lewis, K.D., Ovcharenko, I., Pachter, L. and Rubin, E.M. (2003) Phylogenetic shadowing of primate sequences to find functional regions of the human genome. *Science* 299, 1391–1394.

Bowman, J.L., Bruggemann, H., Lee, J.Y. and Mummenhoff, K. (1999) Evolutionary changes in floral structure within *Lepidium* L. (Brassicaceae). *International Journal of Plant Sciences* 160, 917–929.

Boyce, J.M., Knight, H., Deyholos, M., Openshaw, M.R., Galbraith, D.W., Warren, G. and Knight, M.R. (2003) The *sfr6* mutant of *Arabidopsis* is defective in transcriptional activation via CBF/DREB1 and DREB2 and shows sensitivity to osmotic stress. *Plant Journal* 34, 395–406.

Bradshaw, H. and Schemske, D. (2003) Allele substitution at a flower colour locus produces a pollinator shift in monkeyflowers. *Nature* 426, 176–178.

Bray, E. (1997) Plant responses to water deficit. *Trends in Plant Science* 2, 48–54.

Bressan, R.A., Zhang, C., Zhang, H., Hasegawa, P.M., Bohnert, H.J. and Zhu, J.-K. (2001) Learning from the *Arabidopsis* experience. The next gene search paradigm. *Plant Physiology* 127, 1354–1360.

Brochmann, C., Stedje, B. and Borgen, L. (1992) Gene flow across ploidal levels in *Draba* (Brassicaceae). *Evolutionary Trends in Plants* 6, 125–134.

Bruggemann, H. (2000) Molekulare Systematik und Biogeographie der Gattung *Lepidium* L. und verwanter Sippen (Brassicaceae). PhD thesis, University of Osnabrück, Osnabrück, Germany.

Brumfield, R.T., Beerli, P., Nickerson, D.A. and Edwards, S.V. (2003) The utility of single nucleotide polymorphisms in inferences of population history. *Trends in Ecology and Evolution* 18, 249–256.

Cabin, R.J. (1996) Genetic comparisons of seed bank and seedling populations of a perennial desert mustard, *Lesquerella fendleri*. *Evolution* 50, 1830–1841.

Charlesworth, B., Morgan, M.T. and Charlesworth, D. (1993) The effect of deleterious mutations on neutral molecular variation. *Genetics* 134, 1289–1303.

Charlesworth, D. (2003) Effects of inbreeding on the genetic diversity of populations. *Philosophical Transactions of the Royal Society of London Series B: Biological Sciences* 358, 1051–1070.

Cheong, Y.H., Kim, K.-N., Pandey, G.K., Gupta, R., Grant, J.J. and Luan, S. (2003) *CBL1*, a calcium sensor that differentially regulates salt, drought, and cold responses in *Arabidopsis*. *Plant Cell* 15, 1833–1845.

Clauss, M.J. and Mitchell-Olds, T. (2003) Population genetics of tandem trypsin inhibitor genes in *Arabidopsis* species with contrasting ecology and life history. *Molecular Ecology* 12, 1287–1299.

Comai, L., Tyagi, A.P., Winter, K., Holmes-Davis, R., Reynolds, S.H., Stevens, Y. and Byers, B. (2000) Phenotypic instability and rapid gene silencing in newly formed *Arabidopsis* allotetraploids. *Plant Cell* 12, 1551–1568.

Comai, L., Madlung, A., Josefsson, C. and Tyagi, A. (2003) Do the different parental 'heteromes' cause genomic shock in newly formed allopolyploids? *Philosophical Transactions of the Royal Society of London Series B: Biological Sciences* 358, 1149–1155.

Comes, H.P. and Kadereit, J.W. (1998) The effect of quaternary climatic changes on plant distribution and evolution. *Trends in Plant Science* 3, 432–438.

Cooper, G.M. and Sidow, A. (2003) Genomic regulatory regions: insights from comparative sequence analysis. *Current Opinion in Genetics and Development* 13, 604–610.

Crespo, M.B., Lledo, M.D., Fay, M.F. and Chase, M.W. (2000) Subtribe Vellinae (Brassiceae, Brassicaceae): a combined analysis of ITS nrDNA sequences and morphological data. *Annals of Botany* 86, 53–62.

Dobes, C., Mitchell-Olds, T. and Koch, M. (2003) Multiple hybrid formation in natural populations: concerted evolution of the internal transcribed spacer of nuclear ribosomal DNA (ITS) in North American *Arabis divaricarpa* (Brassicaceae). *Molecular Biology and Evolution* 20, 338–350.

Dobes, C., Mitchell-Olds, T. and Koch, M. (2004) Extensive chloroplast haplotype variation indicates Pleistocene hybridization and radiation of North American *Arabis drummondii*, *A.* × *divaricarpa*, and *A. holboellii* (Brassicaceae). *Molecular Ecology* 13, 349–370.

Dvorák, F. (1973) Importance of indumentum for investigation of evolutional relationship in family Brassicaceae. *Osterreichische Botanische Zeitschrift* 121, 155–164.

Ehrendorfer, F. (1980) Polyploidy and distribution. In: Lewis, W.H. (ed.) *Polyploidy. Biological Relevance.* Plenum Press, New York, pp. 45–60.

Erickson, L.R., Straus, N.A. and Beversdorf, W.D. (1983) Restriction patterns reveal origins of chloroplast genomes in *Brassica* amphiploids. *Theoretical and Applied Genetics* 65, 201–206.

Farmer, E.E., Almeras, E. and Krishnamurthy, V. (2003) Jasmonates and related oxylipins in plant responses to pathogenesis and herbivory. *Current Opinion in Plant Biology* 6, 372–378.

Franzke, A. and Hurka, H. (2000) Molecular systematics and biogeography of the *Cardamine pratensis* complex (Brassicaceae). *Plant Systematics and Evolution* 224, 213–234.

Franzke, A., Pollmann, K., Bleeker, W., Kohrt, R. and Hurka, H. (1998) Molecular systematics of *Cardamine* and allied genera (Brassicaceae): ITS and non-coding chloroplast DNA. *Folia Geobotanica* 33, 225–240.

Galloway, G.L., Malmberg, R.L. and Price, R.A. (1998) Phylogenetic utility of the nuclear gene arginine decarboxylase: an example from Brassicaceae. *Molecular Biology and Evolution* 15, 1312–1320.

Gómez-Campo, C. (1999) Taxonomy. In: Gomez-Campo, C. (ed.) *Biology of* Brassica *coenospecies*. Elsevier, Amsterdam, pp. 3–32.

Hall, A.E., Fiebig, A. and Preuss, D. (2002a) Beyond the *Arabidopsis* genome: opportunities for comparative genomics. *Plant Physiology* 129, 1439–1447.

Hall, J.C., Sytsma, K.J. and Iltis, H.H. (2002b) Phylogeny of Capparaceae and Brassicaceae based on chloroplast sequence data. *American Journal of Botany* 89, 1826–1842.

Harris, S.A. and Ingram, R. (1991) Chloroplast DNA and biosystematics – the effects of intraspecific diversity and plastid transmission. *Taxon* 40, 393–412.

Hauser, L.A. and Crovello, T.J. (1982) Numerical analysis of generic relationships in Thelypodieae (Brassicaceae). *Systematic Botany* 7, 249–268.

Hayek, A. (1911) Entwurf eines Cruciferensystems auf phylogenetischer Grundlage. *Beihefte zum Botanischen Centralblatt* 27, 127–335.

Heenan, P.B. and Mitchell, A.D. (2003) Phylogeny, biogeography and adaptive radiation of *Pachycladon* (Brassicaceae) in the mountains of South Island, New Zealand. *Journal of Biogeography* 30, 1737–1749.

Heenan, P.B., Mitchell, A.D. and Koch, M. (2002) Molecular systematics of the New Zealand *Pachycladon* (Brassicaceae) complex: generic circumscription and relationship to *Arabidopsis* sens. lat. and *Arabis* sens. lat. *New Zealand Journal of Botany* 40, 543–562.

Hoisington, D., Khairallah, M., Reeves, T., Ribaut, J.-M., Skovmand, B., Taba, S. and Warburton, M. (1999) Plant genetic resources: what can they contribute toward increased crop productivity? *Proceedings of the National Academy of Sciences USA* 96, 5937–5943.

Hong, R.L., Hamaguchi, L., Busch, M.A. and Weigel, D. (2003) Regulatory elements of the floral homeotic gene *AGAMOUS* identified by phylogenetic footprinting and shadowing. *Plant Cell* 15, 1296–1309.

Howard, D.J. (1999) Conspecific sperm and pollen precedence and speciation. *Annual Review of Ecology and Systematics* 30, 109–132.

Hurka, H. and Neuffer, B. (1997) Evolutionary processes in the genus *Capsella* (Brassicaceae). *Plant Systematics and Evolution* 206, 295–316.

Inada, D.C., Bashir, A., Lee, C., Thomas, B.C., Ko, C., Goff, S.A. and Freeling, M. (2003) Conserved noncoding sequences in the grasses. *Genome Research* 13, 2030–2041.

Janchen, E. (1942) Das System der Cruciferen. *Osterreichische Botanische Zeitschrift* 91, 1–18.

Judd, W.S., Sanders, R.W. and Donoghue, M.J. (1994) Angiosperm family pairs: preliminary phylogenetic analysis. *Harvard Papers in Botany/Harvard University Herbaria* 5, 1–51.

Kachroo, A., Nasrallah, M.E. and Nasrallah, J.B. (2002) Self-incompatibility in the Brassicaceae: receptor–ligand signaling and cell-to-cell communication. *Plant Cell* 14, S227–238.

Kerkeb, L. and Kramer, U. (2003) The role of free histidine in xylem loading of nickel in *Alyssum lesbiacum* and *Brassica juncea*. *Plant Physiology* 131, 716–724.

Kliebenstein, D.J., Kroymann, J., Brown, P., Figuth, A., Pedersen, D., Gershenzon, J. and Mitchell-Olds, T. (2001) Genetic control of natural variation in *Arabidopsis* glucosinolate accumulation. *Plant Physiology* 126, 811–825.

Kliman, R.M., Andolfatto, P., Coyne, J.A., Depaulis, F., Kreitman, M., Berry, A.J., McCarter, J., Wakeley, J. and Hey, J. (2000) The population genetics of the origin and divergence of the *Drosophila simulans* complex species. *Genetics* 156, 1913–1931.

Koch, M. (2002) Genetic differentiation and speciation in prealpine *Cochlearia*: allohexaploid *Cochlearia bavarica* Vogt (Brassicaceae) compared to its diploid ancestor *Cochlearia pyrenaica* DC. in Germany and Austria. *Plant Systematics and Evolution* 232, 35–49.

Koch, M. (2003) Molecular phylogenetics, evolution and population biology in Brassicaceae. In: Sharma, A.K. and Sharma, A. (eds) *Plant Genome Biodiversity and Evolution*. Vol. 1, part A. *Phanerogams*. Science Publishers, Enfield, New Hampshire, pp. 1–35.

Koch, M. and Al-Shehbaz, I.A. (2000) Molecular systematics of the Chinese *Yinshania* (Brassicaceae): evidence from plastid and nuclear ITS DNA sequence data. *Annals of the Missouri Botanical Garden* 87, 246–272.

Koch, M. and Al-Shehbaz, I. (2002) Molecular data indicate complex intra- and intercontinental differentiation of American *Draba* (Brassicaceae). *Annals of the Missouri Botanical Garden* 89, 88–109.

Koch, M. and Al-Shehbaz, I. (2004) Taxonomic and phylogenetic evaluation of the American '*Thlaspi*' species: identity and relationship to the Eurasian genus *Noccaea* (Brassicaceae). *Systematic Botany* 29, 375–384.

Koch, M. and Bernhardt, K.G. (2004) Comparative biogeography of the cytotypes of annual *Microthlaspi perfoliatum* (Brassicaceae) in Europe using isozymes and cpDNA data: refugia, diversity centers, and postglacial colonization. *American Journal of Botany* 91, 115–124.

Koch, M. and Hurka, H. (1999) Isozyme analysis in the polyploid complex *Microthlaspi perfoliatum* (L.) F. K. Meyer: morphology, biogeography and evolutionary history. *Flora* 194, 33–48.

Koch, M. and Mummenhoff, K. (2001) *Thlaspi* s.str. (Brassicaceae) versus *Thlaspi* s.l.: morphological and anatomical characters in the light of ITS nrDNA sequence data. *Plant Systematics and Evolution* 227, 209–225.

Koch, M., Hurka, H. and Mummenhoff, K. (1996) Chloroplast DNA restriction site variation and RAPD-analyses in *Cochlearia* (Brassicaceae): biosystematics and speciation. *Nordic Journal of Botany* 16, 585–603.

Koch, M., Huthmann, M. and Hurka, H. (1998) Molecular biogeography and evolution of the *Microthlaspi perfoliatum* s.l. polyploid complex (Brassicaceae): chloroplast DNA and nuclear ribosomal DNA restriction site variation. *Canadian Journal of Botany* 76, 382–396.

Koch, M., Bishop, J. and Mitchell-Olds, T. (1999a) Molecular systematics of *Arabidopsis* and *Arabis*. *Plant Biology* 1, 529–537.

Koch, M., Hurka, H. and Mummenhoff, K. (1999b) Molecular phylogenetics of *Cochlearia* L. and allied genera based on nuclear ribosomal ITS DNA sequence analysis contradict traditional concepts of their evolutionary relationship. *Plant Systematics and Evolution* 216, 207–230.

Koch, M.A., Haubold, B. and Mitchell-Olds, T. (2000) Comparative evolutionary analysis of chalcone synthase and alcohol dehydrogenase loci in *Arabidopsis*, *Arabis*, and related genera (Brassicaceae). *Molecular Biology and Evolution* 17, 1483–1498.

Koch, M., Haubold, B. and Mitchell-Olds, T. (2001a) Molecular systematics of the Brassicaceae: evidence from coding plastidic *matK* and nuclear *Chs* sequences. *American Journal of Botany* 88, 534–544.

Koch, M.A., Weisshaar, B., Kroymann, J., Haubold, B. and Mitchell-Olds, T. (2001b) Comparative genomics and regulatory evolution: conservation and function of the *Chs* and *Apetala3* promoters. *Molecular Biology and Evolution* 18, 1882–1891.

Koch, M., Al-Shehbaz, I. and Mummenhoff, K. (2003a) Molecular systematics, evolution, and population biology in the mustard family (Brassicaceae). *Annals of the Missouri Botanical Garden* 90, 151–171.

Koch, M., Bernhardt, K.G. and Kochjarová, J. (2003b) *Cochlearia macrorrhiza* (Brassicaceae): a bridging species between *Cochlearia* taxa from the Eastern Alps and the Carpathians. *Plant Systematics and Evolution* 242, 137–147.

Koch, M., Huthmann, M. and Bernhardt, K.G. (2003c) *Cardamine amara* L. (Brassicaceae) in dynamic habitats: genetic composition and diversity of seed bank and established populations. *Basic and Applied Ecology* 4, 339–348.

Koltunow, A. and Grossniklaus, U. (2003) Apomixis: a developmental perspective. *Annual Review of Plant Biology* 54, 547–574.

Kowalski, S.P., Lan, T.-H., Feldman, K.A. and Paterson, A.H. (1994) Comparative mapping of *Arabidopsis thaliana* and *Brassica oleracea* chromosomes reveals islands of conserved organization. *Genetics* 138, 499–510.

Kropf, M., Kadereit, J.W. and Comes, H.P. (2002) Late Quaternary distributional stasis in the submediterranean mountain plant *Anthyllis montana* L. (Fabaceae) inferred from ITS sequences and amplified fragment length polymorphism markers. *Molecular Ecology* 11, 447–463.

Kropf, M., Kadereit, J.W. and Comes, H.P. (2003) Differential cycles of range contraction and expansion in European high mountain plants during the Late Quaternary: insights from *Pritzelago alpina* (L.) O. Kuntze (Brassicaceae). *Molecular Ecology* 12, 931–949.

Kroymann, J., Donnerhacke, S., Schnabelrauch, D. and Mitchell-Olds, T. (2003) Evolutionary dynamics of an *Arabidopsis* insect resistance quantitative trait locus. *Proceedings of the National Academy of Sciences USA* 100, 14587–14592.

Kuittinen, H., de Haan, A., Vogl, C., Oikarinen, S., Leppala, J., Mitchell-Olds, T., Koch, M., Langley, C. and Savolainen, O. (2004) Comparing the maps of close relatives *Arabidopsis lyrata* and *Arabidopsis thaliana*. *Genetics* (in press).

Kunkel, B.N. and Brooks, D.M. (2002) Cross talk between signaling pathways in pathogen defense. *Current Opinion in Plant Biology* 5, 325–331.

Lagercrantz, U. (1998) Comparative mapping between *Arabidopsis thaliana* and *Brassica nigra* indicates that *Brassica* genomes have evolved through extensive genome replication accompanied by chromosome fusions and frequent rearrangements. *Genetics* 150, 1217–1228.

Lagercrantz, U. and Lydiate, D. (1996) Comparative genome mapping in *Brassica*. *Genetics* 144, 1903–1910.

Lan, T.-H. and Paterson, A.H. (2000) Comparative mapping of quantitative trait loci sculpting the curd of *Brassica oleracea*. *Genetics* 155, 1927–1954.

Lee, J.-Y., Mummenhoff, K. and Bowman, J.L. (2002) Allopolyploidization and evolution of species with reduced floral structures in *Lepidium* L. (Brassicaceae). *Proceedings of the National Academy of Sciences USA* 99, 16835–16840.

Levin, D.A. (1990) The seed bank as a source of genetic novelty in plants. *American Naturalist* 135, 563–572.

Liu, F., Zhang, L. and Charlesworth, D. (1998) Genetic diversity in *Leavenworthia* populations with different inbreeding levels. *Proceedings of the Royal Society of London Series B: Biological Science* 265, 293–301.

Lombi, E., Tearall, K.L., Howarth, J.R., Zhao, F.-J., Hawkesford, M.J. and McGrath, S.P. (2002) Influence of iron status on cadmium and zinc uptake by different ecotypes of the hyperaccumulator *Thlaspi caerulescens*. *Plant Physiology* 128, 1359–1367.

Marhold, K., Lihova, J., Perny, M., Grupe, R. and Neuffer, B. (2002) Natural hybridization in *Cardamine* (Brassicaceae) in the Pyrenees: evidence from morphological and molecular data. *Botanical Journal of the Linnean Society* 139, 275–294.

Matzk, F., Meister, A. and Schubert, I. (2000) An efficient screen for reproductive pathways using mature seeds of monocots and dicots. *Plant Journal* 21, 97–108.

Mayfield, J.A., Fiebig, A., Johnstone, S.E. and Preuss, D. (2001) Gene Families from the *Arabidopsis thaliana* pollen coat proteome. *Science* 292, 2482–2485.

Maynard Smith, J. and Haigh, J. (1974) The hitch-hiking effect of a favourable gene. *Genetical Research Cambridge* 23, 23–35.

McKay, J.K., Bishop, J.G., Lin, J.Z., Richards, J.H., Sala, A. and Mitchell-Olds, T. (2001) Local adaptation across a climatic gradient despite small effective population size in the rare sapphire rockcress. *Proceedings of the Royal Society of London Series B: Biological Sciences* 268, 1715–1721.

McKay, J.K., Richards, J.H. and Mitchell-Olds, T. (2003) Genetics of drought adaptation in *Arabidopsis thaliana*: I. Pleiotropy contributes to genetic correlations among ecological traits. *Molecular Ecology* 12, 1137–1151.

Meyer, F.K. (1973) Conspectus der 'Thlaspi'-Arten Europas, Afrikas und Vorderasiens. *Feddes Repertorium* 84, 449–470.

Meyer, F.K. (1979) Kritische Revision der 'Thlaspi'-Arten Europas, Afrikas und Vorderasiens. I. Geschichte, Morphologie und Chorologie. *Feddes Repertorium* 90, 129–154.

Mitchell, A.D. and Heenan, P.B. (2000) Systematic relationships of New Zealand endemic Brassicaceae inferred from nrDNA ITS sequence data. *Systematic Botany* 25, 98–105.

Mitchell-Olds, T. (2001) *Arabidopsis thaliana* and its wild relatives: a model system for ecology and evolution. *Trends in Ecology and Evolution* 16, 693–700.

Mitchell-Olds, T. and Clauss, M.J. (2002) Plant evolutionary genomics. *Current Opinion in Plant Biology* 5, 74–79.

Mitchell-Olds, T., James, R.V., Palmer, M.V. and Williams, P.H. (1995) Genetics of *Brassica rapa* (syn. *campestris*). 2. Selection for multiple disease resistance to three fungal pathogens: *Peronospora parasitica*, *Albugo candida*, and *Leptosphaeria maculans*. *Heredity* 75, 362–369.

Mummenhoff, K. (1995) Should *Cardaria draba* (L.) Desv. be classified within the genus *Lepidium* L. (Brassicaceae)? Evidence from subunit polypeptide composition of RUBISCO. *Feddes Repertorium* 106, 25–28.

Mummenhoff, K. and Hurka, H. (1990) Evolution of the tetraploid *Capsella bursa-pastoris* (Brassicaceae): isoelectric-focusing analysis of rubisco. *Plant Systematics and Evolution* 172, 205–213.

Mummenhoff, K. and Hurka, H. (1994) Subunit polypeptide composition of rubisco and the origin of allopolyploid *Arabidopsis suecica* (Brassicaceae). *Biochemical Systematics and Ecology* 22, 807–811.

Mummenhoff, K. and Hurka, H. (1995) Allopolyploid origin of *Arabidopsis suecica* (Fries) norrlin – evidence from chloroplast and nuclear genome markers. *Botanica Acta* 108, 449–456.

Mummenhoff, K. and Koch, M. (1994) Chloroplast DNA restriction site variation and phylogenetic relationships in the genus *Thlaspi* sensu lato Brassicaceae. *Systematic Botany* 19, 73–88.

Mummenhoff, K., Franzke, A. and Koch, M. (1997a) Molecular data reveal convergence in fruit characters used in the classification of *Thlaspi* s.l. (Brassicaceae). *Botanical Journal of the Linnean Society* 125, 183–199.

Mummenhoff, K., Franzke, A. and Koch, M. (1997b) Molecular phylogenetics of *Thlaspi* sl (Brassicaceae) based on chloroplast DNA restriction site variation and sequences of the internal transcribed spacers of nuclear ribosomal DNA. *Canadian Journal of Botany* 75, 469–482.

Mummenhoff, K., Bruggemann, H. and Bowman, J.L. (2001) Chloroplast DNA phylogeny and biogeography of *Lepidium* (Brassicaceae). *American Journal of Botany* 88, 2051–2063.

Nasrallah, M.E., Yogeeswaran, K., Snyder, S. and Nasrallah, J.B. (2000) *Arabidopsis* species hybrids in the study of species differences and evolution of amphiploidy in plants. *Plant Physiology* 124, 1605–1614.

Neuffer, B. (1989) Leaf morphology in *Capsella* (Cruciferae): dependency on environments and biological parameters. *Beiträge zur Biologie der Pflanzen* 64, 39–54.

Neuffer, B. (1990) Ecotype differentiation in *Capsella*. *Vegetatio* 89, 165–171.

Neuffer, B. and Hurka, H. (1999) Colonization history and introduction dynamics of *Capsella bursa-pastoris* (Brassicaceae) in North America: isozymes and quantitative traits. *Molecular Ecology* 8, 1667–1681.

Neuffer, B. and Koch, K. (1996) Maternal effects on populations of *Capsella bursa-pastoris* (L.) med from Scandinavia and the Alps – independence of seed weight and the germination behavior of the maturation conditions in lowland and mountains. *Biologisches Zentralblatt* 115, 337–352.

Nordborg, M., Charlesworth, B. and Charlesworth, D. (1996) Increased levels of polymorphism surrounding selectively maintained sites in highly selfing species. *Proceedings of the Royal Society of London Series B: Biological Sciences* 263, 1033–1039.

Nurnberger, T. and Brunner, F. (2002) Innate immunity in plants and animals: emerging parallels between the recognition of general elicitors and pathogen-associated molecular patterns. *Current Opinion in Plant Biology* 5, 318–324.

O'Kane, S.L. Jr and Al-Shehbaz, I.A. (1997) A synopsis of *Arabidopsis* (Brassicaceae). *Novon* 7, 323–327.

O'Kane, S.L. Jr and Al-Shehbaz, I.A. (2003) Phylogenetic position and generic limits of *Arabidopsis* (Brassicaceae) based on sequences of nuclear ribosomal DNA. *Annals of the Missouri Botanical Garden* 90, 603–612.

O'Kane, S.L. Jr, Schaal, B.A. and Al-Shehbaz, I.A. (1997) The origins of *Arabidopsis suecica* (Brassicaceae) as indicated by nuclear rDNA sequences. *Systematic Botany* 21, 559–566.

Oono, Y., Seki, M., Nanjo, T., Narusaka, M., Fujita, M., Satoh, R., Satou, M., Sakurai, T., Ishida, J., Akiyama, K., Iida, K., Maruyama, K., Satoh, S., Yamaguchi-Shinozaki, K. and Shinozaki, K. (2003) Monitoring expression profiles of *Arabidopsis* gene expression during rehydration process after dehydration using ca. 7000 full-length cDNA microarray. *Plant Journal* 34, 868–887.

Palmer, J.D., Shields, C.R., Cohen, D.B. and Orton, T.J. (1983) Chloroplast DNA evolution and the origin of amphidiploid *Brassica* species. *Theoretical and Applied Genetics* 65, 181–189.

Persans, M.W., Nieman, K. and Salt, D.E. (2001) Functional activity and role of cation-efflux family members in Ni hyperaccumulation in *Thlaspi goesingense*. *Proceedings of the National Academy of Sciences USA* 98, 9995–10000.

Pigliucci, M. and Byrd, N. (1998) Genetics and evolution of phenotypic plasticity to nutrient stress in *Arabidopsis* – drift, constraints or selection. *Biological Journal of the Linnean Society* 64, 17–40.

Pigliucci, M. and Marlow, E.T. (2001) Differentiation for flowering time and phenotypic integration in *Arabidopsis thaliana* in response to season length and vernalization. *Oecologia* 127, 501–508.

Pigliucci, M., Cammell, K. and Schmitt, J. (1999) Evolution of phenotypic plasticity: a comparative approach in the phylogenetic neighbourhood of *Arabidopsis thaliana*. *Journal of Evolutionary Biology* 12, 779–791.

Pineros, M.A. and Kochian, L.V. (2003) Differences in whole-cell and single-channel ion currents across the plasma membrane of mesophyll cells from two closely related *Thlaspi* species. *Plant Physiology* 131, 583–594.

Pollack, E. (1987) On the theory of partially inbreeding finite populations. I. Partial selfing. *Genetics* 117, 353–360.

Pollard, H., Cruzan, M. and Pigliucci, M. (2001) Comparative studies of reaction norms in *Arabidopsis*. I. Evolution of response to daylength. *Evolutionary Ecology Research* 3, 129–155.

Prantl, K. (1891) Cruciferae. *Die natürlichen Pflanzenfamilien* III, 145–206.

Price, R.A., Al-Shehbaz, I.A. and Palmer, J.D. (1994) Systematic relationships of *Arabidopsis*: a molecular and morphological approach. In: Meyerowitz, E. and Somerville, C. (eds) *Arabidopsis*. Cold Spring Harbor Press, Cold Spring Harbor, New York, pp. 7–19.

Price, R.A., Al-Shehbaz, I.A. and O'Kane, S.L. (2001) *Beringia* (Brassicaceae), a new genus of Arabidopsoid affinities from Russia and North America. *Novon* 11, 332–336.

Ramos-Onsins, S., Stranger, B., Mitchell-Olds, T. and Aguade, M. (2004) Multilocus analysis of variation and speciation in the closely-related species *Arabidopsis halleri* and *A. lyrata*. *Genetics* 166, 373–388.

Reboud, X. and Zeyl, C. (1994) Organelle inheritance in plants. *Heredity* 72, 132–140.

Regvar, M., Vogel, K., Irgel, N., Wraber, T., Hildebrandt, U., Wilde, P. and Bothe, H. (2003) Colonization of pennycresses (*Thlaspi* spp.) of the Brassicaceae by arbuscular mycorrhizal fungi. *Journal of Plant Physiology* 160, 615–626.

Richards, A.J. (2003) Apomixis in flowering plants: an overview. *Philosophical Transactions of the Royal Society of London Series B: Biological Sciences* 358, 1085–1093.

Rieseberg, L.H., Raymond, O., Rosenthal, D.M., Lai, Z., Livingstone, K., Nakazato, T., Durphy, J.L., Schwarzbach, A.E., Donovan, L.A. and Lexer, C. (2003) Major ecological transitions in wild sunflowers facilitated by hybridization. *Science* 301, 1211–1216.

Rodman, J.E., Karol, K.G., Price, R.A. and Sytsma, K.J. (1996) Molecules, morphology, and Dahlgrens expanded order Capparales. *Systematic Botany* 21, 289–307.

Rodman, J.E., Soltis, P.S., Soltis, D.E., Sytsma, K.J. and Karol, K.G. (1998) Parallel evolution of glucosinolate biosynthesis inferred from congruent nuclear and plastid gene phylogenies. *American Journal of Botany* 85, 997–1006.

Rollins, R.C. (1993) *The Cruciferae of Continental North America*. Stanford University Press, Stanford, California.

Roy, B.A. (1993) Patterns of rust infection as a function of host genetic diversity and host density in natural populations of the apomictic crucifer, *Arabis holboellii*. *Evolution* 47, 111–124.

Roy, B.A. (1995) The breeding systems of six species of *Arabis* (Brassicaceae). *American Journal of Botany* 82, 869–877.

Roy, B.A. (2001) Patterns of association between crucifers and their flower-mimic pathogens: host jumps are more common than coevolution or cospeciation. *Evolution* 55, 41–53.

Sall, T., Jakobsson, M., Lind-Hallden, C. and Hallden, C. (2003) Chloroplast DNA indicates a single origin of the allotetraplold *Arabidopsis suecica*. *Journal of Evolutionary Biology* 16, 1019–1029.

Sarret, G., Saumitou-Laprade, P., Bert, V., Proux, O., Hazemann, J.-L., Traverse, A., Marcus, M.A. and Manceau, A. (2002) Forms of zinc accumulated in the hyperaccumulator *Arabidopsis halleri*. *Plant Physiology* 130, 1815–1826.

Savolainen, O., Langley, C.H., Lazzaro, B.P. and Freville, H. (2000) Contrasting patterns of nucleotide polymorphism at the alcohol dehydrogenase locus in the outcrossing *Arabidopsis lyrata* and the selfing *Arabidopsis thaliana*. *Molecular Biology and Evolution* 17, 645–655.

Schein, M., Yang, Z., Mitchell-Olds, T. and Schmid, K.J. (2004) Rapid evolution of a pollen-specific oleosin-like gene family from *Arabidopsis thaliana* and closely related species. *Molecular Biology and Evolution* 21, 659–669.

Schemske, D.W. and Bradshaw, H.D. (1999) Pollinator preference and the evolution of floral traits in monkeyflowers (*Mimulus*). *Proceedings of the National Academy of Sciences USA* 96, 11910–11915.

Schmid, K.J., Sorensen, T.R., Stracke, R., Torjek, O., Altmann, T., Mitchell-Olds, T. and Weisshaar, B. (2003) Large-scale identification and analysis of genome-wide single-nucleotide polymorphisms for mapping in *Arabidopsis thaliana*. *Genome Research* 13, 1250–1257.

Schranz, M.E., Quijada, P., Sung, S.-B., Lukens, L., Amasino, R. and Osborn, T.C. (2002) Characterization and effects of the replicated flowering time gene *FLC* in *Brassica rapa*. *Genetics* 162, 1457–1468.

Schulz, O.E. (1936) Cruciferae. In: Engler, A. and Harms, H. (eds) *Die natürlichen Pflanzenfamilien*. Verlag von Wilhelm Engelmann, Leipzig, pp. 227–658.

Shah, J. (2003) The salicylic acid loop in plant defense. *Current Opinion in Plant Biology* 6, 365–371.

Sharbel, T.F. and Mitchell-Olds, T. (2001) Recurrent polyploid origins and chloroplast phylogeography in the *Arabis holboellii* complex (Brassicaceae). *Heredity* 87, 59–68.

Sharbel, T.F., Haubold, B. and Mitchell-Olds, T. (2000) Genetic isolation by distance in *Arabidopsis thaliana*: biogeography and post-glacial colonization of Europe. *Molecular Ecology* 9, 2109–2118.

Shu, G., Amaral, W., Hileman, L.C. and Baum, D.A. (2000) *LEAFY* and the evolution of rosette flowering in violet cress (*Jonopsidium acaule*, Brassicaceae). *American Journal of Botany* 87, 634–641.

Singh, G., Dyer, P. and Ashby, A. (1999) Intra-specific and inter-specific conservation of mating-type genes from the discomycete plant-pathogenic fungi *Pyrenopeziza brassicae* and *Tapesia yallundae*. *Current Genetics* 36, 290–300.

Soltis, D.E. and Soltis, P.S. (1999) Polyploidy: recurrent formation and genome evolution. *Trends in Ecology and Evolution* 14, 348–352.

Spielman, M., Vinkenoog, R. and Scott, R.J. (2003) Genetic mechanisms of apomixis. *Philosophical Transactions of the Royal Society of London Series B: Biological Sciences* 358, 1095–1103.

Stahl, E.A., Dwyer, G., Mauricio, R., Kreitman, M. and Bergelson, J. (1999) Dynamics of disease resistance polymorphism at the *Rpm1* locus of *Arabidopsis*. *Nature* 400, 667–671.

Stebbins, G.L. (1940) The significance of polyploidy in plant evolution. *American Naturalist* 74, 54–66.

Strauss, S.Y., Siemens, D.H., Decher, M.B. and Mitchell-Olds, T. (1999) Ecological costs of plant resistance to herbivores in the currency of pollination. *Evolution* 53, 1105–1113.

Sunnucks, P. (2000) Efficient genetic markers for population biology. *Trends in Ecology and Evolution* 15, 199–203.

Swanson, W. and Vacquier, V. (2002a) The rapid evolution of reproductive proteins. *Nature Review Genetics* 3, 137–144.

Swanson, W.J. and Vacquier, V.D. (2002b) Reproductive protein evolution. *Annual Review of Ecology and Systematics* 33, 161–179.

Sweeney, P.W. and Price, R.A. (2000) Polyphyly of the genus *Dentaria* (Brassicaceae): evidence from trnL intron and ndhF sequence data. *Systematic Botany* 25, 468–478.

Takhtajan, A. (1997) *Diversity and Classification of Flowering Plants*. Columbia University Press, New York.

The Arabidopsis Genome Initiative (2000) Analysis of the genome sequence of the flowering plant *Arabidopsis thaliana*. *Nature* 408, 796–815.

Tian, D., Araki, H., Stahl, E., Bergelson, J. and Kreitman, M. (2002) Signature of balancing selection in *Arabidopsis*. *Proceedings of the National Academy of Sciences USA* 99, 11525–11530.

Tian, D., Traw, M., Chen, J., Kreitman, M. and Bergelson, J. (2003) Fitness costs of R-gene-mediated resistance in *Arabidopsis thaliana*. *Nature* 423, 74–77.

Tremetsberger, K., Konig, C., Samuel, R., Pinsker, W. and Stuessy, T.F. (2002) Infraspecific genetic variation in *Biscutella laevigata* (Brassicaceae): new focus on Irene Manton's hypothesis. *Plant Systematics and Evolution* 233, 163–181.

U, N. (1935) Genome-analysis in *Brassica* with special reference to the experimental formation of *B. napus* and peculiar mode of fertilization. *Japanese Journal of Botany* 7, 389–452.

Ungerer, M.C., Halldorsdottir, S.S., Purugganan, M.D. and Mackay, T.F.C. (2003) Genotype-environment interactions at quantitative trait loci affecting inflorescence development in *Arabidopsis thaliana*. *Genetics* 165, 353–365.

Urbanska, K.M., Hurka, H., Landolt, E., Neuffer, B. and Mummenhoff, K. (1997) Hybridization and evolution in *Cardamine* (Brassicaceae) at Urnerboden, central Switzerland – biosystematic and molecular evidence. *Plant Systematics and Evolution* 204, 233–256.

van Dijk, P. and van Damme, J. (2000) Apomixis technology and the paradox of sex. *Trends in Plant Science* 5, 81–84.

Warwick, S.I. and Al-Shehbaz, I.A. (1998) Generic evaluation of *Boleum*, *Euzomodendron*, and *Vella* (Brassicaceae). *Novon* 8, 321–325.

Warwick, S.I. and Black, L.D. (1994) Evaluation of the subtribes Moricandiinae, Savignyinae, Vellinae, and Zillinae (Brassicaceae, Tribe Brassiceae) using chloroplast DNA restriction site variation. *Canadian Journal of Botany* 72, 1692–1701.

Warwick, S.I. and Black, L.D. (1997a) Molecular phylogenies from theory to application in *Brassica* and allies (Tribe Brassiceae, Brassicaceae). *Opera Botanica* 132, 159–168.

Warwick, S.I. and Black, L.D. (1997b) Phylogenetic implications of chloroplast DNA restriction site variation in subtribes Raphaninae and Cakilinae (Brassicaceae, tribe Brassiceae). *Canadian Journal of Botany* 75, 960–973.

Warwick, S.I., Al-Shehbaz, I.A., Price, R.A. and Sauder, C. (2002) Phylogeny of *Sisymbrium* (Brassicaceae) based on ITS sequences of nuclear ribosomal DNA. *Canadian Journal of Botany* 80, 1002–1017.

Weinig, C., Ungerer, M.C., Dorn, L.A., Kane, N.C., Toyonaga, Y., Halldorsdottir, S.S., Mackay, T.F.C., Purugganan, M.D. and Schmitt, J. (2002) Novel loci control variation in reproductive timing in *Arabidopsis thaliana* in natural environments. *Genetics* 162, 1875–1884.

Weinig, C., Dorn, L.A., Kane, N.C., German, Z.M., Halldorsdottir, S.S., Ungerer, M.C., Toyonaga, Y., Mackay, T.F.C., Purugganan, M.D. and Schmitt, J. (2003a) Heterogeneous selection at specific loci in natural environments in *Arabidopsis thaliana*. *Genetics* 165, 321–329.

Weinig, C., Stinchcombe, J.R. and Schmitt, J. (2003b) QTL architecture of resistance and tolerance traits in *Arabidopsis thaliana* in natural environments. *Molecular Ecology* 12, 1153–1163.

Widmer, A. and Baltisberger, M. (1999) Extensive intraspecific chloroplast DNA (cpDNA) variation in the alpine *Draba aizoides* L. (Brassicaceae): haplotype relationships and population structure. *Molecular Ecology* 8, 1405–1415.

Wright, S., Lauga, B. and Charlesworth, D. (2003) Subdivision and haplotype structure in natural populations of *Arabidopsis lyrata*. *Molecular Ecology* 12, 1247–1263.

Xiong, L. and Zhu, J.-K. (2002) Molecular and genetic aspects of plant responses to osmotic stress. *Plant, Cell and Environment* 25, 131–139.

Zunk, K., Mummenhoff, K., Koch, M. and Hurka, H. (1996) Phylogenetic relationships of *Thlaspi* sl (subtribe Thlaspidinae, Lepidieae) and allied genera based on chloroplast DNA restriction-site variation. *Theoretical and Applied Genetics* 92, 375–381.

Zunk, K., Mummenhoff, K. and Hurka, H. (1999) Phylogenetic relationships in tribe Lepidieae (Brassicaceae) based on chloroplast DNA restriction site variation. *Canadian Journal of Botany* 77, 1504–1512.

9 Genetic variation in plant populations: assessing cause and pattern

David J. Coates and Margaret Byrne

*Science Division, Department of Conservation and Land Management, Locked Bag
104, Bentley Delivery Centre, WA 6983, Australia*

Introduction

The assessment of patterns of genetic variation in plant populations has made critical
contributions to many studies in evolutionary biology, conservation genetics, plant
breeding and ecological genetics. The value
in such assessments not only relates to
quantification of the amount and distribution of genetic variation in populations but
also to the investigation of those processes
that influence patterns of genetic variation.
Plants in particular, with their huge diversity in breeding systems and contrasting
life-histories, provide a rich source of information in relation to patterns and
processes that characterize genetic diversity in populations. For example, the relatively recent broad-based interest in
applying population genetic principles in
the conservation of small fragmented populations and management of rare and geographically restricted species has resulted
in a plethora of new studies investigating
patterns of genetic variation in plant populations (see Young and Clarke, 2000). In
attempting to explain the observed patterns, many of these studies have highlighted the complex interactions between
factors such as evolutionary history, breeding system, mode of reproduction and

events associated with recent habitat fragmentation such as reduced population size
and increased isolation.

A major development in plant population
genetics has been the significant advance,
over the last three decades, in technologies
that allow the direct molecular characterization of genes and gene products. In many
respects the development and use of various
molecular markers that track changes in
individual genes has revolutionized population genetics and broadened its applicability
across many fields in biology. The effective
neutrality of molecular markers means they
are ideal for a broad range of plant population genetic, conservation genetic and evolutionary studies such as investigating patterns
of gene flow, mating systems, population
genetic structure, hybridization and effective
population size.

In conjunction with these advances there
has been a dramatic increase in studies of
intraspecific variation often combining population genetics, phylogenetics and biogeography. Such integrated approaches have
seen the development of new fields of study
such as phylogeography (see Avise, 2000),
which have made significant contributions to
our understanding of evolutionary and ecological processes in plant populations (see
Schaal and Olsen, 2000).

Measuring Genetic Variation

Genetic variation in plant populations has been measured using a broad range of approaches. These include: (i) the assessment of quantitative (continuous) characters such as seed set, growth rate and time to flowering; (ii) observable heritable polymorphisms such as flower colour and including recessive lethal alleles; (iii) chromosome rearrangements such as translocations and inversions; (iv) protein variation, in particular isozyme electrophoresis; and (v) nuclear and organelle DNA variation.

The application of molecular markers to investigations of genetic variation in populations started with the development of protein electrophoresis and analysis of isozyme variation, some three decades ago. Subsequently a wide array of DNA-based markers is now available that have allowed an ongoing refinement of approaches to the study of population-based variation and micro-evolutionary change. Extensive data sets are now available on population genetic variation in numerous plant species, for allozyme variation (see Hamrick and Godt, 1989) and more recently DNA-based markers such as RAPDs (see Nybom and Bartish, 2000). Not only has the development of these molecular markers allowed the visualization of locus-specific variation in populations, it has also been accompanied by the development of methods that allow the ready interpretation of this information in the context of population genetic theory (see Weir, 1996).

Quantitative and molecular marker variation

Before reviewing the range of molecular markers now available for population genetic and evolutionary studies in plants it is important to consider whether single locus variation based on these markers is representative of the entire genome. Most studies of genetic variation in populations based on molecular genetic markers consider those markers to be selectively neutral or near neutral. This assumption, although probably correct in most instances, may not necessarily always be the case (see Merila and Crnokrak, 2001). In addition, if they are effectively neutral, measures of population genetic variation based on these markers will not be expected to reflect the actions of selection on other parts of the genome.

Instead of molecular markers, genetic variation in populations can be investigated by assessing quantitative variation that is under polygenic control where many loci, and the environmental effects on those loci, contribute to the quantitative variation in the traits being investigated. Yet analysing patterns of genetic variation from molecular markers has become increasingly popular as molecular techniques become more cost effective and less invasive. Unfortunately evidence for concordance in these two measures of genetic diversity is equivocal, with a number of studies suggesting it is poor (Reed and Frankham, 2001; MacKay and Latta, 2002). Therefore a key question is how well are molecular marker and quantitative variation correlated?

It is often assumed that the various measures of genetic variation are positively correlated, yet there are a number of reasons why there may be a lack of agreement between measures of genetic diversity based on molecular markers and quantitative traits. Molecular markers may not necessarily track quantitative genetic variation due to non-additive effects, differential selection, different mutation rates, environmental effects on quantitative variation, and the influence of genetic variation on gene regulation (see Lynch et al., 1999; Reed and Frankham, 2001). There is increasing evidence that within populations there is little association between levels of genetic variation estimated by molecular marker heterozygosity and life-history trait heritabilities. This implies that neutral molecular markers are unlikely to provide conservation biologists and evolutionary biologists with any clear indication of a population's evolutionary potential. For example, in a meta-analysis Reed and Frankham (2001) found no correlation between molecular markers and quantitative trait variation over 19 animal and plant studies.

In comparison, the relationship between genetic differentiation among populations based on molecular markers and quantitative traits is less clear. The critical question of whether molecular marker differences reflect adaptive divergence among populations has been raised in a number of studies (Lynch *et al.*, 1999; Merila and Crnokrak, 2001; Reed and Frankham, 2001; McKay and Latta, 2002). As pointed out by Lynch (1996) significant molecular divergence provides strong evidence that adaptive divergence has the opportunity to occur but lack of any molecular divergence is likely to be uninformative.

Investigations of population differentiation in plants that have compared estimates from allozymes with quantitative traits have provided contrasting results. In some species there is a strong correlation between the two estimates of differentiation, while in others, such as some forest trees, variation in quantitative traits associates closely with environmental gradients but allozyme variation does not (see Hamrick, 1983). Despite this, a trend does appear to be emerging that supports observations by Hamrick (1983) that quantitative traits show as much or more differentiation among populations than allozyme markers. This trend has been confirmed in recent reviews by Merila and Crnokrak (2001) and McKay and Latta (2002) where population differentiation for quantitative traits (Q_{ST}) is typically higher than estimates of neutral molecular divergence based on F_{ST} (Table 9.1).

Table 9.1. Comparisons of divergence in quantitative traits (Q_{ST}) and divergence in marker genes (F_{ST}) for plant species where Q_{ST} partitions quantitative genetic variation in an analogous fashion to F_{ST} for single gene markers. For neutral traits, F_{ST} and Q_{ST} should be equal, while the level of difference between them can be used to infer directional selection ($Q_{ST} > F_{ST}$) or selection favouring the same phenotype in different populations ($Q_{ST} < F_{ST}$). Note that in most cases Q_{ST} is larger than F_{ST} indicating that natural selection is likely to be a significant force in determining patterns of quantitative trait differentiation among plant populations.

Species	Q_{ST}	F_{ST}	Marker	Reference [a]
Arabidopsis thaliana	0.830	0.890	Allozymes, microsatellites	1
Arabidopsis thaliana	0.885	1.000	Allozymes	2
Arabis fecunda	0.980	0.200	Allozymes	1
Brassica insularis	0.060	0.210	Allozymes	2
Centaurea corymbosa	0.220	0.364	Allozymes	2
Clarkia dudleyana	0.380	0.075	Allozymes	2
Clarkia dudleyana	0.353	0.068	Allozymes	1
Larix laricina	0.490	0.050	Allozymes	2
Larix occidentalis	0.490	0.086	Allozymes	2
Medicago truncatula	0.584	0.330	Allozymes	2
Phlox drummondii	0.250	0.038	Allozymes	2
Picea glauca	0.360	0.035	Allozymes	2
Picea sitchensis	0.290	0.079	Allozymes	2
Pinus brutia	0.250	0.140	Allozymes	2
Pinus contorta	0.120	0.019	Allozymes	1,2
Pinus sylvestris	0.364	0.018	Allozymes, microsatellites, RFLPs	1
Pseudotsuga menziesii	0.420	0.022	Allozymes	2
Quercus petrea	0.310	0.025	Allozymes	2
Salix vimnalis	0.070	0.041	Allozymes	1,2
Scabiosa canescens	0.095	0.164	Allozymes	1,2
Scabiosa columbaria	0.452	0.123	Allozymes	1,2
Sequoiadendron giganteum	0.180	0.097	Allozymes	2
Silene diclinis	0.118	0.052	Allozymes	1,2

[a] 1 and 2 are reviews that contain references to the original studies: 1. McKay and Latta (2002); 2. Merila and Crnokrak (2001).

As emphasized by both Merila and Crnokrak (2001) and McKay and Latta (2002), key issues that can be addressed in studies that compare molecular marker variation with quantitative trait variation are the insights they provide into the relative importance of genetic drift and natural selection as causes of population divergence in specific quantitative traits. Population structure of quantitative traits is not expected to differ from that of single molecular marker loci if the traits determined by both are selectively neutral. Consequently any differences between Q_{ST} and F_{ST} estimates can be attributed to natural selection (see Table 9.1). That is, the extent of local adaptation can be assessed for various traits and the Q_{ST} for those traits experiencing the strongest local selection will be expected to show the largest difference from the molecular marker F_{ST}.

Despite these recent reviews, covering a range of different plant species, there is still no clear answer to the question of how well divergence based on neutral markers predicts that based on quantitative traits. Merila and Crnokrak (2001) found in their meta-analysis based on 27 plant and animal species that the level of differentiation in neutral marker loci is closely predictive of the level of differentiation in loci encoding quantitative traits. In contrast, McKay and Latta (2002) found that differentiation in neutral marker loci and quantitative trait loci was poorly correlated across their sample of 29 species. However, both emphasize the need for further theoretical and empirical studies to address the relationship between Q_{ST} and F_{ST}.

Molecular markers in plant population genetics

Over the last few decades, the use of neutral molecular markers has dominated studies on population genetic structure and geographic patterns of genetic variation in plants. The range of markers now available allows increased flexibility for investigators to utilize more than one marker at varying spatial and temporal scales. These may range from broad historically based geographical variation (see Thompson, 1999; Schaal and Olsen, 2000) to finer scale metapopulation genetic structure (see Manel *et al.*, 2003). Popular molecular marker techniques for population genetic studies include isozymes and an array of DNA-based techniques such as restriction fragment length polymorphism (RFLP), random amplified polymorphic DNA (RAPD), amplified fragment length polymorphism (AFLP) and microsatellites or simple sequence repeats (SSRs). Some of these markers, such as isozymes, RFLPs and microsatellites, are codominant and can be analysed as single locus markers, while others, such as RAPDs and AFLPs, are dominant multilocus markers (Table 9.2).

Isozymes (allozymes)

Isozyme electrophoresis has made an immense contribution to research in plant population genetics, systematics and evolution (see Soltis and Soltis, 1989), and plant conservation biology (see Falk and Holsinger, 1991). Despite the development of a range of DNA-based markers, allozymes continue to be an important and reliable tool for the study of genetic variation and evolutionary processes in plant populations. There seems little doubt that their widespread and continued use stems from their cost effectiveness, technical ease, number of available loci and codominant inheritance (Arnold and Imms, 1998).

Apart from advantages in terms of cost and time, allozymes have a number of other advantages that make them convenient and reliable genetic markers. These include Mendelian inheritance, codominant expression, and similarity of apparently homologous isozyme loci and their allozyme patterns between different species. Other advantages lie not so much in the technique, analytical methods or properties of isozymes, but rather in the large number of studies that have been conducted on a wide range of plant taxa. From this large database of information it is now possible to make quite useful generalizations on the relationships between patterns of genetic variation

Table 9.2. Common molecular techniques used in assessing patterns of genetic variation in plant populations.

Technique	Methodology	Technical difficulty	Level of polymorphism	Resolution	Reliability
Codominant, single-locus markers					
Allozymes/isozymes	Gel electrophoresis and visualization of cellular enzymes and proteins	Easy	Low to moderate	Moderate	Very high
Restriction fragment length polymorphism (RFLP)	Digestion of total genomic DNA with restriction endonucleases followed by Southern blotting and hybridization with specific DNA fragments	Moderate to difficult	Moderate to high	High	Very high
Microsatellites or simple sequence repeats (SSRs)	Specifically developed PCR primers used to amplify hypervariable tandemly repeated units. Variation at these loci can be investigated in both nuclear and chloroplast genomes	Difficult	Very high	High	High
Dominant multilocus markers					
Random amplified polymorphic DNA (RAPD)	Amplification of random DNA segments using arbitrary short sequence primers ($\approx$ 10 nucleotides in length)	Easy	Moderate to high	Moderate	Medium
Amplified fragment length polymorphism (AFLP)	Amplification of total genomic DNA digested with restriction endonucleases, where the subsequent restriction site is then used as a primer binding site for selective amplification using PCR primers that anneal perfectly to target sequences	Moderate	High	High	Medium to high

and factors such as life-history, breeding system, geographic distribution and habitat (see Hamrick and Godt, 1989).

Although isozyme analysis has a number of features that make it an attractive technique, it also has a number of well-known limitations. There is the potential for significant undetected variation given that only about 20–30% of base substitutions in the gene will result in detectable change using standard electrophoresis conditions. Probably the greatest restriction of isozyme markers is the relatively low level of variation. Compared with markers such as microsatellites, allozymes clearly have much less resolving power in, for example, paternity analysis and gene flow studies. The greatest value that isozyme analysis will have in the future is not necessarily as an alternative to more recent DNA-based techniques but as a source of supporting data that may indicate directions for further detailed molecular studies.

Nuclear DNA markers

RESTRICTION FRAGMENT LENGTH POLYMORPHISMS The major use of DNA-based markers in plant population studies commenced with the application of endonuclease methodologies and the development of DNA restriction fragment length polymorphisms. Single-copy nuclear RFLPs provide a large number of highly variable codominant markers. The demonstration that they could be readily used in plant breeding studies and for analysing population genetic variation in plants led to their rapid utilization, particularly in crop species and wild relatives (Clegg, 1989).

Some of the first applications of RFLPs in plant population studies involved investigations of genetic variation at the high-copy ribosomal loci (see Schaal *et al.*, 1991). These investigations were based on rDNA intergenic spacer length and restriction site variation. While some species, such as *Clematis fremontii*, can show significant rDNA variation at the population level, others show little or no length variation or restriction site variation (Schaal *et al.*, 1991). Thus rDNA RFLP data was found to be relatively limited in its application to population studies.

In contrast, single-copy nuclear RFLP markers have proved to be particularly informative in the analysis of patterns of genetic variation within and among plant populations. Comparisons of genetic variation using RFLPs with that of allozymes indicate that the level of polymorphism detected with RFLP loci is generally three to four times higher than the level detected with isozymes (see Byrne *et al.*, 1998). As with other DNA markers, this higher level of variation can be readily attributed, at least theoretically, to the assaying of all mutational variation compared with only a subset of total variation detectable with isozymes. Although most comparisons of RFLPs versus isozymes show similar levels and patterns of divergence among populations, there are notable exceptions such as in *Beta vulgaris* subsp. *maritima* (sea beet) (Raybould *et al.*, 1996). These findings have been generally attributed to selection operating on some isozyme loci, although, as Raybould *et al.* (1996) point out, an alternative scenario is that RFLPs could be under disruptive selection.

Despite their significant resolving power in population genetics, single-copy nuclear RFLPs have become less popular in recent times. Reasons given are that they are time consuming and expensive, they require the use of radiolabelled probes, and relatively large amounts of DNA are needed. Yet, where they have been used, they have proved to be extremely robust and informative in assessing patterns of population genetic variation in a range of species covering a broad range of plant genera and families (see Byrne *et al.*, 1998). In addition, in contrast to microsatellites (see below) they can be readily used across species and genera.

MICROSATELLITES Microsatellites or SSRs are now recognized as potentially one of the most useful genetic markers in plant population studies. Microsatellites consist of tandem repeats of a short 'motif' sequence, usually of one to six bases. These regions occur frequently and randomly in plant and animal genomes and often have large numbers of moderately frequent alleles. Thus they show extensive variation between individuals within populations and have been developed for a wide range of purposes in plant breeding, conservation biology and population genetics including forensics, paternity analysis and gene mapping (Jarne and Lagoda, 1996). In particular, in plant population genetic studies they have proved to be ideal for assessing gene flow among populations (see Chase *et al.*, 1996), and are ideally suited to fine-scale analysis of mating within populations.

Unfortunately the significant benefits of such hypervariable codominant markers are offset by the time and effort involved in their development (see Squirrell *et al.*, 2003). Sequence information is required to design appropriate primers and such information is generally only available for a limited number of commercially important species. For most plant species, microsatellites can only be developed from clones isolated through construction and screening of a genomic library. As pointed out by Squirrell *et al.* (2003), the development of a working primer set is subject to a considerable attrition rate compared with the original number of clones sequenced. This contrasts with, and increases the appeal of, other highly polymorphic markers such as AFLPs and RAPDs (see below) where generic primers are readily available.

One approach that has been successfully

used in some animal groups involves cross-species amplification of microsatellite loci. However, this approach is much less successful in plants where the general patterns indicate that cross-species amplification of plant microsatellites will be restricted to closely related species or at least congeners (Peakall et al., 1998; Butcher et al., 2000), although broader cross-transferability has been demonstrated in Vitaceae (Arnold et al., 2002).

RANDOM AMPLIFIED POLYMORPHIC DNA RAPDs are probably the next most common DNA markers that have been used in plant population genetic studies. These markers were the first of a number of multilocus PCR-based markers that have been widely applied across plant species. The technique uses single arbitrary primer sequences to amplify anonymous regions of the genome and can be used to identify and screen numerous polymorphic loci. Since no sequence information is needed, this technique is particularly applicable in cases where little molecular genetic information is available on the target species. Furthermore, the assay is very simple and fast, and many loci can be identified, often with a single reaction.

Initially RAPDs were extensively used in plant breeding studies and particularly in genome mapping to identify quantitative trait loci. Subsequently they became increasingly popular in studies of genetic variation in natural populations, often in conjunction with, or as an adjunct to, isozyme studies (see Peakall et al., 1995). Patterns of genetic diversity and population genetic structure using RAPDs have now been investigated in a broad range of plant species (Harris, 1999; Nybom and Bartish, 2000). They have proved to be of particular value where they have revealed useful levels of genetic variation within and/or between populations despite the detection of only minimal allozyme variation.

Although RAPDs provide a fast and cost-effective means of investigating genetic variation, they have a number of limitations (see Arnold and Imms, 1998; Harris, 1999). Reproducibility has often been cited as a sig-nificant problem since the lack of specificity associated with the use of short arbitrary primers can lead to increased sensitivity to PCR conditions resulting in erratic amplification. However, this problem can be minimized with strictly controlled and standardized reaction conditions. Homology is also an important issue, given that comigration of products assumes homology, which may not necessarily be correct (see Rieseberg, 1996). This is more likely to be a problem in comparisons where higher levels of genetic divergence are involved, such as between taxa, and is less likely to be an issue in population genetic studies. Another limitation is that the majority of RAPD markers are dominant and there is therefore a significant loss of information content compared with codominant markers such as isozymes, RFLPs and microsatellites. In this regard, comparisons between RAPDs and allozymes need to be treated with caution, given the different approaches used for dealing with monomorphic loci and that dominant RAPD loci are unsuited to estimation of population genetic parameters such as F statistics and G_{ST} unless assumptions are made regarding the breeding system and Hardy–Weinberg equilibrium (Lynch and Milligan, 1994; Bussell, 1999).

AMPLIFIED FRAGMENT LENGTH POLYMORPHISMS AFLPs are a more recently developed molecular marker for plant population genetic studies. Like RAPDs, AFLPs have an advantage over RFLPs because they are applicable to DNA of any origin without prior sequence information, primer synthesis or library construction. Also, like RAPDs, most AFLPs are dominant markers, but unlike RAPD protocols a single AFLP reaction can survey as many as 100–200 loci (see Mueller and Wolfenbarger, 1999). The polymorphisms typically exhibit Mendelian inheritance, enabling their use for typing, identification and mapping of genetic characteristics. The potential to produce hundreds of polymorphic loci per individual and the relative ease of applicability to most organisms makes AFLPs one of the most efficient molecular markers for generating polymorphic loci in plant population genetic studies (Mueller and Wolfenbarger, 1999).

Although AFLP analyses do not suffer from the reproducibility problems that potentially exist in the use of RAPDs, they are still largely dominant markers and therefore pose the same difficulties in reliance on the assumptions of equilibrium for estimations of population genetic parameters. However, their high resolution and the ease with which one can generate large numbers of genetic markers has seen a steady increase in their application in the analysis of population level variation in plant species.

AFLPs and microsatellites are rapidly becoming the DNA-based markers of choice for plant population genetic structure, mating system and gene flow studies. A number of recent investigations have compared the potential level of resolution and efficiency of these markers in plant population studies. As expected, microsatellites are considerably more polymorphic than AFLPs at the locus level, but AFLPs are much more efficient at revealing polymorphic loci. For example, in *Avicennia marina* average expected heterozygosity for AFLPs was only 0.193 compared with 0.780 for microsatellites, but all of the 918 AFLP bands scored were polymorphic (Maguire *et al.*, 2002).

Chloroplast DNA (cpDNA) variation

Whereas the molecular markers discussed previously are based on various portions of the nuclear genome, another marker that has considerable potential in plant population studies is cpDNA. Compared with the nuclear genome of higher plants, which consist of a diploid complement of randomly segregating biparentally inherited chromosomes, the chloroplast genome is predominantly uniparentally inherited and consists of a single circular molecule. In angiosperms the chloroplast genome is generally maternally inherited, while in most gymnosperms it is paternally inherited. A number of different approaches have been utilized to characterize variation in cpDNA. The first and probably most widely used involves RFLP analysis of the entire chloroplast genome. An important factor assisting this approach resides in the relatively high degree of

sequence conservation in the chloroplast genome that allows heterologous probes to be utilized across most plant families. With the development of PCR, a number of RFLP studies have now been carried out based on PCR-amplified products, thus avoiding the time required for Southern hybridization.

A limitation of cpDNA analysis in many species is the relatively low sequence diversity, which can be attributed to the haploid status of the genome, reduced rate of mutation and lack of recombination (see Schaal *et al.*, 1998; Ennos *et al.*, 1999). To address this issue there have been attempts to target sections of the chloroplast genome that have relatively high mutation rates. Useful levels of intraspecific variation have been detected in non-coding regions of the cpDNA genome. For example, Vaillancourt and Jackson (2000) detected significant levels of polymorphism for eucalypts in sequence studies of the J_{LA} junction between the inverted repeats and the large single-copy region, with much of the variation due to complex insertion/deletions. Another useful approach with considerable potential involves the analysis of variation using chloroplast microsatellites (Powell *et al.*, 1995). A recent assessment of the occurrence of microsatellites in six species where the cpDNA genome has been completely sequenced detected a total of 505 cpDNA microsatellites (Provan *et al.*, 1999). Although high levels of polymorphism make cpDNA microsatellites extremely useful markers in plants for gene flow, their high mutation rate, similar to nuclear microsatellites, makes them unlikely to be useful markers in phylogenetic analysis, primarily because of the problems of homoplasy where the same mutation can arise from independent events.

Despite occasional difficulties in accessing appropriate levels of variation in plant population studies, cpDNA has proved to be extremely useful in providing insights into population and evolutionary processes that could not be delivered by nuclear markers.

One of the most fundamental applications of patterns of cpDNA variation has been in the analysis of phylogeographic patterns of population variation (see section on

'Historical associations and phylogeographic patterns', below). In addition to uniparental inheritance and lack of recombination, there are a number reasons why cpDNA may be far more suitable for such studies than nuclear markers. The effective population size of an organelle gene is theoretically half that of a nuclear gene, so cpDNA variation is likely to be more sensitive to population differentiation by genetic drift. Also, because the maternally inherited cpDNA will only be dispersed by seed, the genetic differences existing when populations come into contact will probably break down much more slowly than for nuclear markers, leaving a signature of historical relationships for much longer (see Ennos *et al.*, 1999).

Significant Determinants of Genetic Variation in Plant Populations

Determinants of patterns of genetic variation in plant populations are extremely varied and often involve complex interactions between plant attributes such as life-form, floral architecture, mode of reproduction, incompatibility system, pollination system, and ecological and environmental parameters that may influence pollination events, population size and isolation (see Table 9.3). A further level of complexity can be added when one considers the evolutionary history of the species where events such as climate change and localized extinction, contraction to refugia, range expansion and fluctuations in population size over time can also have substantial influence on the current patterns of genetic variation in a plant population (Schaal *et al.*, 1998). Table 9.3 provides a summary of attributes and an indication of their level of influence on patterns of genetic variation within and between plant populations. Four key themes are explored in the following sections that broadly address those attributes.

Geographical distribution and rarity

Predictions regarding the genetic consequences of restricted geographic range and rarity in plants generally follow genetic the-

ory for small populations occupying a narrow range of environments and have received significant attention over the last decade (see Karron, 1987; Hamrick and Godt, 1989; Barrett and Kohn, 1991; Ellstrand and Elam, 1993; Gitzendanner and Soltis, 2000). Theoretically, geographically restricted and rare plants might be expected to show low levels of genetic variation within both species and populations because of selection under a narrow range of environmental conditions and genetic drift and inbreeding in small, isolated populations (Barrett and Kohn, 1991; Ellstrand and Elam, 1993). Low genetic variation within geographically restricted species may also be due to founder events associated with recent speciation (Gottlieb, 1981; Loveless and Hamrick, 1988). Broad comparisons between geographically restricted and widespread species generally follow predicted trends, with allozyme studies showing that geographically restricted species have less genetic variability than widespread species (Hamrick and Godt, 1989; Table 9.3). However, this association was not evident in a review of plant population-based studies using RAPDs (Nybom and Bartish, 2000).

A useful demonstration of the relationship between geographic range and patterns of genetic variation can be found in comparisons between closely related taxa of a triggerplant (*Stylidium*) species complex in south-west Australia (Coates *et al.*, 2003). The *Stylidium caricifolium* complex taxa show a range of geographic distributions but are phylogenetically closely related and are characterized by the same pollination system, similar seed dispersal mechanisms, self-compatibility and frequent geitonogamous self-pollination. These characteristics make it possible to readily investigate theoretical predictions of lower genetic diversity and reduced genetic structure for rare and geographically restricted taxa in this complex, while at the same time minimizing any confounding effects that may be associated with phylogenetic differences and differences in life-history attributes and breeding system (see Karron, 1987; Karron *et al.*, 1988; Gitzendanner and Soltis, 2000).

Table 9.3. Levels of genetic variation at the population level and differentiation among populations based on allozymes (Hamrick and Godt, 1989) and RAPDs (Nybom and Bartish, 2000) for species with different attributes.

Attributes	Allozymes				RAPDs			
	N	H_{ep}	N	G_{ST}	N	H_{pop}	N	G_{ST}
Taxonomic status		***		***		***		NS
Gymnosperms	56	0.160	80	0.068	5	0.386	6	0.180
Monocotyledons	80	0.144	81	0.231	9	0.190	6	0.310
Dicotyledons	338	0.096	246	0.273	27	0.191	19	0.320
Life form		***		***		NS		NS
Annual	187	0.105	146	0.357	4	0.125	2	0.470
Short-lived perennial					13	0.207	13	0.300
Short-lived perennial (herbaceous)	159	0.096	119	0.233				
Short-lived perennial (woody)	11	0.094	8	0.088				
Long-lived perennial					23	0.242	14	0.230
Long-lived perennial (herbaceous)	4	0.084	2	0.213				
Long-lived perennial (woody)	115	0.149	131	0.076				
Geographic range		***		NS		NS		NS
Endemic	100	0.063	52	0.248	5	0.191	5	0.190
Narrow	115	0.105	82	0.242	4	0.233	7	0.220
Regional	180	0.118	186	0.216	16	0.222	9	0.350
Widespread	85	0.159	87	0.210	15	0.208	9	0.330
Breeding system		***		***		***		***
Selfing	113	0.074	78	0.510	8	0.091	5	0.590
Mixed					6	0.219	5	0.190
Mixed–animal	85	0.090	60	0.216				
Mixed–wind	10	0.198	11	0.100				
Outcrossing					24	0.260	18	0.230
Outcrossing–animal	164	0.124	124	0.197				
Outcrossing–wind	102	0.148	134	0.099				
Seed dispersal		**		***		NS		NS
Gravity	199	0.101	161	0.277	16	0.212	16	0.300
Gravity–attached	12	0.127	11	0.124				
Attached	68	0.137	52	0.257	3	0.165	2	0.470
Explosive	34	0.062	23	0.243				
Ingested	54	0.129	39	0.223	17	0.228	9	0.170
Wind	105	0.123	121	0.143	3	0.261	2	0.230
Mode of reproduction		NS		NS				
Sexual	413	0.114	352	0.225				
Sexual and asexual	56	0.103	54	0.213				
Successional status		NS		***		**		**
Early	198	0.107	165	0.289	10	0.166	8	0.500
Mid	182	0.106	121	0.259	19	0.195	13	0.230
Late	103	0.133	121	0.101	12	0.287	9	0.200

N, number of taxa; H_{ep}, genetic diversity (allozymes); H_{pop}, genetic diversity (RAPDs); G_{ST}, proportion of total genetic diversity among populations. Significance levels: * $P < 0.05$; ** $P < 0.01$; *** $P < 0.001$; NS, not significant.

Population-based estimates of genetic diversity are relatively high for all taxa in the *S. caricifolium* complex when compared with other outcrossing animal-pollinated and short-lived herbaceous perennials, but indicate significant differences in genetic variability between geographically restricted taxa and more widespread taxa. The three most widespread and common species showed consistently, and in a number of cases significantly, higher levels of genetic diversity than four of the six geographically restricted species. Lower levels of genetic diversity in the geographically restricted taxa are likely to be due to a number of factors. Fluctuations in population size and repeated bottlenecks associated with extended Pleistocene climatic instability may be primary determinants. Habitat specificity is also a likely factor, with three taxa confined to breakaways and rocky slopes associated with granite outcrops, banded ironstone and laterites, while the fourth is restricted to coastal dune systems. In addition, the close phylogenetic relationship between three taxa suggests that lower levels of genetic diversity in the two rare and geographically restricted taxa may be due to founder events associated with relatively recent divergence and isolation.

Despite reduced levels of genetic diversity in geographically restricted taxa this trend is not consistent across all such taxa in the complex. Two other rare and geographically restricted taxa show comparable or higher levels of genetic diversity than the three widespread species. A number of explanations have been given for unexpectedly high levels of genetic variation in rare and geographically restricted species. These include recent origin and retention of high levels of variation from a widespread progenitor (Gottlieb *et al.*, 1985), hybridization (see below), maintenance of relatively large populations (Young and Brown, 1996), long-term stability of populations occupying refugia (Lewis and Crawford, 1995), and relatively recent fragmentation of previously widespread species (Karron, 1991). The likely causes of the unexpectedly high levels of genetic variation in these two taxa are large population size prior to recent frag-

mentation and hybridization, respectively. This example reinforces issues raised by various authors in relation to rarity. They stress that rarity has multiple origins and that genetic and ecological consequences of rarity cannot necessarily be generalized in any simplistic fashion (Fiedler and Ahouse, 1992; Gaston, 1997).

Another generalized assumption is that geographically restricted species might also show less genetic structure across their range than widespread species. The *S. caricifolium* complex study appears to confirm this prediction with a significant trend from higher F_{ST} values for the more widespread taxa to lower values for the more geographically restricted taxa. However, this trend has not been observed generally in comparisons between rare and widespread congeners (Gitzendanner and Soltis, 2000), or in broader comparisons between species with contrasting geographic ranges (Table 9.3). Explanations for this include confounding effects due to differences in life-history, taxonomic biases, different evolutionary histories and sampling strategies (see Hamrick and Godt, 1989; Godt and Hamrick, 1999).

Mode of reproduction and clonality

A feature of flowering plants is the vast array of different modes of reproduction both within and across plant families (Richards, 1997). In particular, both asexual and sexual reproduction may vary among species, among populations within species and within populations (Ellstrand and Roose, 1987; Coates, 1988; Kennington and James, 1998; Sydes and Peakall, 1998). In some cases the reproductive systems may change within populations or species in response to environmental factors (Eckert and Barrett, 1993; Richards, 1997) or with time (Stöcklin, 1999). Although sexual reproduction is generally prevalent, the ability to asexually reproduce is common in plants resulting in clonal or partially clonal populations. Mechanisms of asexual reproduction can be divided into vegetative reproduction and apomixis (Richards, 1997). In the latter, a variety of mechanisms can result in the production of seeds with the same diploid geno-type as the maternal genotype. In many cases, fertilization of the endosperm is required (pseudogamy), indicating that polli-nation is just as essential for these species as sexual species. In contrast, vegetative repro-duction is achieved by a propagule other than seed and is often associated with exten-sive clonality in populations of some species.

Assumptions regarding the population genetic consequences of lack of sexual repro-duction are varied; some suggest that asexual populations will be genotypically depauper-ate, while others indicate that asexual popula-tions may be as genetically variable as sexual ones (see Ellstrand and Roose, 1987). In their review, Ellstrand and Roose (1987) found that the vast majority of clonal plants investi-gated were multiclonal both within and between populations, and that they often pos-sess considerable genetic variability. Similarly, Hamrick and Godt (1989) found no differ-ences at the population level between sexual species and species that reproduce by both sexual and asexual means (Table 9.3). Interestingly Hamrick *et al.* (1992) showed that in long-lived woody shrubs genetic diver-sity was significantly higher in populations of species with combined sexual and asexual reproduction compared with entirely sexual species. These findings indicate that clonal species may maintain comparable or some-times higher levels of genetic diversity than sexually producing species.

Geographical patterns of clonal variation within species will be expected to have a sig-nificant influence on the partitioning of genetic variation among populations. Higher levels of differentiation and lower levels of genetic variation would be pre-dicted for clonal populations where rela-tively few clones characterize the populations. This pattern was clearly observed in *Acacia anomala*, a small herba-ceous shrub known from only ten popula-tions occurring in two disjunct areas some 30 km apart (Fig. 9.1). Each population group covers only a few kilometres, with allozymes indicating high genetic divergence between the two groups. Significantly, this level of divergence is associated with a change from sexual reproduction to vegeta-tive reproduction and clonality. The north-ern populations reproduce primarily by seed, and individuals within populations generally show different multilocus geno-types and high levels of genetic diversity. In contrast, southern populations are charac-terized by a few multilocus genotypes, often with fixed heterozygosity at multiple loci and consequently low levels of genetic diver-sity (Coates, 1988).

Mating system

The mating system is expected to be a key factor influencing levels of genetic variation within populations and the population genetic structure of a species. Plant mating systems are influenced by a number of attributes such as flowering phenology, pre- and post-zygotic incompatibility and flower structure, as well as a range of ecological fac-tors. These may include mode of pollina-tion, population size and density, and population position in the landscape. Plant species, therefore, exhibit a wide array of mating systems and this diversity is better thought of as a continuum rather than as specific categories (Schemske and Lande, 1985; Brown, 1989; Barrett and Eckert, 1990). Although this is generally acknowl-edged, it is important to be able to recognize the major mating system types, given the contrasting effects they may have on pat-terns of genetic variation within plant species. A useful summary of mating system modes is given by Brown (1989). These are: predominant self-fertilization, predominant outcrossing, mixed selfing and outcrossing, apomixis and intragametophytic selfing or haploid selfing as may occur in homo-sporous ferns and allied lower plants. Some 20% of higher plant species are predomi-nantly selfing and, as emphasized by Brown (1989), critical factors that will influence the outcomes of mating in such species will be the occurrence and pattern of variation in occasional outcrossing events. Conversely, in predominantly outcrossing species the pres-ence of selfing and biparental inbreeding will have a major influence on genetic struc-ture both within and between populations.

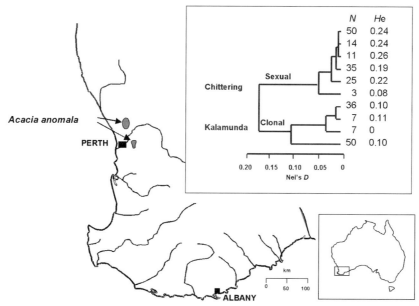

Fig. 9.1. Geographical distribution and unweighted pair-group method using an arithmetic average (UPGMA) of largely sexual and clonal populations of the rare and endangered grass wattle, *Acacia anomala*. The northern populations (Chittering) are sexual while the southern populations (Kalamunda) are clonal. Structuring of genetic diversity among populations, based on allozymes, is shown in the UPGMA. The estimated number of plants in each population (*N*) and gene diversity (*He*) are shown for each population (UPGMA branch). For clonal populations the estimate of *He* is based on the number of ramets (Coates, 1988).

The significance of mating systems as a primary determinant of genetic structure and levels of genetic variation within plant populations has been emphasized in a number of studies. Selfing species are expected to have less genetic variation within populations and greater genetic differentiation between populations than outcrossing species. Reviews comparing allozyme, RAPD and quantitative genetic variation across a wide range of species generally support these expectations (see Table 9.3), with predominantly outcrossed species having significantly higher levels of genetic diversity than selfing species or species with a mixed mating system (Hamrick and Godt, 1989; Schoen and Brown, 1991; Charlesworth and Charlesworth, 1995; Nybom and Bartish, 2000).

It has also been shown in broad comparisons between plant species that the partitioning of genetic variation among populations is generally influenced more by breeding system than by any other factor (Hamrick and Godt, 1989; Nybom and Bartish, 2000). Allozyme data indicate that selfing species have 51% of their genetic variation partitioned among populations, while in outcrossed and mixed mating system species this is reduced to 10–22% (Table 9.3). For example, in a study on two sympatric *Delphinium* species, Williams *et al.* (2001) found that the many-flowered *Delphinium barbeyi* had a lower outcrossing rate through increased geitonogamous self-pollination and a tenfold increase in population subdivision, compared with the few-flowered and more highly outcrossing *Delphinium nuttallianum*. In contrast, Hamrick *et al.* (1992) found that breeding system played a relatively minor role in predicting levels of genetic diversity among populations of woody plants, although they point out that this may be because of the limited range of breeding systems in their sample of woody plants.

Interspecific comparisons of mating systems by various authors have resulted in the recognition of a number of key issues associated with mating system change. For example, plant species investigated by Schemske and Lande (1985) show a strongly bimodal distribution of outcrossing rates with a significant deficit of species with intermediate outcrossing rates. They argue that predominant selfing and predominant outcrossing are the stable endpoints in mating system evolution, suggesting that selection on outcrossing rates is strongly directional. Investigations by Barrett and Eckert (1990) indicate a significant association between outcrossing rate and longevity, suggesting that increased outcrossing in long-lived woody species may be because of an increased genetic load in such species (see also Ledig, 1986). Pollination mode has also been found to influence patterns of outcrossing rates among species; wind-pollinated species show a clear bimodal distribution while animal-pollinated species do not (Aide, 1986; Barrett and Eckert, 1990). Although such comparative studies provide valuable insight into factors that may have significant influences on mating system patterns in different species, it is important to recognize the limitations in these approaches. Comparisons between many unrelated species may be strongly confounded by their different evolutionary histories (Barrett and Eckert, 1990).

Intraspecific variation in outcrossing rates has the potential to be much more valuable in assessing the causes of mating system change and subsequent changes in patterns of genetic variation. Levels of outcrossing within populations are the outcome of a complex interaction of the environmental, demographic, life-history and genetic characteristics of the populations (Barrett and Eckert, 1990). Good examples of this complexity are provided in a number of recent studies that have focused on the patterns and causes of mating system variation in rare and endangered plant populations. A range of threatening processes such as habitat loss, degradation and fragmentation are likely to cause significant changes in population size, density, isolation and pollination biology. Understanding how these changes will influ-

ence the mating system and thus patterns of genetic variation remains a key issue in the conservation of such populations. In addition, populations of such species can provide valuable experimental systems for investigating forces that affect mating systems. For example, in a review of mating system variation in animal-pollinated rare and endangered populations in Western Australia, Sampson et al. (1996) found that outcrossing rates from disturbed populations of mixed mating species often differed substantially from those of undisturbed populations (Table 9.4). Disturbance, associated with loss of understorey species and weed invasion, may influence pollinator type, abundance and activity, and population density and structure. Clearly, as expected, habitat disturbance has a significant effect on the mating system and levels of inbreeding in populations of these species.

In addition to outcrossing rates, another mating system parameter that has proved to be informative in comparative studies among populations is the correlation of outcrossed paternity, that is, the probability that sibs share the same father. For example, studies on bird-pollinated, long-lived woody shrubs in the family Proteaceae have shown that average paternal diversity in open-pollinated sib arrays can be low. In two rare woody shrubs, *Lambertia orbifolia* (Coates and Hamley, 1999) and *Grevillea iaspicula* (Hoebee and Young, 2001), the high levels of correlated paternity and low estimates of neighbourhood size were attributed to mating between small groups of plants. This interpretation was generally consistent with pollinator observations, which indicated that bird movements are frequently restricted to only a few mature plants.

Gene flow

Gene flow is a key factor in shaping gene pools and the population genetic structure of a species, both as a force in maintaining genetic continuity between populations and as a means by which genetic diversity can be enhanced. Thus the gene flow potential of a species will be expected to have a major

Table 9.4. Estimates of outcrossing rate in natural and disturbed plant populations of animal-pollinated, mixed mating, rare and endangered Western Australian flora. t_m is the multilocus outcrossing rate. Outcrossing rates for disturbed populations often differ substantially from those of undisturbed populations. In some species there appears to be a trend to reduced outcrossing in disturbed populations (*Banksia cuneata, Eucalyptus rhodantha, Lambertia orbifolia*) while in another there is no clear trend (*Verticordia fimbrilepis*).

Population/size	Population status	t_m
Banksia cuneata[a]		
1. 56	High disturbance	0.67 (0.04)
2. 40	Undisturbed	0.95 (0.05)
3. 86	Low disturbance	0.76 (0.05)
4. 120	Low disturbance	0.88 (0.07)
Banksia triscuspis[a]		
1. 1	Remnant	0.77 (0.01)
2. 3	Undisturbed	0.74 (0.01)
3. 5	Undisturbed	0.92 (0.01)
4. 350	Undisturbed	0.84 (0.05)
5. 108	Undisturbed	0.79 (0.04)
6. 24	Remnant	1.02 (0.04)
7. 4,140	Burnt	0.69 (0.04)
8. 89	Undisturbed	0.71 (0.02)
Eucalyptus rameliana[a]		
1. 83	Undisturbed	0.84 (0.03)
2. 57	Undisturbed	0.48 (0.05)
3. 200	Undisturbed	0.97 (0.03)
		0.96 (0.02)[d]
Eucalyptus rhodantha[a]		
1. 180	Undisturbed remnant	0.59 (0.04)
		0.67 (0.05)[d]
2. 14	High disturbance	0.26 (0.05)
Lambertia orbifolia[b]		
1. 483	Undisturbed	0.53 (0.08)
2. 250	Undisturbed	0.72 (0.10)
3. 100	Low disturbance	0.57 (0.11)
4. 56	High disturbance	0.41 (0.08)
Verticordia fimbrilepis[c]		
1. 90	High disturbance	0.62 (0.07)
2. 305	Undisturbed	0.52 (0.13)
3. 796	High disturbance	0.70 (0.13)
4. 59,270	Undisturbed	0.73 (0.10)

[a]Data for these species are from Sampson *et al.* (1996).
[b]Data from Coates and Hamley (1999).
[c]Data from Sampson (personal communication).
[d]Within population studies.
Standard errors in parentheses.

influence on the partitioning of genetic variation among plant populations and levels of genetic variation within populations. The review of allozyme data in plants by Hamrick and Godt (1989) clearly supports this expectation based on their analysis of the relationship between G_{ST} and H_{ep}, and species attributes, summarized in Table 9.3.

Limited gene flow may lead to divergence between populations as a consequence of genetic drift or differential selection in local environments. Whereas high levels of gene flow enhance homogenization between gene pools and increase effective population size, they may also result in the establishment of unfavourable alleles where populations are differentially adapted to localized conditions. In the latter case, it has been suggested that gene flow may result in outbreeding depression (see Templeton, 1986; Fenster and Dudash, 1994), where the progeny of distant matings may be less fit than the progeny of near or neighbour matings. Genetic causes of outbreeding depression may relate to the external environment or to intrinsic factors involving critical interactions between groups of genes or coadapted gene complexes. For example, F_1 hybrids may not be adapted to either parental environment because heterozygosity at major single genes, important in local adaptation, results in reduced survivorship or reproduction. Alternatively reduced fitness in F_1 hybrids may relate to the break up of coadapted gene complexes (see Fenster and Dudash, 1994). Waser (1993) provides evidence for outbreeding depression in a number of different plant species following crossing between geographically distant populations. In contrast to these findings other studies have found that gene migration between populations or subpopulations may result in increased fitness or heterosis (Ledig, 1986).

Gene flow, like the mating system, is heavily influenced by the pollination system of the target species but importantly it is also influenced by seed dispersal. Although various indirect measures of gene flow such as pollinator flight distances, seed dispersal, dye and pollen analogue dispersal, genetic diversity statistics and distribution of alleles among populations (see Broyles et al. 1994; Schnabel and Hamrick, 1995) are useful in comparative studies, and as indicators of potential gene flow within and between populations, they do not necessarily provide a true representation of effective gene movement. As pointed out by Levin (1981), there is good reason to believe that dispersal data

underestimate gene flow. For example, in *Chamaecrista fasciculata*, despite the estimates of limited gene dispersal based on pollen and seed dispersal events, decreased fitness of progeny from near-neighbour matings and increased fitness from more distant matings suggest that gene flow will be more widespread (Fenster, 1991). In addition, studies on the distribution of genetic variation based on molecular markers indicate that pollen-mediated gene flow can be far greater than expected from direct estimates of pollen dispersal. For example, in the bee-pollinated *Lupinus texensis*, Schaal (1980) demonstrated the effects of pollen carry over with significantly greater patterns of gene movement estimated from isozyme markers than pollen movement.

More recently, direct estimates of the dispersal of genes via pollen have been developed through the application of molecular markers in paternity analysis and parentage assignment. These approaches involve comparisons of segregating alleles in parental and progeny cohorts and have been developed to determine paternal contributions through exclusion techniques, maximum-likelihood estimates or a combination of these methods using fractional paternity (see Devlin et al., 1988). Evidence from a number of paternity studies indicates that there can be significant gene flow into a population from outside sources (see Broyles et al., 1994; Schnabel and Hamrick, 1995). Estimates of pollen from outside sources based on allozyme studies can be as high as 50% in the perennial herb *Asclepias exaltata* (Broyles et al., 1994) and 30% in the leguminous tree *Gleditsia triacanthos* (Schnabel and Hamrick, 1995). Similarly, estimates based on microsatellite markers indicate comparable levels in some tropical tree species (Chase et al., 1996).

While gene flow in plants has been largely investigated in terms of pollen flow, often using nuclear markers, far less attention has been given to gene flow mediated by seed dispersal. Although seed movement can clearly make a significant contribution to gene flow, it has proved difficult to measure and is not well understood in relation to population genetic variation in plants. The availability of cpDNA markers combined with the development of suitable theoretical methodologies (see Ennos *et al.*, 1999) has provided a basis for addressing this important issue in plant population studies. Ennos *et al.* (1999) show that, by the joint measuring of genetic differentiation for nuclear and cpDNA markers, the ratio of pollen to seed flow among populations can be readily estimated and compared across various plant species.

Geographical Patterns of Genetic Variation among Plant Populations

Geographical patterns of intraspecific genetic variation have been of fundamental interest to plant evolutionary biologists and population geneticists as they are often assumed to represent the initiation of events that will lead to independent evolutionary lineages and allopatric speciation. This information is also of particular importance to those involved in the management and conservation of genetic resources either for species of commercial interest or for species targeted in conservation programmes. For example, strategies for the collection and maintenance of genetic resources of commercial crops place considerable reliance on understanding the pattern and distribution of genetic variation among source populations across different eco-geographic regions. The value of these data in the management of crop genetic resources is reflected in the large number of species for which variation in the genetic structure of landrace populations of crops and wild relatives of domesticated plants has been investigated using isozyme analysis (see Frankel *et al.*, 1995).

Studies of intraspecific geographical patterns of genetic variation have been used to investigate the wide array of factors that may be important as evolutionary determinants. These include historical associations among populations; the role of selection, gene flow and drift; the development of barriers to gene flow among populations; and spatial variation in mating systems (Thompson, 1999). Although studies on intraspecific patterns of genetic variation in plants are numerous there are certain key areas of investigation that highlight the many combinations of historical, demographic and ecological processes that may generate such patterns. These are investigated in the following sections.

Disjunct distributions and differentiation

Disjunct population systems are common in plant species in many parts of the world and may reflect geological and edaphic complexities, localized extinction events, contraction to refugia associated with climate change, and island isolation associated with changes in sea levels. Genetic studies of these population systems can not only give indications of the level and patterns of divergence among populations but can also give valuable clues to the significance of these processes in the evolution of individual plant species and groups of plant species across regions. For example, the significant divergence among populations of intercontinental acacias found in northern Australia, such as *Acacia aulacocarpa* (G_{ST} = 0.626; McGranahan *et al.*, 1997), probably reflects a wider distribution on the Australian geological plate in the Tertiary followed by sea-level changes, geographic separation and contractions due to cycles of aridity during the Quaternary. Climatic fluctuations during the Quaternary, particularly the Pleistocene, have also been important in the evolution of other tree species such as *Abies* firs in Mexico and Guatemala. Here, the significant differentiation among populations appears to be associated with the increased isolation of populations as they retreated upwards during the Pleistocene glaciation and the warming period that followed (Aguirre-Planter *et al.*, 2000).

Some floras found in regions with Mediterranean climates, such as the flora of south-west Australia and the Mediterranean flora, are typically rich with closely related disjunct species or species with disjunct populations (Thompson, 1999; Coates, 2000). In both regions, disjunct distributions appear to be associated with a number of key geohistorical phenomena. In south-west Australia these include the existence of marine, edaphic and climatic barriers that have separated this area from the rest of Australia since the Eocene; the formation of nutrient deficient sands and laterites favouring a shrubland flora that could readily adapt to the increasing aridity of the late Tertiary and Quaternary; and climatic and landscape instability in the transitional rainfall zone (Hopper, 1979). In the Mediterranean flora, however, disjunct distributions have been associated with the geological complexity of the Mediterranean basin; movement and isolation of tectonic microplates during the Tertiary; island isolation with change in sea levels; and dispersal (Thompson, 1999).

Many species in the south-west Australian flora are likely to be relictual and probably had wider, more continuous distributions during favourable climatic regimes up to the early Pleistocene. Following the increased aridity and climatic instability during the Pleistocene, these taxa have become locally extinct but have survived as disjunct remnants, particularly through the semi-arid transitional rainfall zone. Recent gene flow between these disjunct population groups, either by long distance seed dispersal or pollen movement, has probably been limited or absent for long periods. As a consequence, significant genetic differentiation between populations is typical of many species and is particularly evident in rare and geographically restricted species. Relatively high levels of population differentiation have been reported for 22 animal-pollinated, mainly outcrossing, taxa with disjunct population systems. These taxa cover a range of south-west Australian genera including long-lived woody shrubs and trees, and herbaceous perennials (Coates, 2000). Thompson (1999) describes similar patterns of divergence among disjunct populations of species in the Mediterranean flora.

A typical example of these patterns in south-west Australia can be found in *Lambertia orbifolia*, a large, bird-pollinated woody shrub known only from seven populations that have a significant disjunct distribution (Fig. 9.2). Allozyme studies show that the genetic divergence between all popula-

tions is very high ($F_{ST} = 0.441$). Phylogenetic analyses based on either gene frequency data or genetic distance give identical tree topologies and indicate that the two disjunct population groups are separate evolutionary lineages. The analysis of cpDNA variation confirms this conclusion (Byrne *et al.*, 1999). Inferences based on the proposed long-term effects of Pleistocene climate change on the south-west Australian flora suggest that the current population genetic structure in *L. orbifolia* is the result of local extinction of intervening populations, and extended isolation of the two remnants (Coates and Hamley, 1999). This is supported by studies on large endemic forest eucalypts in areas between the two population groups that show patterns of local extinction and range contraction due to climate change (Wardell-Johnson and Coates, 1996).

Historical associations and phylogeographic patterns

Previous mention has been made of the influence of historical events on patterns of genetic variation among disjunct plant populations. In those cases, the opportunity for contemporary gene flow among populations is low and any phylogenetic similarity is probably due to common ancestry rather than any ongoing process of genetic exchange. Yet in more continuous population systems it is far more difficult to assess the significance of contemporary patterns of gene flow versus genetic similarity due to recent common ancestry. Most studies of inter-population genetic variation in plants are based on allele frequency data from markers such as allozymes, nuclear RFLPs, RAPDs, AFLPs and microsatellites where genetic change over time cannot be directly inferred (Schaal and Olsen, 2000). As pointed out by Schaal and Olsen (2000), estimates of genetic exchange and the analysis of population genetic structure are generally based on models that assume equilibrium between genetic drift and gene flow. When investigating determinants of geographical patterns of genetic variation such estimates based on, for example, F statistics (Wright,

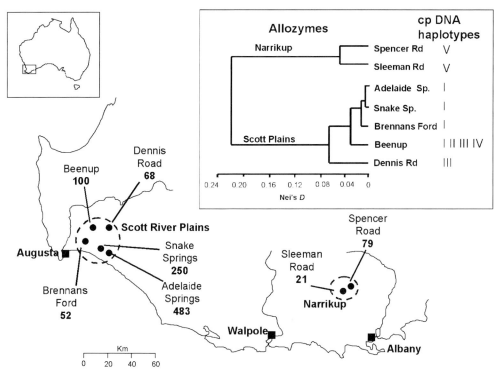

Fig. 9.2. Phylogenetic relationships based on UPGMA of allozyme data and geographical distribution of disjunct populations (Scott River Plains and Narrikup) of *Lambertia orbifolia*. Population size (number of reproductively mature plants) is given after each population name. The analysis of cpDNA variation based on RFLPs detected eight mutations distributed over five haplotypes (I–V) with the Narrikup populations distinguished by a single haplotype characterized by six unique mutations.

1951) do not distinguish between historical effects and contemporary patterns of gene flow. Furthermore, gene flow drift equilibrium is generally considered to be an unlikely scenario for most plant populations, although it may be more prevalent in ancient floras where metapopulation systems within species may have remained relatively stable over long periods of time (see Coates *et al.*, 2003).

Phylogeography provides an approach that potentially allows discrimination between historical and contemporary patterns of gene exchange (Schaal *et al.*, 1998). As a sub-discipline of biogeography, phylogeography involves the analysis of the geographical distribution of genealogical lineages and focuses on the assessment of historical factors as determinants of evolutionary patterns among populations (Avise, 2000). Chloroplast DNA has proved to be an extremely useful source of variation for phylogeographic studies in a number of plant species. The lack of recombination and uniparental mode of inheritance in cpDNA permits genealogical relationships to be followed across populations and the delimitation of phylogeographically distinct populations. Unfortunately, as mentioned previously, cpDNA may have limited application in some cases because insufficient variation is present to allow geographical patterns to be detected. In particular, the variance in cpDNA diversity is likely to be large, with some species showing little diversity due to recent selective sweeps (see Ennos *et al.*, 1999).

A number of studies based primarily on cpDNA variation have now been published describing common plant phylogeographic patterns for certain geographic regions.

Probably the best examples to date are those that show postglacial migration routes from Pleistocene refugia for multiple plant species in both Europe and north-west America. Phylogeographic analyses in European oak species indicate that there are three main haplotype lineages with postglacial migration inferred to start from three distinct southerly refugia in the Balkan, Iberian and Italian peninsulas (see Ferris *et al.*, 1999). Similar patterns of cpDNA haplotype variability have been shown for two other European tree species, beech (*Fagus sylvatica*; Demesure *et al.*, 1996) and alder (*Alnus glutinosa*; King and Ferris, 1998); both studies indicate a single glacial refugium in the Carpathians. Phylogeographic studies based on cpDNA variation in six plant species from the Pacific Northwest also show patterns concordant with postglacial colonization. Here the cpDNA phylogenies for five vascular plants and one fern, representing a range of different life-histories, indicate two clades of populations with a north–south separation. This separation has also been supported by population genetic studies on other plant species (Soltis *et al.*, 1997).

Recent phylogeographic studies on species in south-west Australia indicate some commonality in geographical patterns although the findings here are generally more complex and appear to reflect a much more ancient pattern of population evolution associated with climatic instability since the late Tertiary. Studies on two species based on cpDNA variation, *Santalum spicatum* (Western Australian sandalwood) and *Eucalyptus loxophleba*, indicate two geographically distinct haplotype lineages showing a north–south separation. In both species, a nested clade analysis inferred past fragmentation as the most likely cause of the differentiation between the lineages (Fig. 9.3). The level of sequence divergence between lineages was similar in both species and suggests a mid-Pleistocene timeframe for this divergence. This shared phylogeographic pattern is consistent with a hypothesis of significant climatic fluctuations during the Pleistocene and suggests that such climatic instability has resulted in significant fragmentation events in the flora of this region.

Evolutionarily significant units and conservation

An important factor in assessing priorities for the conservation and management of species involves understanding patterns of intraspecific genetic variation to identify populations which may be critical for the conservation of genetic resources and evolutionary processes (Hopper and Coates, 1990; Newton *et al.*, 1999). Although species are generally accepted as primary units for conservation, existing taxonomies may not adequately consider intraspecific variation. The concept of the 'evolutionarily significant unit' (ESU) has subsequently been introduced to deal with groups of populations that warrant separate management for conservation (Ryder, 1986). The concept is based on a sound understanding of the evolutionary significance of geographically based genetic variation within a species, with ESUs generally considered to be geographically discrete.

Criteria for defining an ESU have included significant divergence of allele frequencies, specific levels of genetic distance and phylogenetic differences based on certain genes. A specific approach outlined by Moritz (1994) defined an ESU as an historically isolated and independently evolving set of populations. In animals this was regarded as populations showing reciprocal monophyly for mitochondrial DNA (mtDNA) alleles with significant divergence of alleles at nuclear loci. As mentioned previously, using organelle DNA, such as mtDNA or cpDNA, permits genealogical relationships to be followed across populations and the delimitation of phylogeographically distinct population groups and ESUs. For example, the array of different cpDNA lineages found in various European tree species has been considered indicative of separate ESUs. Perhaps more importantly, however, these studies have highlighted the value of refugial areas in southern Europe. These are considered key areas for genetic resource conservation because they contain many unique haplotypes (Newton *et al.*, 1999).

Although defining ESUs based on cpDNA variation and phylogeographic analysis has potential in plant conservation

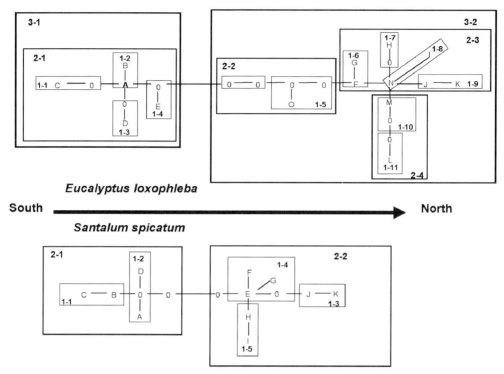

Fig. 9.3. Haplotype network showing nested clades for *Eucalyptus loxophleba* and *Santalum spicatum*. Haplotypes are represented by letters. Interior haplotypes not detected in the sample are represented by 0. Each line connecting haplotypes represents a single mutational change. One-step clades are indicated by thin-lined boxes, two- and three-step clades by heavier-lined boxes. Differentiation of the three-step clades in *E. loxophleba* and two-step clades in *S. spicatum* represent distinct lineages in each species and show a shared phylogeographic pattern indicating a north–south geographic separation. Inferences from a nested clade analysis identified past fragmentation as the most likely cause of the significant geographic association of genealogical lineages for each species. Application of a molecular clock indicates a mid-Pleistocene split and that the past fragmentation is most likely due to significant climate instability during that period in the south-west Australian region (Byrne *et al.*, 2003; Byrne and Hines, 2004).

genetics (see Newton *et al.*, 1999), it may well be limited, given the low level of cpDNA variation found in some plant groups. Where cpDNA is restricted, it will probably be necessary to define conservation units in terms of contemporary population genetic structure based on polymorphic nuclear markers, such as allozymes, RAPDs or microsatellites, rather than historical population structure inferred from phylogeographic analyses. Whilst these data lack genealogical information and therefore have limitations in phylogeographic analyses, significant divergence in allele frequencies at nuclear loci, combined with phylogenetic analysis of gene frequency data, may still be extremely valuable in identifying conservation units in plant taxa (Coates, 2000).

The example given previously (*L. orbifolia*, Fig. 9.2) clearly indicates the value of such information in setting conservation priorities. Significant genetic divergence among disjunct populations based on nuclear genes (allozymes) and cpDNA indicates two ESUs. *L. orbifolia* has an International Union for Conservation of Nature and Natural Resources (IUCN) and Western Australian ranking of endangered, but with the large number of critically endangered plants in that State it has a relatively low priority for conservation and recovery actions. Recently, the two ESUs were recognized as distinct conservation units and listed separately under Western Australian State legislation. Small population size, increased inbreeding and threats posed by disease and habitat degradation indicate that the Narrikup ESU (Fig. 9.2) is critically endangered (Coates and Hamley, 1999). Subsequently, this ESU was targeted for a major re-introduction programme with a new population established on a nearby nature reserve. This population of some 500 plants is now successfully recruiting from seed produced by the plants originally established on the site (L. Monks, Perth, 2003, personal communication).

Despite the value of population genetic and phylogeographic studies in informing genetic resource conservation and setting priorities for conservation, caution is needed in interpreting the results from such studies. Genetic differences based on molecular markers may be indicative of adaptive differ-entiation among populations but this will certainly not always be the case. Likewise, lack of any molecular marker differentiation does not necessarily allow one to infer a lack of adaptive differences. In addition, as pointed out by Newton et al. (1999), each marker has its limitations. While cpDNA has proved to be extremely valuable, its relatively slow mutation rate may mean that intraspecific patterns of variation based on shared haplotypes may be confounded by much more ancient relationships perhaps covering timeframes of millions of years. In that case, phylogeographic patterns may not be readily interpretable in relation to the recent history of the populations and geographical patterns (see Ennos et al., 1999). It would be more appropriate in these situations to combine the cpDNA analyses with studies of genetic relatedness based on nuclear markers that may help elucidate the influence of recent population history on current population genetic structure.

References

Aguirre-Planter, E., Furnier, G.R. and Eguiarte, L.E. (2000) Low levels of genetic variation and high levels of genetic differentiation among populations of species of *Abies* from southern Mexico and Guatemala. *American Journal of Botany* 87, 362–371.

Aide, T.M. (1986) The influence of wind and animal pollination on variation in outcrossing rates. *Evolution* 40, 434–435.

Arnold, C., Rosetto, M., McNally, J. and Henry, R.J. (2002) The application of SSRs characterized for grape (*Vitis vinifera*) to conservation studies in Vitaceae. *American Journal of Botany* 89, 22–28.

Arnold, M.L. and Imms, S.K. (1998) Molecular markers, gene flow, and natural selection. In: Soltis, D.E., Soltis, P.S. and Doyle, J.J. (eds) *Molecular Systematics of Plants. II. DNA Sequencing.* Kluwer Academic Publishers, Boston, Massachusetts, pp. 442–458.

Avise, J. (2000) *Phylogeography.* Harvard University Press, Harvard, Massachusetts.

Barrett, S.C.H. and Eckert, C.G. (1990) Variation and evolution of mating systems in seed plants. In: Kawano, S. (ed.) *Biological Approaches and Evolutionary Trends in Plants.* Academic Press, London, pp. 229–254.

Barrett, S.C.H. and Kohn, J.R. (1991) Genetics and evolutionary consequences of small population size in plants: implications for conservation. In: Falk, D.A. and Holsinger, K.E. (eds) *Genetics and Conservation of Rare Plants.* Oxford University Press, New York, pp. 3–30.

Brown, A.H.D. (1989) Genetic characterization of mating systems. In: Brown, A.H.D., Clegg, M.T., Kahler, A.L. and Weir, B.S. (eds) *Plant Population Genetics, Breeding and Genetic Resources.* Sinauer Associates, Sunderland, Massachusetts, pp. 145–162.

Broyles, S.B., Schnabel, A. and Wyatt, R. (1994) Evidence for long-distance pollen dispersal in milkweeds (*Asclepias exaltata*). *Evolution* 48, 1032–1040.

Bussell, J.D. (1999) The distribution of random amplified polymorphic DNA (RAPD) diversity amongst populations of *Isotoma petraea* (Lobeliaceae). *Molecular Ecology* 8, 775–789.

Butcher, P.A., Decroocq, S., Gray, Y. and Moran, G.F. (2000) Development, inheritance and cross species amplification of microsatellite markers from *Acacia mangium*. *Theoretical and Applied Genetics* 101, 1282–1290.

Byrne, M. and Hines, B. (2004) Phylogeographical analysis of cpDNA variation in *Eucalyptus loxophleba* (Myrtaceae). *Australian Journal of Botany* 52, 459–470.

Byrne, M., Parrish, T.L. and Moran, G.F. (1998) Nuclear RFLP diversity in *Eucalyptus nitens*. *Heredity* 81, 225–233.

Byrne, M., Macdonald, B. and Coates, D.J. (1999) Divergence in the chloroplast genome and nuclear rDNA of the rare Western Australian plant *Lambertia orbifolia* (Proteaceae). *Molecular Ecology* 8, 1789–1796.

Byrne, M., Macdonald, B. and Brand, J. (2003) Phylogeography and divergence in the chloroplast genome of Western Australian Sandalwood (*Santalum spicatum*). *Heredity* 91, 389–395.

Charlesworth, D. and Charlesworth, B. (1995) Quantitative genetics in plants: the effect of the breeding system on genetic variability. *Evolution* 49, 911–920.

Chase, M.R., Moller, C., Kesseli, R. and Bawa, K.S. (1996) Distant gene flow in tropical trees. *Nature* 383, 398–399.

Clegg, M.T. (1989) Molecular diversity in plant populations. In: Brown, A.H.D., Clegg, M.T., Kahler, A.L. and Weir, B.S. (eds) *Plant Population Genetics, Breeding and Genetic Resources*. Sinauer Associates, Sunderland, Massachusetts, pp. 98–115.

Coates, D.J. (1988) Genetic diversity and population genetic structure in the rare Chittering grass wattle, *Acacia anomala* Court. *Australian Journal of Botany* 36, 273–286.

Coates, D.J. (2000) Defining conservation units in a rich and fragmented flora: implications for the management of genetic resources and evolutionary processes in south-west Australia. *Australian Journal of Botany* 48, 329–339.

Coates, D.J. and Hamley, V.L. (1999) Genetic divergence and mating systems in the endangered and geographically restricted species *Lambertia orbifolia* Gardner (Proteaceae). *Heredity* 83, 418–427.

Coates, D.J., Carstairs, S. and Hamley, V.L. (2003) Evolutionary patterns and genetic structure in localized and widespread species in the *Stylidium caricifolium* complex (Stylideaceae). *American Journal of Botany* 90, 997–1008.

Demesure, B., Comps, B. and Petit, R.J. (1996) Chloroplast DNA phylogeography of the common beech (*Fagus sylvatica* L.) in Europe. *Evolution* 50, 2515–2520.

Devlin, B., Roeder, K. and Ellstrand, N.C. (1988) Fractional paternity assignment: theoretical development and comparison to other methods. *Theoretical and Applied Genetics* 76, 369–380.

Eckert, C.G. and Barrett, S.C.H. (1993) Clonal reproduction and patterns of genotypic diversity in *Decodon verticillatus* (Lythraceae). *American Journal of Botany* 80, 1175–1182.

Ellstrand, N.C. and Elam, D.R. (1993) Population genetic consequences of small population size: implications for plant conservation. *Annual Review of Ecology and Systematics* 24, 217–242.

Ellstrand, N.C. and Roose, M.L. (1987) Patterns of genotypic diversity in clonal plant species. *American Journal of Botany* 74, 123–131.

Ennos, R.A., Sinclair, W.T., Hu, X.-S. and Langdon, A. (1999) Using organelle markers to elucidate the history, ecology and evolution of plant populations. In: Hollingsworth, P.M., Bateman, R.M. and Gornall, R.J. (eds) *Molecular Systematics and Plant Evolution*. Taylor & Francis, London, pp. 1–19.

Falk, D.A. and Holsinger, K.E. (1991) *Genetics and Conservation of Rare Plants*. Oxford University Press, New York.

Fenster, C.B. (1991) Gene flow in *Chamaecrista fasciculata* (Leguminosae) II. Gene establishment. *Evolution* 45, 410–422.

Fenster, C.B. and Dudash, M.R. (1994) Genetic considerations for plant population restoration and conservation. In: Bowles, M.L. and Whelan, C.J. (eds) *Restoration of Endangered Species*. Cambridge University Press, Cambridge, pp. 34–62.

Ferris, C., King, R.A. and Hewitt, G.M. (1999) Isolation within species and the history of glacial refugia. In: Hollingsworth, P.M., Bateman, R.M. and Gornall, R.J. (eds) *Molecular Systematics and Plant Evolution*. Taylor & Francis, London, pp. 20–34.

Fiedler, P.L. and Ahouse, J.J. (1992) Hierarchies of cause: toward an understanding of rarity in vascular plant species. In: Fiedler, P.L. and Jain, S.K. (eds) *Conservation Biology: the Theory and Practice of Nature Conservation*. Chapman & Hall, New York, pp. 23–48.

Frankel, O.H., Brown, A.H.D. and Burdon, J.J. (1995) *The Conservation of Plant Biodiversity*. Cambridge University Press, Cambridge.

Gaston, K.J. (1997) What is rarity? In: Kunin, W.E. and Gaston, K.J. (eds) *The Biology of Rarity*. Chapman & Hall, London, pp. 30–41.

Gitzendanner, M.A. and Soltis, P.S. (2000) Patterns of genetic variation in rare and widespread congeners. *American Journal of Botany* 87, 783–792.

Godt, M.J.W. and Hamrick, J.L. (1999) Genetic divergence among infraspecific taxa of *Sarracenia purpurea*. *Systematic Botany* 23, 427–438.

Gottlieb, L.D. (1981) Electrophoretic evidence and plant populations. *Progress in Phytochemistry* 7, 1–46.

Gottlieb, L.D., Warwick, S.I. and Ford, V.S. (1985) Morphological and electrophoretic divergence between *Layia discoidea* and *L. glandulosa*. *Systematic Botany* 10, 133–148.

Hamrick, J.L. (1983) The distribution of genetic variation within and among natural plant populations. In: Schonewald-Cox, C.M., Chambers, S.M., MacBryde, B. and Thomas, L. (eds) *Genetics and Conservation*. Benjamin-Cummings, London, pp. 343–348.

Hamrick, J.L. and Godt, M.J.W. (1989) Allozyme diversity in plant species. In: Brown, A.H.D., Clegg, M.T., Kahler, A.L. and Weir, B.S. (eds) *Plant Population Genetics, Breeding and Genetic Resources*. Sinauer Associates, Sunderland, Massachusetts, pp. 43–63.

Hamrick, J.L., Godt, M.J.W. and Sherman-Broyles, S.L. (1992) Factors influencing levels of genetic diversity in woody plant species. *New Forests* 6, 95–124.

Harris, S.A. (1999) RAPDs in systematics – a useful methodology? In: Hollingsworth, P.M., Bateman, R.M. and Gornall, R.J. (eds) *Molecular Systematics and Plant Evolution*. Taylor & Francis, London, pp. 211–228.

Hoebee, S.E. and Young, A.G. (2001) Low neighbourhood size and high interpopulation differentiation in the endangered shrub *Grevillea iaspicula* McGill (Proteaceae). *Heredity* 86, 489–496.

Hopper, S.D. (1979) Biogeographical aspects of speciation in the southwest Australian flora. *Annual Review of Systematics and Ecology* 10, 399–422.

Hopper, S.D. and Coates, D.J. (1990) Conservation of genetic resources in Australia's flora and fauna. *Proceedings of the Ecological Society of Australia* 16, 567–577.

Jarne, P. and Lagoda, P.J.L. (1996) Microsatellites, from molecules to populations and back. *Trends in Ecology and Evolution* 11, 424–429.

Karron, J.D. (1987) A comparison of levels of genetic polymorphism and self-compatibility in geographically restricted and widespread plant congeners. *Evolutionary Ecology* 1, 47–58.

Karron, J.D. (1991) Patterns of genetic variation and breeding systems in rare plant species. In: Falk, D.A. and Holsinger, K.E. (eds) *Genetics and Conservation of Rare Plants*. Oxford University Press, New York, pp. 87–98.

Karron, J.D., Linhart, Y.B., Chaulk, C.A. and Robertson, C.A. (1988) Genetic structure of populations of geographically restricted and widespread species of *Astragalus* (Fabaceae). *American Journal of Botany* 75, 1114–1119.

Kennington, W.J. and James, S.H. (1998) Allozyme and morphometric variation in two closely related mallee species from Western Australia, *Eucalyptus argutifolia* and *E. obtusifolia* (Myrtaceae). *Australian Journal of Botany* 46, 173–186.

King, R.A and Ferris, C. (1998) Chloroplast DNA phylogeography of *Alnus glutinosa* (L.) Gaertn. *Molecular Ecology* 7, 1151–1161.

Ledig, F.T. (1986) Heterozygosity, heterosis, and fitness in outbreeding plants. In: Soulé, M.E. (ed.) *Conservation Biology*. Sinauer Associates, Sunderland, Massachusetts, pp. 77–104.

Levin, D.A. (1981) Dispersal versus gene flow in plants. *Annals of the Missouri Botanical Gardens* 68, 233–253.

Lewis, P.O. and Crawford, D.J. (1995) Pleistocene refugium endemics exhibit greater allozymic diversity than widespread congeners in the genus *Polygonella* (Polygonaceae). *American Journal of Botany* 82, 141–149.

Loveless, M.D. and Hamrick, J.L. (1988) Genetic organization and evolutionary history in two American species of *Cirsium*. *Evolution* 42, 254–265.

Lynch, M. (1996) A quantitative-genetic perspective on conservation issues. In: Avise, J.C. and Hamrick, J.L. (eds) *Conservation Genetics: Case Histories from Nature*. Chapman & Hall, New York, pp. 471–501.

Lynch, M. and Milligan, B.G. (1994) Analysis of population genetic structure with RAPD markers. *Molecular Ecology* 3, 91–99.

Lynch, M., Pfrender, M., Spitze, K., Lehman, N., Hicks, J., Allen, D., Latta, L., Ottene, M., Bogue, F. and Colbourne, J. (1999) The quantitative and molecular genetic architecture of a subdivided species. *Evolution* 53, 100–110.

McGranahan, M., Bell, J.C., Moran, G.F. and Slee, M. (1997) High genetic divergence between geographic regions in the highly outcrossing species *Acacia aulacocarpa* (Cunn. ex Benth.). *Forest Genetics* 4, 1–13.

McKay, J.K. and Latta, R.G. (2002) Adaptive population divergence: markers, QTL and traits. *Trends in Ecology and Evolution* 17, 285–291.

Maguire, T.L., Peakall, R. and Saenger, P. (2002) Comparative analysis of genetic diversity in the mangrove species *Avicennia marina* (Forsk.) Vierh. (Avicenniaceae) detected by AFLPs and SSRs. *Theoretical and Applied Genetics* 104, 388–398.

Manel, S., Schwartz, M.K., Luikart, G. and Taberlet, P. (2003) Landscape genetics: combining landscape ecology and population genetics. *Trends in Ecology and Evolution* 18, 189–197.

Merila, J. and Crnokrak, P. (2001) Comparison of genetic differentiation at marker loci and quantative traits. *Journal of Evolutionary Biology* 14, 892–903.

Moritz, C. (1994) Defining 'evolutionarily significant units' for conservation. *Trends in Ecology and Evolution* 9, 373–375.

Mueller, U.G. and Wolfenbarger, L.L. (1999) AFLP genotyping and fingerprinting. *Trends in Ecology and Evolution* 14, 389–393.

Newton, A.C., Allnutt, T.R., Gillies, A.C.M., Lowe, A.J. and Ennos, R.A. (1999) Molecular phylogeography, intraspecific variation and the conservation of tree species. *Trends in Ecology and Evolution* 14, 140–145.

Nybom, H. and Bartish, I.V. (2000) Effects of life history traits and sampling strategies on genetic diversity estimates obtained with RAPD markers in plants. *Perspectives in Plant Ecology, Evolution and Systematics* 3, 93–114.

Peakall, R., Smouse, P.E. and Huff, D.R. (1995) Evolutionary implications of allozyme and RAPD variation in diploid populations of dioecious buffalograss *Buchloë dactloides*. *Molecular Ecology* 4, 135–147.

Peakall, R., Gilmore, S., Keys, W., Morgante, M. and Rafalski, A. (1998) Cross-species amplification of soybean (*Glycine max*) simple sequence repeats (SSRs) within the genus and other legume genera: implications for the transferability of SSRs in plants. *Molecular Biology and Evolution* 15, 1275–1287.

Powell, W., Morgante, M., NcDevitt, R., Vendramin, G.G. and Rafalski, J.A. (1995) Polymorphism simple sequence repeat regions in chloroplast genomes: applications to population genetics of pines. *Proceedings of the National Academy of Sciences USA* 92, 7759–7763.

Provan, J., Soranzo, N., Wilson, N.J., McNicol, J.W., Morgante, M. and Powell, W. (1999) The use of uni-parentally inherited simple sequence repeat markers in plant population studies and systematics. In: Hollingsworth, P.M., Bateman, R.M. and Gornall, R.J. (eds) *Molecular Systematics and Plant Evolution*. Taylor & Francis, London, pp. 35–50.

Raybould, A.F., Mogg, R.J. and Clarke, R.T. (1996) The genetic structure of *Beta vulgaris* ssp. *maritima* (sea beet) populations: RFLPs and isozymes show different patterns of gene flow. *Heredity* 77, 245–250.

Reed, D.H. and Frankham, R. (2001) How closely correlated are molecular and quantitative measures of genetic variation? A meta-analysis. *Evolution* 55, 1095–1103.

Richards, A.J. (1997) *Plant Breeding Systems*, 2nd edn. Chapman & Hall, London.

Rieseberg, L.H. (1996) Homology among RAPD fragments in interspecific comparisons. *Molecular Ecology* 5, 99–105.

Ryder, O.A. (1986) Species conservation and systematics: the dilemma of subspecies. *Trends in Ecology and Evolution* 1, 9–10.

Sampson, J.F., Coates, D.J. and van Leeuwen, S.J. (1996) Mating system variation in animal-pollinated rare and endangered plant populations in Western Australia. In: Hopper, S.D., Chappill, J.A., Harvey, M.S. and George, A.S. (eds) *Gondwanan Heritage: Past, Present and Future of the Western Australian Biota*. Surrey Beatty and Sons, Chipping Norton, Australia, pp. 187–195.

Schaal, B.A. (1980) Measurement of gene flow in *Lupinus texensis*. *Nature* 284, 450–451.

Schaal, B.A. and Olsen, K.M. (2000) Gene genealogies and population variation in plants. *Proceedings of the National Academy of Sciences USA* 97, 7024–7029.

Schaal, B.A., Leverich, W.J. and Rogstad, S.H. (1991) A comparison of methods for assessing genetic variation in plant conservation biology. In: Falk, D.A. and Holsinger, K.E. (eds) *Genetics and Conservation of Rare Plants*. Oxford University Press, New York, pp. 123–134.

Schaal, B.A., Hayworth, D.A., Olsen, K.M., Rauscher, J.T. and Smith, W.A. (1998) Phylogeographic studies in plants: problems and prospects. *Molecular Ecology* 7, 465–475.

Schemske, D.W. and Lande, R. (1985) The evolution of self-fertilization and inbreeding depression in plants. II. Empirical observations. *Evolution* 39, 41–52.

Schnabel, A. and Hamrick, J.L. (1995) Understanding the population genetic structure of *Gleditsia triacanthos* L.: the scale and pattern of pollen gene flow. *Evolution* 49, 921–931.

Schoen, D.J. and Brown, A.H.D. (1991) Intraspecific variation in population gene diversity and effective population size correlates with the mating system in plants. *Proceedings of the National Academy of Sciences USA* 88, 4494–4497.

Soltis, D.E. and Soltis, P.S. (1989) *Isozymes in Plant Biology*. Dioscorides Press, Portland, Oregon.

Soltis, D.E., Gitzendanner, M.A., Strenge, D.D. and Soltis, P.S. (1997) Chloroplast DNA intraspecific phylogeography of plants from the Pacific Northwest of North America. *Plant Systematics and Evolution* 206, 353–373.

Squirrell, J., Hollingsworth, P.M., Woodhead, M., Russell, J., Lowe, A.J., Gibby, M. and Powell, W. (2003) How much effort is required to isolate microsatellites from plants? *Molecular Ecology* 12, 1339–1348.

Stöcklin, J. (1999) Differences in life history traits of related *Epilobium* species: clonality, seed size and seed number. *Folia Geobotanica and Phytotaxonomica Praha* 34, 7–18.

Sydes, M.A. and Peakall, R. (1998) Extensive clonality in the endangered shrub *Haloragodendron lucasii* (Haloragaceae) revealed by allozymes and RAPDs. *Molecular Ecology* 7, 87–93.

Templeton, A.R. (1986) Coadaptation and outbreeding depression. In: Soulé, M.E. (ed.) *Conservation Biology*. Sinauer Associates, Sunderland, Massachusetts, pp. 105–116.

Thompson, J.D. (1999) Population differentiation in Mediterranean plants: insights into colonization history and the evolution and conservation of endemic species. *Heredity* 82, 229–236.

Vaillancourt, R.E. and Jackson, H.D. (2000) A chloroplast DNA hypervariable region in eucalypts. *Theoretical and Applied Genetics* 101, 473–477.

Wardell-Johnson, G. and Coates, D.J. (1996) Links to the past: local endemism in four species of forest eucalypts in southwestern Australia. In: Hopper, S.D., Chappill, J.A., Harvey, M.S. and George, A.S. (eds) *Gondwanan Heritage: Past, Present and Future of the Western Australian Biota*. Surrey Beatty & Sons, Chipping Norton, Australia, pp. 137–154.

Waser, N.M. (1993) Population structure, optimal outbreeding, and assortative mating in angiosperms. In: Thornhill, N.W. (ed.) *The Natural History of Inbreeding and Outbreeding*. University of Chicago Press, Chicago, Illinois, pp. 173–199.

Weir, B.S. (1996) *Genetic Data Analysis II*. Sinauer Associates, Sunderland, Massachusetts.

Williams, C.F., Ruvinsky, J., Scott, P.E. and Hews, D.K. (2001) Pollination, breeding system, and genetic structure in two sympatric *Delphinium* (Ranunculaceae) species. *American Journal of Botany* 88, 1623–1633.

Wright, S. (1951) The genetical structure of populations. *Annals of Eugenics* 15, 323–354.

Young, A.G. and Brown, A.H.D. (1996) Comparative population genetic structure of the rare woodland shrub *Daviesia suaveolens* and its common congener *D. mimosoides*. *Conservation Biology* 10, 374–381.

Young, A.G. and Clarke, G.M. (2000) *Genetics, Demography and Viability of Fragmented Populations*. Cambridge University Press, Cambridge.

10   Evolution of the flower

Douglas E. Soltis,[1] Victor A. Albert,[2] Sangtae Kim,[1] Mi-Jeong Yoo,[1] Pamela S. Soltis,[3] Michael W. Frohlich,[4] James Leebens-Mack,[5] Hongzhi Kong,[5,6] Kerr Wall,[5] Claude dePamphilis[5] and Hong Ma[5]

[1]*Department of Botany and the Genetics Institute, University of Florida, Gainesville, FL 32611, USA;* [2]*The Natural History Museums and Botanical Garden, University of Oslo, NO-0318 Oslo, Norway;* [3]*Florida Museum of Natural History and the Genetics Institute, University of Florida, Gainesville, FL 32611, USA;* [4]*Department of Botany, Natural History Museum, London SW7 5BD, UK;* [5]*Department of Biology, The Huck Institutes of the Life Sciences and Institute of Molecular Evolutionary Genetics, The Pennsylvania State University, University Park, PA 16802, USA;* [6]*Laboratory of Systematic and Evolutionary Botany, Institute of Botany, The Chinese Academy of Sciences, Beijing 100093, China*

Introduction

The origin and evolution of the flower have been intensively studied not only because of the great importance of flowers (and especially the fruits they produce) in providing human food, but also because of their crucial role in angiosperm sexual reproduction and many plant–animal interactions. The past centuries of morphologically and taxonomically based studies of flowers generated much information, but left some of the most critical questions of flower origin and evolution unresolved. Recent progress in understanding angiosperm (and seed plant) phylogeny provides a solid framework for evaluating evolutionary innovation, and identifies the taxa that provide the best insights into key innovations. The recent growth of developmental genetics provides exciting new data for understanding flower evolution; the interplay of developmental genetics with focused studies of morphology, development and phylogeny has generated a new field of study: the evolution of

development (evo-devo). Evo-devo offers the best hope for rapid advance in the understanding of flower evolution. To appreciate this potential one must be cognizant of recent advances in all of these fields – phylogeny, morphology and developmental genetics – that are merging to create evo-devo.

Here we describe recent progress in the study of floral evolution, beginning with advances in phylogeny and the reconstruction of trends in floral evolution. We include a brief comparative review of some of the genes known to regulate flower development, with an emphasis on recent studies relevant to the classic ABC model of flower development. We conclude with a perspective on future research on floral biology at the genomic level. Throughout our discussion we describe how experimental genetic and phylogenetic analyses are together improving our understanding of the evolution of floral architecture and the molecules regulating floral development.

Trends in Floral Evolution Inferred from Phylogeny

Background

A clear understanding of angiosperm phylogeny has recently emerged (e.g. Qiu *et al.*, 1999; Soltis *et al.*, 1999, 2000; Barkman *et al.*, 2000; Zanis *et al.*, 2002, 2003). These well-resolved and highly concordant DNA-based phylogenies have important implications for interpreting the morphology of early angiosperms and subsequent patterns of floral evolution.

Before the application of explicit phylogenetic methods, several investigators proposed that the first angiosperms had large, *Magnolia*-like flowers (Arber and Parkin, 1907; Bessey, 1915; Takhtajan, 1969; Cronquist, 1981). Stebbins (1974), in contrast, suggested that the earliest flowers were moderate in size. Endress (1987) proposed that the earliest angiosperm was bisexual, but that the transition to unisexuality was relatively easy, the perianth was undifferentiated and could be easily lost, and that the number of floral parts was labile.

Early phylogenetic studies focused attention on several herbaceous lineages (e.g. Nymphaeaceae, Piperaceae and Chloranthaceae; Fig. 10.1) as possible first-branching extant angiosperms (Donoghue and Doyle, 1989; Doyle *et al.*, 1994). Based on these results, it was suggested that early flowers were small, with a trimerous perianth, and with few stamens and carpels. However, more recent analyses (e.g. Mathews and Donoghue, 1999; Parkinson *et al.*, 1999; Qiu *et al.*, 1999; Soltis *et al.*, 1999, 2000; Barkman *et al.*, 2000; Doyle and Endress, 2000; Graham and Olmstead, 2000; Zanis *et al.*, 2002, 2003; Borsch *et al.*, 2003; Hilu *et al.*, 2003) place *Amborella*, Nymphaeaceae (including Cabombaceae; see APG II, 2003) and Austrobaileyales as basal to other extant angiosperms (Fig. 10.2). This topology suggests instead that the earliest flowers were small to moderate in size, with an undifferentiated perianth, stamens lacking a well-differentiated filament, and a gynoecium composed of one or more distinct carpels.

Fossils are critical for inferring the origin and early diversification of angiosperms, but fossil flowers of the earliest angiosperms are scarce. None the less, early Cretaceous angiosperm fossils are consistent with the hypothesis that the first flowers were small to moderate in size, with an undifferentiated perianth (Crane, 1985; Friis *et al.*, 1994, 2000; Crane *et al.*, 1995), although *Magnolia*-like forms also occurred during the same geological time (e.g. *Archaeanthus*; Dilcher and Crane, 1984). In addition, some early angiosperms lacked a perianth (e.g. *Archaefructus*; Sun *et al.*, 2002), but these may not be basal within angiosperms (Friis *et al.*, 2003). There are no known fossils representing unequivocal stem-group angiosperms (i.e. angiosperms that attach below the basal node leading to *Amborella*, Nymphaeaceae and all other living angiosperms).

One way to infer ancestral states is to employ character-state reconstruction with phylogenetic trees and programs such as MacClade (Maddison and Maddison, 1992). Using this approach, the evolution of specific floral characters in basal angiosperms has been reconstructed (e.g. Albert *et al.*, 1998; Doyle and Endress, 2000; Ronse De Craene *et al.*, 2003; Zanis *et al.*, 2003; Soltis *et al.*, 2004). We review some of the findings of these character-state reconstructions below using the most conservative optimization method (all most parsimonious states; Maddison and Maddison, 1992). Other reconstructions, using other optimization methods and tree topologies, are provided in the references noted above. Most of the same general conclusions are supported regardless of optimization.

Perianth differentiation

A differentiated or bipartite perianth has an outer whorl of sepals clearly differentiated from the inner whorl(s) of petals. In contrast, an undifferentiated perianth lacks clear differentiation between the outer and inner whorls, or the perianth may consist of undifferentiated spirally arranged parts. These undifferentiated perianth organs

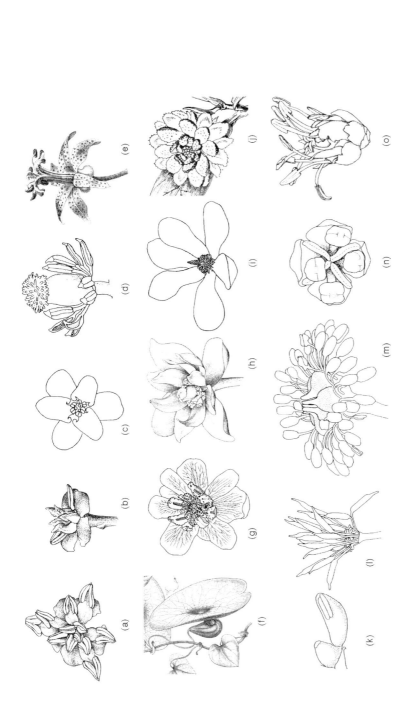

Fig. 10.1. Floral diversity in basal angiosperms (a–i) and early-diverging eudicots (j–m). (a) *Amborella trichopoda* (Amborellaceae), staminate flower (from Endress and Igersheim, 2000). (b) *A. trichopoda* (Amborellaceae), pistillate flower (from Endress and Igersheim, 2000). (c) *Cabomba aquatica* (Nymphaeaceae) (from Endress, 1994b). (d) *Trimenia papuana* (Trimeniaceae) (from Endress and Sampson, 1983). (e) *Tricyrtis pilosa* (Liliaceae), flower (from Engler in Engler and Prantl, 1887–1915). (f) *Aristolochia* (Aristolochiaceae) flower (from Solereder in Engler and Prantl, 1887–1915). (g) *Austrobaileya scandens* (Austrobaileyaceae) (from Endress, 1980). (h) *Takhtajania perrieri* (Winteraceae; Canellales) (from Endress et al., 2000). (i) *Magnolia* × *soulangiana* (Magnoliaceae; Magnoliales) (from Endress, 1987). (j) *Eupomatia* (Eupomatiaceae) flowering shoot (from Uphof in Engler and Prantl, 1959). (k) *Sarcandra chloranthoides* (Chloranthaceae) (from Endress, 1987). (l) *Euptelea polyandra* (Eupteleaceae) (from Endress, 1986). (m) *Trochodendron* (Trochodendraceae) (from Endress, 1986). (n) *Tetracentron* (Trochodendraceae) (from Endress, 1986). (o) *Buxus balearica* (Buxaceae), inflorescence with lateral staminate flowers and terminal carpellate flower (from Von Balthazar and Endress, 2002).

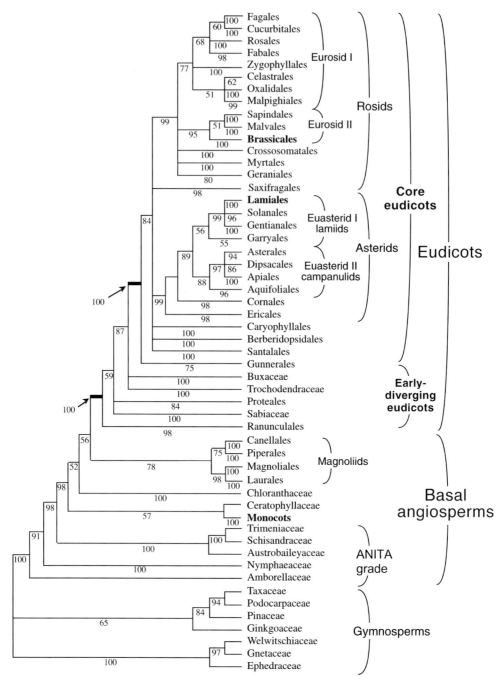

Fig. 10.2. Summary topology for angiosperms showing general positions of model organisms (in bold). Modified from Soltis *et al.* (2003).

have traditionally been referred to as tepals. The term tepal was coined by De Candolle (1827) to describe perianth organs (sepals and petals) that are not clearly differentiated morphologically; thus, the entire perianth may be petaloid. Takhtajan (1969), in contrast, used the term 'tepal' in a phylogenetic sense such that all monocots have tepals. Takhtajan's definition limits the application of tepal to specific groups of angiosperms and requires different terms for an undifferentiated perianth in other groups. Following other recent investigators, we will use the term tepal as defined by De Candolle.

Distinguishing sepals from petals is not always straightforward (Endress, 1994a; Albert *et al.*, 1998). Whereas sepals and petals are readily distinguished in most eudicots (~ 75% of all angiosperms), this is often not the case in basal angiosperms (Fig. 10.1), many of which have numerous undifferentiated perianth parts arranged in spirals, rather than in distinct whorls, a condition long considered ancestral (e.g. Bessey, 1915; Cronquist, 1968; Takhtajan, 1969).

The origin of a differentiated perianth of sepals and petals has long been of interest (e.g. Eames, 1931; Hiepko, 1965; Kosuge, 1994; Albert *et al.*, 1998; Kramer and Irish, 1999, 2000). It has been proposed that petals evolved first and that sepals evolved later (e.g. Albert *et al.*, 1998) and that petals have evolved multiple times from different floral organs in different groups (e.g. Eames, 1961; Takhtajan, 1969; Kosuge, 1994; Albert *et al.*, 1998; Zanis *et al.*, 2003).

Takhtajan (1969, 1997) suggested two origins of petals, one from stamens and one from bracts. Support for multiple, independent origins of petals has come from morphological studies showing that 'petals' of various angiosperms exhibit major differences and can be grouped into two basic classes (e.g. Endress, 1994a; Kramer and Irish, 2000). In one group are petals that resemble stamens. The petals are developmentally delayed and are similar in appearance to stamen primordia at inception (Endress, 1994a). These petals have sometimes been termed andropetals. The second type of petaloid organ (conventionally termed tepals; Cronquist, 1981) is found in undifferentiated perianths and is more leaf-like in general characteristics. These petals initiate and mature much earlier than do the stamens and are generally more leaf-like in appearance than are other petals (Smith, 1928; Tucker, 1960; Takhtajan, 1969, 1997).

Following Albert *et al.* (1998), two or more whorls of perianth parts must be present for an unambiguous interpretation of sepals and petals. If only a single perianth whorl is present, it may be difficult to interpret as 'sepals' or 'petals' (see also Endress, 1994a,b). Is the single whorl an undifferentiated perianth, composed of neither sepals nor petals, or is the single whorl composed of either sepals or petals with the other perianth whorl absent? A single-whorled perianth has traditionally been referred to as being composed of 'sepals' as a matter of convention (e.g. Cronquist, 1968). Families of basal angiosperms that contain taxa with a single-whorled perianth include nearly all Aristolochiaceae (except *Saruma*), all Myristicaceae and Chloranthaceae (*Hedyosmum*). In some cases, however, the nature of a single-whorled perianth can be determined through comparison with the perianths of closely related taxa. In Aristolochiaceae, most taxa have a single-whorled perianth that is considered a calyx (Cronquist, 1968, 1981; Tucker and Douglas, 1996; Takhtajan, 1997). In contrast, *Saruma* has two perianth whorls that are differentiated into sepals and petals. Furthermore, in some species of *Asarum*, petals apparently begin to develop, but the only traces are small, thread-like structures (Leins and Erbar, 1985).

In recent reconstructions (Ronse De Craene *et al.*, 2003; Zanis *et al.*, 2003; Soltis *et al.*, 2004) (Fig. 10.3), the ancestral state for the angiosperms is an undifferentiated perianth. *Amborella* and Austrobaileyales have an undifferentiated perianth. In contrast, the ancestral state for Nymphaeaceae is reconstructed as equivocal because some Nymphaeaceae (e.g. *Cabomba*, *Brasenia*, *Nuphar*) have a differentiated perianth whereas more derived waterlilies (*Victoria*, *Nymphaea*) have an undifferentiated perianth. Above the basal angiosperm grade, the undifferentiated perianth continues to be ancestral for the remaining angiosperms (Fig. 10.3). Importantly, all reconstructions indicate that a differentiated perianth evolved multiple times (see Albert *et al.*, 1998). Separate origins include some Nymphaeaceae, monocots, some Magnoliaceae, Annonaceae, Canellaceae, some

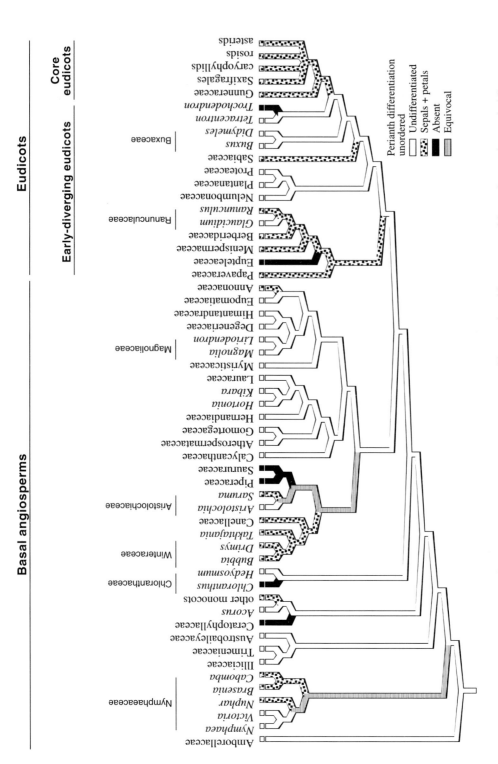

Fig. 10.3. MacClade reconstruction using all most parsimonious states optimization (Maddison and Maddison, 1992) of the evolution of perianth differentiation in angiosperms, with an emphasis on basal angiosperms and early-diverging eudicots. Topology is based on Zanis *et al.* (2002, 2003) and Soltis *et al.* (2000, 2003). Data are from Zanis *et al.* (2003) and Ronse De Craene *et al.* (2003). Modified from Soltis *et al.* (2004). For impact of other optimization methods (ACCTRAN, DELTRAN) see Zanis *et al.* (2003), Ronse De Craene *et al.* (2003) and Soltis *et al.* (2004).

Aristolochiaceae and Siparunaceae with additional origins in early-diverging eudicots (e.g. Papaveraceae, Menispermaceae, *Ranunculus*, Sabiaceae) and core eudicots. Comparative developmental studies are required to test whether multiple origins of perianth differentiation were driven by similar changes in gene function and regulation.

Phyllotaxis

Amborella has spiral phyllotaxis (Fig. 10.1), as do members of Austrobaileyales. In some basal families, phyllotaxis is complex. For example, in some Nymphaeaceae, phyllotaxis has been considered spiral, but it now appears to be primarily whorled, or in some cases irregular (Endress, 2001). In some Winteraceae (*Drimys* and *Pseudowintera*), phyllotaxis is primarily whorled, but occasionally spiral (Doust, 2000). In *Drimys winteri*, flowers within one tree vary between spiral and whorled (Doust, 2001).

The distinction between spiral and whorled is not always clear. In *Amborella*, recent developmental studies indicate that some floral organs (e.g. carpels) are initiated in a nearly whorl-like manner, although they are commonly described as spirally arranged (Buzgo *et al.*, 2004b). Studies of other basal angiosperms reveal that in some cases floral organs that appear to be whorled in mature flowers actually result from spiral initiation of primordia and a bimodal distribution of long and short time intervals between the initiation of consecutive organ primordia (Tucker, 1960; Leins and Erbar, 1985; Endress, 1994a). Thus, both spiral and whorled phyllotaxis of mature flowers result from the organs developing in a spiral sequence (Endress, 1987). For example, *Illicium* has spiral phyllotaxis in developing buds, but in mature flowers the carpels have an apparently whorled arrangement. Furthermore, even in some eudicots the sepals initiate in a spiral sequence, with the later-arising sepals positioned slightly inside the earliest to originate, as reflected in their imbricate arrangement at maturity. The inner organs arise in precise whorls, and even the sepals have traditionally been considered whorled, because of their close apposition at maturity.

Although spiral phyllotaxis is present in Amborellaceae and Austrobaileyales, the presence of whorled (and irregular) phyllotaxis in Nymphaeaceae makes the ancestral reconstruction for perianth phyllotaxis for the angiosperms dependent on the coding of the outgroup. However, outgroup coding is problematic because the immediate sister group of the angiosperms is unknown. Furthermore, no fossil group is known to have possessed flowers. If the outgroup is coded as lacking a perianth, then either a spiral or whorled phyllotaxis is reconstructed as equally parsimonious for the base of the angiosperms. If the outgroup is coded as having a spirally arranged perianth, then a spiral perianth is reconstructed as ancestral for the angiosperms. If the outgroup is coded as having a whorled perianth, then a whorled perianth is ancestral for the angiosperms with a spiral perianth evolving several times.

Above the Amborellaceae, Nymphaeaceae, Austrobaileyales grade, whorled perianth phyllotaxis is reconstructed as ancestral for all remaining angiosperms with multiple shifts to a spiral perianth occurring in basal lineages, including Calycanthaceae, Atherospermataceae, Gomortegaceae, some Monimiaceae, Degeneriaceae and some Magnoliaceae (Fig. 10.4). A possible transformation from whorled to spiral phyllotaxis may have occurred in *Drimys* and *Pseudowintera* (Winteraceae), which have a complex phyllotaxis involving spirals and multiple whorls (Doust, 2000, 2001; Endress *et al.*, 2000). Still additional reversals to a spiral perianth are found in the early-diverging eudicots *Nelumbo* (Proteales) and *Xanthorhiza*, *Caltha* and *Ranunculus* (Ranunculaceae). Thus, perianth phyllotaxis is highly labile in basal angiosperms and in basal eudicots (Endress, 1994b; Albert *et al.*, 1998; Ronse De Craene *et al.*, 2003; Zanis *et al.*, 2003; Soltis *et al.*, 2004). Again, comparative developmental studies are necessary to determine whether unrelated taxa with convergent phyllotaxis share common regulatory networks for organ initiation.

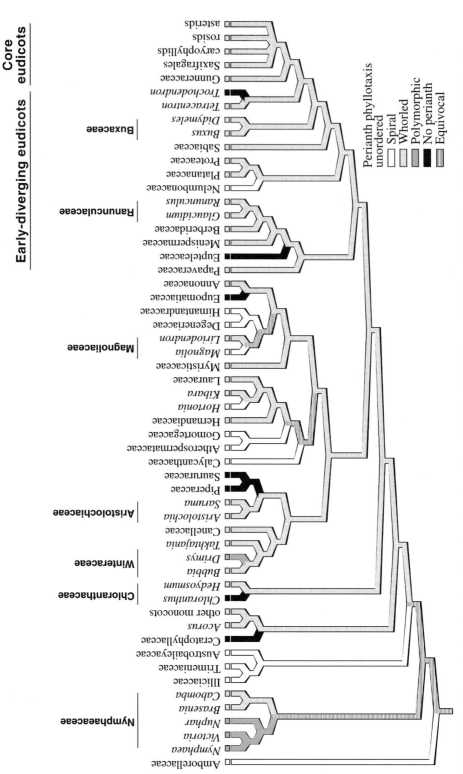

Fig. 10.4. MacClade reconstruction using all most parsimonious states optimization (Maddison and Maddison, 1992) of the evolution of perianth phyllotaxis in angiosperms, with an emphasis on basal angiosperms and early-diverging eudicots. Topology is based on Zanis *et al.* (2002, 2003) and Soltis *et al.* (2000, 2003). Data are from Zanis *et al.* (2003) and Ronse De Craene *et al.* (2003). Modified from Soltis *et al.* (2004). For impact of other optimization methods (ACCRAN, DELTRAN) see Zanis *et al.* (2003), Ronse De Craene *et al.* (2003) and Soltis *et al.* (2004).

Merosity

Among basal angiosperms, many lineages have numerous parts, some clades are trimerous, and others defy simple coding of merosity. In Winteraceae, the outermost floral organs are in dimerous whorls, followed by a switch to tetramerous whorls, and finally (in *Takhtajania*) a change to pentamerous whorls (Endress *et al.*, 2000). Similarly, in Magnoliaceae, the perianth of some species of *Magnolia* is an indeterminate spiral, whereas that of *Liriodendron* and other species of *Magnolia* is in three trimerous whorls and may represent a transition from spiral to whorled phyllotaxis (Tucker, 1960; Erbar and Leins, 1981, 1983).

Amborella and Austrobaileyales have an indeterminate spiral (Fig. 10.1). However, within Nymphaeaceae, *Cabomba*, *Brasenia* and *Nuphar* they are trimerous; other genera (e.g. *Victoria*, *Nymphaea*) are trimerous or tetramerous (Endress, 2001). As found for phyllotaxis (above), reconstruction of the ancestral merosity of extant angiosperms is dependent on the coding of merosity for the outgroup. If the outgroup of the angiosperms is coded as having an indeterminate number of perianth parts, then an indeterminate number is also ancestral for the angiosperms. Alternatively, if the ancestor of the angiosperms is considered to lack a perianth, then it is equally parsimonious for the base of the angiosperms to be either trimerous or indeterminate in perianth merosity (see Zanis *et al.*, 2003; Soltis *et al.*, 2004).

However, regardless of outgroup coding, above the basal grade of *Amborella*, Nymphaeaceae and Austrobaileyales, the ancestral character state for all remaining angiosperms is a trimerous perianth (Fig. 10.5) (e.g. Ronse De Craene *et al.*, 2003; Zanis *et al.*, 2003; Soltis *et al.*, 2004). Thus, although the trimerous condition is typically associated with monocots, these results indicate that trimery played a major role in the early evolution and diversification of the flower (Kubitzki, 1987).

Following the origin of a trimerous perianth, there was a return to an indeterminate spiral perianth in several basal lineages, including Calycanthaceae (e.g. *Calycanthus*), the clade of Atherospermataceae and Gomortegaceae, Himantandraceae, some Monimiaceae (e.g. *Hortonia*) and some Magnoliaceae (*Magnolia*). A perianth has also been lost several times (e.g. Eupomatiaceae (see below), Piperaceae, most Chloranthaceae, and Ceratophyllaceae) (Fig. 10.5).

These reconstructions indicate that perianth merosity is labile in basal angiosperms (see also Endress, 1987, 1994b; Albert *et al.*, 1998; Zanis *et al.*, 2003), a condition that continues through the early-diverging eudicots (Fig. 10.5). Dimery is often seen in early-diverging eudicots. However, trimery is also prevalent (Ranunculales), and pentamery is seen in some taxa. In contrast, in core eudicots, pentamery predominates. Interestingly, dimery is found in *Gunnera*, sister group to all other core eudicots. Thus, reconstructions not only indicate that perianth merosity is labile in basal angiosperms and early-diverging eudicots, but also suggest that a dimerous perianth could be the immediate precursor to the pentamery characteristic of eudicots (Soltis *et al.*, 2003). Once more, comparative developmental studies are required to elucidate the molecular basis of changes in merosity throughout angiosperm history.

Genes Controlling Early Floral Development

The models

Developmental genetic analyses have provided unprecedented insights into the molecular mechanisms that determine identities of the principal floral organs, at least in the eudicot model organisms used for these studies. *Arabidopsis thaliana* and *Antirrhinum majus*, two derived eudicots, were the first models studied, and are still the best understood. Investigations of these models have resulted in the identification and understanding of over 80 genes critical for normal floral development, including genes involved in flower initiation; however, the true number is bound to be much larger (Zhao *et al.*, 2001a; Ni *et al.*, 2004) (Fig. 10.6). Careful morphological developmental

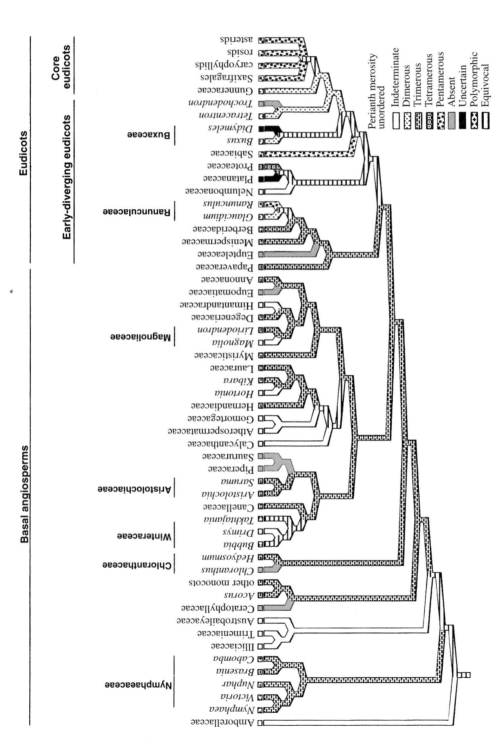

Fig. 10.5. MacClade reconstruction using all most parsimonious states optimization (Maddison and Maddison, 1992) of the evolution of perianth merosity (merism) in angiosperms, with an emphasis on basal angiosperms and early-diverging eudicots. Topology is based on Zanis *et al.* (2002, 2003) and Soltis *et al.* (2000, 2003). Data are from Zanis *et al.* (2003) and Ronse De Craene *et al.* (2003). Modified from Soltis *et al.* (2004). For impact of other optimization methods (ACCTRAN, DELTRAN) see Zanis *et al.* (2003), Ronse De Craene *et al.* (2003) and Soltis *et al.* (2004).

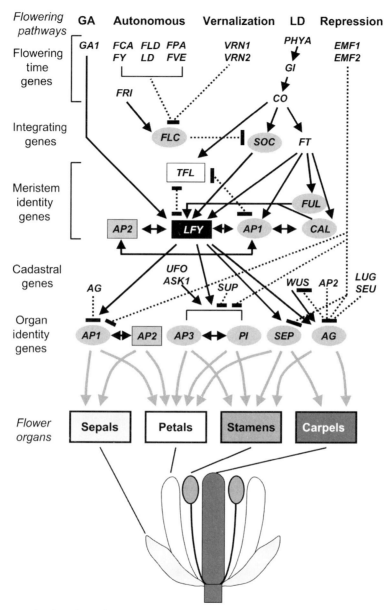

Fig. 10.6. Genes that have been demonstrated genetically to regulate flowering time, floral meristem and organ identities in *Arabidopsis*. MADS-box genes are shown in ovals. For genes that encode other types of proteins, only those that play critical roles in floral meristem and organ identities are shown in boxes. The black lines and arrows indicate positive genetic interaction; the dotted lines with a short bar at the end represent negative genetic interactions. The arrows indicate that the specific organ identity gene(s) is (are) required for the identity of the corresponding organ. See Fig. 10.7 for an illustration of the ABC model. Although few genes have been identified that function downstream of the organ identity genes (Sablowski and Meyerowitz, 1998), a number of putative downstream genes for *LFY* and *AP3/PI* have been reported recently from microarray analysis (Schmid *et al.*, 2003; Zik and Irish, 2003). Modified from a figure in Soltis *et al.* (2002), with recent information on the regulation of floral meristem identity genes by *CO* and *FT* (Schmid *et al.*, 2003) and regulation of floral organ identity genes by *EMF1*, *EMF2*, *LUG* and *SEU* (Franks *et al.*, 2002; Moon *et al.*, 2003; Schmid *et al.*, 2003).

studies (Smyth *et al.*, 1990) provided a foundation for evaluating the effects of mutations and defining gene functions. This integration of morphological and developmental genetic investigations has characterized the work on several other model systems as well, including the derived monocots *Zea mays* and *Oryza sativa* (Poaceae), and to a lesser extent *Petunia hybrida* and *Lycopersicon esculentum* (= *Solanum lycopersicum*), both of Solanaceae (Coen and Meyerowitz, 1991; Meyerowitz *et al.*, 1991; Ma, 1994, 1998; Weigel and Meyerowitz, 1994; Weigel, 1995; Yanofsky, 1995; Ma and dePamphilis, 2000; Zhao *et al.*, 2001a; Irish, 2003).

The best-known genes controlling floral organ identity are the A, B and C function genes (Coen and Meyerowitz, 1991; Meyerowitz *et al.*, 1991). According to the ABC model, three overlapping gene functions, A, B and C, act alone or in combination to specify the four types of floral organs (Fig. 10.7). In 1990, the genes representing *deficiens* (B class) and *agamous* (C class) mutants were cloned from *Antirrhinum* and *Arabidopsis*, respectively.

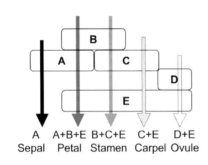

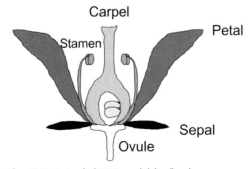

Fig. 10.7. Extended ABC model for floral organ specification (modified from Theißen, 2001).

Homologous genes from these two models sometimes have different names, creating some confusion for newcomers to the field; we therefore often provide both names in our overview. The protein products of *DEFICIENS* (*DEF* = *APETALA3* (*AP3*) in *Arabidopsis*) and *AGAMOUS* (*AG* = *PLENA* (*PLE*) in *Antirrhinum*) were found to be from the same family of transcription factors, which are regulators of the expression of other genes (Schwarz-Sommer *et al.*, 1990). This family was named MADS-box genes after a DNA-binding amino acid domain present in $\underline{M}CM1$ (mini-chromosome maintenance-1; from yeast), $\underline{A}G$, $\underline{D}EF$ and $\underline{S}RF$ (serum response factor; from humans). MADS-box genes encode a conserved domain that constitutes most of the DNA-binding domain.

It had been hypothesized from mutant phenotypes that the *DEF* (= *AP3*) and *AG* (= *PLE*) genes control floral organ identity in a combinatorial, whorl-specific fashion: A function directs sepal identity; B function together with A specifies petals; B plus C function designates stamens; and C alone promotes carpel development (Meyerowitz *et al.*, 1991; Ma, 1994; Weigel and Meyerowitz, 1994; see below; Fig. 10.7). The *DEF* and *AG* gene products were assigned to the B and C functions, respectively.

As noted, in *Arabidopsis*, the A function genes are *AP1* and *AP2* (Fig. 10.7), the B function genes are *AP3* (= *DEF*) and *PISTILLATA* (*PI* = *GLO* in *Antirrhinum*), genes that resulted from an ancient duplication event (discussed below), and the C function is specified by *AG* (= *PLE*) (reviewed in Ma, 1994; Ma and dePamphilis, 2000). Genetic studies were crucial for the identification of these gene functions, with mutations in each of these genes affecting two adjacent whorls. For example, *ap3* mutants produce sepals and carpels instead of petals and stamens, respectively. Double- and triple-mutant analyses in *Arabidopsis* have further clarified the genetic interactions among A, B, C class genes. Expression studies have also been important in confirming aspects of the ABC model. All of the ABC MADS-box genes are expressed in the regions of the floral meristem that they help specify. The model is supported by over-expression studies of the ABC genes in *Arabidopsis*, which can place any of the four flower organs in any of the four whorls.

Recently, in *Arabidopsis*, the role of the class E genes, *SEPALLATA1*, *SEPALLATA2* and *SEPALLATA3*, has been demonstrated: they act redundantly to specify petals, stamens and carpels (Pelaz *et al.*, 2000, 2001; Theißen, 2001). These genes were identified through their sequence similarity to *AG*, rather than through individual mutant phenotypes. Triple-null mutants of *SEP1-3* produce 'flowers' consisting only of sepal-like organs, suggesting that these related genes have redundant functions in controlling the identity of petals and reproductive organs (Pelaz *et al.*, 2000, 2001). Floral MADS-domain proteins can form homodimers, heterodimers and tetramers, providing a mechanism for the interaction of genes within and between the A, B, C and E functions (Theißen, 2001) (see Fig. 10.10).

In addition to A, B, C and E function genes, numerous other genes are also regulators of normal floral development. Furthermore, not all floral regulators are MADS-box genes. In *Arabidopsis*, the non-MADS *APETALA2* (*AP2*) confers A function along with the MADS gene *APETALA1* (*AP1= SQUAMOSA* (*SQUA*) in *Antirrhinum*). *LEAFY* (*LFY*), which controls the entire floral developmental programme, codes for a previously unknown type of transcriptional regulator (Weigel *et al.*, 1992). Space does not permit review of all of the numerous genes involved in floral development here. Readers are encouraged to consult recent reviews (e.g. Ma, 1998; Zhao *et al.*, 2001a; Ni *et al.*, 2004; Fig. 10.6). Because most genes with known functions in flower development have been detected through their single-gene mutant phenotypes, genes such as the *sepallata* genes with redundant function (Pelaz *et al.*, 2000; Theißen, 2001), or genes that are lethal when disrupted, are not usually discovered except through detailed follow-up analysis. As a result, even this rapidly growing collection of genes of known function must be considered an underestimate of the genes with critical roles in flower development.

Genes are also known that specify the floral character of the apical meristem that forms the flower. The genes *FLORICAULA* (*FLO*) and *LEAFY* (*LFY*) of *Antirrhinum* and *Arabidopsis*, respectively, are transcription factors of a family unique to land plants. *FLO/LFY* is single copy in diploid angiosperms. *FLO/LFY* is expressed in a graded manner and acts synergistically with the MADS-box gene *SQUAMOSA/AP1* to specify the floral character of the apex. These genes integrate signals from multiple pathways involved in the transition to flowering. Some of the additional genes involved in floral specification are shown in Fig. 10.6 (e.g. Coen *et al.*, 1990; Weigel *et al.*, 1992; Weigel, 1995; Riechmann and Meyerowitz, 1997; Ma, 1998; Theißen *et al.*, 2000; Theißen, 2001; Zhao *et al.*, 2001a).

New model plants

Exhaustive studies of a few key model plants, chiefly *Arabidopsis* and *Antirrhinum*, have provided enormous insights into the genetic control of flower development. However, a key question is, are the models of the genetic control of floral development in these derived eudicots applicable to all angiosperms? Interestingly, the conservation of A function is unclear in angiosperms other than Brassicaceae. Another floral developmental model emphasizing the B and C functions alone (called at that time A and B) was developed even before the ABC model, and this focus might be more broadly applicable (Schwarz-Sommer *et al.*, 1990) (Fig. 10.8). The genetic architecture of floral development in angiosperms other than the well-known models should also be investigated (e.g. Albert *et al.*, 1998; Kramer and Irish, 2000; Soltis *et al.*, 2002). To obtain maximal benefit from the enormous resources afforded by well-developed models for floral developmental genetics, it is imperative that researchers expand their emphasis to include additional species representing a wider phylogenetic coverage of angiosperms.

The rapid increase in interest in the evolutionary developmental biology ('evo-devo') of the flower has stimulated the investiga-

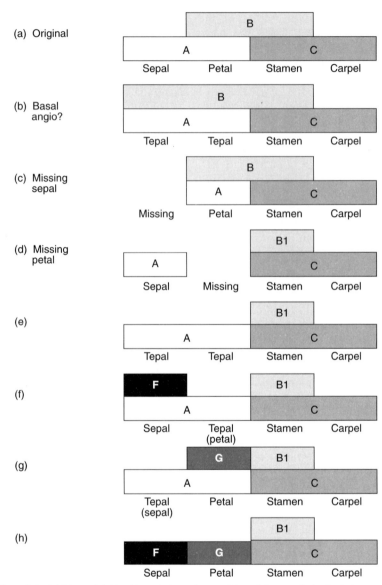

Fig. 10.8. The original ABC model (a) with variations that could explain morphological changes. Versions (b–d) simply allow the change of the domains of A and B functions to account for the diversity in the perianth. Version (e) makes the control of the tepal identity similar to that of the sepal identity in derived eudicots, although tepals are often morphologically similar to petals. Versions (f–h) propose 'F and G functions' different from the ABC functions in distribution and in consequence to control perianth identities. B1 is used instead of B when the function is only used to control the stamen identity.

tion of a number of new 'model' plants, and many of these are under investigation as part of genomics initiatives (Soltis *et al.*, 2002; De Bodt *et al.*, 2003). New models have typically been chosen based on their significant phylogenetic positions (Fig. 10.2). *Amborella* (Amborellaceae) and waterlilies (Nymphaeaceae) were chosen because they represent the sister groups to all other angiosperms. Other basal angiosperms (e.g.

Lauraceae and Magnoliaceae) are also the focus of study, as is *Acorus* (Acoraceae), the sister to all other monocots. Poppies (*Papaver* and *Eschscholzia*) are important choices because they represent an early-diverging eudicot lineage (Fig. 10.2) and provide a critical link between derived eudicot models (e.g. *Arabidopsis* and *Antirrhinum*) and basal angiosperms.

The growing list of new models not only expands the phylogenetic diversity under study, but also the diversity of floral form that is currently under molecular and genetic investigation (Soltis *et al.*, 2002; De Bodt *et al.*, 2003). In addition to the standard whorled arrangements of parts, new models such as *Amborella* exhibit a spiral perianth that is undifferentiated. *Gerbera*, a derived asterid in the sunflower family, is also a useful model because of its divergent inflorescence format: multiple flowers of different phenotypes borne together in a dense head (Yu *et al.*, 1999; Kotilainen *et al.*, 2000).

The new models also have important limitations. For most, genetic studies are not yet possible. Although developmental morphological and molecular studies can lead to the formulation of useful hypotheses regarding the evolution of gene functions, these await testing using genetic studies. For basal angiosperms that are woody (e.g. *Amborella*, *Persea*) and not readily analysed genetically, definitive conclusions about gene functions will be difficult to achieve. Therefore, herbaceous basal angiosperms (e.g. the waterlily *Cabomba*) and herbaceous basal eudicots (e.g. *Papaver* or *Eschscholzia*) may have the greatest potential as new models because of their short life cycles and the transformability of *Papaver* (Baum *et al.*, 2002).

New technologies might provide effective methods for reverse genetic analysis of new genetic models. Methods that use viruses to generate small, interfering RNA and to posttranscriptionally silence a gene of interest (Lu *et al.*, 2003) might be applicable in mature plants, even in long-lived perennials. Such new methods, if perfected, that allow easy elucidation of gene function in diverse plants by mutation or by gene silencing, could become as important for evo-devo studies as PCR has been for molecular phylogenetics.

Limits on the generality of floral developmental genetics

Through molecular evolutionary and gene exchange studies, it was determined that *AP3* represents the *Arabidopsis* homologue of *DEF* from *Antirrhinum*. Similarly, sequence and functional homologies were found between *AG* and *LFY* and their *Antirrhinum* counterparts (*PLE* and *FLO*). However, the situation with A function genes is more complex. Mutations in the *Antirrhinum* *SQUAMOSA* gene, a likely orthologue of *AP1* from *Arabidopsis*, cause floral meristem defects similar to *ap1* mutants. Flowers of *squamosa* mutants also exhibit defects in petal development, although the role of *SQUAMOSA* in controlling petal identity is thought to be less important than that of *AP1* in *Arabidopsis*. Recently, two *Antirrhinum* homologues, *LIPLESS1* and *LIPLESS2* (*LIP1* and *LIP2*), of the *Arabidopsis AP2* gene, have been shown to have redundant functions in controlling sepal and petal development (Keck *et al.*, 2003) in a manner similar to that of *AP2*. However, unlike *AP2*, *LIP1/2* do not seem to be involved in the negative regulation of the C-function gene *PLENA* (*PLE* = *AG*). Therefore, results from *Antirrhinum* support a critical role for A function in determining perianth identities, but the interactions between known genes involved in A and C functions seem to be different between *Arabidopsis* and *Antirrhinum*. *AP1*- and *AP2*-like genes have been identified from a diverse array of angiosperms (Litt and Irish, 2003); however, whether they play a role in A function is not known.

Therefore, the existence of a conserved A function in angiosperm flower development is still uncertain, although there should be gene functions that specify sepal identity in species that produce a differentiated perianth. It is possible that different genes serve this function in different flowering plant lineages, or that the determination of sepal and petal identity is more complex than depicted in the ABC model. Based on the presence of B- and C-function MADS-box genes in gymnosperms (which lack flowers), it has been hypothesized that determination of flower organ identity has evolved from a more ancient role of these genes in sex determination (Hahn and Somerville, 1988; Münster *et al.*, 1997; Winter *et al.*, 1999). It has also been hypothesized that 'true' sepals of the kind expressed by *Arabidopsis* and *Antirrhinum* are a relatively recent evolutionary innovation, because basal eudicots and monocots characteristically lack discrete sepals and petals and bear tepals instead.

Conservation of control over floral specification of the flower apex by *FLO/LFY* has been shown for several eudicots, but additional functions are also known. For example, the *LFY* homologue of *Pisum* (Fabaceae) controls compound leaf development in addition to the transition to flowering. In grasses, *LFY* homologues probably direct the development of inflorescence meristems rather than only floral meristems, as in *Arabidopsis*.

As noted above, the *SEPALLATA* (*SEP*) genes of *Arabidopsis* provide redundant function required for floral organ identity. However, research on other organisms such as *Gerbera* has shown that *SEP*-class homologues play divergent roles in development of the condensed, head inflorescence as well as in the different floral forms that are borne by it. Specifically, one *SEP*-like gene confers the C function only in the staminal whorl of flowers borne at the periphery of the inflorescence, whereas another *SEP* homologue (the probable duplication partner of the first) appears to confer the C function only in carpels (Teeri *et al.*, 2002). This partitioning of genetic function has probably had morphological evolutionary consequences because the outer flowers of *Gerbera* inflorescences are male-sterile and highly asymmetrical, with fused, elongate petals, whereas the inner flowers are bisexual and very close to symmetrical, with non-elongate petals. The *Gerbera* inflorescence looks very much like a single flower, and probably attracts pollinating insects in the same capacity.

The ABC Model: New Data, New Views

ABCs of basal angiosperms

Recent investigations of basal angiosperms have provided an important assessment of the applicability of the ABC model to all angiosperms. Certainly much of the ABC framework is conserved in a number of eudicots and grasses, but there are important variations on the ABC theme in some flowering plants (Fig. 10.8). For example, in contrast to the well-differentiated sepal and petal whorls of eudicots such as *Arabidopsis* and *Antirrhinum*, the two outer floral whorls in many members of the monocot family Liliaceae (lily family) are petaloid and almost identical in morphology (Fig. 10.1). Importantly, in *Tulipa* (tulip), the B-class genes are expressed in both petaloid whorls, as well as in stamens (Kanno *et al.*, 2003). This situation supports the idea that petals and petal-like organs require B function, regardless of the position of these organs within the flower.

Similarly, in some Nymphaeaceae (waterlilies) such as *Nuphar*, the outer whorl of the flower, sometimes referred to as sepals, exhibits B class gene expression, as do the petals, stamens and staminodes (Kim *et al.*, 2004) (Fig. 10.8). In *Amborella*, which has a spirally arranged perianth with parts that are not differentiated into sepals and petals (Fig. 10.1), a similar pattern is observed, with B class genes expressed throughout the perianth, as well as in the stamens (Fig. 10.8). Similar expression data have been forthcoming for basal angiosperms in the magnoliid clade. In *Magnolia* (Magnoliaceae), B class gene expression has been documented throughout the perianth whorls, as well as in stamens and staminodes (Kramer and Irish, 2000; Kim *et al.*, unpublished). A similar pattern of B-class gene expression has been observed in basal eudicots such as *Papaver* (Papaveraceae) and various members of Ranunculaceae (Kramer and Irish, 2000; Kramer *et al.*, 2003). The expression of B-function genes in sepal-like organs suggests that these B-function genes are not sufficient to specify petal identity.

The expression of C-class genes has also been examined in several basal angiosperms, and the results for this gene match the predictions of the ABC model. For example, homologues of *AGAMOUS* have been isolated from *Amborella*, and these are expressed in carpels, stamens and staminodes (Kim *et al.*, unpublished). Data for the expression of A-class genes from the basal-most angiosperms remain fragmentary.

Thus, recent data suggest a modified ABC model for basal angiosperms, with B-class genes expressed and presumably functioning throughout the perianth and stamens (Fig. 10.7) (see Van Tunen *et al.*, 1993; Albert *et al.*, 1998) following the original 'BC model' idea put forth by Schwarz-Sommer *et al.* (1990). From a phylogenetic standpoint, the ABC model may reflect a more recent programme that is important in *Arabidopsis* and possibly other eudicots. The specification of sepals, which may have evolved more than once (Albert *et al.*, 1998), may well be encoded by different genes in different angiosperm lineages. The pattern of B-class gene expression observed in basal angiosperms and basal eudicots probably represents the ancestral condition, with the model originally proposed for *Arabidopsis* and *Antirrhinum* a derived modification (Fig. 10.8).

An important evolutionary question now becomes: at which node in the angiosperm tree did the switch from the more general BC model occur? Functional studies in phylogenetically critical taxa are required before this question can be answered, but the switch probably coincided with the evolution of the core eudicots (Fig. 10.2). Other important changes in floral genes similarly appear to coincide with the origin of core eudicots, including duplication of *AP3* yielding the eu*AP3* gene lineage, as well as the origin of *AP1* (Kramer *et al.*, 1998; Litt and Irish, 2003).

Molecular phylogenetic analyses of the gene families involved in floral development are elucidating the important role that gene duplication has played in the evolution of flower development. The gene duplications and losses evident in gene family phylogenies can confuse discussions of functional evolution when the genes with equivalent function in different species are not orthologous. At the same time, orthology does not always coincide with strict functional equivalence (e.g. the discussion of *LIP* genes in Keck *et al.*, 2003). Given the lack of perfect correspondence between gene function and phylogeny, a clear distinction should be made between functional and phylogenetically based classifications of gene relationships (Becker and Theißen, 2003).

AP3/PI-like genes: an ancient duplication

The evolution of MADS-box genes has involved a series of gene duplications and subsequent diversification, as well as losses. Several investigators have conducted phylogenetic analyses of the floral MADS-box genes (e.g. Purugganan, 1997; Theißen *et al.*, 2000; Johansen *et al.*, 2002; Nam *et al.*, 2003; Becker and Theißen, 2003). For example, a duplication yielding the A and E + *AGL6* class genes occurred approximately 413 million years ago (mya) (Nam *et al.*, 2003), and the ages of several other prominent MADS-box gene duplications have also been estimated (e.g. Purugganan *et al.*, 1995; Purugganan, 1997; Nam *et al.*, 2003).

Whereas angiosperms possess two B-class paralogues (*AP3* = *DEF*, and *PI* = *GLO*), only one certain B-class homologue has been found in gymnosperms, suggesting that an ancient duplication led to the presence of the *AP3* and *PI* homologues. However, the accelerated rate of evolution of *AP3* and *PI* relative to other MADS-box genes precluded estimation of the age of the *AP3/PI* duplication by molecular clock-based substitution rate methods (Purugganan *et al.*, 1995; Purugganan, 1997; Kramer *et al.*, 1998; Nam *et al.*, 2003). Tree-based methods using a data set of over 20 new *AP3* and *PI* gene sequences for basal angiosperms estimated that the *AP3/PI* duplication occurred approximately 260 mya (range of 230–290 mya) (Kim *et al.*, 2004). This date places the duplication shortly after the split between extant gymnosperms and angiosperms and on the 'stem' lineage of extant flowering plants. This indicates that the *AP3/PI* duplication occurred perhaps 100 million years before the oldest fossil flowers (generally placed at 125–131.8 mya; Hughes, 1994). Thus, this suggests that the joint expression of *AP3* and *PI* did not immediately result in the formation of petals, structures for which they control development in extant angiosperms, because no such structures are present in the fossil record at that time. This raises the question: what was the early (pre-angiosperm) role of the *AP3* and *PI* homologues? The co-expression of *AP3* and *PI* homologues could reflect an evolutionary innovation of animal-attractive, petal-like organs well before the appearance of angiosperms in the fossil record. Indeed, some fossil, non-angiosperm seed plants from the appropriate timeframe, such as the glossopterids, had sterile spathe-like organs attached to male or female reproductive structures (e.g. Crane, 1985).

Transcription-factor complexes: early flexibility?

A striking result of Kim *et al.* (2004) is the strong similarity between *Amborella* AP3 and PI C-domain amino acid sequences (Fig. 10.9). The C domains, as well as K- and MADS-domains, signal the assembly of multimers for several MADS proteins in core eudicots (Egea-Cortines *et al.*, 1999; Ferrario *et al.*, 2003). Indeed, higher-order multimers are probably the active state of B-function MADS-box proteins (Egea-Cortines *et al.*, 1999; Honma and Goto, 2000; Theißen, 2001; Ferrario *et al.*, 2003).

Heterodimerization of AP3 and PI proteins is required for DNA binding in the core eudicots that have been studied. However, PI/PI homodimers are possible in some monocots, at least *in vivo*, although it is not clear whether these can bind DNA (Fig. 10.10). (Even the *Arabidopsis* PI proteins can form homodimers, but these cannot bind DNA (Riechmann *et al.*, 1996).) Selective fix-

ation of heterodimerization has been hypothesized for the morphologically more stereotyped core eudicots (Winter *et al.*, 2002). However, the phylogenetic point at which heterodimerization became enforced is not yet clear. Hints from sequence comparison suggest that *Amborella*, and perhaps some other basal lineages (e.g. Nymphaeaceae), may have retained some capacity for B-class homodimerization.

Because *Amborella* proteins may have K-domain heterodimerization signals that differ from those in *Arabidopsis* and other well-studied angiosperms, the data suggest that *Amborella* B-function proteins may have different dimerization dynamics from monocots and core eudicots. Two different *AP3* genes are present in *Amborella* (*Amborella AP3-1* and *AP3-2*). *Amborella* may be capable of forming PI/PI, AP3-1/AP3-1 and AP3-2/AP3-2 homodimers and perhaps AP3-1/AP3-2 heterodimers. Furthermore, if the amino acid residues in the K1 subdomain of *Amborella* AP3 and PI are not sufficient to prevent heterodimerization, but only weaken it, perhaps *Amborella* can also form AP3-1/PI and AP3-2/PI heterodimers (Fig. 10.10). Recent studies using transgenic *Arabidopsis* plants indicate that the C terminus of AP3 is sufficient to confer AP3 functionality on the paralogous PI protein (Lamb and Irish, 2003). This finding, when considered in the light of *Amborella* and its indistinct AP3 and PI C domains, also supports the possibility of AP3-1/AP3-2 heterodimerization.

A simple extension of the *Arabidopsis* 'quartet model' for MADS protein function (Fig. 10.10; Theißen, 2001) can accommodate both

(a)

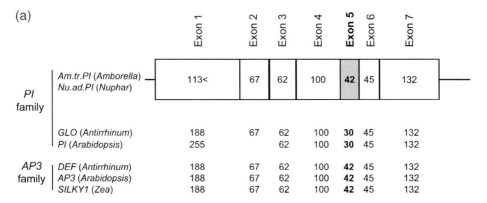

		Exon 1	Exon 2	Exon 3	Exon 4	**Exon 5**	Exon 6	Exon 7
PI family	*Am.tr.PI (Amborella)* / *Nu.ad.PI (Nuphar)*	113<	67	62	100	42	45	132
	GLO (Antirrhinum)	188	67	62	100	**30**	45	132
	PI (Arabidopsis)	255		62	100	**30**	45	132
AP3 family	*DEF (Antirrhinum)*	188	67	62	100	**42**	45	132
	AP3 (Arabidopsis)	188	67	62	100	**42**	45	132
	SILKY1 (Zea)	188	67	62	100	**42**	45	132

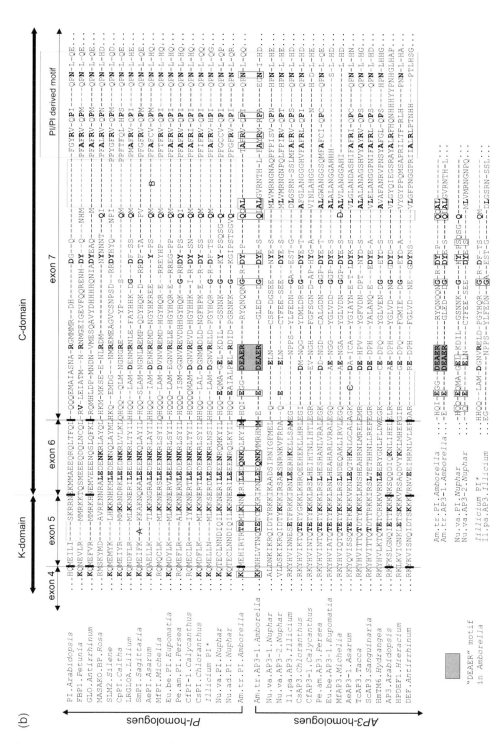

Fig. 10.9. *AP3/PI* gene structure in basal angiosperms. (a) (Opposite.) The size of exon 5 in *Amborella* and *Nuphar* compared with that observed in *Arabidopsis* and other eudicots. (b) Comparison of *AP3/PI* domain similarities in *Amborella* and other basal angiosperms (from Kim *et al.*, 2004).

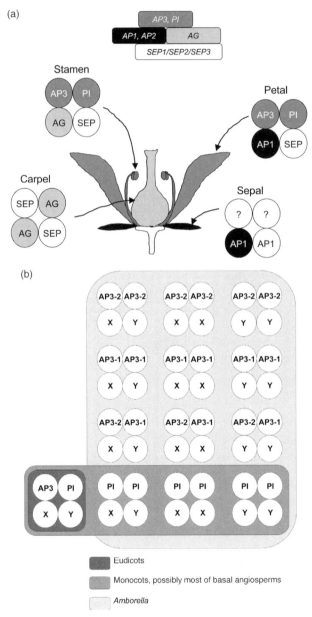

Fig. 10.10. Transcription-factor complexes. (a) Quartet model of floral organ specification in *Arabidopsis* (Theißen, 2001). (b) Extension of the quartet model for determination of floral organ identity to include *Amborella*. MADS protein tetramers are shown schematically (as in (a); see Theißen, 2001). One possible model is presented with the following assumptions: (i) AP3/PI obligate heterodimerization occurs in core eudicots; (ii) additional MADS proteins X and Y are available in cells; (iii) PI/PI dimers can tetramerize with all three configurations of X and Y in monocots (Winter *et al.*, 2002) and possibly other basal angiosperms, whereas only the XY configuration is possible in core eudicots; (iv) *Amborella* AP3/AP3 and PI/PI dimers possess similar capacities for C-domain tetramerization specification; and (v) *Amborella* AP3-1 and AP3-2 proteins are able heterodimerize. Tetramer potential would be 12:4:1 for *Amborella*, monocots (and perhaps other basal angiosperms) and core eudicots, respectively. If *Amborella* AP3 and PI can also heterodimerize to some extent, the ratio of possible quartets in *Amborella*:monocots:core eudicots becomes 18:4:1.

the monocot and *Amborella* cases. In this hypothetical example, the MADS protein tetramer AP3/PI/X/Y specifies a particular organ identity in *Arabidopsis*. Assuming that homodimerization is the ancestral state for B-function proteins (Winter *et al.*, 2002), we can invoke a model whereby PI/PI homodimers, as known from and argued to have functional significance in monocots (Münster *et al.*, 2001; Winter *et al.*, 2002), are more flexible in their protein partnerships. This scenario could call for the possibility of PI/PI/X/Y, PI/PI/X/X and PI/PI/Y/Y tetramers in monocots and most other basal angiosperms (Fig. 10.10).

Although this hypothesis must be tested using gel shift and yeast 2-, 3- and 4-hybrid assays (Winter *et al.*, 2002; Ferrario *et al.*, 2003), the implications of this model (Fig. 10.10; Kim *et al.*, 2004) are that *Amborella* would have 12 times more tetramer possibilities than a core eudicot and three times greater tetramer potential than a monocot or other basal angiosperm with limited homodimerization potential. Given that *Amborella* may have the capacity to form more different protein quartets for a given number of genes, it should possess more distinct controls (and therefore flexibility) over organ identity and development than any other flowering plant. The waterlily *Nuphar* also has considerable C-domain similarity for the AP3 and PI proteins, and this may be sufficient to provide the Nymphaeaceae with at least some extra tetramerization possibilities. By contrast, *Illicium*, which represents the next most basal clade of angiosperms after the Nymphaeaceae (Austrobaileyales; Fig. 10.2), has lost most of the C-domain AP3/PI similarity. Furthermore, a deletion in the K domain of *PI* (Fig. 10.9) first appears in *Illicium* (Austrobaileyales) and is fixed in all other angiosperms. Although most flowering plants are canalized in their possibilities for heterodimerization and multimer formation (Theißen, 2001), several eudicots (e.g. Ranunculales, Kramer *et al.*, 2003; *Petunia*, Ferrario *et al.*, 2003) and monocots (Münster *et al.*, 2001) may have regained some potential for developmental flexibility by a different mechanism involving later duplications of *AP3* homologues, *PI* homologues, or both.

The data suggest that the evolution of the control of B-function MADS-box genes in the development of the earliest flowers was dynamic, with different 'experiments' tried. *Amborella*, which may be the most flexible living angiosperm in its developmental genetics, is the sole surviving representative of its clade. Some of this same biochemical flexibility may also be present in waterlilies. These are testable hypotheses, to be pursued with more rigorous molecular investigations. None the less, *Amborella* B-function proteins would have represented a considerable increase in complexity over the demonstrated B-protein homodimerization known for conifers (Sundström *et al.*, 1999) and *Gnetales* (Winter *et al.*, 2002). However, the amino acid structural evidence suggests that this flexibility was rapidly lost before the bulk of the angiosperm radiation occurred. The unique phylogenetic position of *Amborella* and waterlilies, coupled with their apparently ancestral and flexible mode of B-gene function, make them model organisms that should be studied more intensively.

The Early Floral Genome

Rice and *Arabidopsis*: similarity in gene copy number

Early angiosperms clearly had the basic framework of B- and C-function genes in place. However, these genes are only a few of those involved in floral organ development and identity (Fig. 10.6) (Zhao *et al.*, 2001a). Complete sequencing of the rice and *Arabidopsis* genomes has made it possible to conduct comparisons of floral gene homologues shared by a derived monocot and a derived eudicot. These comparisons reveal a striking similarity in the number of homologues of genes involved in floral identity in the two species (Fig. 10.11). The similarity in gene family sizes is surprising given that the genome of rice is four times larger than that of *Arabidopsis* and the predicted number of protein-coding genes is just over twice as large in rice (Goff *et al.*, 2002; Yu *et al.*, 2002). Similarities in gene family size may be due to conservation of orthologous sets of rice and *Arabidopsis* genes or conservation of

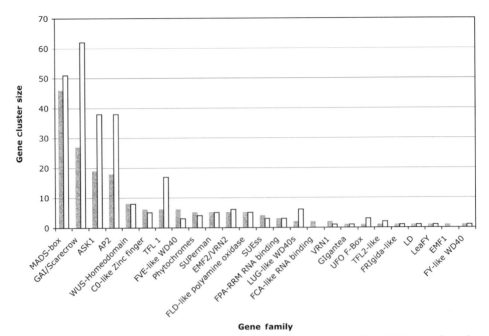

Fig. 10.11. Similarity of size of gene families that contain key floral regulators (Fig. 10.6) in two distantly related flowering plant species, *Arabidopsis* (shaded bars) and rice (open bars). The proteomes of *Arabidopsis* (26,993 proteins) and rice (62,657 proteins) were gathered into 20,934 'tribes' or putative gene families (Wall *et al.*, unpublished) using the Markov-clustering method of Enright *et al.* (2002).

gene number following independent gene duplications and losses after the split of the monocot and eudicot lineages at least 125 mya. Phylogenetic analyses of gene families allow us to test these hypotheses and investigate the evolution of gene function.

Basal angiosperms: a diverse tool kit of floral genes

As more data have emerged from major EST (expressed sequence tag) projects on angiosperms, it has become possible to make broader comparisons of some of the numerous genes and gene families that are involved in normal floral development. Particularly useful have been ESTs obtained for several basal angiosperms (www. floralgenome.org). Many genes identified in rice and *Arabidopsis* have clear homologues in basal angiosperms. Given that extant basal angiosperms represent old lineages (the waterlily lineage, for example, is among the oldest in the fossil record of

angiosperms; Friis *et al.*, 2001), the data suggest that early angiosperms possessed a diverse tool kit of floral genes.

As more genes are examined phylogenetically, it is also clear that there are different types of floral gene histories. In some cases, the gene phylogenies roughly track organismal phylogeny. This is the case for the B-class genes, *PI* and *AP3*. The single-copy gene *Gigantea* also appears to track organismal phylogeny (Chanderbali *et al.*, unpublished). However, several gene families present in rice and *Arabidopsis* exhibit an array of different evolutionary patterns (see below).

AP3/PI-*like genes*

Phylogenetic analyses of *AP3* and *PI* homologues (Kim *et al.*, 2004) resulted in gene trees that generally track the organismal phylogeny (Fig. 10.12) inferred from analyses of large data sets of plastid, mitochondrial and nuclear rDNA sequences. *Amborella* and *Nuphar* (Nymphaeaceae) appear as sisters to all other angiosperms, in complete

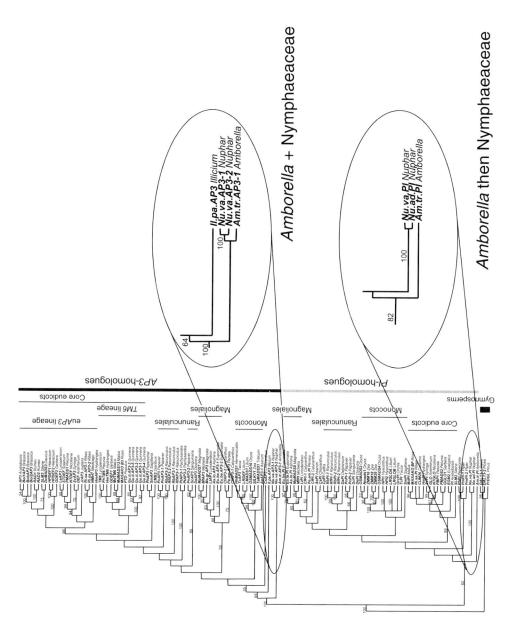

Fig. 10.12. *AP3* and *PI* gene trees. Strict consensus of 72 equally most parsimonious trees (shown as a phylogram) using M-, I-, K- and C-domain regions of amino acid sequences. Numbers above branches are bootstrap values; only values above 50% are indicated (from Kim *et al.*, 2004).

agreement with the organismal phylogeny (see references in 'Background'). Several clades of *AP3* and *PI* homologues that correspond to well-supported organismal clades were consistently recognized by Kim *et al.* (2004), including Magnoliales and monocots. The eu*AP3* gene clade, which was previously described (Kramer *et al.*, 1998), was recovered in most analyses.

SHAGGY-*like kinases*

The *SHAGGY/GSK3*-like kinases are non-receptor Ser-Thr kinases that play numerous roles in plants and animals (Kim and Kimmel, 2000). The rice and *Arabidopsis* proteomes include 69 and 79 *SHAGGY*-like kinases, respectively, but these genes can be subdivided into smaller gene families. For example, ten *Arabidopsis* genes were identified as forming a clade with *SHAGGY* itself (*AtSK* genes; Dornelas *et al.*, 2000). The *AtSK* family was shown to form four subclades in a phylogenetic analysis (Charrier *et al.*, 2003): (i) *AtSK41* and *AtSK42* formed a subclade sister to the remaining genes, which were weakly supported as a clade (bootstrap support < 50%); (ii) *AtSK31* and *AtSK32* formed a second subclade sister to the remaining genes, which formed a well-supported (88% bootstrap) clade (this clade was composed of the two remaining subclades, each of which received strong bootstrap support); (iii) *AtSK21*, *AtSK22* and *AtSK23* (100%); and (iv) *AtSK11*, *AtSK12*, *AtSK13* (98%). The *AtSK* loci appear to have diverse functions. Mutant-based analyses indicate that *AtSK11* and *AtSK12* have a role in floral development; expression analyses suggest that *AtSK31* is flower-specific (Charrier *et al.*, 2003). The *SHAGGY*-like kinases are also involved in plant responses to stress.

The Floral Genome Project research consortium has obtained ESTs for a number of *SHAGGY*-like kinase genes in basal angiosperms. Yoo *et al.* (2005) conducted a phylogenetic analysis of all *SHAGGY*-like kinase genes available in public databases, as well as the ESTs from basal angiosperms (Fig. 10.13). Plant *SHAGGY*-like kinase genes form a well-supported clade distinct from those of animals (see also Charrier *et al.*, 2003). Across all angiosperms, Yoo *et al.* identified four

clades of *SHAGGY*-like kinase genes that mirrored the *AtSK* subgroups reported for *Arabidopsis*. Importantly, *SHAGGY*-like kinase genes from rice and from basal angiosperms were also represented in these four clades. For example, *SHAGGY*-like kinase ESTs from basal angiosperms appeared in all four of the subclades noted (Fig. 10.13). These data indicate that *SHAGGY*-like kinase genes diversified into these four well-marked clades early in angiosperm evolution.

SKP1-*like proteins*

Gene duplications within the angiosperms are also important in the history of the *SKP1* gene family. SKP1 (S-phase kinase-associated protein 1) is a core component of Skp1-Cullin-F-box protein (SCF) ubiquitin ligases and mediates protein degradation, thereby regulating many fundamental processes in eukaryotes such as cell-cycle progression, transcriptional regulation and signal transduction (Hershko and Ciechanover, 1998; Callis and Vierstra, 2000). Among the four components of the SCF complexes, Rbx1 and Cullin form a core catalytic complex, an F-box protein acts as a receptor for target proteins, and SKP1 links one of the variable F-box proteins with a Cullin (Zheng *et al.*, 2002). There is only one known functional SKP1 protein in human and yeast, and this unique protein is able to interact with different F-box proteins to ubiquinate different substrates (Ganoth *et al.*, 2001). In some plant and invertebrate species, however, there are multiple *SKP1* genes, which have evolved at highly heterogeneous rates (Farras *et al.*, 2001; Nayak *et al.*, 2002; Yamanaka *et al.*, 2002; Kong *et al.*, 2004). The extreme rate of heterogeneity observed among the 38 rice and 19 *Arabidopsis* *SKP1* homologues raised concerns that long-branch attraction may obscure true relationships in phylogenetic analyses of the entire gene family. For this reason, Kong *et al.* (2004) partitioned the original data set into subsets of genes with slow, medium and rapid rates of evolution and analysed each group separately. Most *SKP1* homologues observed in EST databases were included in the set of slowly evolving genes. In *Arabidopsis*, the slowly evolving *SKP1* genes were expressed more widely (in more tissues and more develop-

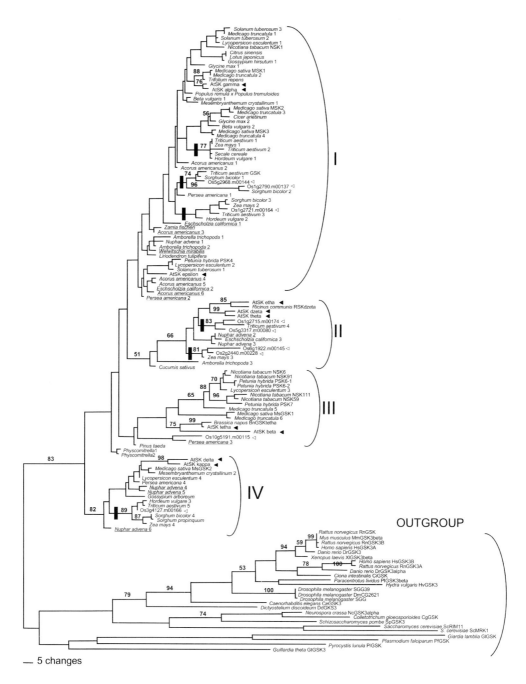

Fig. 10.13. *SHAGGY*-like kinase tree. Strict consensus of equally most parsimonious trees (shown as a phylogram) based on phylogenetic analysis of amino acid sequences. Closed triangle represents *GSK/SHAGGY*-like protein kinase from *Arabidopsis* and open triangle represents *Oryza*. Clade designations (I–IV) follow those given to *Arabidopsis* sequences (see text). ESTs provided by the Floral Genome Project (www.floralgenome.org) are underlined; monocot-specific clades are indicated by a vertical bar. Numbers above branches are bootstrap values; only values above 50% are indicated (from Yoo *et al.*, 2005).

mental stages) and at higher levels than the more rapidly evolving rice and *Arabidopsis SKP1* homologues. In addition, the strength of purifying selection was found to be significantly greater in the slowly evolving *Arabidopsis SKP1*-like genes (Kong *et al.*, 2004). Taken together, these results suggest that the slowly evolving *SKP1* homologues serve the most fundamental function(s) to interact with Cullin and F-box proteins.

The two slowly evolving *Arabidopsis SKP1*-like *1* and *2* genes, *ASK1* and *ASK2*, are important for vegetative and flower development and essential for male meiosis (Samach *et al.*, 1999; Yang *et al.*, 1999; Zhao *et al.*, 1999, 2001b, 2003). Slowly evolving *SKP1* homologues from other plant species usually have very similar sequences, suggesting that they may also serve similar fundamental functions (Kong *et al.*, 2004). Multiple slowly evolving *SKP1* homologues have been sampled in EST studies for a variety of angiosperm species, including *Liriodendron*, *Persea*, *Mesembryanthemum*, *Vitis*, *Medicago*, *Lotus*, *Rosa*, *Arabidopsis*, *Brassica*, *Gossypium*, *Helianthus* and *Solanum*. While relationships are poorly resolved across the angiosperm *SKP1* phylogeny, it is clear that gene duplication events that have occurred throughout angiosperm history have contributed to this set of conserved genes (Fig. 10.14). Recent duplication has increased *SKP1* gene diversity in *Brassica*, *Helianthus*, *Medicago* and *Triticum*. In contrast, conserved paralogues in *Liriodendron*, *Persea*, *Mesembryanthemum*, *Vitis*, *Lotus*, *Rosa*, *Arabidopsis*, *Gossypium* and Solanaceae are the products of ancient duplication events. Interestingly, the basal position of the sole *Amborella SKP1* homologue sampled from a set of 10,000 ESTs suggests that all of these duplications occurred after the origin of the angiosperms (Fig. 10.14).

Homology of Floral Organs: Extending Out from New Models

Can we use expression data to determine organ identity?

The homology of characters leading to the assessment of organ identity can be inferred from the mature phenotype, from the posi-
tions and function of organs within a flower, from developmental morphology, from phylogeny, from developmental genetics, or a combination of these approaches (Albert *et al.*, 1998; Buzgo *et al.*, 2004a). Albert *et al.* (1998) were among the first to explore the topic of using gene expression data as one means of determining floral organ identity, and this application of expression information continues to be a topic of debate. As an example, there are now divergent definitions of perianth organs and interpretations of organ identity. Sepals typically are the outermost organs of the flower, whereas petals are conspicuous organs, typically of the second perianth whorl. The two outer floral whorls in *Tulipa* may be positionally homologous to sepals and petals, respectively. However, both whorls are morphologically petaloid, and, as noted, patterns of B-class gene expression in both whorls resemble those of eudicot petals (Kanno *et al.*, 2003). If gene expression patterns are conserved across the broad phylogenetic distances from *Arabidopsis* to tulip, then these data suggest homology of both whorls to petals. Alternatively, changes in expression patterns of B-function genes may have occurred during angiosperm evolution (e.g. 'shifting borders'; Kramer *et al.*, 2003); if so, similar expression patterns may not indicate homology. Extension of expression and functional data to homology assessment of the lodicules of grasses is even more challenging. Although lodicules occur in the position of petals and exhibit B-class gene expression (e.g. Schmidt and Ambrose, 1998), as predicted for petals, their unique morphology suggests that they may not be 'petals', despite their position and gene expression patterns. Thus, morphology, developmental data and genetic data may provide conflicting evidence of homology (organ identity) and yet ultimately a more complete, and complex, view of a structure (Buzgo *et al.*, 2004a).

Eupomatia: a case study

Eupomatia (Eupomatiaceae; Fig. 10.1) is a genus of two species that possess an unusual structure (a calyptra) that encloses and presumably protects the flower in bud. The origin of the calyptra has been debated. Some

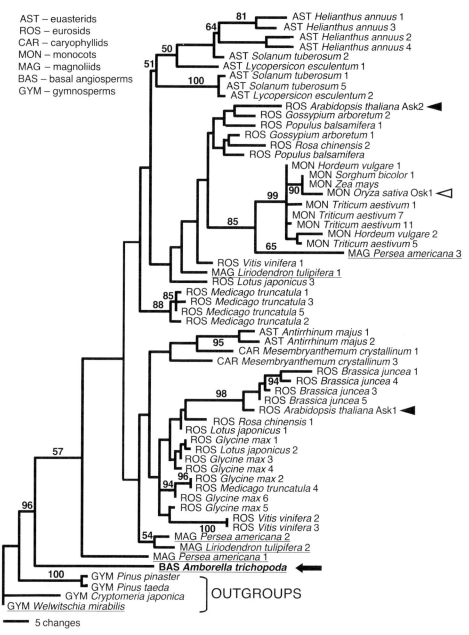

Fig. 10.14. Phylogenetic relationships of slowly evolving SKP1 proteins from select rosid (ROS, ROS1, ROS2), euasterid (AST), monocot (MON), magnoliid (MAG), basal angiosperm (BAS) and gymnosperm (GYM) species suggest that gene duplications have occurred both before and after the origin of major taxa within the angiosperms. The maximum likelihood tree shown is one of >2000 most parsimonious trees. Maximum parsimony bootstrap values higher than 50% are shown above or below the branch. Support for each branch was tested with 500 replicates of bootstrap analysis using random input order for each replicate. Note that most nodes on the tree are not well supported because the regions used for analysis are rather short (149 amino acids) and highly conserved. The taxonomic categories for these species follow Soltis *et al.* (2000). ESTs generated by the Floral Genome Project are underlined. Closed triangles indicate SKP1 homologues from *Arabidopsis* and open triangles those from rice. Modified from Kong *et al.* (2004).

have considered it to represent a modified perianth (Cronquist, 1981), but developmental data indicate that it represents a modified bract (Endress, 2003). Eupomatiaceae are closely related to the *Magnolia* family, members of which have a well-developed perianth of showy tepals, as well as bracts that enclose the flower.

Kim *et al.* (unpublished) examined the expression of A and B class genes in the calyptra. The B-function genes (*AP3* and *PI*) isolated from staminodes of *Eupomatia* species were strongly expressed in developing stamens, staminodes and carpels, but either not expressed or expressed weakly in the calyptra, at a level consistent with expression in leaves. As reviewed, recent studies suggest that in basal angiosperms and monocots an 'ancestral' ABC model (compared with *Arabidopsis*) operates with B-function genes expressed throughout the perianth. Following this model (Fig. 10.8), the pattern of expression of floral genes in the calyptra of *Eupomatia* generally matches the expectations for a non-floral organ (such as a leaf or bract) rather than predictions for perianth, consistent with Endress's (2003) interpretation based on developmental morphology.

Thus, in some situations such as a clade of related taxa, floral gene expression may be useful in addressing the origins of enigmatic structures. There are caveats, however. In the case of a basal angiosperm such as *Eupomatia*, comparisons are better made in light of the proposed 'ancestral' ABC model, rather than the classic ABC model of *Arabidopsis* and other core eudicots.

Gene Evidence and the Origin of Flowers

An understanding of the nature of the flower in basal angiosperms should help in elucidating the evolutionary origin of the flower itself. Flowers differ so greatly from the reproductive structures of living and fossil gymnosperms that the origin of the flower has long been a famous question in evolutionary biology. Numerous hypotheses based on morphological, developmental and palaeobotanical studies have been proposed (reviewed in Stebbins, 1974; Crane, 1985;

Hughes, 1994), but each typically accounts for only a limited range of observations, and none is testable, unless revealing fossils happen to be discovered.

The Mostly Male Theory (Frohlich and Parker, 2000; Frohlich, 2001, 2002, 2003) arose through studies of the *LFY* gene, in particular from the observations that *LFY* is single copy in diploid angiosperms, but that there are two copies present in all extant gymnosperm groups, owing to an ancient duplication predating the divergence of angiosperms from extant gymnosperms. Angiosperms would have inherited both copies of *LFY*, but one of these has been lost. Expression of the gymnosperm *LFY*s in pine, coupled with the role of *LFY* in angiosperms, suggests that one gymnosperm *LFY* helps to specify the male reproductive unit while the other helps to specify the female unit. Angiosperms retain only the male-specifying unit, suggesting that the angiosperm flower may derive more from the male reproductive structure of the gymnosperm ancestor, rather than from the female unit. Other data from extant plants and from fossils are consistent with this view, and together suggest that bisexuality of the angiosperm reproductive structure may have arisen when the ovule antecedent became ectopic upon microsporophylls of a reproductive unit resembling that of the fossil gymnosperm group, Corystospermales. The theory is consistent with the flower antecedent originally consisting of stamen- and carpel-like structures, but without a perianth, and with insect-attractive features and insect pollination long predating the full elaboration of the flower (Frohlich, 2001), as suggested by the timing of the *AP3/PI* gene duplication described above.

Other recent gene-based hypotheses also focus on the origin of angiosperm bisexual reproductive structures from the unisexual structures typical of gymnosperms. Albert *et al.* (2002) proposed an alternative to Mostly Male. Their hypothesis stresses the possible functional replacement of one copy of *LFY* by the other, resulting in expression of both male- and female-specific genes in the reproductive unit. Theißen *et al.* (2002) suggest that changed expression patterns of B-class genes could have generated bisexual reproductive structures from either male or female cones of gymnosperms. Both of these hypotheses, but especially that of Theißen *et al.* (2002), suggest relatively equal participation of male- and female-derived gymnosperm genes in the organization of the flower. If the distinction between gymnosperm male and female structures is fully determined by differences in B-gene expression, then changes in B-gene expression patterns should bring the full panoply of male- or female-specific genes into the (formerly) unisexual cone of the other gender.

The relative contribution of male- and female-specific gymnosperm genes to those active in the flower constitutes a direct test of these theories. The gene discovery, gene phylogeny and gene expression studies of the Floral Genome Project (see below) will provide this test.

Future Prospects

The evolution of flower morphology is being elucidated through research on the genetic mechanisms of reproductive development in diverse angiosperms. *Arabidopsis* has figured prominently in these studies. Although *Arabidopsis* was the first plant to have its nuclear genome completely sequenced (in 2000), earlier genetic studies of *Arabidopsis* (beginning in the late 1980s) paved the way for evolutionary interpretations of the molecular processes underlying floral diversity.

The evolutionary genetics of floral morphology

While the simplicity of the ABC model made it seem that diversions from 'normal' sepal/petal/stamen/carpel identity among angiosperms might be explored through comparative expression studies of A-, B- and C-function genes (as has been done analogously with homeobox genes in the segmental evolution of animals), the greater genetic complexity now recognized behind flower development indicates that this view requires revision. For example, one author of this contribution once felt that explaining the homeotic evolution of a second corolla (fused petal) whorl in the Hawaiian genus *Clermontia* (Campanulaceae) would be a simple issue of demonstrating ectopic expression of B-function genes in the first, normally sepalar, whorl. However, with the generality of the A function now in question, the mechanistic basis for the double-corolla phenotype in *Clermontia* might be other than simply out-of-place B-function gene expression. Furthermore, if altered B-gene expression is the cause, various mechanisms could generate such modified expression. The naturally occurring mutation could be within a B-function coding sequence, or perhaps in its transcriptional promoter, which might have elements that fine-tune spatio-temporal expression. It could equally well be that the double-corolla lesion is in a different gene, the product of which normally interacts with a B-function gene's promoter to regulate where it expresses. In other words, the gene that normally excludes B-function genes from the first whorl of *Clermontia*, or any gene upstream of it in its developmental regulatory cascade, could be the culprit. Analysing this problem will not be simple, because *Clermontia*, unlike weedy *Arabidopsis*, is a small tree that is much less tractable to genetic studies that require progeny analysis. Such 'forward' genetic studies, starting with a phenotypically recognizable mutation and culminating with the gene linked to it, may be difficult to accomplish outside of the model plants. Therefore, investigators are turning more and more to 'reverse' genetic approaches that start with a gene sequence that is suspected to have a particular function (e.g. through a molecular evolutionary relationship to genes of known function) and work backwards to establish this function through transgenic experiments that over- and/or underexpress the gene's protein product. In this way the *Gerbera SEP*-like genes were characterized.

The Floral Genome Project, a large-scale effort to identify genes specific to flower development and those linked to floral diversification, is under way (www.floralgenome.org; overview in Soltis *et al.*, 2002). The Floral Genome Project is sequencing genes expressed during the earliest stages of floral development in diverse lineages of angiosperms, particularly basal angiosperms such as *Amborella*, waterlilies, tulip tree (Magnoliaceae), avocado (Lauraceae) and *Acorus* (sister to all other monocots), plus *Eschscholzia* (poppy, a basal eudicot), and will obtain expression data for genes in a subset of these taxa. Genes in common with or distinct to these species should provide valuable molecular tools for the next generation of plant evolutionary developmental research.

Acknowledgements

This research was supported in part by an NSF Plant Genome Grant, DBI-0115684. We thank Matyas Buzgo, William Farmerie, Lena Landherr, Yi Hu, Marlin Druckenmeyer, Sheila Plock and John Carlson for technical assistance in building cDNA libraries and generating EST data.

References

Albert, V.A., Gustafsson, M.H.G. and Di Laurenzio, L. (1998) Ontogenetic systematics, molecular developmental genetics, and the angiosperm petal. In: Soltis, D.E., Soltis, P.S. and Doyle, J.J. (ed.) *Molecular Systematics of Plants* II. Kluwer, New York.

Albert, V.A., Oppenheimer, D. and Lindqvist, C. (2002) Pleiotropy, redundancy and the evolution of flowers. *Trends in Plant Science* 7, 297–301.

APG II (2003) An update of the Angiosperm Phylogeny Group classification for the orders and families of flowering plants. *Botanical Journal of the Linnean Society* 141, 399–436.

Arber, E.A.N. and Parkin, J. (1907) On the origin of angiosperms. *Journal of the Linnean Society, Botany* 38, 29–80.

Barkman, T.J., Chenery, G., McNeal, J.R., Lyons-Weiler, J. and DePamphilis, C.W. (2000) Independent and combined analysis of sequences from all three genomic compartments converge to the root of flowering plant phylogeny. *Proceedings of the National Academy of Sciences USA* 97, 13166–13171.

Baum, D.A., Doebley, J., Irish, V.F. and Kramer, E.M. (2002) Response: Missing links: the genetic architecture of flower and floral diversification. *Trends in Plant Science* 7, 31–34.

Becker, A. and Theißen, G. (2003) The major clades of MADS-box genes and their role in the development and evolution of flowering plants. *Molecular Phylogenetics and Evolution* 29, 464–489.

Bessey, C.E. (1915) The phylogenetic taxonomy of flowering plants. *Annals of the Missouri Botanical Garden* 2, 109–164.

Borsch, T., Hilu, K.W., Quandt, D., Wilde, V., Neinhuis, C. and Barthlott, W. (2003) Non-coding plastid *trnT-trnF* sequences reveal a well resolved phylogeny of basal angiosperms. *Journal of Evolutionary Biology* 16, 558–576.

Buzgo, M., Soltis, D.E., Soltis, P.S. and Hong, M. (2004a) Toward a comprehensive integration of morphological and genetic studies of floral development. *Trends in Plant Science* 9, 164–173.

Buzgo, M., Soltis, P.S. and Soltis, D.E. (2004b) Floral developmental morphology of *Amborella trichopoda* (Amborellaceae). *International Journal of Plant Sciences* (in press).

Callis, J. and Vierstra, R.D. (2000) Protein degradation in signaling. *Current Opinion in Plant Biology* 3, 381–386.

Charrier, B., Champion, A., Henry, Y. and Kreis, M. (2003) Expression profiling of the whole *Arabidopsis* Shaggy-like kinase multigene family by real-time reverse transcriptase-polymerase chain reaction. *Plant Physiology* 130, 577–590.

Coen, E.S. and Meyerowitz, E.M. (1991) The war of the whorls: genetic interactions controlling flower development. *Nature* 353, 31–37.

Coen, E.S., Momero, J.M., Doyle, S., Elliot, R., Murphy, G. and Carpenter, R. (1990) *Floricaula*: a homeotic gene required for flower development in *Antirrhinum majus. Cell* 63, 1311–1322.

Crane, P.R. (1985) Phylogenetic analysis of seed plants and the origin of angiosperms. *Annals of the Missouri Botanical Garden* 72, 716–793.

Crane, P.R., Friis, E.M. and Pedersen, K.R. (1995) The origin and early diversification of angiosperms. *Nature* 374, 27–33.

Cronquist, A. (1968) *The Evolution and Classification of Flowering Plants.* Houghton Mifflin, Boston, Massachusetts.

Cronquist, A. (1981) *An Integrated System of Classification of Flowering Plants.* Columbia University Press, New York.

De Bodt, S., Raes, J., Van de Peer, Y. and Theißen, G. (2003) And then there were many: MADS goes genomic. *Trends in Plant Science* 8, 475–483.

De Candolle, A.P. (1824–1873) *Prodomus systematis naturalis regni vegetabilis.* Déterville, Paris.

Dilcher, D.L. and Crane, P.R. (1984) *Archaeanthus:* an early angiosperm from the Cenomanian of the western interior of North America. *Annals of the Missouri Botanical Garden* 71, 351–383.

Donoghue, M.J. and Doyle, J.A. (1989) Phylogenetic analysis of angiosperms and the relationships of Hamamelidae. In: Crane, P.R. and Blackmore, S. (eds) *Evolution, Systematics, and Fossil History of Hamamelidae.* Clarendon Press, Oxford, pp. 17–45.

Dornelas, M.C., Lejeune, B., Dron, M. and Kreis, M. (2000) The *Arabidopsis SHAGGY*-related protein kinase (*ASK*) gene family: structure, organization and evolution. *Gene* 212, 249–257.

Doust, A.N. (2000) Comparative floral ontogeny in Winteraceae. *Annals of the Missouri Botanical Garden* 87, 366–379.

Doust, A.N. (2001) The developmental basis of floral variation in *Drimys winteri* (Winteraceae). *International Journal of Plant Sciences* 162, 697–717.

Doyle, J.A. and Endress, P.K. (2000) Morphological phylogenetic analysis of basal angiosperms: comparison and combination with molecular data. *International Journal of Plant Sciences* 161, S121–S153.

Doyle, J.A., Donoghue, M.J. and Zimmer, E. (1994) Integration of morphological and ribosomal RNA data on the origin of angiosperms. *Botanical Review* 52, 321–431.

Eames, A.J. (1931) The vascular anatomy of the flower with refutation of the theory of carpel polymorphism. *American Journal of Botany* 18, 147–188.

Egea-Cortines, M., Saedler, H. and Sommer, H. (1999) Ternary complex formation between the MADS-box proteins *SQUAMOSA, DEFICIENS* and *GLOBOSA* is involved in the control of floral architecture in *Antirrhinum majus. EMBO Journal* 18, 5370–5379.

Endress, P.K. (1980) Ontogeny, function and evolution of extreme floral construction in Monimiaceae. *Plant Systematics and Evolution* 134, 79–120.

Endress, P.K. (1986) Floral structure, systematics and phylogeny in Trochodendrales. *Annals of the Missouri Botanical Garden* 73, 297–324.

Endress, P.K. (1987) The early evolution of the angiosperm flower. *Trends in Ecology and Evolution* 2, 300–304.

Endress, P.K. (1994a) *Diversity and Evolutionary Biology of Tropical Flowers.* Cambridge University Press, Cambridge.

Endress, P.K. (1994b) Floral structure and evolution of primitive angiosperms: recent advances. *Plant Systematics and Evolution* 192, 79–97.

Endress, P.K. (2001) The flowers in extant basal angiosperms and inferences on ancestral flowers. *International Journal of Plant Sciences* 162, 1111–1140.

Endress, P.K. (2003) Early floral development in Eupomatiaceae. *International Journal of Plant Sciences* 164, 489–503.

Endress, P.K. and Igersheim, A. (2000) Gynoecium structure and evolution in basal angiosperms. *International Journal of Plant Sciences* 161, S211–S223.

Endress, P.K. and Sampson, F.B. (1983) Floral structure and relationships of the Trimeniaceae (Laurales). *Journal of the Arnold Arboretum* 64, 447–473.

Endress, P.K., Igersheim, A., Sampson, F.B. and Schatz, G.E. (2000) Floral structure of *Takhtajania* and its systematic position in Winteraceae. *Annals of the Missouri Botanical Garden* 87, 347–365.

Engler, A. and Prantl, K. (eds) (1887–1915) *Die Natürlichen Pflanzenfamilien,* 1st edn. W. Engelmann, Leipzig.

Engler, A. and Prantl, K. (eds) (1959) *Die Natürlichen Pflanzenfamilien,* 2nd edn. W. Engelmann, Leipzig.

Enright, A.J., Van Dongen, S. and Ouzounis, C.A. (2002) An efficient algorithm for large scale detection of protein families. *Nucleic Acids Research* 30, 1575–1584.

Erbar, C. and Leins, P. (1981) Zur Spirale in Magnolienbluten. *Beitrage zur Biologie der Pflanzen* 56, 225–241.

Erbar, C. and Leins, P. (1983) Zur Sequenz von Blutenorganen bei enigen Magnoliiden. *Botanische Jahrbucher für Systematik* 103, 433–449.

Farras, R., Ferrando, A., Jasik, J., Kleinow, T., Okresz, L., Tiburcio, A., Salchert, K., Del Pozo, C., Schell, J. and Koncz, C. (2001) SKP1–SnRK protein kinase interactions mediate proteasomal binding of a plant SCF ubiquitin ligase. *EMBO Journal* 20, 2742–2756.

Ferrario, S., Immink, R.G., Shchennikova, A., Busscher-Lange, J. and Angenent, G.C. (2003) The MADS box gene *FBP2* is required for the *SEPALLATA* function in *Petunia. Plant Cell* 15, 914–925.

Franks, R.G., Wang, C., Levin, J.Z. and Liu, Z. (2002) SEUSS, a member of a novel family of plant regulatory proteins, represses floral homeotic gene expression with LEUNIG. *Development* 129, 253–263.

Friis, E.M., Pedersen, K.R. and Crane, P.R. (1994) Angiosperm floral structures from the Early Cretaceous of Portugal. *Plant Systematics and Evolution, Supplement* 8, 31–49.

Friis, E.M., Pedersen, K.R. and Crane, P.R. (2000) Reproductive structure and organization of basal angiosperms from the early Cretaceous (Barremian or Aptian) of western Portugal. *International Journal of Plant Science* 161, S169–S182.

Friis, E.M., Pedersen, K.R. and Crane, P.R. (2001) Fossil evidence of waterlilies (Nymphaeales) in the Early Cretaceous. *Nature* 410, 357–360.

Friis, E.M., Doyle, J.A., Endress, P.K. and Leng, Q. (2003) *Archaefructus* – angiosperm precursor or specialized early angiosperm? *Trends in Plant Science* 8, 369–373.

Frohlich, M.W. (2001) A detailed scenario and possible tests of the Mostly Male theory of flower evolutionary origins. In: Zelditch, M.L. (ed.) *Beyond Heterochrony: the Evolution of Development.* Wiley-Liss, New York, pp. 59–104.

Frohlich, M.W. (2002) The Mostly Male theory of flower origins: summary and update regarding the Jurassic pteridosperm *Pteroma.* In: Cronk, Q.C.B., Bateman, R.M. and Hawkins, J.A. (eds) *Developmental Genetics and Plant Evolution*, Systematics Association Special Volume Series 65. Taylor & Francis, New York, pp. 85–108.

Frohlich, M.W. (2003) An evolutionary scenario for the origin of flowers. *Nature Reviews Genetics* 4, 559–566.

Frohlich, M.W. and Parker, D.S. (2000) The Mostly Male theory of flower evolutionary origins: from genes to fossils. *Systematic Botany* 25, 155–170.

Ganoth, D., Borstein, G., Ko, Y.K., Larsen, B., Tyers, M., Pagano, M. and Hershko, A. (2001) The cell-cycle regulatory protein Cks1 is required for SCF (Skp2)-mediated ubiquitinylation of p27. *Nature Cell Biology* 3, 321–324.

Goff, S.A., Ricke, D., Lan, T.H., Presting, G., Wang, R., Dunn, M., Glazebrook, J., Sessions, A., Oeller, P., Varma, H., Hadley, D., Hutchison, D., Martin, C., Katagiri, F., Lange, B.M., Moughamer, T., Xia, Y., Budworth, P., Zhong, J., Miguel, T., Paszkowski, U., Zhang, S., Colbert, M., Sun, W.L., Chen, L., Cooper, B., Park, S., Wood, T.C., Mao, L., Quail, P., Wing, R., Dean, R., Yu, Y., Zharkikh, A., Shen, R., Sahasrabudhe, S., Thomas, A., Cannings, R., Gutin, A., Pruss, D., Reid, J., Tavtigian, S., Mitchell, J., Eldredge, G., Scholl, T., Miller, R.M., Bhatnagar, S., Adey, N., Rubano, T., Tusneem, N., Robinson, R., Feldhaus, J., Macalma, T., Oliphant, A. and Briggs, S. (2002) A draft sequence of the rice genome (*Oryza sativa* L. ssp. *japonica*). *Science* 296, 92–100.

Graham, S. and Olmstead, R. (2000) Utility of 17 chloroplast genes for inferring the phylogeny of the basal angiosperms. *American Journal of Botany* 87, 1712–1730.

Hahn, G.W. and Sommerville, C.R. (1988) Genetic control of morphogenesis in *Arabidopsis. Developmental Genetics* 9, 73–89.

Hershko, A. and Ciechanover, A. (1998) The ubiquitin system. *Annual Review of Biochemistry* 67, 425–479.

Hiepko, P. (1965) Vergleichend-morphologische und entwicklungs-geschichtliche Untersuchungen über das Perianth bei den Polycarpicae. *Botanische Jarbucher für Systematik* 84, 359–508.

Hilu, K.W., Borsch, T., Muller, K., Soltis, D.E., Soltis, P.S., Savolainen, V., Chase, M.W., Powell, M., Alice, L., Evans, R., Sauquet, H., Neinhuis, C., Slotta, T., Rohwer, J. and Catrou, L. (2003) Inference of angiosperm phylogeny based on *matK* sequence information. *American Journal of Botany* 90, 1758–1776.

Honma, T. and Goto, K. (2000) Complexes of MADS-box proteins are sufficient to convert leaves into floral organs. *Nature* 409, 525–529.

Hughes, N.F. (1994) *The Enigma of Angiosperm Origins.* Cambridge University Press, Cambridge.

Irish, V.F. (2003) The evolution of floral homeotic gene function. *BioEssays* 25, 637–646.

Johansen, B., Pedersen, L.B., Skipper, M. and Frederiksen, S. (2002) MADS-box gene evolution–structure and transcription patterns. *Molecular Phylogenetics and Evolution* 23, 458–480.

Kanno, A., Saeki, H., Kameya, T., Saedler, H. and Theißen, G. (2003) Heterotropic expression of class B floral homeotic genes supports a modified ABC model for tulip (*Tulipa gesneriana*). *Plant Molecular Biology* 52, 831–841.

Keck, E., McSteen, P., Carpenter, R. and Coen, E. (2003) Separation of genetic functions controlling organ identity in flowers. *EMBO Journal* 22, 1058–1066.

Kim, I. and Kimmel, A.R. (2000) GSKs, a master switch regulating cell-fate specification and tumorigenesis. *Current Opinion in Genetics and Development* 10, 508–514.

Kim, S., Yoo, M.-J., Albert, V.A., Farris, J.S., Soltis, P.S. and Soltis, D.E. (2004) Phylogeny and diversification of B-function MADS-box genes in angiosperms: evolutionary and functional implications of a 260-million-year-old duplication. *American Journal of Botany* 91 (in press).

Kong, H., Leebens-Mack, J., Ni, W., DePamphilis, C.W. and Ma, H. (2004) Highly heterogeneous rates of evolution in the SKP1 gene family in plants and animals: functional and evolutionary implications. *Molecular Biology and Evolution* 21, 117–128.

Kosuge, K. (1994) Petal evolution in Ranunculaceae. *Plant Systematics and Evolution* 152, 185–191.

Kotilainen, M., Elomaa, P., Uimari, A., Albert, V.A., Yu, D. and Terri, T.H. (2000) GRCD1, an AGL2-like MADS box gene, participates in the C function during stamen development in *Gerbera hybrida. Plant Cell* 12, 1893–1902.

Kramer, E.M. and Irish, V.F. (1999) Evolution of genetic mechanisms controlling petal development. *Nature* 399, 144–148.

Kramer, E.M. and Irish, V.F. (2000) Evolution of the petal and stamen developmental programs: evidence from comparative studies of the lower eudicots and basal angiosperms. *International Journal of Plant Sciences* 161, S29–S40.

Kramer, E.M., Dorit, R.L. and Irish, V.F. (1998) Molecular evolution of genes controlling petal and stamen development: duplication and divergence within the *APETALA3* and *PISTILLATA* MADS-box gene lineages. *Genetics* 149, 765–783.

Kramer, E.M., Di Stillio, V.S. and Schluter, P.M. (2003) Complex patterns of gene duplication in the *Apetala3* and *Pistillata* lineages of the Ranunculaceae. *International Journal of Plant Sciences* 164, 1–11.

Kubitzki, K. (1987) Origin and significance of trimerous flowers. *Taxon* 36, 21–28.

Lamb, R.S. and Irish, V.F. (2003) Functional divergence within the *APETALA3/PISTILLATA* floral homeotic gene lineages. *Proceedings of the National Academy of Sciences USA* 100, 6558–6563.

Leins, P. and Erbar, C. (1985) Ein Beitrag zur Blutenentwicklung der Aristolochiaceen, einer Vermittlergruppe zu den Monokotylen. *Botanische Jahrbucher für Systematik* 107, 343–368.

Litt, A. and Irish, V.F. (2003) Duplication and diversification in the *APETALA1/FRUITFULL* floral homeotic gene lineage: implications for the evolution of floral development. *Genetics* 165, 821–833.

Lu, R., Martin-Hernandez, A.M., Peart, J.R., Malcuit, I. and Baulcombe, D.C. (2003) Virus-induced gene silencing in plants. *Methods* 30, 296–303.

Ma, H. (1994) The unfolding drama of flower development: recent results from genetic and molecular analyses. *Genes and Development* 8, 745–756.

Ma, H. (1998) To be, or not to be, a flower – control of floral meristem identity. *Trends in Genetics* 14, 26–32.

Ma, H. and dePamphilis, C. (2000) The ABCs of floral evolution. *Cell* 101, 5–8.

Maddison, W.P. and Maddison, D.R. (1992) *MacClade: analysis of Phylogeny and Character Evolution.* Sinauer Associates, Sunderland, Massachusetts.

Mathews, S. and Donoghue, M. (1999) The root of angiosperm phylogeny inferred from duplicate phytochrome genes. *Science* 286, 947–950.

Meyerowitz, E.M., Bowman, J.L., Brockman, L.L., Drews, G.N., Jack, T., Sieburth, L.E. and Weigel, D. (1991) A genetic and molecular model for flower development in *Arabidopsis thaliana. Development Supplement* 1, 157–167.

Moon, Y.H., Chen, L., Pan, R.L., Chang, H.S., Zhu, T., Maffeo, D.M. and Sung, Z.R. (2003) EMF genes maintain vegetative development by repressing the flower program in *Arabidopsis. Plant Cell* 15, 681–693.

Münster, T., Pahnke, J., Di Rosa, A., Kim, J.T., Martin, W., Saedler, H. and Theißen, G. (1997) Floral homeotic genes were recruited from homologous MADS-box genes preexisting in the common ancestor of ferns and seed plants. *Proceedings of the National Academy of Sciences USA* 94, 2415–2420.

Münster, T., Wingen, L.U., Faigl, W., Werth, S., Saedler, H. and Theißen, G. (2001) Characterization of three *GLO-BOSA*-like MADS-box genes from maize: evidence for ancient parology in one class of floral homeotic B-function genes of grasses. *Gene* 262, 1–13.

Nam, J., dePamphilis, C.W., Ma, H. and Nei, M. (2003) Antiquity and evolution of the MADS-box gene family controlling flower development in plants. *Molecular Biology and Evolution* 20, 1435–1447.

Nayak, S., Santiago, S.F., Jin, H., Lin, D., Schedl, T. and Kipreos, E.T. (2002) The *Caenorhabditis elegans Skp1*-related gene family: diverse functions in cell proliferation, morphogenesis, and meiosis. *Current Biology* 12, 277–287.

Ni, W., Li, W., Zhang, X., Hu, W., Zhang, W., Wang, G., Han, T., Zahn, L.M., Zhao, D. and Ma, H. (2004) Genetic control of reproductive development in flowering plants. In: Sharma, R.P. (ed.) *Molecular Plant Physiology.* Haworth Press, New York (in press).

Parkinson, C.L., Adams, K.L. and Palmer, J.D. (1999) Multigene analyses identify the three earliest lineages of extant flowering plants. *Current Biology* 9, 1485–1488.

Pelaz, S., Ditta, G.S., Baumann, E., Wisman, E. and Yanofsky, M.F. (2000) B and C floral organ identity functions require *SEPALLATA* MADS-box genes. *Nature* 405, 200–203.

Pelaz, S., Tapia-Lopez, R., Alvarez-Buylla, E.R. and Yanofsky, M.F. (2001) Conversion of leaves into petals in *Arabidopsis. Current Biology* 11, 182–184.

Purugganan, M.D. (1997) The MADS-box floral homeotic gene lineages predate the origin of seed plants: phylogenetic and molecular clock estimates. *Journal of Molecular Evolution* 45, 392–396.

Purugganan, M.D., Sounsley, S.D., Schmidt, R.J. and Yanofsky, M.F. (1995) Molecular evolution of flower development: diversification of the plant MADS-box regulatory gene family. *Genetics* 140, 345–356.

Qiu, Y.-L., Lee, J., Bernasconi-Quadroni, F., Soltis, D.E., Soltis, P.S., Zanis, M., Chen, Z., Savolainen, V. and Chase, M.W. (1999) The earliest angiosperms: evidence from mitochondrial, plastid and nuclear genomes. *Nature* 402, 404–407.

Riechmann, J.L. and Meyerowitz, E.M. (1997) MADS domain proteins in plant development. *Biological Chemistry* 378, 1079–1101.

Riechmann, J.L., Krizek, B.A. and Meyerowitz, E.M. (1996) Dimerization specificity of *Arabidopsis* MADS domain homeotic proteins Apetala1, Apetala3, Pistillata and Agamous. *Proceedings of the National Academy of Sciences USA* 93, 4793–4798.

Ronse De Craene, L.P., Louis, P., Soltis, P.S. and Soltis, D.E. (2003) Evolution of floral structure in basal angiosperms. *International Journal of Plant Sciences* 164, S329–S363.

Sablowski, R.W. and Meyerowitz, E.M. (1998) A homolog of NO APICAL MERISTEM is an immediate target of the floral homeotic genes *APETALA3/PISTILLATA. Cell* 92, 93–103.

Samach, A., Klenz, J.E., Kohalmi, S.E., Risseeuw, E., Haughn, G.W. and Crosby, W.L. (1999) The *UNUSUAL FLORAL ORGANS* gene of *Arabidopsis thaliana* is an F-box protein required for normal patterning and growth in the floral meristem. *Plant Journal* 20, 433–445.

Schmid, M., Uhlenhaut, N.H., Godard, F., Demar, M., Bressan, R., Weigel, D. and Lohmann, J.U. (2003) Dissection of floral induction pathways using global expression analysis. *Development* 130, 6001–6012.

Schmidt, R.J. and Ambrose, B.A. (1998) The blooming of grass flower development. *Current Opinion in Plant Biology* 1, 60–67.

Schwarz-Sommer, Z., Huijser, P., Nacken, W., Saedler, H. and Sommer, H. (1990) Genetic control of flower development by homeotic genes in *Antirrhinum majus. Science* 250, 931–936.

Smith, G.H. (1928) Vascular anatomy of Ranalian flowers. II. Ranunculaceae (continued), Menispermaceae, Calycanthaceae, Annonaceae. *Botanical Gazette* 85, 152–177.

Smyth, D.R., Bowman, J.L. and Meyerowitz, E.M. (1990) Early flower development in *Arabidopsis. Plant Cell* 2, 755–768.

Soltis, D.E., Soltis, P.S., Chase, M.W., Mort, M.E., Albach, D.C., Zanis, M., Savolainen, V., Hahn, W.H., Hoot, S.B., Fay, M.F., Axtell, M., Swensen, S.M., Prince, L.M., Kress, W.J., Nixon, K.C. and Farris, J.S. (2000) Angiosperm phylogeny inferred from 18S rDNA, *rbcL,* and *atpB* sequences. *Botanical Journal of the Linnean Society* 133, 381–461.

Soltis, D.E., Soltis, P.S., Albert, V.A., dePamphilis, C.W., Frohlich, M., Ma, H. and Theißen, G. (2002) Missing links: the genetic architecture of the flower and floral diversification. *Trends in Plant Sciences* 7, 22–30.

Soltis, D.E., Senters, A.E., Zanis, M.J., Kim, S., Thompson, J.D., Soltis, P.S., Ronse De Craene, L.P., Endress, P.K. and Farris, J.S. (2003) Gunnerales are sister to other core eudicots: implications for the evolution of pentamery. *American Journal of Botany* 90, 463–470.

Soltis, D.E., Endress, P.K., Chase, M.W. and Soltis, P.S. (2004) *Phylogeny and Evolution of Angiosperms.* Smithsonian Institution Press, Washington, DC.

Soltis, P.S., Soltis, D.E. and Chase, M.W. (1999) Angiosperm phylogeny inferred from multiple genes as a research tool for comparative biology. *Nature* 402, 402–404.

Stebbins, G.L. (1974) *Flowering Plants: Evolution Above the Species Level.* Belknap Press, Cambridge, Massachusetts.

Sun, G., Li, Q., Dilcher, D.L., Zheng, S., Nixon, K.C. and Wang, X. (2002) Archaefructaceae, a new basal angiosperm family. *Science* 296, 899–904.

Sundström, J., Carlsecker, A., Svensson, M.E., Svenson, M., Johanson, U., Theißen, G. and Engström, P. (1999) MADS-box genes active in developing pollen cones of Norway spruce (*Picea abies*) are homologous to the B-class floral homeotic genes in angiosperms. *Developmental Genetics* 25, 253–266.

Takhtajan, A. (1969) *Flowering Plants: Origin and Dispersal.* Smithsonian Institution Press, Washington, DC.

Takhtajan, A. (1997) *Diversity and Classification of Flowering Plants.* Columbia University Press, New York.

Teeri, T.H., Albert, V.A., Elomaa, P., Hämäläinen, J., Kotilainen, M., Pöllänen, E. and Uimari, A. (2002) Involvement of non-ABC MADS-box genes in determining stamen and carpel identity in *Gerbera hybrida* (Asteraceae). In: Cronk, Q.C.B., Bateman, R.M. and Hawkins, J.A. (eds) *Developmental Genetics and Plant Evolution.* Taylor & Francis, London, pp. 220–232.

Theißen, G. (2001) Development of floral organ identity: stories from the MADS house. *Current Opinion in Plant Biology* 4, 75–86.

Theißen, G., Becker, A., Di Rosa, A., Kanno, A., Kim, J.T., Münster, T., Winter, K.-U. and Saedler, H. (2000) A short history of MADS-box genes in plants. *Plant Molecular Biology* 42**,** 115–149.

Theißen, G., Becker, A., Winter, K.-U., Münster, T., Kirchner, C. and Saedler, H. (2002) How the land plants learned their floral ABCs: the role of MADS box genes in the evolutionary origin of flowers. In: Cronk, Q.C.B., Bateman, R.M. and Hawkins, J.A. (eds) *Developmental Genetics and Plant Evolution,* Systematics Association Special Volume Series 65. Taylor & Francis, New York, pp. 173–206.

Tucker, S.C. (1960) Ontogeny of the floral apex of *Michelia fuscata. American Journal of Botany* 47, 266–277.

Tucker, S.C. and Douglas, A.W. (1996) Floral structure, development, and relationships of Paleoherbs: *Saruma, Cabomba, Lactoris,* and selected Piperales. In: Taylor, D.W. and Hickey, L.J. (eds) *Flowering Plant Origin, Evolution, and Phylogeny.* Chapman & Hall, New York, pp. 141–175.

Van Tunen, A.J., Eikelboom, W. and Angenet, G.C. (1993) Floral organogenesis in *Tulipa. Flower Newsletter* 16, 33–38.

Von Balthazar, M. and Endress, P.K. (2002) Development of inflorescences and flowers in Buxaceae and the problem of perianth interpretation. *International Journal of Plant Sciences* 163, 847–876.

Weigel, D. (1995) The genetics of flower development: from floral induction to ovule morphogenesis. *Annual Review Genetics* 29, 19–39.

Weigel, D. and Meyerowitz, E.M. (1994) The ABCs of floral homeotic genes. *Cell* 78, 203–209.

Weigel, D., Alvarez, J., Smyth, D.R., Yanofsky, M.F. and Meyerowitz, E.M. (1992) *LEAFY* controls floral meristem identity in *Arabidopsis. Cell* 69, 843–859.

Winter, K.-U., Becker, A., Münster, T., Kim, J., Saedler, H. and Theißen, G. (1999) MADS-box genes reveal that gnetophytes are much more closely related to conifers than to flowering plants. *Proceedings of the National Academy of Sciences USA* 96, 7342–7347.

Winter, K.-U., Weiser, C., Kaufmann, K., Bohne, A., Kirchner, C., Kanno, A., Saedler, H. and Theißen, G. (2002) Evolution of class B floral homeotic proteins: obligate heterodimerization originated from homodimerization. *Molecular Biology and Evolution* 19, 587–596.

Yamanaka, A., Yada, M., Imaki, H., Koga, M., Ohshima, Y. and Nakayama, K. (2002) Multiple *Skp1*-related proteins in *Caenorhabditis elegans*: diverse patterns of interaction with Cullins and F-box proteins. *Current Biology* 12, 267–275.

Yang, M., Hu, Y., Lodhi, M., McCombie, W.R. and Ma, H. (1999) The *Arabidopsis SKP1-LIKE1* gene is essential for male meiosis and may control homolog separation. *Proceedings of the National Academy of Sciences USA* 96, 11416–11421.

Yanofsky, M.F. (1995) Floral meristems to floral organs: genes controlling early events in *Arabidopsis* flower development. *Annual Review of Plant Physiology and Plant Molecular Biology* 46, 167–188.

Yoo, M.-J., Albert, V.A., Soltis, P.S. and Soltis, D.E. (2005) Diversification of glycogen synthase kinase 3/*SHAGGY-like* kinase genes in plants. *BMC Evolutionary Biology* (submitted).

Yu, D.Y., Kotilainen, M., Pollanen, E., Mehto, M., Elomaa, P., Helariutta, Y., Albert, V.A. and Teeri, T.H. (1999) Organ identity genes and modified patterns of flower development in *Gerbera hybrida. Plant Journal* 17, 51–62.

Yu, J., Hu, S., Wang, J., Wong, G.K., Li, S., Liu, B., Deng, Y., Dai, L., Zhou, Y., Zhang, X., Cao, M., Liu, J., Sun, J., Tang, J., Chen, Y., Huang, X., Lin, W., Ye, C., Tong, W., Cong, L., Geng, J., Han, Y., Li, L., Li, W., Hu, G., Li, J., Liu, Z., Qi, Q., Li, T., Wang, X., Lu, H., Wu, T., Zhu, M., Ni, P., Han, H., Dong, W., Ren, X., Feng, X., Cui, P., Li, X., Wang, H., Xu, X., Zhai, W., Xu, Z., Zhang, J., He, S., Xu, J., Zhang, K., Zheng, X., Dong, J., Zeng, W., Tao, L., Ye, J., Tan, J., Chen, X., He, J., Liu, D., Tian, W., Tian, C., Xia, H., Bao, Q., Li, G., Gao, H., Cao, T., Zhao, W., Li, P., Chen, W., Zhang, Y., Hu, J., Liu, S., Yang, J., Zhang, G., Xiong, Y., Li, Z., Mao, L., Zhou, C., Zhu, Z., Chen, R., Hao, B., Zheng, W., Chen, S., Guo, W., Tao, M., Zhu, L., Yuan, L. and Yang, H. (2002) A draft sequence of the rice genome (*Oryza sativa* L. ssp. *indica). Science* 296, 79–92.

Zanis, M.J., Soltis, D.E., Soltis, P.S., Mathews, S. and Donoghue, M.J. (2002) The root of the angiosperms revisited. *Proceedings of the National Academy of Sciences USA* 99, 6848–6853.

Zanis, M.J., Soltis, P.S., Qiu, Y.-L., Zimmer, E. and Soltis, D.E. (2003) Phylogenetic analyses and perianth evolution in basal angiosperms. *Annals of the Missouri Botanical Garden* 90, 129–150.

Zhao, D., Yang, M., Solava, J. and Ma, H. (1999) The *ASK1* gene regulates development and interacts with the *UFO* gene to control floral organ identity in *Arabidopsis. Developmental Genetics* 25, 209–223.

Zhao, D., Yu, Q., Chen, C. and Ma, H. (2001a) Genetic control of reproductive meristems. In: McManus, M.T. and Veit, B. (eds) *Meristematic Tissues in Plant Growth and Development*. Sheffield Academic Press, Sheffield, pp. 89–142.

Zhao, D., Yu, Q., Chen, C. and Ma, H. (2001b) The *ASK1* gene regulates B function gene expression in cooperation with *UFO* and *LEAFY* in *Arabidopsis*. *Development* 128, 2735–2746.

Zhao, D., Ni, W., Feng, B., Han, T., Petrasek, M.G. and Ma, H. (2003) Members of the *Arabidopsis-SKP1*-like gene family exhibit a variety of expression patterns and may play diverse roles in *Arabidopsis*. *Plant Physiology* 133, 203–217.

Zheng, N., Schulman, B.A., Song, L., Miller, J.J., Jeffrey, P.D., Wang, P., Chu, C., Koepp, D.M., Elledge, S.J., Pagano, M., Conaway, R.C., Conaway, J.W., Harper, J.W. and Pavletich, N.P. (2002) Structure of the Cul1-Rbx1-Skp1-F boxSkp2 SCF ubiquitin ligase complex. *Nature* 416, 703–709.

Zik, M. and Irish, V.F. (2003) Flower development: initiation, differentiation and diversification. *Annual Review of Cell and Developmental Biology* 19, 119–140.

11 Diversity in plant cell walls

Philip J. Harris

School of Biological Sciences, The University of Auckland, Private Bag 92019, Auckland, New Zealand

Introduction

The cell walls of vascular plants account for much of the carbon fixed during photosynthesis and make up much of their biomass. In addition to determining the size, shape and mechanical strength of plant cells, the walls of many species of seed plants (angiosperms and gymnosperms) are economically important. They form the bulk of wood for timber or paper, as well as textile and other fibres; in forages, they are a major source of energy for ruminants; they make up most of the fibre in human diets; and they play a key role in determining the texture of fruits and vegetables. Because of their economic importance, much is known about the structure and compositions of the walls of many species of angiosperms and some species of coniferous gymnosperms. However, because pteridophytes are of only minor economic importance, information about the composition of their walls is only fragmentary.

All the walls in vascular plants have a similar construction, consisting of two phases: a fibrillar phase of cellulose microfibrils set in a matrix phase that contains a high proportion of non-cellulosic polysaccharides that vary in structure (Figs 11.1–11.3) (Bacic *et al.*, 1988; Carpita and Gibeaut, 1993). Structural proteins and glycoproteins, as well as phenolic components including lignin, may also be present in the wall matrix. However, before

comparing wall compositions among different taxa, it is important to appreciate that because wall composition varies with wall type, there is diversity in wall composition even within an individual plant. Two major groups of wall types are commonly recognized: primary and secondary (Bacic *et al.*, 1988; Carpita and Gibeaut, 1993). Primary walls are deposited while the cells are still enlarging. By contrast, secondary walls are laid down on the primary wall after cell expansion has stopped, and are usually very much thicker than primary walls. When mature, the different cell types can be grouped according to whether they have only primary walls or both primary and secondary walls. For example, parenchyma cells frequently have only a thin primary wall, whereas sclerenchyma fibre cells have both a primary wall and a very much thicker secondary wall. In the latter cell type, the secondary wall is laid down over the whole surface of the primary wall. However, in xylem tracheary elements, the secondary wall is laid down in a variety of patterns that may cover only a portion of the surface area of the primary wall. Cell types with secondary walls can be further classified into those with lignified and those with non-lignified walls, those with lignified walls being the more common group. In lignified walls, the lignin occurs in both the primary and the secondary walls. Cell types with lignified secondary walls

Cellulose [(1→4)-β-D-glucan]

→4)-β-D-Glc*p*-(1→4)-β-D-Glc*p*-(1→4)-β-D-Glc*p*-(1→4)-β-D-Glc*p*-(1→

Callose [(1→3)-β-D-glucan]

→3)-β-D-Glc*p*-(1→3)-β-D-Glc*p*-(1→3)-β-D-Glc*p*-(1→3)-β-D-Glc*p*-(1→

(1→3),(1→4)-β-D-glucan

→3)-β-D-Glc*p*-(1→4)-β-D-Glc*p*-(1→4)-β-D-Glc*p*-(1→3)-β-D-Glc*p*-(1→

Xyloglucans

    A common subunit of fucogalactoxyloglucans:

→4)-β-D-Glc*p*-(1→4)-β-D-Glc*p*-(1→4)-β-D-Glc*p*-(1→4)-β-D-Glc*p*-(1→

	6	6	6	
	↑	↑	↑	
	α-D-Xyl*p*	α-D-Xyl*p*	α-D-Xyl*p*	
		2	2	
		↑	↑	
		β-D-Gal*p*	β-D-Gal*p*	
			2	
			↑	
			α-L-Fuc*p*	
X	L	F	G

    A common subunit of arabinoxyloglucans of the Solanaceae

→4)-β-D-Glc*p*-(1→4)-β-D-Glc*p*-(1→4)-β-D-Glc*p*-(1→4)-β-D-Glc*p*-(1→

	6	6	
	↑	↑	
	α-D-Xyl*p*	α-D-Xyl*p*	
		2	
		↑	
		α-L-Ara*f*	
X	S	G	G

Fig. 11.1. Structures of the wall polysaccharides cellulose, callose and xyloglucans. The letters G, X, S, L and F refer to an unambiguous nomenclature for xyloglucan structure developed by Fry *et al.* (1993) (see text). Xyloglucans in walls are often acetylated, but, for simplicity, the sites of possible acetylation are not shown.

(Galacto-) glucomannan

$$\rightarrow4)\text{-}\beta\text{-}\text{D-Man}p\text{-}(1\rightarrow4)\text{-}\beta\text{-}\text{D-Man}p\text{-}(1\rightarrow4)\text{-}\beta\text{-}\text{D-Glc}p\text{-}(1\rightarrow4)\text{-}\beta\text{-}\text{D-Glc}p\text{-}(1\rightarrow$$

6   6
↑   ↑
α-D-Gal$p$ α-D-Gal$p$

(Galacto-) mannan

$$\rightarrow4)\text{-}\beta\text{-}\text{D-Man}p\text{-}(1\rightarrow4)\text{-}\beta\text{-}\text{D-Man}p\text{-}(1\rightarrow4)\text{-}\beta\text{-}\text{D-Man}p\text{-}(1\rightarrow4)\text{-}\beta\text{-}\text{D-Man}p\text{-}(1\rightarrow$$

6   6
↑   ↑
α-D-Gal$p$ α-D-Gal$p$

Heteroxylans

4-O-methylglucuronoxylan

$$\rightarrow4)\text{-}\beta\text{-}\text{D-Xyl}p\text{-}(1\rightarrow4)\text{-}\beta\text{-}\text{D-Xyl}p\text{-}(1\rightarrow4)\text{-}\beta\text{-}\text{D-Xyl}p\text{-}(1\rightarrow4)\text{-}\beta\text{-}\text{D-Xyl}p\text{-}(1\rightarrow$$

2
↑
4-O-Me-α-D-Glc$p$A

Glucuronoarabinoxylan (GAX)

$$\rightarrow4)\text{-}\beta\text{-}\text{D-Xyl}p\text{-}(1\rightarrow4)\text{-}\beta\text{-}\text{D-Xyl}p\text{-}(1\rightarrow4)\text{-}\beta\text{-}\text{D-Xyl}p\text{-}(1\rightarrow4)\text{-}\beta\text{-}\text{D-Xyl}p\text{-}(1\rightarrow$$

3   2
↑   ↑
α-L-Ara$f$ α-D-Glc$p$A
5
↑
Ferulic acid

4-O-methylglucuronoarabinoxylan

$$\rightarrow4)\text{-}\beta\text{-}\text{D-Xyl}p\text{-}(1\rightarrow4)\text{-}\beta\text{-}\text{D-Xyl}p\text{-}(1\rightarrow4)\text{-}\beta\text{-}\text{D-Xyl}p\text{-}(1\rightarrow4)\text{-}\beta\text{-}\text{D-Xyl}p\text{-}(1\rightarrow$$

3   2
↑   ↑
α-L-Ara$f$ 4-O-Me-α-D-Glc$p$A

Fig. 11.2. Structures of the wall polysaccharides (galacto-) glucomannan, (galacto-) mannan and heteroxylans. For simplicity, sites of possible acetylation are not shown.

include sclerenchyma fibres and xylem tracheary elements. Cell types with non-lignified secondary walls include cotton-seed hairs, pollen tubes and the thickened parenchyma walls of endosperms and cotyledons in some seeds. The walls of cotton-seed hairs contain about 94% cellulose and the walls of pollen tubes contain callose, a $(1\rightarrow3)$-β-D-glucan (Fig. 11.1) (Bacic *et al.*, 1988; Stone and Clarke, 1992). The secondary walls of transfer cells, which form ingrowths with a diversity of morphologies, are also non-lignified.

Homogalacturonan

$$\rightarrow4)\text{-}\alpha\text{-}D\text{-}GalpA\text{-}(1\rightarrow4)\text{-}\alpha\text{-}D\text{-}GalpA\text{-}(1\rightarrow4)\text{-}\alpha\text{-}D\text{-}GalpA\text{-}(1\rightarrow4)\text{-}\alpha\text{-}D\text{-}GalpA\text{-}(1\rightarrow$$

Xylogalacturonan

$$\rightarrow4)\text{-}\alpha\text{-}D\text{-}GalpA\text{-}(1\rightarrow4)\text{-}\alpha\text{-}D\text{-}GalpA\text{-}(1\rightarrow4)\text{-}\alpha\text{-}D\text{-}GalpA\text{-}(1\rightarrow4)\text{-}\alpha\text{-}D\text{-}GalpA\text{-}(1\rightarrow$$

$$\begin{array}{ccc} 3 & & 3 \\ \uparrow & & \uparrow \\ \beta\text{-}D\text{-}Xylp & & \beta\text{-}D\text{-}Xylp \end{array}$$

Apiogalacturonan

$$\rightarrow4)\text{-}\alpha\text{-}D\text{-}GalpA\text{-}(1\rightarrow4)\text{-}\alpha\text{-}D\text{-}GalpA\text{-}(1\rightarrow4)\text{-}\alpha\text{-}D\text{-}GalpA\text{-}(1\rightarrow4)\text{-}\alpha\text{-}D\text{-}GalpA\text{-}(1\rightarrow$$

$$\begin{array}{ccc} 2 & & 2 \\ \uparrow & & \uparrow \\ \beta\text{-}D\text{-}Apif & & \beta\text{-}D\text{-}Apif \\ & & 3' \\ & & \uparrow \\ & & \beta\text{-}D\text{-}Apif \end{array}$$

Rhamnogalacturonan I (RG-I)

$$\rightarrow2)\text{-}\alpha\text{-}L\text{-}Rhap\text{-}(1\rightarrow4)\text{-}\alpha\text{-}D\text{-}GalpA\text{-}(1\rightarrow2)\text{-}\alpha\text{-}L\text{-}Rhap\text{-}(1\rightarrow4)\text{-}\alpha\text{-}D\text{-}GalpA\text{-}(1\rightarrow$$

$$\begin{array}{c} 4 \\ \uparrow \end{array}$$

Mostly pectic arabinan, galactan and
arabinogalactans (Type I)

Arabinan

$$\alpha\text{-}L\text{-}Araf$$
$$\downarrow$$
$$2$$
$$\rightarrow5)\text{-}\alpha\text{-}L\text{-}Araf\text{-}(1\rightarrow5)\text{-}\alpha\text{-}L\text{-}Araf\text{-}(1\rightarrow5)\text{-}\alpha\text{-}L\text{-}Araf\text{-}(1\rightarrow5)\text{-}\alpha\text{-}L\text{-}Araf\text{-}(1\rightarrow$$

$$\begin{array}{ccc} 3 & & 3 \\ \uparrow & & \uparrow \\ \alpha\text{-}L\text{-}Araf & & \alpha\text{-}L\text{-}Araf \end{array}$$

Galactan

$$\rightarrow4)\text{-}\beta\text{-}D\text{-}Galp\text{-}(1\rightarrow4)\text{-}\beta\text{-}D\text{-}Galp\text{-}(1\rightarrow4)\text{-}\beta\text{-}D\text{-}Galp\text{-}(1\rightarrow4)\text{-}\beta\text{-}D\text{-}Galp\text{-}(1\rightarrow$$

Thus, in making comparisons of wall compositions among different taxa, it is important to compare the compositions of the same wall types. Many of the early studies of wall compositions are difficult to interpret and use in comparisons among taxa because the wall preparations were from whole organs, or even whole plants, and contained a mixture of wall types. Ideally, comparisons should be made using wall preparations from the same

Arabinogalactan (Type I)

$$\rightarrow 4)\text{-}\beta\text{-}D\text{-}Gal}p\text{-}(1\rightarrow4)\text{-}\beta\text{-}D\text{-}Gal}p\text{-}(1\rightarrow4)\text{-}\beta\text{-}D\text{-}Gal}p\text{-}(1\rightarrow4)\text{-}\beta\text{-}D\text{-}Gal}p\text{-}(1\rightarrow$$

Rhamnogalacturonan II (RG-II)

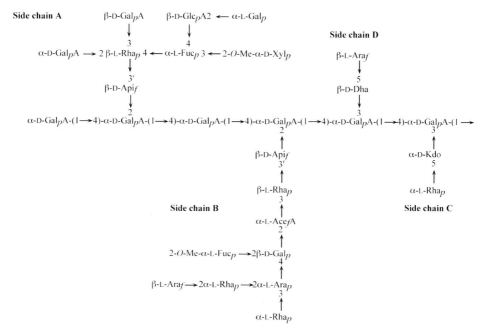

Fig. 11.3. (Above and opposite.) Structures of the pectic polysaccharides that occur in walls of vascular plants. For simplicity, sites of possible acetylation are not shown. The rhamnogalacturonan II (RG-II) structure is that which occurs in grape (wine) (*Vitis vinifera*); however, the substituents on the arabinopyranose residue of the B side chain vary with taxon, resulting in chains containing 6–9 monosaccharides.

cell type. Methods have been described for isolating walls from several different cell types, but these are often difficult to carry out (Harris, 1983). Therefore, more commonly, wall preparations that contain mostly one wall type, such as primary walls or lignified secondary walls, are isolated and chemically analysed. Although so far little used, another way of making comparisons of wall compositions among different taxa is to use immunocytochemistry. A range of monoclonal and polyclonal antibodies are now available that specifically recognize particular wall components, particularly polysaccharides (Willats *et al.*, 2000). These antibodies can be used in conjunction with secondary antibodies labelled either with fluorochromes for fluorescence light microscopy, or with colloidal gold for transmission electron microscopy. Because it is known that environmental conditions and pathogens can also affect wall compositions, plants used for comparative studies should be pathogen free and ideally grown in the same environment. Stage of development of an organ can also affect wall composition: for example, striking changes in wall composition often occur during fruit ripening, and this should be recognized in making comparisons among taxa.

Recognition of the considerable diversity in wall composition that occurs among different vascular plants is particularly important as increasingly attention focuses on the wall compositions of just a few model species, such as *Arabidopsis thaliana* and rice (*Oryza sativa*), which have sequenced genomes. Studies of wall diversity among vascular plant taxa have been, and will increasingly be, helped by the enormous impact that gene sequences have had on our understanding of phylogeny. This has been particularly striking for the angiosperms and has resulted in a new classification (APG, 1998; APG II, 2003) that provides an excellent basis for investigating whether variation in a particular wall component discovered in one taxon is also present in phylogenetically related taxa (Figs 11.5 and 11.6).

In this review of diversity in the walls of vascular plants, I have focused particularly on the compositions of primary walls, lignified secondary walls and non-lignified secondary walls of seeds, because these are particularly common wall types and most is known about their diversity of composition. In doing so, I have concentrated on the structures of their component non-cellulosic polysaccharides and lignins, because most information is available about these components. Where information is available, structural proteins and glycoproteins will be discussed. However, walls containing other polymers, including suberin, cutin, cutan and sporopollenin, will not be discussed. Walls of the dicotyledons are discussed first, followed by the monocotyledons, gymnosperms and pteridophytes. The reason for beginning with the walls of the dicotyledons, a phylogenetically advanced group, and concluding with the pteridophytes, the least advanced group, is twofold. First, more research has been done on the walls of a greater variety of families in the dicotyledons than on the walls of any of the other

groups. However, most of this research has been done on the walls of eudicotyledons, rather than the other dicotyledons, often referred to as basal angiosperms, comprising the Amborellaceae, Nymphaeaceae, Austrobaileyales, Chloranthaceae, magnoliids and Ceratophyllales (APG II, 2003) (Fig. 11.5). Second, most of the polymers present in dicotyledon walls are also present in the other groups, although often with different fine structures and in different proportions, and thus most of the wall polymers discussed throughout the review are introduced in this section.

Angiosperm Walls

Dicotyledon walls

Primary walls of dicotyledons

The primary walls of the eudicotyledons are probably the most thoroughly investigated walls in vascular plants and have broadly similar compositions in all taxa so far examined. They contain large proportions of pectic polysaccharides, smaller proportions of xyloglucans, and minor proportions of heteroxylans, glucomannans and/or galactoglucomannans. They also contain structural proteins and glycoproteins and may contain phenolic components (Bacic *et al.*, 1988; Carpita and Gibeaut, 1993). Although much less research has been done on the walls of the other dicotyledons (basal angiosperms), there is so far no evidence that their wall compositions differ from those of the eudicotyledons.

PECTIC POLYSACCHARIDES   Pectic polysaccharides are highly complex polymers comprising four domains: homogalacturonan (HG), rhamnogalacturonan I (RG-I), rhamnogalacturonan II (RG-II) and xylogalacturonan (XGA) (Ridley *et al.*, 2001; Willats *et al.*, 2001) (Fig. 11.3). HG is composed of linear chains of galacturonic acid residues that may be methyl-esterified and acetylated. The extent of acetylation apparently varies with taxon. The HG of sugarbeet (*Beta vulgaris*) (Chenopodiaceae) and potato (*Solanum tuberosum*) (Solanaceae) is highly acetylated

(Willats *et al.*, 2001). Furthermore, the HG in walls of two other Chenopodiaceae species, *Salicornia ramosissima* and *Chenopodium quinoa*, is also highly acetylated (Renard *et al.*, 1993, 1999). Variation has also been found in the pattern of acetylation of HG; this variation may be taxonomic, developmental or both (Perrone *et al.*, 2002). The XGA domain has single β-D-xylosyl residues attached to C-3 of galacturonic acid residues of an HG backbone, and is apparently particularly abundant in the walls of seeds and fruits (Ridley *et al.*, 2001; Willats *et al.*, 2001).

The RG-I domain is composed of alternating galacturonic acid and rhamnose residues. Many of the rhamnose residues have polysaccharide or oligosaccharide side chains rich in arabinose and/or galactose that include arabinans, galactans and Type I arabinogalactans (Fig. 11.3); small amounts of Type II arabinogalactans possibly also occur. The structures of the side chains vary with taxon. For example, RG-I from cabbage (*Brassica oleracea*) walls has mostly side chains of arabinans, whereas RG-I from walls of potato (*S. tuberosum*) tubers has mostly galactan side chains (Harris *et al.*, 1997). However, whether phylogenetically related taxa have similar side chains is unknown. Moreover, immunolabelling studies using monoclonal antibodies that recognize arabinans (LM6) and galactans (LM5) have shown that the structures of these side chains can also vary during development. In suspension-cultured carrot (*Daucus carota*) (Apiaceae) cells, arabinans predominate in the walls of meristematic cells, whereas galactans predominate in the walls of elongating cells (Willats *et al.*, 1999).

Small proportions of the RG-II domain are also present in these walls. This low molecular weight, but highly complex domain, which is attached to HG, contains 11 different monosaccharide residues, including some that are not found elsewhere in plant wall polysaccharides. It has a backbone of at least seven galacturonic acid residues, which has attached to it four structurally different side chains: A, B, C and D. C and D are disaccharides, A, which contains 2-*O*-methyl-α-D-xylose, is an octasaccharide and B, which contains 2-*O*-methyl-α-L-fucose, varies in size (Glushka *et al.*, 2003) (Fig. 11.3). The B side chain contains nine monosaccharides in ginseng (*Panax ginseng*) (Araliaceae) and grape (wine) (*Vitis vinifera*) (Vitaceae), seven in *A. thaliana* and six in sugarbeet and beetroot (*B. vulgaris*). In sycamore (*Acer pseudoplatanus*) (Aceraceae), this side chain varies in length, containing seven, eight or nine monosaccharides (Glushka *et al.*, 2003). The differences in the B side chain occur at the non-reducing end and appear to result from the presence or absence of substituents linked to C-2 and/or C-3 of the arabinopyranose residue (Fig. 11.3). Although these differences probably represent real variation among species, at least some could also result from differences in the isolation procedures. Except for these differences in the B side chain, RG-II has a highly conserved structure and occurs as a dimer cross-linked by 1:2 borate-diol esters involving the apiosyl residue in only side chain A (Ishii *et al.*, 1999).

FERULIC ACID ESTER-LINKED TO PECTIC POLYSACCHARIDES The hydroxycinnamic acid ferulic acid (Fig. 11.4) is attached to pectic polysaccharides in the primary walls of spinach (*Spinacia oleracea*) and sugarbeet (*B. vulgaris*) (Chenopodiaceae). It is attached to the arabinan and galactan side chains of RG-I, where it is esterified via its carboxyl group to the C(O)2 hydroxyl of arabinofuranosyl residues in the arabinans and to the C(O)6 hydroxyl of galactopyranosyl residues in the galactans (Ishii and Tobita, 1993; Colquhoun *et al.*, 1994). The primary walls of spinach also contain small amounts of ester-linked *p*-coumaric acid (Fig. 11.4) that may be linked to RG-I in the same way (Fry, 1988). In addition to ferulic and *p*-coumaric acids, early research showed that the walls of spinach also contained 5-5′ dehydrodiferulic acid, which was then known simply as 'diferulic acid' (Hartley and Harris, 1981; Fry, 1983) (Fig. 11.4). More recently, a range of other dehydrodiferulic acids has been found in the walls of sugarbeet and beetroot (*B. vulgaris*) (Waldron *et al.*, 1997). The most abundant dimers are the 8-*O*-4′ and 8-5′ dimers, with smaller proportions of the 5-5′ and 8-8′

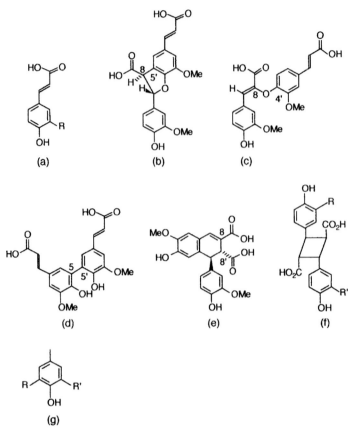

Fig. 11.4. Structures of some of the phenolic components that occur in walls of vascular plants: (a) *p*-coumaric acid (R = H) and ferulic acid (R = OCH₃); (b–e) examples of dehydrodiferulic acids, (b) 8-5′ dimer (benzofuran form), (c) 8-*O*-4′ dimer, (d) 5-5′ dimer, (e) 8-8′ dimer (aryltetralin form); (f) examples of substituted cyclobutanes, 4,4′-dihydroxy-α-truxillic acid (R = R′ = H), 4,4′-dihydroxy-3,3′-dimethoxy-α-truxillic acid (R = R′ = OCH₃), and 4,4′-dihydroxy-3-methoxy-α-truxillic acid (R = OCH₃, R′ = H); (g) structural units in lignin, *p*-hydroxyphenyl (R = R′ = H), guaiacyl (R = H, R′ = OCH₃) and syringyl (R = R′ = OCH₃).

dimers (Fig. 11.4). Dehydrodiferulic acids have also been found in the walls of another species of Chenopodiaceae, *Chenopodium quinoa*, where the 8-*O*-4′ dimer also predominates (Renard *et al.*, 1999). If the two ferulic acid residues of such dimers are attached to different RG-I molecules, then the dimer forms a cross-link between the RG-I molecules.

Ester-linked ferulic acid, however, is not confined to the primary walls of the Chenopodiaceae. In a survey of the primary walls of 251 species in 150 families of dicotyledons, using UV fluorescence microscopy at two pH values, Hartley and Harris (1981) found ester-linked ferulic acid

in all ten families they examined of the order Caryophyllales as defined by Cronquist (1981), but in no other families (Fig. 11.5). These families were as follows: Aizoaceae, Amaranthaceae, Basellaceae, Cactaceae, Caryophyllaceae, Chenopodiaceae, Didiereaceae, Nyctaginaceae, Phytolaccaceae and Portulacaceae. Species of the two other families in this order, Achatocarpaceae and Molluginaceae, were not examined. It should be noted that the order Caryophyllales as defined by Cronquist (1981) is the core of the much larger order Caryophyllales as defined by APG II (2003). Chemical analyses were also carried out on wall preparations, and *p*-coumaric acid

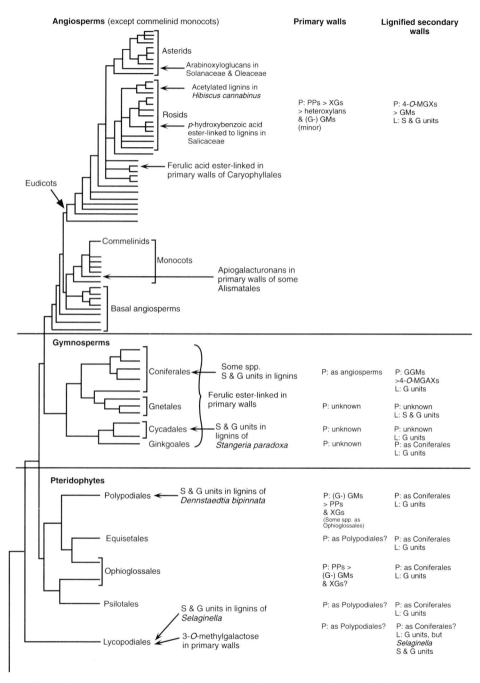

Fig. 11.5. The phylogeny of vascular plants based on nucleotide sequences of genes. The figure is a combination of trees adapted from Pryer *et al.* (2001), Soltis *et al.* (2002) (for the gymnosperms) and APG II (2003) (for the angiosperms). On the right are the non-cellulosic polysaccharides of the primary and lignified secondary walls and the lignin monomers of the lignified secondary walls. Selected wall features of particular taxa are shown arrowed. Abbreviations: 4-*O*-MGX = 4-*O*-methylglucuronoxylan; 4-*O*-MGAX = 4-*O*-methylglucuronoarabinoxylan; G = guaiacyl; (G-) GM = (galacto-) glucomannans; L = lignins; P = non-cellulosic polysaccharides; PP = pectic polysaccharides; S = syringyl; XG = xyloglucans.

and/or 5-5′ dehydrodiferulic acid were detected in some of these wall preparations (Hartley and Harris, 1981). However, it is not known if the ferulic, *p*-coumaric and 5-5′ dehydrodiferulic acid in the walls of these other species are ester-linked to arabinans and galactans as in the walls of spinach and sugarbeet.

More recently, small amounts of ferulic and 8-*O*-4′ dehydrodiferulic acids have also been found ester-linked to the walls of carrots (Parr *et al.*, 1997). These small amounts of hydroxycinnamic acids, which were not detected in the study of Hartley and Harris (1981), nevertheless could have an important influence on the properties of the walls. Small amounts of such acids may be of widespread occurrence in dicotyledon primary walls. The walls of carrots also contain significant amounts of ester-linked *p*-hydroxybenzoic acid but, as with the ferulic acid, the polysaccharide to which it is attached is unknown (Hartley and Harris, 1981; Parr *et al.*, 1997).

XYLOGLUCANS   Xyloglucans have a linear backbone of (1→4)-linked β-D-glucopyranosyl residues substituted at C(O)6 with α-D-xyopyranosyl residues. Other substituents can also occur on the xylose residues and an unambiguous nomenclature has been developed by Fry *et al.* (1993) to describe the structures of xyloglucans. The letters G, X, S, L and F are used to refer to the following structures: G = unsubstituted β-Glc*p*; X = α-D-Xyl*p*-(1→6)-β-D-Glc*p*; S and L = X with α-L-Ara*f*-(1→2)- and β-D-Gal*p*-(1→2)- attached to the non-reducing end, respectively; and F = L with α-L-Fuc*p*-(1→2)- attached to the non-reducing end (Fig. 11.1).

The xyloglucans in the primary walls of most dicotyledon species examined so far are substituted with galactose and fucose, and are referred to as fucogalactoxyloglucans. They have an XXXG core structure: three successive glucose residues are substituted with xylose residues, but the fourth glucose is unsubstituted (Fig. 11.1) (Vincken *et al.*, 1997). An excellent way of analysing their structures is by treating them with a (1→4)-endo-β-glucanase and then identifying and quantifying the oligosaccharides

released. Such analyses have shown that fucogalactoxyloglucans contain three major subunits: XXXG, XXFG and XLFG. For example, the xyloglucan from leaves of *A. thaliana* yielded 45% XXXG, 24% XXFG, 16% XLFG, 8% XXLG, 4% XLLG and 3% XLXG (Vanzin *et al.*, 2002). The structure of XLFG is shown in Fig. 11.1. However, the proportions of the different xyloglucan subunits vary somewhat with the organ from which the walls are isolated (Pauly *et al.*, 2001). The xyloglucan from walls of sycamore is unusual in that about 3% of the backbone glucose residues have attached to them the unusual side chains β-D-Xyl*p*-(1→2)-, α-L-Ara*f*-(1→2)- or α-L-Ara*f*-(1→3)-β-D-Xyl*p*-(1→2); these glucose residues also have an α-D-Xyl*p*-(1→6)- residue attached to them (Vincken *et al.*, 1997).

Nevertheless, not all dicotyledons have fucogalactoxyloglucans: so far, two families, Solanaceae and Oleaceae, have been identified that have arabinoxyloglucans. These contain no fucose but have S side chains (Fig. 11.1). In contrast to the fucogalactoxyloglucans, the arabinoxyloglucans of the Solanaceae have an XXGG core structure. Galactose occurs attached to some of the xylose residues in the xyloglucans of tomato (*Lycopersicon esculentum*) and potato (*S. tuberosum*), but not in the xyloglucan of tobacco (*Nicotiana tabacum*) (Vincken *et al.*, 1996; York *et al.*, 1996; Jia *et al.*, 2003). Another side chain (T = tomato) β-Ara*f*-(1→3)-α-Ara*f*-(1→2)-α-Xyl*p* occurs in the xyloglucan of tomato, but not tobacco or potato. The major subunits of tobacco xyloglucan are XSGG and XXGG and of tomato are XXGG, XSGG, LSGG and XTGG (Vierhuis *et al.*, 2001). The structure of XSGG is shown in Fig. 11.1. Unlike the arabinoxyloglucans of the Solanaceae, that in the walls of olive (*Olea europaea*) (Oleaceae) has an XXXG core structure, with XXXG, XXSG and XLSG as the major subunits (Vierhuis *et al.*, 2001).

Both the families Solanaceae and Oleaceae are in the asterids (euasterid I) (Fig. 11.5), the Solanaceae in the order Solanales and the Oleaceae in the Lamiales (APG II, 2003). It would be interesting to know the structures of xyloglucans of species in other families in both of these orders and in the other two orders of the euasterid I clade: Garryales and Gentianales. There is already evidence that the xyloglucan of a mint (*Mentha arvensis* × *Mentha spicata*) (Lamiaceae) (Lamiales) is probably also an arabinoxyloglucan (Maruyama *et al.*, 1996). However, the xyloglucan of burdock, *Arctium lappa* (Asteraceae) (Asterales), in the euasterid II clade is a fucogalactoxyloglucan (Vierhuis *et al.*, 2001).

HETEROXYLANS AND GLUCOMANNANS AND/OR GALACTOGLUCOMANNANS Small proportions of heteroxylans occur in these walls, and the heteroxylan from the walls of sycamore suspension-cultured cells has been characterized as a glucuronoarabinoxylan (GAX) (Darvill *et al.*, 1980). This has a similar structure to the GAXs of commelinid monocotyledons (Fig. 11.2), but the side chains are all linked at C(O)2 to the xylose residues of the (1→4)-β-D-xylan backbone. Small proportions of glucomannans and/or galactoglucomannans (Fig. 11.2) also occur in primary cell walls (Bacic *et al.*, 1988; Schröder *et al.*, 2001).

PROTEINS AND GLYCOPROTEINS Three groups of structural proteins and glycoproteins are often recognized: extensins, proline-rich proteins and glycine-rich proteins (Cassab and Varner, 1988; Showalter, 1993). In addition, arabinogalactan proteins (AGPs) occur in the apoplast and are secreted in slimes and mucilages; some of these AGPs may be ionically bound to walls. However, it is now apparent that proline-rich proteins, extensins and AGPs form a family of proline-/hydroxyproline-rich glycoproteins (P/HRGPs) with a continuum of molecules from AGPs, containing 99% carbohydrate, to proline-rich proteins, which are either not or only minimally glycosylated (Gaspar *et al.*, 2001).

Lignified secondary walls of dicotyledons

As for the primary walls, there are no indications so far that the compositions of the lignified secondary walls of the eudicotyledons differ from those of the rest of the dicotyledons (basal angiosperms).

POLYSACCHARIDES In contrast to the primary walls, the major non-cellulosic polysaccharides are usually 4-*O*-methylglucuronoxylans, with smaller proportions of glucomannans (Fig. 11.2). There are exceptions, however; for example in the walls of bast fibres of sunn hemp (*Crotalaria juncea*) (Fabaceae), glucomannans are the major non-cellulosic polysaccharides (Bacic *et al.*, 1988).

PROTEINS AND GLYCOPROTEINS In addition to occurring in primary walls, extensins and proline-rich proteins occur in lignified secondary walls. For example, extensins occur in the walls of sclerenchyma cells in the seed coats of soybean (*Glycine max*) (Fabaceae) (Cassab and Varner, 1988). Proline-rich proteins may be involved in the process of lignification (Showalter, 1993). Although the best-characterized glycine-rich protein, GRP 1.8, is present in protoxylem vessels, which have lignified secondary-wall thickenings, immunocytochemistry showed that it is in fact located in the primary walls between the wall thickenings (Ringli *et al.*, 2001).

LIGNINS Lignins are unique to the walls of vascular plants; there is no good evidence for lignins in bryophyte walls (Sarkanen and Hergert, 1971; Chen, 1991). Lignins are conventionally defined as complex polymers resulting from the oxidative polymerization of three types of hydroxycinnamyl alcohol precursors (monolignols) that result in *p*-hydroxyphenyl, guaiacyl and syringyl units in lignins (Fig. 11.4). These units are joined by a series of inter-unit linkages, the relative abundance of which appears to depend largely on the proportions of the different types of unit. The most frequent is the β-*O*-4 (β-aryl ether) linkage, constituting 50% or more of all linkages. Dicotyledon lignins consist mostly of guaiacyl and syringyl units, with trace amounts of *p*-hydroxyphenyl units (Sarkanen and Hergert, 1971; Chen, 1991; Boerjan *et al.*, 2003). They usually have a syringyl:guaiacyl ratio of 2–2.5:1, which varies with taxon. Early research indicated low proportions of syringyl units in the lignins of some species of the Winteraceae, although considerable variation apparently occurs in the syringyl:guaiacyl ratio within this family (Sarkanen and Hergert, 1971). A lower proportion of syringyl units occurs in the lignin of tension wood, formed in hardwoods in regions held under tension (Chen, 1991). The proportions of syringyl and guaiacyl units also vary with cell type: lignin in the walls of sclerenchyma fibres is enriched in syringyl units, whereas that in the walls of xylem vessels is enriched in guaiacyl units (Saka and Goring, 1985). Moreover, even within the wall of a particular cell type, the lignin is not homogeneous: *p*-hydroxyphenyl units are laid down first, then guaiacyl and finally syringyl (Donaldson, 2001).

The lignins of some taxa are acylated. The lignins of the Salicaceace, which comprises the willows (*Salix* spp.) and poplars (*Populus* spp.), are *p*-hydroxybenzoylated (Landucci *et al.*, 1992), and the lignin of kenaf (*Hibiscus cannabinus*) (Malvaceae) is acetylated (Lu and Ralph, 2002) (Fig. 11.5). Recent evidence indicates that these acylated lignins result from the polymerization of acylated monolignols; acylation does not occur after the polymerization of the lignin, as was first believed (Lu and Ralph, 2002; Boerjan *et al.*, 2003).

Lignins also contain a variety of minor components, many of which are produced in much greater quantities by mutants or transgenics involving lignin pathway genes (Boerjan *et al.*, 2003). Such studies have shown that these components are also synthesized by the polymerization of non-conventional monolignols. For example, transgenic tobacco (*Nicotiana tabacum*) that had the gene encoding the enzyme cinnamyl alcohol dehydrogenase (CAD) downregulated, incorporated significant amounts of hydroxycinnamyl aldehyde into its lignin. CAD converts hydroxycinnamyl aldehydes to hydroxycinnamyl alcohol, and when this enzyme is downregulated, the hydroxycinnamyl aldehydes build up and are incorporated into lignin (Ralph *et al.*, 2001). Hydroxycinnamyl aldehydes are minor components of normal lignins and react with the histochemical reagent phloroglucinol-HCl to give a red colour. These studies illustrate the plasticity of lignification and

the need to broaden the conventional definition of lignin to include these non-conventional monomers.

Secondary walls of dicotyledon seeds

Thick, non-lignified secondary walls occur in the cotyledons and endosperms of many dicotyledon seeds. These walls usually contain large amounts of one of four polysaccharides that function as reserve carbohydrates and are mobilized during germination (Bacic *et al.*, 1988; Buckeridge *et al.*, 2000). Two of these polysaccharides, galactomannans and mannans (Fig. 11.2), occur only in seed walls. Galactomannans occur in the endosperms of many species of Fabaceae and their structures are of chemotaxonomic value: species of the three subfamilies (Caesalpinioideae, Mimosoideae and Faboideae) can be distinguished by their ratio of mannose to galactose (Buckeridge *et al.*, 2000). Galactomannans also occur in the seeds of some other families, for example the Annonaceae and Convolvulaceae. Mannans, with little or no galactose, have been recorded in the seeds of several species including *Coffea arabica* (Rubiaceae) and *Carum carvi* (Apiaceae) (Bacic *et al.*, 1988). The other two polysaccharides, (1→4)-β-galactans and xyloglucans, occur as normal components of primary walls. (1→4)-β-galactans, similar in structure to those in side chains of RG-I (Fig. 11.3), occur in the cotyledons of some species of *Lupinus* (Fabaceae). However, the xyloglucans, which occur in the seeds of the Fabaceae (subfamily Caesalpinioideae) and many other families (Kooiman, 1960), have been particularly extensively studied.

The xyloglucans of seeds, like most non-seed xyloglucans, have the XXXG core structure. However, seed xyloglucans, with one exception, lack fucose. A comparative study of seed xyloglucans of the three species *Tropaeolum majus* (Tropaeolaceae), *Tamarindus indica* and *Copaifera langsdorffi* (Fabaceae) showed that they all contained the same major structural units (XXXG, XLXG, XXLG and XLLG), but the proportions varied with the species (Buckeridge *et al.*, 1992). For the xyloglucan from *C. langsdorffi*, there was also evidence for slight differences in the proportions between different plant populations.

Other variations in the structures of seed xyloglucans have also been reported. The xyloglucan of *Hymenaea courbaril* (Fabaceae) has a core structure based on XXXG and XXXXG, in about equal proportions (Buckeridge *et al.*, 1997). Unusually, the xyloglucan from the cotyledon walls of jojoba (*Simmondsia chinensis*) (Simmondsiaceae) seeds contains the same major structural units as dicotyledon fucogalactoxyloglucans: XXXG, XXFG and XLFG (Hantus *et al.*, 1997). However, even more remarkable is the occurrence in this xyloglucan of two further major structural units: XXJG and XLJG, where J (jojoba) is α-L-Gal*p*-(1→2)-β-D-Gal*p*-(1→2)-α-D-Xyl*p*-(1→6)-, which is unusual because it contains both L-Gal and D-Gal. An identical side chain occurs in the xyloglucan of the walls of the *mur1* mutant of *A. thaliana* where XXJG and XLJG structural units occur (Zablackis *et al.*, 1996). In this mutant, the α-L-Gal*p*-(1→2)- residues replace the α-L-Fuc*p*-(1→2)- residues of the wild-type.

Monocotyledon walls

In a survey of the primary walls of 104 species in 52 families of monocotyledons using UV fluorescence microscopy at two pH values, Harris and Hartley (1980) showed that ester-linked ferulic acid was confined to those species belonging to a group that was subsequently also identified using DNA sequences and named the commelinoid group (Chase *et al.*, 1993). However, this name has recently been changed to the commelinid group to avoid confusion with the Commelinoideae, a subfamily of the Commelinaceae (APG II, 2003). The commelinid group, which is the most highly evolved major group in the monocotyledons, comprises the Arecales (palms), Commelinales, Zingiberales and Poales (Fig. 11.6). The other, non-commelinid monocotyledons comprise the Acorales, Alismatales and lilioids (Asparagales, Dioscorales, Liliales and Pandanales) (Chase *et al.*, 2000; APG II, 2003). The commelinid and non-commelinid groups have quite

different wall compositions, especially of their primary walls, and will be discussed separately.

Primary walls of monocotyledons

NON-COMMELINID PRIMARY WALLS   Surveys of the monosaccharide compositions of the primary walls of a range of non-commelinid monocotyledons indicated that the non-cellulosic polysaccharide compositions of these walls are similar to those of the primary walls of dicotyledons and contain a large proportion of pectic polysaccharides (Jarvis *et al.*, 1988; Harris *et al.*, 1997; Harris, 2000). The neutral monosaccharide profiles indicated that RG-I galactans are more abundant than RG-I arabinans (Harris *et al.*, 1997; Harris, 2000).

Detailed analyses of the non-cellulosic polysaccharides, including linkage analyses, of the primary walls of non-commelinid monocotyledons have been done on only a few species. These include onion (*Allium cepa*) (Alliaceae) (Redgwell and Selvendran, 1986) and asparagus (*Asparagus officinalis*) (Asparagaceae) (Waldron and Selvendran, 1990). In both species, the walls contain large proportions of pectic polysaccharides, comprising homogalacturonans and rhamnogalacturonan I. Although RG-II has not been identified in the walls of non-commelinid monocotyledons, its widespread occurrence in other seed plants suggests that it also occurs in this group. The structures of the xyloglucans in onion and garlic (*Allium sativa*) walls have been investigated and found to be fucogalactoxyloglucans, identical to those in dicotyledon walls (Ohsumi and Hayashi, 1994).

In addition, the walls of some species of the Alismatales contain pectic polysaccharides rich in the monosaccharide apiose (Harris, 2000; Ridley *et al.*, 2001) (Fig. 11.5). Apiogalacturonan, which is a homogalacturonan bearing mono- and di-apiosyl side chains attached to C-2 and C-3 of the backbone (Fig. 11.3), has been isolated from the walls of *Lemna minor* (Araceae). Apiose-rich pectic polysaccharides also occur in the walls of vegetative parts of the sea grasses *Heterozostera tasmanica*, *Zostera marina*, *Zostera pacifica* and *Phyllospadix* sp. (Zosteraceae).

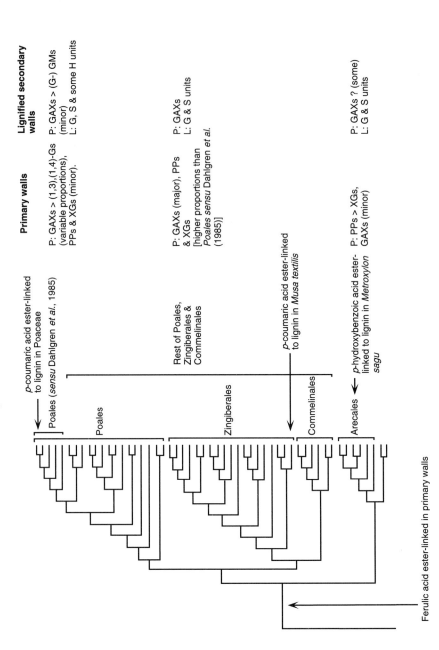

Fig. 11.6. The phylogeny of the commelinid monocotyledons based on nucleotide sequences of three genes (re-drawn from Chase *et al.*, 2000). On the right are the non-cellulosic polysaccharides of the primary and lignified secondary walls and the lignin monomers of the lignified secondary walls. Selected wall features of particular taxa are shown arrowed. Abbreviations: as for Fig. 11.5 and (1,3),(1,4)-G = (1→3),(1→4)-β-D-glucan; GAX = glucuronoarabinoxylan; H = *p*-hydroxyphenyl.

Small amounts of ester-linked ferulic and *p*-coumaric acids, together with a range of dehydrodiferulic acids, have recently been reported in the walls of asparagus (Rodriguez-Arcos *et al.*, 2002). As in the dicotyledons, these small amounts of hydroxycinnamic acids, which were not detected in the study of Harris and Hartley (1980), may be of widespread occurrence in the walls of non-commelinid monocotyledons.

Although virtually nothing is known about the structural proteins and glycoproteins in the walls of this group, there is a report of extensins being isolated from the walls of asparagus (Kieliszewski *et al.*, 1992).

COMMELINID PRIMARY WALLS: POACEAE Most research on the walls of the commelinid group has focused on the walls of the Poaceae (grasses and cereals) because of their enormous economic importance. Indeed, for many years the composition of Poaceae primary walls, which is quite different from those of dicotyledon and non-commelinid walls, was sometimes falsely regarded as being representative of all monocotyledon primary walls. In Poaceae walls, the most abundant non-cellulosic polysaccharides are GAXs rather than pectic polysaccharides. GAXs have mostly single α-L-arabinosyl and α-D-glucuronosyl (or its 4-*O*-methyl derivative) residues linked at C(O)3 and C(O)2 respectively to the xylose residues of the (1→4)-β-D-xylan backbone (Fig. 11.2) (Carpita, 1996; Harris, 2000). However, more complex side chains also occur (Bacic *et al.*, 1988; Wende and Fry, 1997). Ferulic acid and small amounts of *p*-coumaric acid (Fig. 11.4) are esterified by their carboxyl groups to the C(O)5 hydroxyl of some of the arabinosyl residues.

Poaceae walls also contain variable proportions of (1→3),(1→4)-β-D-glucans, which are linear polysaccharides containing both (1→3)- and (1→4)-glycosidic linkages; 70% of the linkages are usually (1→4) (Bacic *et al.*, 1988; Stone and Clarke, 1992; Smith and Harris, 1999) (Fig. 11.1). These polysaccharides occur in only small proportions in some primary walls, for example those of mesophyll cells of ryegrass (*Lolium multiflorum* and *Lolium perenne*) leaves contain <3%,

but in large proportions in others, for example those of barley (*Hordeum vulgare*) starchy endosperm contain 75%. In the walls of coleoptiles, the proportions vary with stage of development.

Pectic polysaccharides and xyloglucans also occur in Poaceae walls, but only in small proportions. The pectic polysaccharides have similar structures to those in the walls of dicotyledons (Carpita, 1996; Harris, 2000). For example, RG-Is isolated from the walls of suspension-cultured cells of maize (*Zea mays*) and rice (*O. sativa*) have similar structures to the RG-I from the walls of suspension-cultured sycamore (Thomas *et al.*, 1989). RG-II from the walls of bamboo (*Phyllostachys edulis*) has an identical structure to RG-IIs from dicotyledons, with the variable B side chain containing eight monosaccharides (Glushka *et al.*, 2003) (Fig. 11.3). The xyloglucans are less branched than those of dicotyledons and non-commelinid monocotyledons; they usually contain no fucose, less xylose and much less galactose (Vincken *et al.*, 1997). However, a fucosylated xyloglucan has been reported in the walls of suspension-cultured tall fescue (*Festuca arundincea*) cells (McDougall and Fry, 1994). It has recently been shown that the xyloglucan in the walls of barley coleoptiles has a core structure based on XXGG and XXGGG (Gibeaut *et al.*, 2002).

In addition to ferulic and *p*-coumaric acids, Poaceae walls contain two series of dimers of hydroxycinnamic acids: dehydrodimers, formed by oxidation, and substituted cyclobutanes, formed photochemically (Fig. 11.4). 5-5′ Dehydrodiferulic acid ('diferulic acid') was the first dehydrodimer discovered and, for many years, was thought to be the only dehydrodimer present (Harris and Hartley, 1980). However, the full range of dehydrodimers of ferulic acid that can theoretically be formed has now been discovered (Ralph *et al.*, 1994); the most abundant are the 8-5′, 8-*O*-4′ and 8-8′ dimers (Fig. 11.4). Additionally, dehydrodimers of sinapic acid and heterodimers of ferulic and sinapic acids have been found in the walls of several cereal grains, especially wild rice (*Zizania aquatica*) (Bunzel *et al.*, 2003), and a 4-*O*-8′, 5′-5″ dehydrotrimer of ferulic acid has been found in the walls of maize (*Z. mays*) bran (Rouau *et al.*, 2003). Substituted cyclobutanes formed from ferulic and/or *p*-coumaric acid also occur; the most abundant are 4,4'-dihydroxy-α-truxillic acid, 4,4'-dihydroxy-3,3'-dimethoxy-α-truxillic acid and 4,4'-dihydroxy-3-methoxy-α-truxillic acid (Fig. 11.4) (Hartley and Morrison, 1991). Both the dehydrodimers and the substituted cyclobutanes are believed to be esterified to GAXs at both carboxyl groups thus cross-linking the GAXs. It is also theoretically possible for the dehydrotrimer of ferulic acid to cross-link up to three GAXs (Rouau *et al.*, 2003).

The same three groups of structural proteins and glycoproteins described in dicotyledon walls also occur in Poaceae walls; however, most is known about the extensins, which occur in smaller proportions than in dicotyledon walls (Carpita, 1996). Two extensins, one rich in histidine and the other rich in threonine, have been isolated from *Z. mays* walls (Kieliszewski and Lamport, 1994). Interestingly, the histidine-rich extensin shares sequence homology and glycosylation patterns with both extensins and AGPs.

COMMELINID PRIMARY WALLS: OTHER FAMILIES
The composition of the Poaceae primary wall is not unique within commelinid monocotyledons. The walls of species in other families within the order Poales (*sensu* Dahlgren *et al.*, 1985) have a similar composition, with large proportions of GAXs, variable amounts of (1→3),(1→4)-β-glucans, and small proportions of pectic polysaccharides and xyloglucans (Smith and Harris, 1999; Harris, 2000). The Poales (*sensu* Dahlgren *et al.*, 1985) is part of the much larger Poales order as defined by Chase *et al.* (2000) and APG II (2003) (Fig. 11.6).

In contrast to the walls of the Poales (*sensu* Dahlgren *et al.*, 1985), the walls of the Arecaceae (the palms), which is a basal family in the commelinid group, have a similar composition to the walls of dicotyledons and non-commelinid monocotyledons, with large proportions of pectic polysaccharides and smaller amounts of xyloglucans (Fig. 11.6). However, they do contain small proportions of GAXs to which ferulic acid is probably

ester-linked. There is also some evidence for the xyloglucans being fucogalactoxyloglucans, as occur in the dicotyledons and non-commelinid monocotyledons (Carnachan and Harris, 2000a).

The walls of the rest of the commelinid group have compositions intermediate between that of the Poales (*sensu* Dahlgren *et al.*, 1985) and that of the Arecaceae (Fig. 11.6). As classified by APG II (2003) and Chase *et al.* (2000), these plants comprise the Commelinales, the Zingiberales and the rest of the Poales. As in the walls of the Poales (*sensu* Dahlgren *et al.*, 1985), GAXs are major components, but more pectic polysaccharides and probably more xyloglucans are usually present (Harris *et al.*, 1997). A detailed analysis, including linkage analysis of the polysaccharides, has been done on the walls of pineapple (*Ananas comosus*) (Bromeliaceae) (Poales *sensu* APG II, 2003) (Smith and Harris, 1995). The ferulic acid in these walls is esterified to GAXs in exactly the same way as in the walls of the Poaceae (Smith and Harris, 2001). The pineapple walls contained only a small amount of pectic polysaccharides, but the walls were isolated from a ripe fruit and the proportion of pectic polysaccharides may have decreased during ripening. As in the walls of palms, there is some evidence that the xyloglucans in these walls are fucogalactoxyloglucans (Smith and Harris, 1995).

In addition to ester-linked ferulic acid, the walls of commelinid monocotyledons contain 5-5' dehydrodiferulic acid ('diferulic acid') (Harris and Hartley, 1980). More recently, other dehydrodimers of ferulic acid have been found in the walls of Chinese water chestnut (*Eleocharis dulcis*) (Parr *et al.*, 1996) and chufa (*Cyperus esculentus*) (Parker *et al.*, 2000), both in the Cyperaceae.

Lignified secondary walls of monocotyledons

NON-COMMELINID LIGNIFIED SECONDARY WALLS
Only fragmentary information is available about these walls. 4-*O*-Methylglucuronoxylans, similar to those in the equivalent walls of dicotyledons, have been characterized from the walls of sisal (*Agave sisalana*), *Sansevieria trifasciata* and *Cordyline indivisa*

(Asparagaceae) (Bacic *et al.*, 1988). The lignins of non-commelinid monocotyledons are apparently similar to those of dicotyledons, although acylated lignins have not been reported (Sarkanen and Hergert, 1971; Chen, 1991).

COMMELINID LIGNIFIED SECONDARY WALLS
The major non-cellulosic polysaccharides in the lignified walls of Poaceae are GAXs with similar structures to those in their primary walls, but with a lower degree of substitution by glycosyl residues (Smith and Harris, 1999). Ferulic acid and its dimers, which are ester-linked to GAXs, appear to act as nucleation sites for lignin polymerization and become incorporated into lignin (Boerjan *et al.*, 2003). In addition to GAXs, small proportions of xyloglucans, $(1\rightarrow3),(1\rightarrow4)$-β-glucans, and glucomannans and/or galactoglucomannans also occur in these walls (Smith and Harris, 1999; Trethewey and Harris, 2002).

Poaceae lignins are similar to dicotyledon lignins in containing guaiacyl and syringyl units, but the proportions of *p*-hydroxyphenyl units are slightly higher (Sarkanen and Hergert, 1971; Chen, 1991; Boerjan *et al.*, 2003). However, in contrast to dicotyledon lignins, Poaceae lignins are acylated with *p*-coumaric acid (Lu and Ralph, 1999). As with acylated lignins in dicotyledons, recent evidence indicates that these are synthesized by the polymerization of acylated monolignols (Boerjan *et al.*, 2003). Both ester-linked *p*-coumaric acid and the *p*-hydroxyphenyl units of lignins are oxidized to *p*-hydroxybenzaldehyde with alkaline nitrobenzene. However, for many years, it was often assumed that the *p*-hydroxybenzaldehyde originated only from *p*-hydroxyphenyl units of the lignins and hence the lignins had a high content of these units.

There is some evidence that the non-cellulosic polysaccharide compositions of the lignified secondary walls of all other commelinid monocotyledons are similar to those in the Poaceae. The walls of species of other families in the Poales (*sensu* Dahlgren *et al.*, 1985) that have been examined have similar polysaccharide compositions to the Poaceae (Smith and Harris, 1999). GAXs have also been identified in the walls of *Cyperus papyrus* (Cyperaceae), pineapple (*A. comosus*) (Bromeliaceae) (Bacic *et al.*, 1988) and sago palm (*Metroxylon sagu*) (Arecaceae) (Ozawa *et al.*, 1998).

The lignins of all other commelinid monocotyledons are probably also composed of mostly guaiacyl and syringyl units. As in Poaceae lignin, *p*-coumaric acid is ester-linked to the lignin of abaca (*Musa textilis*) (Musaceae) (Sun *et al.*, 1999), but in sago palm (*M. sagu*), *p*-hydroxybenzoic acid is ester-linked to lignin (Kuroda *et al.*, 2001) (Fig. 11.6).

Secondary walls of monocotyledon seeds

As in dicotyledon seeds, the seeds of many monocotyledons have thick, non-lignified walls that contain mostly one structural type of polysaccharide that is mobilized as a reserve carbohydrate during germination (Bacic *et al.*, 1988; Buckeridge *et al.*, 2000). In the non-commelinid monocotyledons, the endosperms of many species of Asparagales and Liliales contain glucomannans. Among the commelinid monocotyledons, the endosperms of various species of the Arecaceae (palms) contain mannans or galactomannans, and the aleurone walls of the Poaceae contain arabinoxylans and $(1\rightarrow3),(1\rightarrow4)$-β-glucans. As in the GAXs of primary walls of Poaceae, the arabinoxylans of the aleurone walls of wheat (*Triticum aestivum*) have ferulic acid and small amounts of *p*-coumaric acid esterified by their carboxyl groups to the C(O)5 hydroxyl of some of the arabinosyl residues; however, only minor amounts of dehydrodimers of these hydroxycinnamic acids and no substituted cyclobutanes occur in these walls (Rhodes *et al.*, 2002). In addition to proline-rich and glycine-rich proteins, wheat aleurone walls contain proteins with up to 23% serine (Rhodes and Stone, 2002).

Gymnosperm Walls

Compared with the 249,000 species of angiosperms, there are only 840 extant species of gymnosperms, which form a monophyletic group divided into four orders: the Coniferales (conifers); the Ginkgoales, which contains only one species, *Ginkgo biloba*; the Cycadales (cycads); and the Gnetales, which contains *Ephedra* (Ephedraceae), *Gnetum* (Gnetaceae) and *Welwitschia* (Welwitschiaceae) (Mabberley, 1997; Judd *et al.*, 2002; Soltis *et al.*, 2002) (Fig. 11.5). Most research has been done on the walls of the conifers because of the economic importance of many of these as sources of timber and wood pulp.

Primary walls of gymnosperms

Polysaccharides

The walls of coniferous gymnosperms have similar non-cellulosic polysaccharide compositions to those of dicotyledons; the most abundant non-cellulosic polysaccharides are pectic polysaccharides, followed by xyloglucans (Fig. 11.5). The pectic polysaccharides are similar to those of dicotyledons and comprise homogalacturonan, RG-I and RG-II domains. In addition, an XGA (Fig. 11.3) has been characterized from the walls of *Pinus mugo* (Pinaceae) pollen (Bouveng, 1965). RG-Is from the walls of suspension-cultured *Pseudotsuga menziesii* (Pinaceae) cells (Thomas *et al.*, 1987) and *Cryptomeria japonica* (Taxodiaceae) cambium (Edashige and Ishii, 1997) differ only in structural details from RG-Is from dicotyledon walls. RG-II in the walls of *Pinus densiflora* (Pinaceae) hypocotyls occurs as a dimer cross-linked by 1:2 borate-diol esters and has an identical structure to RG-IIs from dicotyledons, with the variable B side chain containing six monosaccharide residues (Shimokawa *et al.*, 1999; Glushka *et al.*, 2003) (Fig. 11.3). The xyloglucans also have identical structures to those from dicotyledon walls. They are fucogalactoxyloglucans and have been best characterized in the walls of *C. japonica* cambium (Kakegawa *et al.*, 1998) (Fig. 11.1). Small amounts of a heteroxylan, probably similar in structure to heteroxylans in dicotyledon primary walls, have been reported from the walls of suspension-cultured *P. menziesii* cells (Thomas *et al.*, 1987) and the cambium of *C. japonica* (Edashige *et al.*, 1995). Significant

amounts of glucomannans and/or galactoglucomannans have also been reported in the walls of suspension-cultured cells (Edashige and Ishii, 1996) and cambium of *C. japonica* (Edashige *et al.*, 1995). However, it is possible that at least some of these glucomannans and/or galactoglucomannans could result from contaminating lignified secondary walls in the wall preparations. None of these polysaccharides was found in the walls of suspension-cultured *P. menziesii* cells (Thomas *et al.*, 1987). The non-cellulosic polysaccharide compositions of the primary walls of the other three classes of gymnosperms are unknown.

Ester-linked ferulic acid

A survey using UV fluorescence microscopy at two pH values showed that the primary walls of species from all 17 extant families of gymnosperms contain ester-linked ferulic acid (Carnachan and Harris, 2000b) (Fig. 11.5). Analyses of primary wall preparations showed that ester-linked *p*-coumaric acid was also present. In addition, three dehydrodimers of ferulic acid, two 8-8′ dimers and a 8-5′ dimer, have been reported from hypocotyl walls of *Pinus pinaster* (Pinaceae) (Sánchez *et al.*, 1996). However, the wall polysaccharides to which these hydroxycinnamic acid monomers and dimers are ester-linked are unknown.

Proteins and glycoproteins

Structural proteins and glycoproteins, similar to those in dicotyledon walls, have been identified in gymnosperm walls. Hydroxyproline arabinosides have been isolated from the walls of suspension cultures of the coniferous gymnosperm *Cupressus* sp. (Cupressaceae) and from *G. biloba* (Ginkgoaceae) and *Ephedra* sp. (Ephedraceae), indicating the presence of extensins (Lamport and Miller, 1971). However, unlike dicotyledon extensins, more hydroxyproline tri-arabinoside was isolated than the tetra-arabinoside. More recently, an extensin and a proline-rich protein have been characterized from the walls of suspension-cultured cells of *P. menziesii* (Kieliszewski *et al.*, 1992; Kieliszewski and

Lamport, 1994). In addition, expressed sequence tags encoding proline-rich and glycine-rich proteins have been identified in a zone including the cambium and developing xylem of *Pinus taeda* (Pinaceae) (Allona *et al.*, 1998). However, it is unclear whether the proteins occur in the primary walls of the cambium or in the lignified secondary walls of the tracheids.

Lignified secondary walls of gymnosperms

Polysaccharides

The major non-cellulosic polysaccharides in the lignified secondary walls of most species of coniferous gymnosperms examined are galactoglucomannans (*O*-acetyl-galactogluco-mannans), together with smaller amounts of 4-*O*-methylglucuronoarabinoxylans (Whistler and Chen, 1991) (Fig. 11.2). However, there are exceptions: in the walls of *Calocedrus decurrens* (Cupressaceae) (Whistler and Chen, 1991) and *Podocarpus lambertii* (Podocarpaceae) (Bochicchio and Reicher, 2003), the 4-*O*-methylglucuronoarabinoxylans predominate over the galactoglucomannans. The polysaccharide composition of the lignified secondary walls of *G. biloba* (Ginkgoaceae) is similar to that of the walls of most coniferous gymnosperms (Timell, 1960).

Lignins

In the walls of *G. biloba* and most coniferous gymnosperms and cycads, the lignins contain almost all guaiacyl units (Sarkanen and Hergert, 1971; Logan and Thomas, 1985; Chen, 1991) (Fig. 11.5). In addition to these units, *p*-hydroxyphenyl units occur in compression wood formed in regions of coniferous gymnosperms under compression (Chen, 1991). Significant proportions of syringyl units have also been reported in the lignins of some coniferous gymnosperms, including *Tetraclinis articulata* (Cupressaceae) and certain species of *Podocarpus* (Podocarpaceae), and in the lignin of the cycad *Stangeria paradoxa* (Stangeriaceae) (Towers and Gibbs, 1953) (Fig. 11.5). Additionally, both syringyl and guaiacyl

units have been reported in the lignins of all three families of the Gnetales: Gnetaceae, Ephedraceae and Welwitschiaceae (Sarkanen and Hergert, 1971; Logan and Thomas, 1985; Chen, 1991) (Fig. 11.5).

Proteins and glycoproteins

An extensin-like protein has been characterized from the tracheid walls of *P. taeda* (*Pinaceae*) (Bao *et al.*, 1992).

Pteridophyte Walls

The traditionally recognized group, the pteridophytes, which contains 9800 extant species (Mabberley, 1997), is now recognized as paraphyletic. A phylogenetic analysis using a combination of morphological data and DNA sequences from four genes identified two monophyletic lineages: the Lycopodiales (lycophytes or lycopsids), comprising the club mosses and their allies that form the most basal group of vascular plants; and a lineage comprising the Psilotales (whisk ferns), the Equisetales (horsetails or sphenopsids), the Polypodiales (the leptosporangiate ferns) and the Ophioglossales (the eusporangiate ferns, comprising the Marattiaceae and Ophioglossaceae) (Pryer *et al.*, 2001; Judd *et al.*, 2002) (Fig. 11.5). Compared with the walls of other vascular plants little is known about the compositions of pteridophyte walls.

Primary walls of pteridophytes

Polysaccharides

In contrast to the primary walls of seed plants, acid hydrolysates of the primary walls of many pteridophyte taxa contain high concentrations of mannose, indicating the presence of large proportions of mannose-containing polysaccharides, probably glucomannans and/or galactoglucomannans (Bailey and Pain, 1971; Popper and Fry, 2003). Pteridophytes with large proportions of such polysaccharides in their walls include the lycophytes, the whisk fern

Psilotum nudum (Psilotaceae) and the horse-tail *Equisetum debile* (Equisetaceae) (Popper and Fry, 2003). Bailey and Pain (1971) examined the walls of a wide range of species of ferns (both leptosporangiate and eusporangiate) and divided the species into three groups depending on whether their walls contained low (0–3%), medium or high (>10%) concentrations of mannose. The eusporangiate ferns contained only low concentrations of mannose in their walls, but the leptosporangiate ferns included species in all three groups, although species with low mannose concentrations in their walls appear to be confined to more derived species. Interestingly, however, there is evidence that walls with different concentrations of these mannose-containing polysaccharides can occur even within a single species (White *et al.*, 1986). Acid hydrolysates of walls from suspension cultures of the leptosporangiate fern *Pteridium aquilinum* (Dennstaedtiaceae) contained different concentrations of mannose depending on whether the cultures were generated from the sporophyte or the gametophyte. The walls of sporophyte cultures contained only 1% mannose, but the walls of the gametophyte culture contained 11%. Overall, however, the walls of most pteridophytes probably contain substantial proportions of glucomannans and/or galactoglucomannans.

There is also evidence that the walls of all pteridophytes contain pectic homogalacturonans (Popper and Fry, 2003). RG-I has so far not been reported in pteridophyte walls, but Matsunaga *et al.* (2004) have recently shown that RG-II is present in the primary walls of all extant groups of pteridophytes (lycophytes, whisk ferns, horsetails, eusporangiate and leptosporangiate ferns), where, as in seed plants, it occurs as a dimer cross-linked by 1:2 borate-diol esters. RG-IIs from the lycophyte *Selaginella kraussiana* (Selaginellaceae) and the horsetail *Equisetum hyemale* have identical monosaccharide compositions to RG-IIs from seed plants, but RG-IIs from another lycophyte, *Lycopodium tristachyum* (Lycopodiaceae), the whisk fern *Psilotum nudum* and four species of leptosporangiate ferns also contain the unusual monosaccharide 3-*O*-methyl-rhamnose (trivial name acofriose). Remarkably, the detailed structures of the RG-IIs of *S. kraussiana* and *E. hyemale*, including those of the B side chains, are apparently identical to that of the RG-II of the eudicotyledon *A. thaliana* (Matsunaga *et al.*, 2004). In the RG-IIs of *L. tristachyum, P. nudum* and the leptosporangiate ferns, *Ceratopteris thalictroides* (Pteridaceae) and *Platycerium bifurcatum* (Polypodiaceae), the 3-*O*-methyl-rhamnose is linked to C-3, and sometimes also to C-2, of the arabinopyranose residue at the non-reducing end of the B side chain (Fig. 11.3); otherwise, the structures of these RG-IIs are identical to that of *A. thaliana*. Popper and Fry (2003) also found 3-*O*-methyl-rhamnose in acid hydrolysates of primary walls of lycophyte species in the family Lycopodiaceae, but not in the family Selaginellaceae. However, they did not find this monosaccharide in acid hydrolysates of primary walls of other species of vascular plants, including other pteridophytes. As the proportions of RG-II in the primary walls of lycophytes and other pteridophytes are similar (Matsunaga *et al.*, 2004), it is possible that in lycophyte walls other polysaccharides, in addition to RG-II, contain 3-*O*-methyl-rhamnose.

Interestingly, acid hydrolysates of lycophyte primary walls, but the walls of no other vascular plants, contain appreciable quantities of another unusual monosaccharide, 3-*O*-methyl-D-galactose (Popper *et al.*, 2001; Popper and Fry, 2003) (Fig. 11.5). The wall polysaccharide or polysaccharides containing this monosaccharide are unknown, although there are reports in angiosperms of small quantities of 3-*O*-methyl-galactose in the side chains of RG-I.

Xyloglucans have also been found in the walls of a range of pteridophytes, including lycophytes, leptosporangiate ferns and a eusporangiate fern, but the structures of these are unknown (Popper and Fry, 2003).

Proteins and glycoproteins

Little is also known about the structural proteins and glycoproteins in pteridophyte walls. However, hydroxyproline arabinosides have been isolated from the walls of *Equisetum* sp. and the leptosporangiate fern *Onoclea sensibilis* (Dryopteridaceae), indicating the presence of extensins (Lamport and Miller, 1971).

Lignified secondary walls of pteridophytes

Polysaccharides

There is no published information on the non-cellulosic polysaccharides of wall preparations solely obtained from cells with lignified secondary walls. However, galacto-glucomannans and 4-*O*-methylglu-curonoarabinoxylans have been isolated and characterized from wall preparations obtained from whole rachides of the leptosporangiate ferns *Osmunda cinnamomea* (Osmundaceae) (Timell, 1962a,b) and *Pteridium aquilinum* (Bremner and Wilkie, 1966, 1971). These rachides contain a high proportion of lignified secondary walls and so it is likely that most of these polysaccharides were from these walls. In both species, the 4-*O*-methylglucuronoarabinoxylans have similar structures to those in the lignified secondary walls of coniferous gymnosperms, but with a lower proportion of arabinose residues. Galactoglucomannans have also been isolated from wall preparations from whole organs of *Lycopodium clavatum*, *P. nudum* and *Equisetum arvense* (Timell, 1964). Such wall preparations would have contained both primary and lignified secondary walls. Nevertheless, because of their thickness, it is likely that lignified secondary walls would have made a substantial gravimetric contribution to the isolated walls, and at least some of the galactoglucomannans were probably from these walls. The limited evidence available indicates that most species in both lineages of the pteridophytes have lignified secondary walls in which the glucomannans and/or galactoglucomannans occur either in greater amounts than the 4-*O*-methylglucuronoarabinoxylans or in about equal amounts. However, it is possible that 4-*O*-methylglucuronoarabinoxylans predominate in the walls of some leptosporangiate ferns (Timell, 1962c).

Lignins

Although few studies have been done, it appears that most pteridophytes have lignins similar to those of coniferous gymnosperms, containing almost all guaiacyl units (Sarkanen and Hergert, 1971; Logan

and Thomas, 1985; Chen, 1991) (Fig. 11.5). Taxa with lignins of this type include the lycophyte genus *Isoetes* (Isoetaceae), the whisk fern *Psilotum*, the horsetail *Equisetum* and both eusporangiate and leptosporangiate ferns. However, exceptions have been reported in which the lignins also contain a significant proportion of syringyl units. Such lignins occur in the lycophyte *Selaginella* (although not in *Lycopodium*) and in the leptosporangiate fern *Dennstaedtia bipinnata* (Dennstaedtiaceae) (White and Towers, 1967; Logan and Thomas, 1985).

Wall Evolution

It is possible to trace the evolution of the compositions of primary and lignified secondary walls (Figs 11.5 and 11.6). From what little is known about the compositions of the walls of the lycophytes, the most basal extant group of vascular plants, it is evident that most of the non-cellulosic polysaccharides found in angiosperm walls, as well as lignins, had already evolved at the base of the vascular plant lineage. These include the highly complex, pectic polysaccharide RG-II (Matsunaga *et al.*, 2004). However, the relative proportions of the different polysaccharides and the structures of the lignins changed during vascular plant evolution. In both primary and lignified secondary walls, there was apparently a marked decrease in the proportions of glucomannans and/or galactoglucomannans, but this decrease occurred more rapidly in the primary walls than in the lignified secondary walls. Thus, in the primary walls of the lycophytes, these polysaccharides are predominant, but in the walls of coniferous gymnosperms, they are only a minor component and pectic polysaccharides are the predominant non-cellulosic polysaccharides, as they are in most angiosperms (dicotyledons and non-commelinid monocotyledons). Even within the pteridophytes, the eusporangiate and some leptosporangiate ferns have primary walls with only a small proportion of glucomannans and/or galactoglucomannans. However, these polysaccharides persisted in lignified secondary walls until the evolution of angiosperms, when heteroxylans became predominant in these walls.

Within the angiosperms, the most striking evolutionary changes in wall composition occurred in the primary walls of the commelinid monocotyledons. At the base of this group, the Arecaceae (palms) have walls similar to those of dicotyledons and non-commelinid monocotyledons, except for the presence of ester-linked ferulic acid, which is probably linked to GAXs that occur in these walls in small amounts. During the evolution of the commelinid group, the proportions of these GAXs, with ferulic acid, esterified to them, apparently increased, reaching their highest proportions in the walls of the Poales (*sensu* Dahlgren *et al.*, 1985), where the $(1\rightarrow3),(1\rightarrow4)$-$\beta$-glucans probably first appeared. At what stage in commelinid evolution the simple xyloglucans found in the walls of the Poaceae evolved from fucogalactoxyloglucans is unknown.

The evolution of lignin structure can also be traced from lignins with almost all guaiacyl units in the pteridophytes and coniferous gymnosperms, cycads and *G. biloba* to lignins with both guaiacyl and syringyl units that occur in the angiosperms. However, reports of syringyl units in the lignins of some lycophytes, a leptosporangiate fern, the Gnetales, a cycad and a few species of coniferous gymnosperms suggest that syringyl units in lignin probably evolved a number of times.

Despite the major differences in proportions of different non-cellulosic polysaccharides in primary and lignified secondary walls and in lignin structure among different vascular plants, all these walls presumably perform all the functions necessary for plant growth and development. This suggests that different non-cellulosic polysaccharides and lignins with different structures play similar and essential roles that have been conserved during evolution (Harris, 2000).

Conclusions

Although much is known about the compositions of the walls of many vascular plants, there are many gaps in our knowledge. In particular, information about the walls of pteridophytes and non-coniferous gymnosperms is only fragmentary. Phylogenetic trees, based on nucleotide sequences of genes, which are now available for all groups of vascular plants, provide powerful guides for selecting taxa for wall analyses.

Acknowledgements

I thank Professor B.A. Stone and J.A.K. Trethewey for critically reading the manuscript, and Professor A.G. Darvill for providing a pre-print of a paper.

References

Allona, I., Quinn, M., Shoop, E., Swope, K., Cyr, S.S., Carlis, J., Reidl, J., Retzel, E., Campbell, M.M., Sederoff, R. and Whetten, R.W. (1998) Analysis of xylem formation in pine by cDNA sequencing. *Proceedings of the National Academy of Sciences USA* 95, 9693–9698.

APG (1998) An ordinal classification for the families of flowering plants. *Annals of the Missouri Botanical Garden* 85, 531–553.

APG II (2003) An update of the Angiosperm Phylogeny Group classification for the orders and families of flowering plants: APG II. *Botanical Journal of the Linnean Society* 141, 399–436.

Bacic, A., Harris, P.J. and Stone, B.A. (1988) Structure and function of plant cell walls. In: Preiss, J. (ed.) *The Biochemistry of Plants*. Academic Press, San Diego, California, pp. 297–371.

Bailey, R.W. and Pain, V. (1971) Polysaccharide mannose in New Zealand ferns. *Phytochemistry* 10, 1065–1073.

Bao, W.L., Omalley, D.M. and Sederoff, R.R. (1992) Wood contains a cell-wall structural protein. *Proceedings of the National Academy of Sciences USA* 89, 6604–6608.

Bochicchio, R. and Reicher, F. (2003) Are hemicelluloses from *Podocarpus lambertii* typical of gymnosperms? *Carbohydrate Polymers* 53, 127–136.

Boerjan, W., Ralph, J. and Baucher, M. (2003) Lignin biosynthesis. *Annual Review of Plant Biology* 54, 519–546.

Bouveng, H.O. (1965) Polysaccharides in pollen. II. Xylogalacturonan from mountain pine pollen. *Acta Chemica Scandinavica* 19, 953–963.

Bremner, I. and Wilkie, K.C.B. (1966) The hemicelluloses of bracken. Part I. An acidic xylan. *Carbohydrate Research* 2, 24–34.

Bremner, I. and Wilkie, K.C.B. (1971) The hemicelluloses of bracken. Part II. A galactoglucomannan. *Carbohydrate Research* 20, 193–203.

Buckeridge, M.S., Rocha, D.C., Reid, J.S. and Dietrich, S.M.C. (1992) Xyloglucan structure and post-germinative metabolism in seeds of *Copaifera langsdorfii* from savanna and forest populations. *Physiologia Plantarum* 86, 145–151.

Buckeridge, M.S., Crombie, H.J., Mendes, C.J.M., Reid, J.S.G., Gidley, M.J. and Vieira, C.C.J. (1997) A new family of oligosaccharides from the xyloglucan of *Hymenaea courbaril* L. (Leguminosae) cotyledons. *Carbohydrate Research* 303, 233–237.

Buckeridge, M.S., dos Santos, H.P. and Tine, M.A.S. (2000) Mobilisation of storage cell wall polysaccharides in seeds. *Plant Physiology and Biochemistry* 38, 141–156.

Bunzel, M., Ralph, J., Kim, H., Lu, F.C., Ralph, S.A., Marita, J.M., Hatfield, R.D. and Steinhart, H. (2003) Sinapate dehydrodimers and sinapate–ferulate heterodimers in cereal dietary fiber. *Journal of Agricultural and Food Chemistry* 51, 1427–1434.

Carnachan, S.M. and Harris, P.J. (2000a) Polysaccharide compositions of primary cell walls of the palms *Phoenix canariensis* and *Rhopalostylis sapida*. *Plant Physiology and Biochemistry* 38, 699–708.

Carnachan, S.M. and Harris, P.J. (2000b) Ferulic acid is bound to the primary cell walls of all gymnosperm families. *Biochemical Systematics and Ecology* 28, 865–879.

Carpita, N.C. (1996) Structure and biogenesis of the cell walls of grasses. *Annual Review of Plant Physiology and Plant Molecular Biology* 47, 445–476.

Carpita, N.C. and Gibeaut, D.M. (1993) Structural models of primary cell walls in flowering plants: consistency of molecular structure with the physical properties of the walls during growth. *The Plant Journal* 3, 1–30.

Cassab, G.I. and Varner, J.E. (1988) Cell wall proteins. *Annual Review of Plant Physiology and Plant Molecular Biology* 39, 321–353.

Chase, M.W., Soltis, D.E., Olmstead, R.G., Morgan, D., Les, D.H., Mishler, B.D., Duvall, M.R., Price, R.A., Hills, H.G., Qiu, Y.-L., Kron, K.A., Rettig, J.H., Conti, E., Palmer, J.D., Manhart, J.R., Sytsma, K.J., Michaels, H.J., Kress, W.J., Karol, K.G., Clark, W.D., Hedren, M., Gaut, B.S., Jansen, R.K., Kim, K.-J., Wimpee, C.F., Smith, J.F., Furnier, G.R., Strauss, S.H., Xiang, Q.-Y., Plunkett, G.M., Soltis, P.S., Swensen, S.M., Williams, S.E., Gadek, P.A., Quinn, C.J., Eguiarte, L.E., Golenberg, E., Learn, G.H. Jr, Graham, S.W., Barrett, S.C.H., Dayanandan, S. and Albert, V.A. (1993) Phylogenetics of seed plants: an analysis of nucleotide sequences from the plastid gene *rbc*L. *Annals of the Missouri Botanical Garden* 80, 528–580.

Chase, M.W., Soltis, D.E., Soltis, P.S., Rudall, P.J., Fay, M.F., Hahn, W.H., Sullivan, S., Joseph, J., Molvray, M., Kores, P.J., Givnish, T.J., Sytsma, K.J. and Pires, J.C. (2000) Higher-level systematics of the monocotyledons: an assessment of current knowledge and a new classification. In: Wilson, K.L. and Morrison, D.A. (eds) *Monocots: Systematics and Evolution*. CSIRO, Melbourne, pp. 3–16.

Chen, C.-L. (1991) Lignins: occurrence in woody tissues, isolation, reactions, and structure. In: Lewin, M. and Goldstein, I.S. (eds) *Wood Structure and Composition*. Marcel Dekker, New York, pp. 183–261.

Colquhoun, I.J., Ralet, M.-C., Thibault, J.-F., Faulds, C.B. and Williamson, G. (1994) Structure and identification of feruloylated oligosaccharides from sugar-beet pulp by NMR spectroscopy. *Carbohydrate Research* 263, 243–256.

Cronquist, A. (1981) *An Integrated System of Classification of Flowering Plants*. Columbia University Press, New York.

Dahlgren, R.M.T., Clifford, H.T. and Yeo, P.F. (1985) *The Families of the Monocotyledons*. Springer-Verlag, New York.

Darvill, J.E., McNeil, M., Darvill, A.G. and Albersheim, P. (1980) Structure of plant cell walls XI. Glucuronoarabinoxylan. A second hemicellulose in the primary cell walls of suspension-cultured sycamore cells. *Plant Physiology* 66, 1135–1139.

Donaldson, L.A. (2001) Lignification and lignin topochemistry – an ultrastructural view. *Phytochemistry* 17, 859–873.

Edashige, Y. and Ishii, T. (1996) Structures of cell-wall polysaccharides from suspension-cultured cells of *Cryptomeria japonica*. *Mokuzai Gakkaishi* 42, 895–900.

Edashige, Y. and Ishii, T. (1997) Rhamnogalacturonan I from xylem differentiating zones of *Cryptomeria japonica*. *Carbohydrate Research* 304, 357–365.

Edashige, Y., Ishii, T., Hiroi, T. and Fujii, T. (1995) Structural analysis of polysaccharides of primary walls from xylem differentiating zones of *Cryptomeria Japonica* D. Don. *Holzforschung* 49, 197–202.

Fry, S.C. (1983) Feruloylated pectins from the primary cell wall: their structure and possible functions. *Planta* 157, 111–123.

Fry, S.C. (1988) *The Growing Plant Cell Wall: Chemical and Metabolic Analysis*. Longman, Harlow, UK.

Fry, S.C., York, W.S., Albersheim, P., Darvill, A., Hayashi, T., Joseleau, J.-P., Kato, Y., Lorences, E.P., Maclachlan, G.A., Mort, A.J., Reid, J.S., Seitz, H.U., Selvendran, R.R., Voragen, A.G.J. and White, A.R. (1993) An unambiguous nomenclature for xyloglucan-derived oligosaccharides. *Physiologia Plantarum* 89, 1–3.

Gaspar, Y., Johnson, K.L., McKenna, J.A., Bacic, A. and Schultz, C.J. (2001) The complex structures of arabinogalactan-proteins and the journey towards understanding function. *Plant Molecular Biology* 47, 161–176.

Gibeaut, D.M., Pauly, M., Bacic, A. and Fincher, G.B. (2002) Cell wall polysaccharide composition of barley coleoptiles. *Plant Polysaccharide Workshop 2002*, Palm Cove, Queensland, Australia.

Glushka, J.N., Terrell, M., York, W.S., O'Neill, M.A., Gucwa, A., Darvill, A.G., Albersheim, P. and Prestegard, J.H. (2003) Primary structure of the 2-*O*-methyl-α-L-fucose-containing side chain of the pectic polysaccharide, rhamnogalacturonan II. *Carbohydrate Research* 338, 341–352.

Hantus, S., Pauly, M., Darvill, A.G., Albersheim, P. and York, W.S. (1997) Structural characterization of novel L-galactose-containing oligosaccharide subunits of jojoba seed xyloglucans. *Carbohydrate Research* 304, 11–20.

Harris, P.J. (1983) Cell walls. In: Hall, J.L. and Moore, A.L. (eds) *Isolation of Membranes and Organelles from Plant Cells*. Academic Press, London, pp. 25–53.

Harris, P.J. (2000) Composition of monocotyledon cell walls: implications for biosystematics. In: Wilson, K.L. and Morrison, D.A. (eds) *Monocots: Systematics and Evolution*. CSIRO, Melbourne, pp. 114–126.

Harris, P.J. and Hartley, R.D. (1980) Phenolic constituents of the cell walls of monocotyledons. *Biochemical Systematics and Ecology* 8, 153–160.

Harris, P.J., Kelderman, M.R., Kendon, M.F. and McKenzie, R.J. (1997) Monosaccharide composition of unlignified cell walls of monocotyledons in relation to the occurrence of wall-bound ferulic acid. *Biochemical Systematics and Ecology* 25, 167–179.

Hartley, R.D. and Harris, P.J. (1981) Phenolic constituents of the cell walls of dicotyledons. *Biochemical Systematics and Ecology* 9, 189–203.

Hartley, R.D. and Morrison, W.H. (1991) Monomeric and dimeric phenolic acids released from cell walls of grasses by sequential treatment with sodium hydroxide. *Journal of the Science of Food and Agriculture* 55, 365–375.

Ishii, T. and Tobita, T. (1993) Structural characterization of feruloyl oligosaccharides from spinach-leaf cell walls. *Carbohydrate Research* 248, 179–190.

Ishii, T., Matsunaga, T., Pellerin, P., O'Neill, M.A., Darvill, A. and Albersheim, P. (1999) The plant cell wall polysaccharide rhamnogalacturonan II self-assembles into a covalently cross-linked dimer. *Journal of Biological Chemistry* 274, 13098–13104.

Jarvis, M.C., Forsyth, W. and Duncan, H.J. (1988) A survey of the pectic content of non-lignified monocot cell walls. *Plant Physiology* 88, 309–314.

Jia, Z., Qin, Q., Darvill, A.G. and York, W.S. (2003) Structure of the xyloglucan produced by suspension-cultured tomato cells. *Carbohydrate Research* 338, 1197–1208.

Judd, W.S., Campbell, C.S., Kellogg, E.A., Stevens, P.F. and Donoghue, M.J. (2002) *Plant Systematics: a Phylogenetic Approach*. Sinauer, Sunderland, Massachusetts.

Kakegawa, K., Edashige, Y. and Ishii, T. (1998) Xyloglucan from xylem-differentiating zones of *Cryptomeria japonica*. *Phytochemistry* 47, 767–771.

Kieliszewski, M. and Lamport, D.T.A. (1994) Extensin: repetitive motifs, functional sites, post-translational codes, and phylogeny. *The Plant Journal* 5, 157–172.

Kieliszewski, M., de Zacks, R., Leykam, J.F. and Lamport, D.T.A. (1992) A repetitive proline-rich protein from the gymnosperm Douglas fir is a hydroxyproline-rich glycoprotein. *Plant Physiology* 98, 919–926.

Kooiman, P. (1960) On the occurrence of amyloids in plant seeds. *Acta Botanica Neerlandica* 9, 208–219.

Kuroda, K., Ozawa, T. and Ueno, T. (2001) Characterization of sago palm (*Metroxylon sagu*) lignin by analytical pyrolysis. *Journal of Agricultural and Food Chemistry* 49, 1840–1847.

Lamport, D.T.A. and Miller, D.H. (1971) Hydroxyproline arabinosides in the plant kingdom. *Plant Physiology* 48, 454–456.

Landucci, L.L., Deka, G.C. and Roy, D.N.A. (1992) ^{13}C NMR study of milled wood lignins from hybrid *Salix* clones. *Holzforschung* 46, 505–511.

Logan, K.J. and Thomas, B.A. (1985) Distribution of lignin derivatives in plants. *New Phytologist* 99, 571–585.

Lu, F. and Ralph, J. (1999) Detection and determination of *p*-coumaroylated units in lignin. *Journal of Agricultural and Food Chemistry* 47, 1888–1992.

Lu, F. and Ralph, J. (2002) Preliminary evidence for sinapyl acetate as a lignin monomer in kenaf. *Chemical Communications* 1, 90–91.

Mabberley, D.J. (1997) *The Plant-book: a Portable Dictionary of the Vascular Plants.* Cambridge University Press, Cambridge.

Maruyama, K., Goto, C., Numata, M., Suzuki, T., Nakagawa, Y., Hoshino, T. and Uchiyama, T. (1996) *O*-Acetylated xyloglucan in extracellular polysaccharides from cell-suspension cultures of *Mentha*. *Phytochemistry* 41, 1309–1314.

Matsunaga, T., Ishii, T., Matsumoto, S., Higuchi, M., Darvill, A., Albersheim, P. and O'Neill, M.A. (2004) Occurrence of the primary cell wall polysaccharide rhamnogalacturonan II in pteridophytes, lycophytes, and bryophytes. Implications for the evolution of vascular plants. *Plant Physiology* 134, 1–13.

McDougall, G.J. and Fry, S.C. (1994) Fucosylated xyloglucans in suspension-cultured cells of the graminaceous monocotyledon, *Festuca arundinacea. Journal of Plant Physiology* 143, 591–595.

Ohsumi, C. and Hayashi, T. (1994) The oligosaccharide units of the xyloglucans in the cell walls of bulbs of onion, garlic and their hybrid. *Plant Cell Physiology* 35, 963–967.

Ozawa, T., Ueno, T., Negishi, O. and Masaki, S. (1998) Chemical characteristics of hemicelluloses in the fibrous residue of sago palm. *Nettai Nogyo* 42, 172–178.

Parker, M.L., Ng, A., Smith, A.C. and Waldron, K.W. (2000) Esterified phenolics of the cell walls of chufa (*Cyperus esculentus* L.) tubers and their role in texture. *Journal of Agricultural and Food Chemistry* 48, 6284–6291.

Parr, A.J., Waldron, K.W., Ng, A. and Parker, M.L. (1996) The wall-bound phenolics of Chinese water chestnut (*Eleocharis dulcis*). *Journal of the Science of Food and Agriculture* 71, 501–507.

Parr, A.J., Ng, A. and Waldron, K.W. (1997) Ester-linked phenolic components of carrot cell walls. *Journal of Agricultural and Food Chemistry* 45, 2468–2471.

Pauly, M., Qin, Q., Greene, H., Albersheim, P., Darvill, A. and York, W.S. (2001) Changes in the structure of xyloglucan during cell elongation. *Planta* 212, 842–850.

Perrone, P., Hegwage, C.M., Thomson, A.R., Bailey, K., Sadler, I.H. and Fry, S.C. (2002) Patterns of methyl and *O*-acetyl esterification in spinach pectins: new complexity. *Phytochemistry* 60, 67–77.

Popper, Z.A. and Fry, S.C. (2003) Primary cell wall composition of bryophytes and charophytes. *Annals of Botany* 91, 1–12.

Popper, Z.A., Sadler, I.H. and Fry, S.C. (2001) 3-*O*-Methyl-D-galactose residues in lycophyte primary cell walls. *Phytochemistry* 57, 711–719.

Pryer, K.M., Schnelder, H., Smith, A.R., Cranfill, R., Wolf, P.G., Hunt, J.S. and Sipes, S.D. (2001) Horsetails and ferns are a monophyletic group and the closest living relatives to seed plants. *Nature* 409, 618–622.

Ralph, J., Quideau, S., Grabber, J.H. and Hatfield, R.D. (1994) Identification and synthesis of new ferulic acid dehydrodimers present in grass cell walls. *Journal of the Chemical Society, Perkin Transactions 1* 23, 3485–3498.

Ralph, J., Lapierre, C., Marita, J.M., Kim, H., Lu, F., Hatfield, R.D., Ralph, S., Chapple, C., Franke, R., Hemm, M.R., Van Doorsselaere, J., Sederoff, R.R., O'Malley, D.M., Scott, J.T., MacKay, J.J., Yahiaoui, N., Boudet, A.-M., Pean, M., Pilate, G., Jouanin, L. and Boerjan, W. (2001) Elucidation of new structures in lignins of CAD- and COMT-deficient plants by NMR. *Phytochemistry* 57, 993–1003.

Redgwell, R.J. and Selvendran, R.R. (1986) Structural features of cell-wall polysaccharides of onion *Allium cepa. Carbohydrate Research* 157, 183–199.

Renard, C.M.G.C., Champenois, Y. and Thibault, J.-F. (1993) Characterisation of the extractable pectins and hemicelluloses of the cell walls of glasswort, *Salicornia ramosissima. Carbohydrate Polymers* 22, 239–245.

Renard, C.M.G.C., Wende, G. and Booth, E. (1999) Cell wall phenolics and polysaccharides in different tissues of quinoa (*Chenopodium quinoa* Willd). *Journal of the Science of Food and Agriculture* 79, 2029–2034.

Rhodes, D.I. and Stone, B.A. (2002) Proteins in walls of wheat aleurone cells. *Journal of Cereal Science* 36, 83–101.

Rhodes, D.I., Sadek, M. and Stone, B.A. (2002) Hydroxycinnamic acids in walls of wheat aleurone cells. *Journal of Cereal Science* 36, 67–81.

Ridley, B.L., O'Neill, M.A. and Mohnen, D. (2001) Pectins: structure, biosynthesis, and oligogalacturonide-related signaling. *Phytochemistry* 57, 929–967.

Ringli, C., Keller, B. and Ryser, U. (2001) Glycine-rich proteins as structural components of plant cell walls. *Cellular and Molecular Life Sciences* 58, 1430–1441.

Rodriguez-Arcos, R.C., Smith, A.C. and Waldron, K.W. (2002) Effect of storage on wall-bound phenolics in green asparagus. *Journal of Agricultural and Food Chemistry* 50, 3197–3203.

Rouau, X., Cheynier, V., Surget, A., Gloux, D., Barron, C., Meudec, E., Louis-Montero, J. and Criton, M. (2003) A dehydrotrimer of ferulic acid from maize bran. *Phytochemistry* 63, 899–903.

Saka, S. and Goring, D.A.I. (1985) Localization of lignins in wood cell walls. In: Higuchi, T. (ed.) *Biosynthesis and Biodegradation of Wood Components.* Academic Press, Orlando, Florida, pp. 57–62.

Sánchez, M., Pena, M.J., Revilla, G. and Zarra, I. (1996) Changes in dehydrodiferulic acids and peroxidase activity against ferulic acid associated with cell walls during growth of *Pinus pinaster* hypocotyl. *Plant Physiology* 111, 941–946.

Sarkanen, K.V. and Hergert, H.L. (1971) Classification and distribution. In: Sarkanen, K.V. and Ludwig, C.H. (eds) *Lignins: Occurrence, Formation, Structure and Reactions.* Wiley-Interscience, New York, pp. 43–93.

Schröder, R., Nicholas, P., Vincent, S.J.F., Fischer, M., Reymond, S. and Redgwell, R.J. (2001) Purification and characterisation of a galactoglucomannan from kiwifruit (*Actinidia deliciosa*). *Carbohydrate Research* 331, 291–306.

Shimokawa, T., Ishii, T. and Matsunaga, T. (1999) Isolation and structural characterization of rhamnogalacturonan II–borate complex from *Pinus densiflora*. *Journal of Wood Science* 45, 435–439.

Showalter, A.M. (1993) Structure and function of plant cell wall proteins. *The Plant Cell* 5, 9–23.

Smith, B.G. and Harris, P.J. (1995) Polysaccharide composition of unlignified cell walls of pineapple [*Ananas comosus* (L.) Merr.] fruit. *Plant Physiology* 107, 1399–1409.

Smith, B.G. and Harris, P.J. (1999) The polysaccharide composition of Poales cell walls: Poaceae cell walls are not unique. *Biochemical Systematics and Ecology* 27, 33–53.

Smith, B.G. and Harris, P.J. (2001) Ferulic acid is esterified to glucuronoarabinoxylans in pineapple cell walls. *Phytochemistry* 56, 513–519.

Soltis, D.E., Soltis, P.S. and Zanis, M.J. (2002) Phylogeny of seed plants based on evidence from eight genes. *American Journal of Botany* 89, 1670–1681.

Stone, B.A. and Clarke, A.E. (1992) *Chemistry and Biology of (1→3)-β-Glucans.* La Trobe University Press, La Trobe University, Victoria, Australia.

Sun, R.C., Fang, J.M., Goodwin, A., Lawther, J.M. and Bolton, A.J. (1999) Fractionation and characterization of ball-milled and enzyme lignins from abaca fibre. *Journal of the Science of Food and Agriculture* 79, 1091–1098.

Thomas, J.R., McNeil, M., Darvill, A.G. and Albersheim, P. (1987) Structure of plant cell walls XIX. Isolation and characterization of wall polysaccharides from suspension-cultured Douglas fir cells. *Plant Physiology* 83, 659–671.

Thomas, J.R., Darvill, A.G. and Albersheim, P. (1989) Rhamnogalacturonan I, a pectic polysaccharide that is a component of moncot cell-walls. *Carbohydrate Research* 185, 279–305.

Timell, T.E. (1960) Studies on *Ginkgo biloba*, L. 1. General characteristics and chemical composition. *Svensk Papperstidning* 63, 652–657.

Timell, T.E. (1962a) Studies on ferns (Filicineae). Part 1. The constitution of a xylan from cinnamon fern (*Osmunda cinnamomea* L.). *Svensk Papperstidning* 65, 122–125.

Timell, T.E. (1962b) Studies on ferns (Filicineae). Part 2. The properties of a galactoglucomannan and a cellulose from cinnamon fern (*Osmunda cinnamomea* L.). *Svensk Papperstidning* 65, 173–177.

Timell, T.E. (1962c) Studies on ferns (Filicineae) Part 3. General chemical characteristics and comparison with gymnosperms and angiosperms. *Svensk Papperstidning* 65, 266–272.

Timell, T.E. (1964) Studies on some ancient plants. *Svensk Papperstidning* 67, 356–363.

Towers, G.H.N. and Gibbs, R.D. (1953) Lignin chemistry and taxonomy of higher plants. *Nature* 172, 25–26.

Trethewey, J.A.K. and Harris, P.J. (2002) Location of (1→3)- and (1→3), (1→4)-β-D-glucans in vegetative cell walls of barley (*Hordeum vulgare*) using immunolabelling. *New Phytologist* 154, 347–358.

Vanzin, G.F., Madson, M., Carpita, N.C., Raikhel, N.V., Keegstra, K. and Reiter, W.-D. (2002) The *mur2* mutant of *Arabidopsis thaliana* lacks fucosylated xyloglucan because of a lesion in fucosyltransferase AtFUT1. *Proceedings of the National Academy of Sciences USA* 99, 3340–3345.

Vierhuis, E., York, W.S., Kolli, V.S.K., Vincken, J.-P., Schols, H.A., Van Alebeek, G.-J.W.M. and Voragen, A.G.J. (2001) Structural analyses of two arabinose containing oligosaccharides derived from olive fruit xyloglucan: XXSG and XLSG. *Carbohydrate Research* 332, 285–297.

Vincken, J.-P., Wijsman, A.J.M., Beldman, G., Niessen, W.M.A. and Voragen, A.G.J. (1996) Potato xyloglucan is built from XXGG-type subunits. *Carbohydrate Research* 288, 219–232.

Vincken, J.-P., York, W.S., Beldman, G. and Voragen, A.G.J. (1997) Two general branching patterns of xyloglucan, XXG and XXGG. *Plant Physiology* 114, 9–13.

Waldron, K.W. and Selvendran, R.R. (1990) Composition of the cell walls of different asparagus (*Asparagus officinalis*) tissues. *Physiologia Plantarum* 80, 568–575.

Waldron, K.W., Ng, A., Parker, M.L. and Parr, A.J. (1997) Ferulic acid dehydrodimers in the cell walls of *Beta vulgaris* and their possible role in texture. *Journal of the Science of Food and Agriculture* 74, 221–228.

Wende, G. and Fry, S.C. (1997) 2-*O*-β-D-Xylopyranosyl-(5-*O*-feruloyl)-L-arabinose, a widespread component of grass cell walls. *Phytochemistry* 44, 1019–1030.

Whistler, R.L. and Chen, C.-C. (1991) Hemicelluloses. In: Lewin, M. and Goldstein, I.S. (eds) *Wood Structure and Composition*. Marcel Dekker, New York, pp. 287–319.

White, A.R., Elmore, H.W. and Watson, M.B. (1986) Cell wall polysaccharides and carbohydrate composition of fern tissue cultures. *13th International Carbohydrate Symposium*, Abstract B 129, Cornell University, Ithaca, New York.

White, E. and Towers, G.H.N. (1967) Comparative biochemistry of the lycopods. *Phytochemistry* 6, 663–667.

Willats, W.G.T., Steele-King, C.G., Marcus, S.E. and Knox, J.P. (1999) Side chains of pectic polysaccharides are regulated in relation to cell proliferation and cell differentiation. *The Plant Journal* 20, 619–628.

Willats, W.G.T., Steele-King, C.G., McCartney, L., Orfila, C., Marcus, S.E. and Knox, J.P. (2000) Making and using antibody probes to study plant cell walls. *Plant Physiology and Biochemistry* 38, 27–36.

Willats, W.G.T., McCartney, L., Mackie, W. and Knox, J.P. (2001) Pectin: cell biology and prospects for functional analysis. *Plant Molecular Biology* 47, 9–27.

York, W.S., Kolli, V.S.K., Orlando, R., Albersheim, P. and Darvill, A.G. (1996) The structures of arabinoxyloglucans produced by solanaceous plants. *Carbohydrate Research* 285, 99–128.

Zablackis, E., York, W.S., Pauly, M., Hantus, S., Reiter, W.-D., Chapple, C.C.S., Albersheim, P. and Darvill, A. (1996) Substitution of L-fucose by L-galactose in cell walls of *Arabidopsis mur1*. *Science* 272, 1808–1810.

12 Diversity in secondary metabolism in plants

Peter G. Waterman

Centre for Phytochemistry, Southern Cross University, Lismore, NSW 2480, Australia

---

Introduction

In all higher plants discrete organic compounds are found that are not intimately involved in the primary processes of physiology and development. Collectively such compounds are often referred to as secondary metabolites, presumably because they have no obvious impact on primary metabolism. One of the most striking features of secondary metabolites is their enormous structural diversity. This is well illustrated by the *Dictionary of Natural Products* (2003), the most recent edition of which lists over 140,000 discrete compounds, most of which originate from higher plant sources. They have given rise to a branch of chemistry (natural products chemistry) and have been a significant stimulant for the development of chromatographic, spectroscopic and synthetic techniques that are necessary for their isolation and characterization. Secondary metabolites are also noteworthy for exhibiting a wide range of biological activities and they have played a critical role in the development of our current pharmaceutical and agrochemical armoury.

Why produce secondary metabolites?

Unsurprisingly there is no simple answer to why plants invest a significant, although variable, amount of resources in producing secondary metabolites. However, a consensus view has now evolved among most interested scientists that their presence does, in some way, confer some advantage to the producer, usually through their interaction with extrinsic factors in the environment.

Paramount among these is that they, to a degree, defend the producer against the attacks of predators and pathogens. Such a function would certainly explain why there would be selection for structural diversity in as much as a 'defence role' would need to be viewed in terms of a war of adaptation and counter-adaptation between plants and their pests. The evidence for this role is now compelling (Rosenthal and Berenbaum, 1991).

Other secondary metabolites are implicated in attraction of pollinators (colour and scent) while the complex issue of arranging seed dispersal by an animal mediator can require appreciable manipulation in secondary metabolite profiles in unripe fruit, which needs to be protected, and ripe fruit, where the mediator needs to be encouraged to take the fruit but avoid eating the seed.

Such a view of secondary metabolites differs somewhat from that described by Jarvis (2000). He recognizes compounds that act as internal messengers as secondary metabolites, although he expressly avoids the use of that term because of a blurred distinction from primary metabolites. In this discussion emphasis is placed on those metabolites that are of limited distribution, and conserved messenger molecules, plant hormones and so on are excluded. However, that does not eliminate the possibility, indeed the likelihood, that some secondary metabolites are multifunctional and have both intrinsic and extrinsic roles to play.

The Building Blocks of Secondary Metabolism

It is one of the wonders of evolution that, despite the enormous diversity in end products, the carbon skeletons of most secondary metabolites can be traced back to three building blocks: 'acetate', 'mevalonate' and 'shikimate' (Fig. 12.1). Furthermore, none of these building blocks is used specifically for

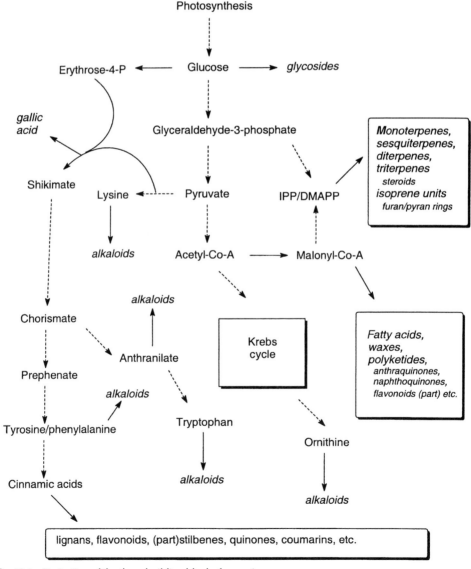

Fig. 12.1. Derivation of the three building blocks from primary processes.

secondary metabolites but they are critical components of primary processes. Some amino acids feature as precursors for nitrogen-containing secondary metabolites, notably the alkaloids. Simple sugars are often involved, giving rise to compounds known collectively as glycosides.

Acetate

The acetate or polyketide pathway is based on a two-carbon unit attached to co-enzyme A (acetyl-coA; Fig. 12.1). Polymerization of the acetate units builds carbon chains that can either remain linear or cyclize. By variation in the length and the mode of cyclization of the polyketide chain, the array of products that can be formed is vast. The highpoint of polyketide secondary metabolite biosynthesis is probably found in the macrolide and polyether antibiotics produced in *Streptomyces*, fungi and some algae (Tsantrizos and Yang, 2000), but it remains a vital player in secondary metabolism of higher plants.

Mevalonate

Mevalonic acid is a six-carbon intermediate that, by decarboxylation, gives rise to a five-carbon branched chain (isoprene) unit that occurs in the activated (through phosphorylation) isomeric forms isopentenyl diphosphate (IPP) and dimethylallyl diphosphate (DMAPP) (see Fig. 12.1). Mevalonic acid is itself formed by the combination of three acetate units. The activated five-carbon units are able to polymerize to give rise to a class of compounds generally referred to as terpenes. The products of the terpenoid pathway are the most numerous and, even when small (C_{15} or C_{20}), can be exceedingly complex.

Quite recently an alternative pathway to IPP and DMAPP has been found to occur in plastids (Rohmer, 1999), whereas the mevalonate pathway occurs in the cytosol. The penultimate product in the new route is deoxyxylulose phosphate (DXP) and the primary precursors are pyruvate and glyceraldehyde-3-phosphate. The DXP route to

IPP is clearly independent of mevalonic acid and it is probably better in future to refer to the collective pathway and its outcomes as the terpenoid pathway.

Shikimate

The shikimic acid pathway, originating in the formation of the cyclic (six-membered ring) acid from erythrose-4-phosphate and phosphoenolpyruvate, spins off relatively few secondary metabolites. The addition of a further three-carbon side chain by the inclusion of further phosphoenolpyruvate to yield the C_6–C_3 phenylpropane is, in contrast, the starting point for many classes of secondary metabolite (Fig. 12.1). Seemingly there are once again parallel pathways to be found in plastids and cytosol, but in this case the precursors appear to be identical (Hrazdina and Jensen, 1992).

Amino acids

There is an interesting dichotomy among natural amino acids with a few (phenylalanine, tyrosine, tryptamine, lysine, ornithine) commonly incorporated into secondary metabolites whilst others are rarely involved, particularly in higher plants. The alkaloids are the largest group of amino acid-derived metabolites and are often classified in respect of the precursor amino acid, which is commonly one of the five amino acids noted above. Other amino acid-derived metabolites include cyanogenic compounds and glucosinolate.

Sugars

A range of hexose and pentose sugars are often added to a preformed carbon skeleton (an aglycone) to form a class of compounds known collectively as glycosides. The most widely exploited is β-D-glucose in the pyranose form linked to the aglycone through an oxygen (ether) bridge. Rhamnose, galactose, xylose and arabinose also occur widely, while many others are recorded less commonly.

On occasion disaccharides right through to decasaccharides made up of mixtures of different sugars can be added.

The addition of the sugar profoundly changes the solubility profile of the glycoside in comparison with the aglycone, making it far more water soluble, more readily transportable, and capable of storage in the cell vacuole.

Integration and ornamentation

In the sections above, the various key building blocks have been introduced in isolation. However, we cannot view them in isolation as, in reality, many secondary metabolites are formed by linking subunits derived from different pathways. Take, for example, the flavonoid derivative rotenone, an insecticidal product found widely in some genera of the Leguminosae–Papilionaceae, notably *Derris* and *Tephrosia*. In Fig. 12.2, the origins of the carbon skeleton of rotenone are highlighted. The main events are as follows:

1. Linkage of a phenylpropane (shikimate) and polyketide (acetate) units to form the flavonoid skeleton.
2. Modifications to the structure of that skeleton to generate an isoflavone with the appropriate oxygenation pattern for formation of the rotenoid nucleus.
3. Addition of methyl (from methionine?) and subsequent cyclization to the rotenoid.
4. Addition of IPP to the rotenoid skeleton followed by cyclization to give the furan ring.

Thus the rotenone incorporates elements from acetate, terpenoid and phenylpropanoid pathways. It also involves addition of methyl units from a methyl donor and modification of the isoflavone skeleton, one methyl and the

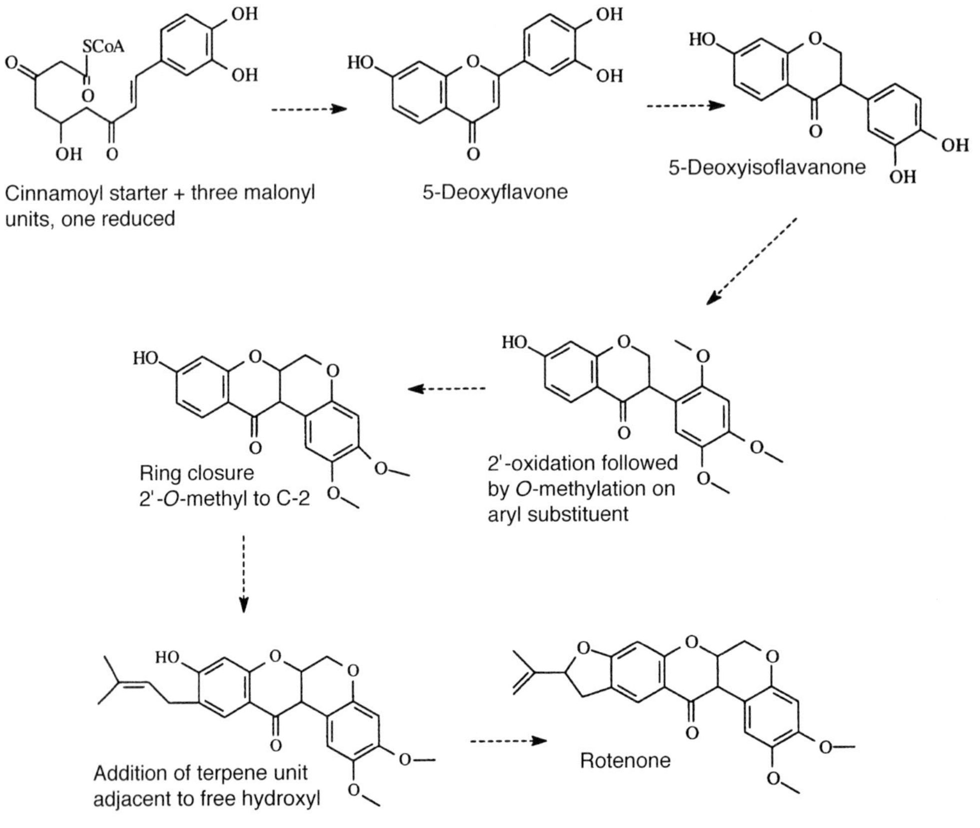

Cinnamoyl starter + three malonyl units, one reduced

5-Deoxyflavone

5-Deoxyisoflavanone

2'-oxidation followed by O-methylation on aryl substituent

Ring closure 2'-O-methyl to C-2

Addition of terpene unit adjacent to free hydroxyl

Rotenone

Fig. 12.2. Steps in the formation of rotenone.

terpenoid element through oxidative and/or reductive processes. This process involves a series of specific enzymes to guide the various steps and is clearly controlled by the genome.

Role of Acetate

The acetate pathway extends by the addition of two-carbon units to an existing 'starter' unit and this will progress until a predetermined number of units have been added. Current evidence points to the adding unit being malonyl-CoA (HOOC-CH_2-CO-CoA) rather than acetyl-CoA (CH_3-CO-CoA) with decarboxylation assisting in driving the condensation reaction.

The simplest outcome of this process is the building up of long chains to produce the ubiquitous fatty acids that are involved in cell membranes and in the production of fixed oils (Forkmann and Heller, 1999). The starter unit is most often 'acetate', leading to linear chains with an even number of carbon atoms, but other options can occur. Between each addition the carbonyl closest to the starter unit is usually eliminated and replaced with either a fully saturated carbon or an olefinic group (unsaturated fatty acids).

Far more important in terms of a contribution to secondary metabolite diversity and also very instructive with respect to the evolutionary steps involved in creating diversity are the cyclic (polyketide) products of the acetate pathway. These are characteristically generated from chains of between three (triketide) and seven (heptaketide) 'acetate' units. In Fig. 12.3, an example of a simple tetraketide is shown, where the starter unit is acetyl-CoA. This simple eight-carbon

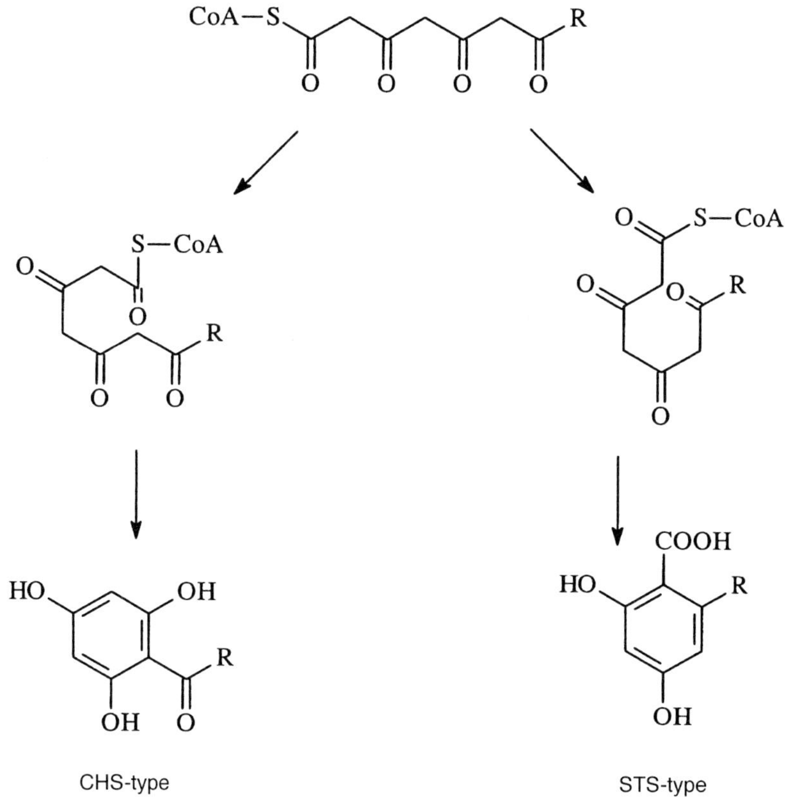

CHS-type   STS-type

Fig. 12.3. Alternative cyclizations of a tetraketide (R = CH_3) bound on to chalcone synthase (CHS)- and stilbene synthase (STS)-related proteins.

chain can yield two distinctly different aromatic (homocyclic) products depending on which carbons are involved in the cyclization reactions. Which of these two cyclizations actually happens depends on a series of proteins that are responsible for the condensations to the tetraketide and then folding that tetraketide in space so that the required ring closure reactions can occur.

The chalcone synthase (CHS; Fig. 12.3) family of proteins is a particularly important one in higher plants because if the starter unit (i.e. R in Fig. 12.3) is the CoA enzyme of a cinnamic acid variously substituted on the aromatic nucleus then the resulting cyclized product is of the class of compounds known as chalcones. Chalcones are the precursors of a huge group of metabolites that are known collectively as the flavonoids and which have been reported to serve a number of functions that are typical of secondary metabolites (for reviews, see Harborne, 1989, 1996).

Flavonoids are known to occur in all major plant phyla back to the Bryophyta. The initial steps in their formation, which are mediated by CHS proteins, appear to have remained the same throughout the timescale that implies for the existence of these enzymes. Diversity in flavonoids has arisen through post-CHS changes that have modified the skeleton by a series of oxidation/reduction reactions (Petersen *et al.*, 1999; Stafford, 2000) and through supplemental additional oxidation of the skeleton, *O*-methylation and glycosylation, addition of terpenoid units (usually five-carbon) that may subsequently cyclize.

The one major modification that has occurred with CHS and which is seen particularly in parts of the Leguminosae is the absence of one of the hydroxyls of the phloroglucinol ring. This occurs through the intervention of an additional enzyme that uses NADPH (nicotinamide adenine dinucleotide phosphate, reduced) to reduce the keto group of the first 'malonyl'-derived unit to an alcohol (Schröder, 2000). Subsequent to ring closure, that hydroxyl is lost in the formation of the aromatic ring (Fig. 12.4). The isoflavonoid compound rotenone (Fig. 12.2) will have been formed through the involvement of this modified

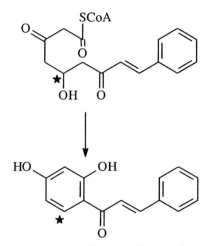

Fig. 12.4. Formation of 6′-deoxychalcone from a modified CHS cluster; * = extender carbons.

CHS. Both normal CHS and the 6′-deoxy-CHS systems can operate in the same species.

While the CHS system has been strongly conserved in chalcone synthesis it has, it seems, proved to be adaptable, in an evolutionary sense, to substitution of the building blocks, in particular the starter unit. For example, Fig. 12.5a gives the structure for a typical acridone alkaloid, a type found widely in the Rutaceae. From classical feeding experiments it has long been known that this originates from anthranilic acid and a triketide. It has now been demonstrated that the combination is catalysed in a manner analogous to that of chalcones but with an enzyme system capable of accepting anthranoyl-CoA as the starter and which has a homology with CHS > 65% (Junghanns *et al.*, 1995). Evidence that homologous systems exist for the biosynthesis of the xanthone skeleton (Fig. 12.5b) and β-triketone derivatives such as humulone (Fig. 12.5c) is significant but as yet not incontrovertible. A number of other likely starter units can be considered in the formation of derivatives of the β-triketone type. Among both the acridone and the xanthone compounds isolated are a few examples devoid of oxygenation in the anticipated positions. It is not unreasonable to presume that in such cases a modification similar to that proposed for 6′-deoxychalcones (Fig. 12.4) is in operation.

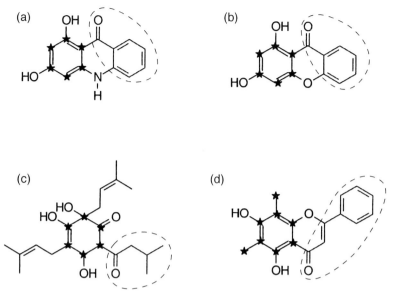

Fig. 12.5. Exploitation of the CHS system to produce structurally diverse secondary metabolites: (a) acridone alkaloids (starter = anthranilic acid); (b) xanthones (starter = benzoic acid); (c) β-triketones of the humulone type (starter = isovaleric acid); (d) C-methyl flavones (extender = methylmalonyl-CoA). * = extender carbons, starter is circled.

C-methyl flavonoids occur in significant numbers. It has been demonstrated (Schröder *et al.*, 1998), at least in *Pinus strobus*, that these are generated using the CHS assemblage with methylmalonyl-CoA as the substrate for chain extension rather than malonyl-CoA (Fig. 12.5d). There is every reason to suspect that this will be true for most if not all C-methylated flavonoids and could also be envisaged for other groups of ketides where C-methylation is found, such as the β-triketones.

In Fig. 12.3, a second cyclization of the tetraketide was demonstrated that gives rise to a dihydroxy aromatic acid. The combination of enzymes giving rise to these condensations and subsequent cyclization is known collectively as the stilbine synthase (STS) assemblage. In the STS mechanism of cyclization, the terminal carbon of the extension is not included and is left as a carbonyl, which is generally lost as carbon dioxide. As with the CHS system, there is now evidence that diversity is achieved in the STS system through the use of a range of different starter units (Fig. 12.6) and loss of oxygena-

tion through the intervention of a reduction stage in chain condensation paralleling the formation of 6′-deoxychalcones.

In CHS and STS systems, the cyclization of the polyketide involves formation of a carbon–carbon bond. Comparable cyclizations of polyketides formed in the same way can also involve condensation involving the terminal carbon (visualized as a carboxylic acid) and one of the internal oxygens of the ketide. Such condensations lead to the formation of lactones, such as parasorboside (Fig. 12.7).

In assessing the acetate pathway, it is striking that the various systems that have evolved are well suited for achieving high diversity. First, each system seems to be able to modify in comparable ways to accept a range of starter units, most of which are generated from other metabolic pathways. Secondly, each assemblage seems to have developed some capacity to vary the number of condensation units, use different condensation units, and perform reduction reactions on condensation units. The way in which these enzyme assemblages parallel each other in their capa-

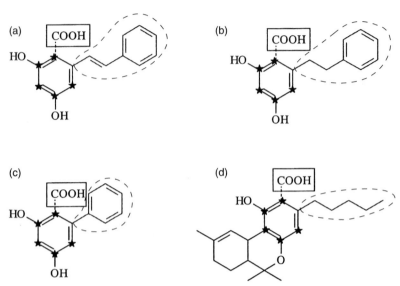

Fig. 12.6. Exploitation of the STS system to produce structurally diverse secondary metabolites: (a) stilbene alkaloids (starter = cinnamic acid); (b) bibenzyl (starter = dihydrocinnamic acid); (c) biphenyl (starter = benzoic acid); (d) cannabinoids (starter = hexanoic acid). * = extender carbons, starter is circled. The carboxylic acid (in rectangular box) is always lost during the biosynthesis.

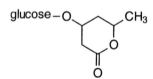

Fig. 12.7. Parasorboside, a 'lactone cyclization product' derived from a triketide. Glucose is added subsequent to cyclization.

bilities suggests that they have to be viewed as maximizing the synthetic potential of a single theme using a framework that can evolve with a minimum of need to invent or reinvent the wheel. As we shall see, this 'mechanistic parallelism' repeats itself in the exploitation of each of the major building blocks of secondary metabolism.

Role of Mevalonate

Secondary metabolites arising from the 'mevalonate' pathway are condensates of the five-carbon precursors IPP and DMAPP (see above) and as a consequence they, for the most part, exhibit a carbon skeleton made up of a multiple of C_5 and the branched-chain isopentene units. Condensation occurs through a series of enzymes known as prenyl transferases, which add IPP units to a starter or intermediate chain that is structurally comparable to DMAPP. Most of the metabolites generated fall into the following four categories: monoterpenes (C_{10}), sesquiterpenes (C_{15}), diterpenes (C_{20}) and triterpenes (C_{30}), the latter giving rise to phytosterols (Fig. 12.8). It is apparent that both the classical mevalonate and the DXP pathways contribute IPP and DMAPP for the synthesis of secondary metabolites but there is some evidence that monoterpenes and diterpenes originate predominantly from DXP and sesquiterpenes and triterpenes from the mevalonate route (Bohlmann *et al.*, 2000).

Conversion of the diphosphate intermediates into secondary metabolites is relatively easy to rationalize from a chemical perspective as it can be envisaged in terms of classical carbonium ion chemistry (Geissman and Crout, 1969). A simple example of this is shown in Fig. 12.9 in which the geranyl diphosphate (GPP)-derived monoterpenyl cation is converted by

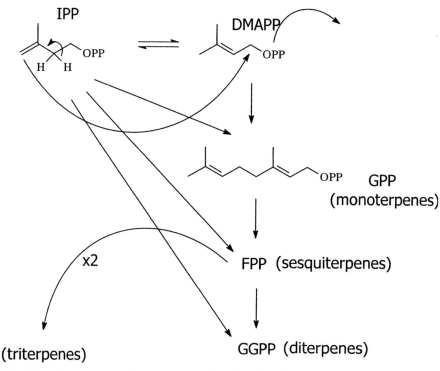

Fig. 12.8. Major classes of secondary metabolites originating from IPP and DMAPP. GPP = geranyl diphosphate; FPP = farnesyl diphosphate; GGPP = geranylgeranyl diphosphate.

a series of carbonium ion shifts and subsequent elimination of the carbonium ion through either hydration or deprotonation to give a series of well-known monoterpenes. Of course, in real life, each of these interactions is governed in a stereospecific manner by a terpene synthase. In Fig. 12.9 just a few of the many possibilities for carbonium ion-induced structural modification of the GPP-type precursor are shown. In the vast majority of cases, the eventual elimination of the carbonium ion is based on one of the two mechanisms shown in Fig. 12.9.

The larger farnesyl diphosphate (FPP) and geranylgeranyl diphosphate (GGPP) intermediates leave even more scope for carbonium ion-driven structural diversity and this is reflected in the numbers of sesquiterpene and diterpene compounds that have been recorded (>11,900 sesquiterpenes in 149 structural classes; >11,100 diterpenes in 119 structural classes; *Dictionary of Natural Products*, 2003).

The triterpenes are formed by the initial condensation of two units of FPP to give squalene that, in the form of squalene 2,3-epoxide, initiates through rupture of the epoxide and subsequent carbonium ion formation a train of rearrangements that eventually leads to the tetracyclic and pentacyclic triterpene skeletons and the plant sterols (Fig. 12.10). In the triterpenes the carbonium ion-driven ring closures generally involve methyl migrations so that a number of different methylation patterns can occur on the triterpene skeleton.

With both diterpenes and triterpenes much of the structural diversity arises through subsequent modification of the terpene skeleton, usually through oxidative and reductive reactions. One example of this is to be found in the *Citrus* family (Rutaceae) and its close allies the Meliaceae, Simaroubaceae and Cneoraceae where the tirucallol tetracyclic triterpene skeleton has been subjected to a series of oxidative ring

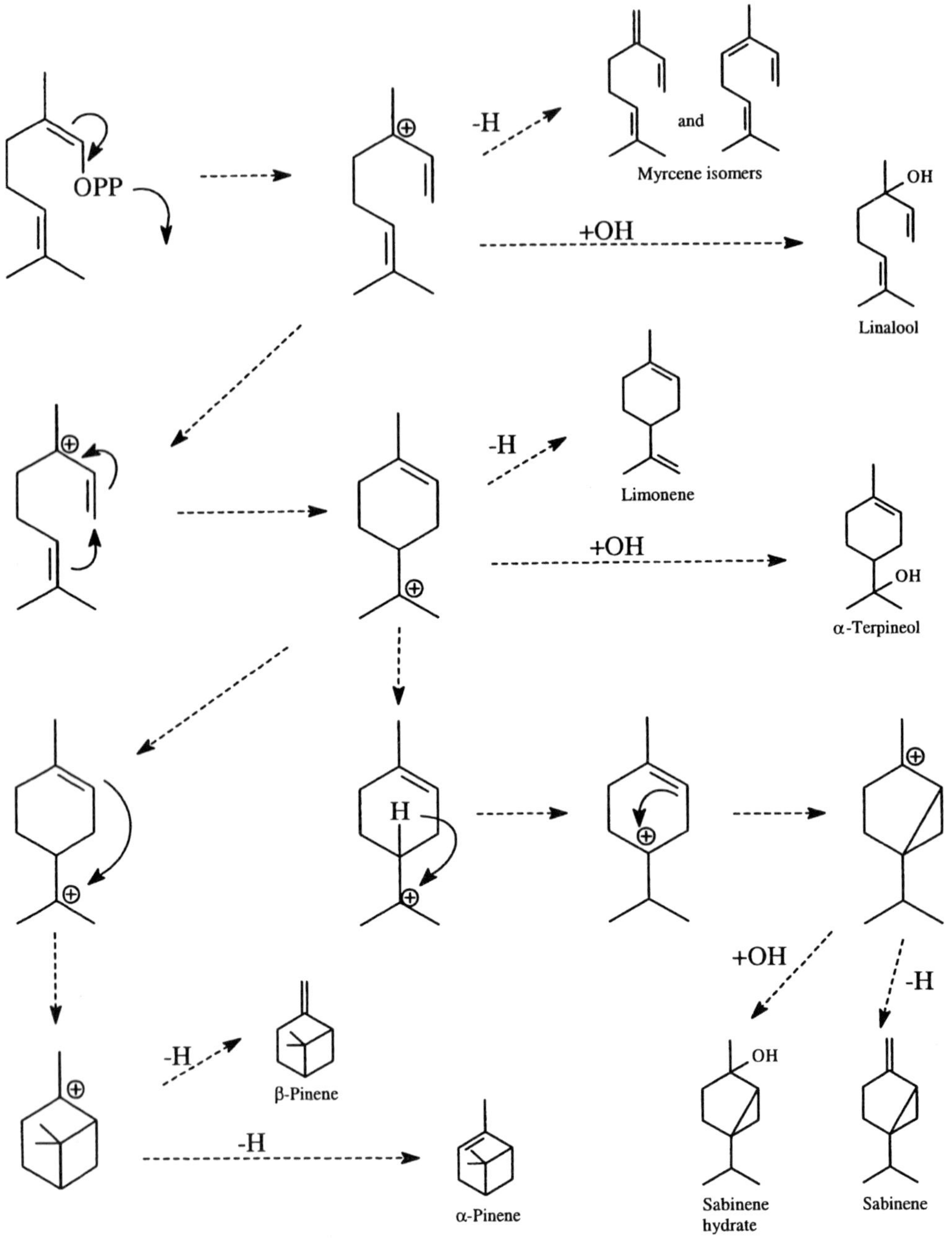

Fig. 12.9. Proliferation of monoterpene structures from a single precursor facilitated by carbonium ion formation and subsequent rearrangements.

openings and ring closures to lead to the limonoid and quassinoid bitter principles. In these plants, enzymatic processes to oxidize, ring open and then reclose as lac- tones (Fig. 12.11) have evolved and adapted to allow changes to be made to each of the homocyclic rings of the triter- pene skeleton. These can lead to far-rang-

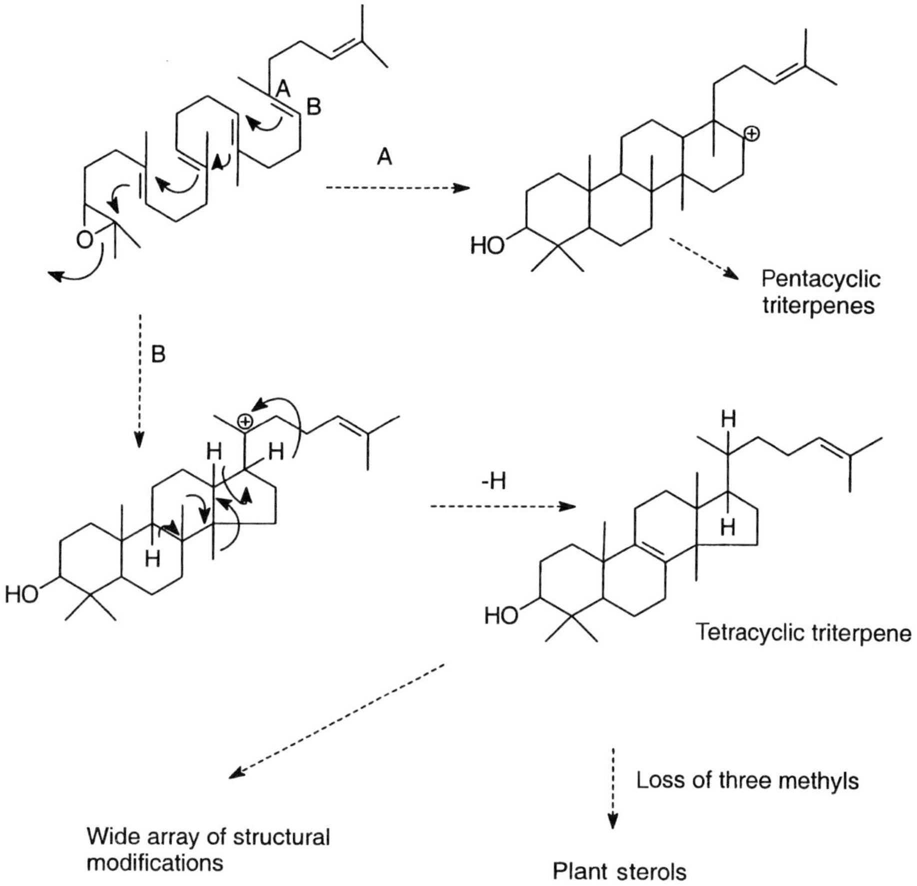

Fig. 12.10. Some key points in the synthesis of triterpenes and plant sterols.

ing changes in the final structure to the extent that some compounds are difficult to recognize as having arisen from a triterpene. Oxidation processes are widespread but rarely achieve the level seen in the Rutaceae and its allies (see review in Waterman and Grundon, 1983).

Alternative modes of cyclization

While the 'head-to-tail' condensation of DMAPP- and IPP-type diphosphates is by far the most common method of skeleton building in terpenes there are alternative pathways that are found in taxonomically restricted groups. For example, the iridoid group of monoterpenes, exemplified by loganin and *seco*-loganic acid, are generated

by a 'head-to-head' condensation, while the pyrethroids require a 'head-to-body' condensation (Fig. 12.12). The iridoid-producing families are now widely considered to be monophyletic and it can be presumed that their evolution in higher plants has been a unique event. Iridoids differ from the classical mainstream monoterpenes arising from GPP in that they tend to be quite highly oxidized and often occur as glycosides. They are also an integral part of one of the major groups of plant alkaloids (see section 'Alkaloids', below).

Terpenes as alkylating agents

The diphosphate unit, with its capacity to give a positively charged species through

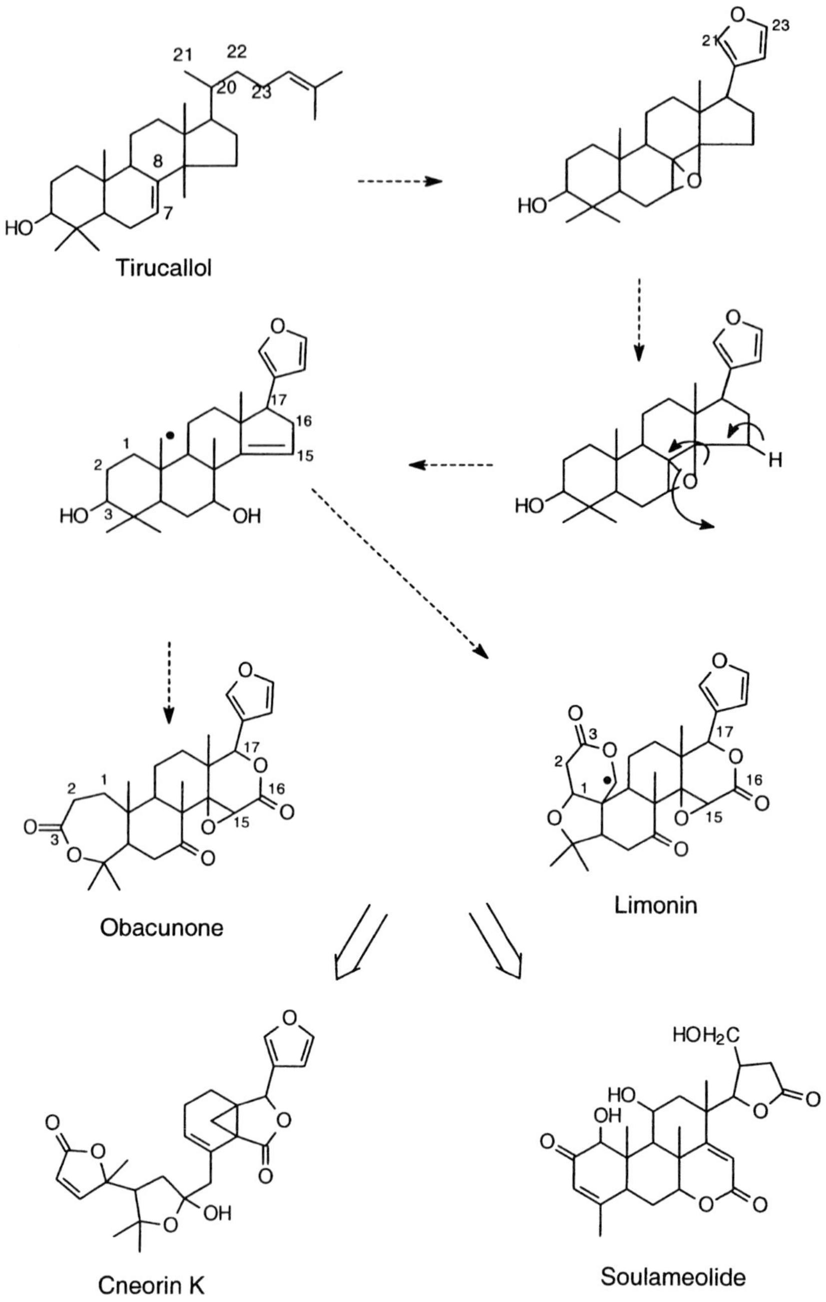

Fig. 12.11. Structural modification of tetracyclic triterpenes involving multiple oxidative ring openings and closures.

elimination of the phosphate, is predisposed to act as an alkylating agent on to electrophilic carbons and other elements. In nature these are usually either aromatic hydroxyl groups or carbons adjacent to aromatic hydroxyls. This environment is met perfectly in the alternately oxygenated rings that originate from the CHS and STS sys-

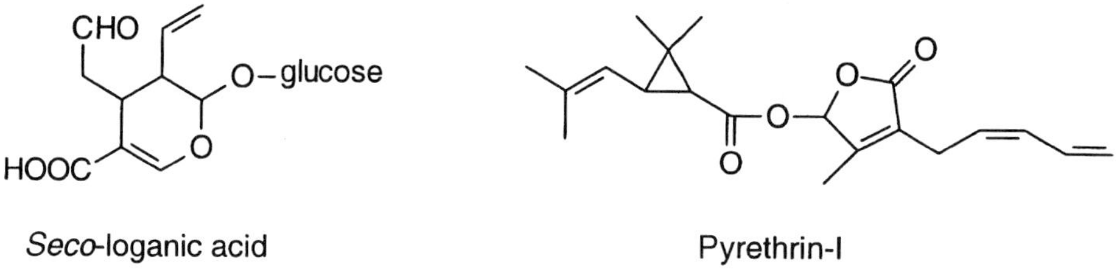
CHO
O–glucose
HOOC
O
Seco-loganic acid
O
O
O
O
O
Pyrethrin-I

Fig. 12.12. Products of unusual terpenoid pathways.

tems in polyketides and the aromatic rings of oxygenated phenylpropanes (see section 'Role of shikimate', below). The normal addition unit is the 3,3-dimethylallyl (DMA) moiety and there is now substantial evidence that specific enzymes direct the DMA unit on to the carbon skeleton. This DMA unit is often subsequently oxidized to give the corresponding epoxide (Fig. 12.13). Isolation studies on plants containing 2′,3′-epoxyDMA will often produce a series of diols and related dehydration products. It is quite likely that these are due to the acid-catalysed opening of the epoxide ring rather than to *in vivo* synthesis by the producer. Notwithstanding this issue, the addition of the DMA unit leads to an increase not only in structural diversity but also in solubility profile within the skeletal type, because of the lipophilic nature of the DMA unit.

Further modifications often occur through cyclization of the DMA side chain leading to either furan or pyran ring systems. These units are very widely distributed in higher plants and, like the open-chain precursor, can modify the properties of the structure to which they are added.

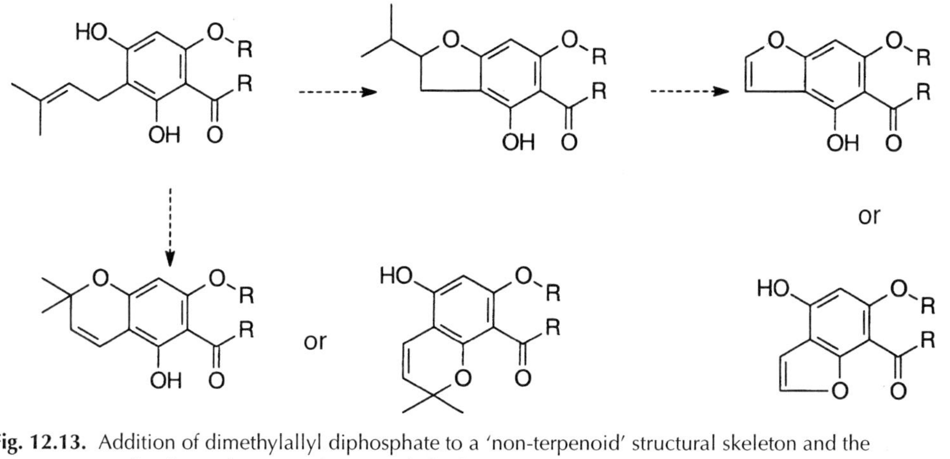
HO
O
R
R
OH O
or
or

Fig. 12.13. Addition of dimethylallyl diphosphate to a 'non-terpenoid' structural skeleton and the consequent cyclization to furano and pyrano derivatives.

While use of the simple five-carbon DMA unit is by far the most common, there are significant numbers of metabolites in which GPP, FPP and even GGPP are added. The greater length of the resulting side chain and the presence of more sights for possible oxidation can lead to more complex ring closures. This is exemplified by a recent report (Sultana *et al.*, 2003) on the structural diversity found in the coumarins of the Australian genus *Philotheca* (Rutaceae).

Role of Shikimate

The formation of the phenylpropanes (compounds with a C_6C_3 skeleton) via the shikimic acid pathway and the aromatic amino acid phenylalanine is well documented (Fig. 12.1). Once again there is evidence for separate pathways operating in plastids and cytosol, but in this case they appear to rely on identical resources and comparable enzyme assemblages (Hrazdina and Jensen, 1992). A spin-off from the main phenylpropanoid route occurs at the shikimic acid stage (C_6C_1) via the formation of anthranilic acid and subsequently the amino acid tryptophan, both of which play a role in the formation of alkaloids (see section 'Alkaloids', below). It is still not entirely clear whether the C_6C_1 product, gallic acid, important in hydrolysable tannin formation (see section 'Tannins', below), is also spun off at the shikimic acid stage or occurs via β-oxidation and chain reduction of a phenylpropane. Current evidence indicates that gallic acid could originate from both sources (Petersen *et al.*, 1999).

Metabolites arising from phenylpropanoid precursors

Secondary metabolites with a carbon skeleton attributable solely to phenylpropanoid precursors are comparatively few in number. The first products are the series of simple *trans*-cinnamic acids that proliferate from cinnamic acid itself, which has no aromatic ring substitution and arises from phenylalanine via the intervention of the pivotal enzyme phenylalanine ammonia-lyase (PAL). A series of enzymes lead to the step-by-step oxidation and *O*-methylation of cinnamic acid to coumaric acid (4-hydroxy), caffeic acid (3,4-dihydroxy), ferulic acid (3-methoxy, 4-hydroxy), 5-hydroxyferulic acid (3-methoxy, 4,5-dihydroxy) and finally sinapic acid (3,5-dimethoxy, 4-methoxy). A family of coumarate:CoA ligase enzymes occurs that have been shown to be able to convert most of these acids to the corresponding CoA product, which is ready to add to a wide range of other products, while a second series of hydroxylases and methyltransferases have been found that will catalyse the same oxidation/methylation reactions on a cinnamic acid-CoA precursor. These two options lead to a very versatile system to allow changes in ring ornamentation to occur at different stages of metabolite production (Petersen *et al.*, 1999).

The two most common groups of secondary metabolites with wholly phenylpropanoid skeletons are the coumarins and lignans/neolignans. Despite the relative simplicity of the coumarin structure and the widespread occurrence of simple coumarins based on umbelliferone (Fig. 12.14), the exact mechanism of formation from phenylalanine remains uncertain (Petersen *et al.*, 1999). The simple coumarin skeleton does not offer great scope for diversification other than through oxidation and *O*-methylation on the homocyclic ring and most of the considerable structural diversity that is found in some families, notably the Apiaceae and Rutaceae, arises through addition of terpenyl units and subsequent cyclization, usually to the furan or pyran derivatives (cf. Fig. 12.13). Furthermore, there is structural convergence to the coumarin skeleton, which can also be achieved from acetate (as in C-methyl coumarins) or from a combination of phenylpropane and acetate (neoflavonoids, 3- or 4-arylcoumarins in which the coumarin skeleton is partly derived from both pathways). A full range of coumarins, most of which originate from a phenylpropanoid precursor, is given by Murray (1997).

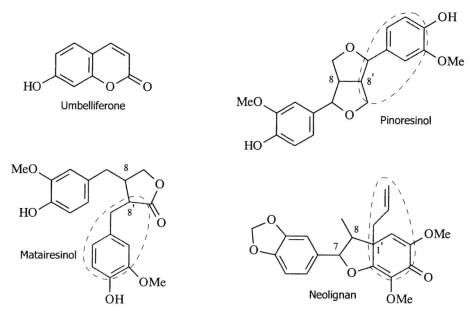

Fig. 12.14. Phenylpropanoid metabolites. In the dimeric lignans (pinoresinol and matairesinol) and neolignans one of the monomers is shown circled.

Lignans and neolignans are dimers of cinnamyl alcohols, usually coniferyl alcohol. The name lignan is generally restricted to compounds where there is a link between C-8 of the participating units (see for example pinoresinol and matairesinol; Fig. 12.14), while in neolignans the primary linkages are most often 3-3′, 8-3′, 8-O-4′ or, as 8-1′ (shown in Fig. 12.14). While lignans are quite widespread in nature and a number of skeletal structures arise from the 8-8′ bond (Ayres and Loike, 1990), the diversification that has occurred in this group of metabolites is relatively unspectacular. Neolignans are of a somewhat restricted distribution but where they do occur they are prone to somewhat greater skeletal diversification (Davin and Lewis, 1992).

Products from phenylpropanoid and acetate pathways

The incorporation of a cinnamoyl-CoA starter into a tetraketide by means of the CHS and STS enzyme groupings has already been discussed above. The class of metabolites known as flavonoids, formed by the CHS route, are among the most widespread compounds in the plant kingdom, present in all major phyla other than algae. The number of flavonoids identified is now in excess of 4000 (Petersen *et al.*, 1999).

The enzymology and genetics of almost all of the 'flavonoid' network of compounds is now well established (Forkmann and Heller, 1999) and there is, through current distribution, evidence to show when each enzyme evolved. Most evolutionary steps involve changes in oxidation and reduction levels (see Fig. 4.13 in Peterson *et al.*, 1999). The most radical skeletal modification is that in which the phenyl group migrates from C-2 to C-3 (Fig. 12.2) to form the isoflavone skeleton. Isoflavone synthase seems to have evolved in the Bryophyta (Stafford, 2000) but it is in the Fabaceae that it is developed to form a major class of metabolites that undergo a series of further cyclizations to form, for example, rotenones, pterocarpans and 3-phenyl-coumarins. An excellent review of structural variation in flavonoids and isoflavonoids can be found in Bohm (1998).

Tannins

The term tannin embraces compounds of different structural types but with a common function: the capacity, under some conditions, to form insoluble complexes with proteins and polysaccharides (Waterman and Mole, 1994). The three most important tannin types are:

1. Phlorotannins: polymers of cyclic triketides in which individual rings are linked to one another by C–C bonds. Thus, phlorotannins are entirely acetate in origin. They are found only in algae.
2. Condensed tannins: polymers of polyhydroxy flavan-3-ol flavonoid units, which are found in pteridophytes, gymnosperms and in woody species of angiosperms. Based on the flavonoid nucleus, they are of mixed phenylpropane/acetate origin (see section above).
3. Hydrolysable tannins: glycosides in which a central monosaccharide unit, most commonly glucose, is esterified by gallic acid (3,4,5-trihydroxybenzoic acid) or ellagic acid (a cyclized dimeric gallic acid). A single hexose sugar can link with up to five of these acid units. Hydrolysable tannins occur within a fairly restricted group of families in higher plants.

In comparison with most classes of secondary metabolites, the tannins are structurally somewhat conservative. This is to some extent inevitable because there are fewer avenues open for structural modification that do not reduce their capacity to bind to their target molecules. However, if you accept that their prime roles are as digestibility reducing and perhaps anti-oxidant compounds and that in the former role their target is primarily a general class of compounds (proteins), then there is probably far less evolutionary pressure to maximize diversification than is found with most other classes of secondary metabolite. The last statement is a simplification of the situation, both in terms of tannin structural diversity and the variability of the tannin–protein interaction (Waterman and Mole, 1994), but holds true for comparison with structural diversification in many groups of secondary metabolites.

Alkaloids

Of all the various classes of secondary metabolites, it is alkaloids that have received by far the most attention. The dominating reason for this is that so many of them exhibit profound biological activities and a significant number have found their way into traditional and modern materia medica (e.g. morphine, codeine, vincristine, cocaine, ergotamine, atropine, hyoscine, physostigmine, strychnine, nicotine).

The structures of alkaloids contain, by definition, nitrogen and the vast majority of nitrogenous secondary metabolites would be classified as alkaloids. The key defining factor for a metabolite to be included in the class of alkaloids is that the source of the nitrogen is an amino acid and that all or most of the amino acid is incorporated into the structure. Essentially terpenoid (steroidal alkaloids, diterpene alkaloids) or polyketide (coniine) metabolites into which nitrogen is inserted share the likelihood for significant biological activity but are referred to as pseudoalkaloids. Other classes of nitrogen-containing secondary metabolites that are usually not considered to be alkaloids are the isothiocyanates and cyanogenic glycosides. All of these various groups are reviewed in Waterman (1993). None of them shares the coherent biosynthetic themes that tie the true alkaloids together as a subset of secondary metabolites, which are derived through a 'shared' route in which the parts may change but the mechanisms uniting them to form alkaloids basically remain the same. As noted previously (Waterman, 1998), this 'shared' theme in alkaloid formation requires the bringing together of a nitrogen source (the NH_2 of an amino acid) and a ketone source (usually in the form of an aldehyde) and the formation of a C–N (imino) bond involving the two entities, together with subsequent cyclization involving this imino group. How this can occur is demonstrated in the formation of the pyrrolizidine alkaloids (Fig. 12.15). In this example, ornithine contributes both of the key functional groups through different modifications. The components contribut-

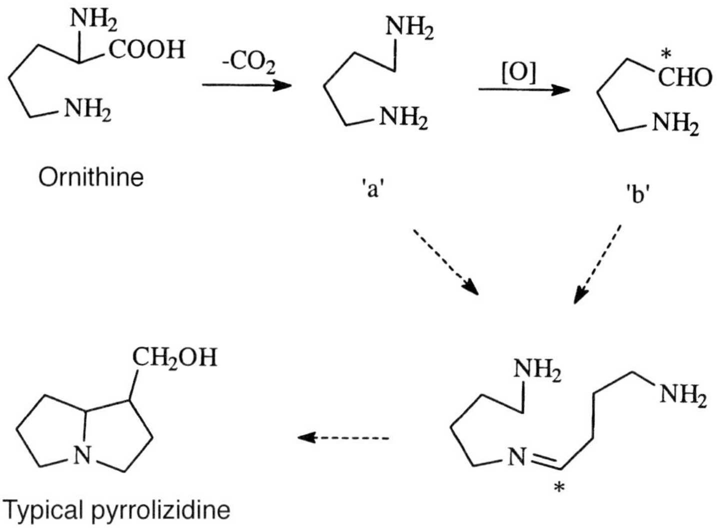

Fig. 12.15. Formation of pyrrolizidine alkaloids from two ornithine precursor units. * = the ketone-bearing carbon.

ing to the basic heterocyclic skeletons of the more widespread classes of alkaloids are listed in Table 12.1.

In the biosynthesis of alkaloids, therefore, the key process would appear to be bringing the two participating functional groups into close proximity in an environment where formation of the C=N (imine) bond can be generated. The ability to diversify the alkaloid skeleton then depends on the evolution of modifications that allow the substrate to be changed. Unfortunately identification of the genes involved in alkaloid biosynthesis has been slow (Saito and Murakoshi, 1998). When we eventually

access the genes responsible for the key imino bond-generating step it is going to be fascinating to see how it differs between such diverse higher plants as the daffodil, opium poppy and lupin.

Once the initial alkaloid skeleton is generated, then further diversification uses much the same types of modification already described for other classes (according to the *Dictionary of Natural Products* (2003) there are currently 224 structural classes of alkaloids known). In particular the 1-benzylisoquinoline and indole-monoterpene types diversify through a wide range of possible secondary cyclizations (see Chapters 9 and 10 in

Table 12.1. Origins of major subclasses of alkaloids.

Alkaloid type	N source	Ketone source
Amaryllidaceae	Tyrosine/phenylalanine	Phenylbenzaldehyde derived from tyrosine/phenylalanine
1-Benzylisoquinoline	Tyrosine/phenylalanine	Phenylacetaldehyde derived from tyrosine/phenylalanine
Emetine type	Tyrosine	Seco-loganin (see Fig. 12.12)
Indole-monoterpene	Tryptophan	Seco-loganin
Anthranilate	Anthranilic acid	'Acetate' chains
Betalains	Tyrosine or proline	Tyrosine (betalamic acid)
Tropane	Ornithine	Ornithine (same molecule) and then diketide
Pyrrolizidine	Ornithine	Aminoaldehyde derived from ornithine
Quinolizidine	Lysine	Lysine-derived aldehyde(s)

From Hrazdina and Jensen, 1992.

Waterman, 1993). Subsequent to these refinements in the actual carbon skeleton there are then all the classic oxidation/methylation type ornamentations that produce further diversity, which, while relatively subtle, can have significant impact on biological activity (Wink, 1998).

Evolution of Secondary Metabolites: Some Concluding Comments

The most striking feature of secondary metabolism in higher plants is not that there is such staggering diversity in end products but that the level of diversity is attained on the back of so few building blocks and on the use of a relatively small number of chemical reactions to bring about skeletal formation and modification and a similarly small number for skeleton ornamentation. It seems apparent that, in the process of evolution, individual enzymes have proved adaptable in taking on new substrates and have been integrated into pathways at different points to assist in the generation of structurally altered end products.

In the 1960s, with the advent of chemical taxonomy as a science, there was real hope among phytochemists that we would be able to predict phylogenetic relationships through the array of secondary metabolites produced by any given species. Subsequently, we have found this not to be the case and that, while it is certainly true that we can link some classes of compounds with individual families or related groups of families, secondary metabolites generally show distributions which contain disjunct elements.

This may well be due, in part, to parallelism and convergence in biosynthetic pathways, given the apparent adaptability of many of the individual enzymes involved and the limited number of building blocks. Probably more important is that the genes for secondary metabolite synthesis are highly conserved in the genomes of higher plants and may become dormant for considerable phylogenetic distances before being reactivated. That this is the case for quinolizidine alkaloids was demonstrated some 20 years ago (Wink and Witte, 1983) but, unfortunately, the phenomenon remains largely explored. Confirming conservation and establishing how the genes are involved in the generation of secondary metabolites is, today, a major challenge facing molecular biology.

References

Ayres, D.C. and Loike, J.D. (1990) *Lignans: Chemical, Biological and Clinical Properties.* Cambridge University Press, Cambridge.

Bohlmann, J., Gershenzon, J. and Aubourg, S. (2000) Biochemical, molecular genetic and evolutionary aspects of defense-related terpenoid metabolism in conifers. In: Romeo, J.T., Ibrahim, R., Varin, L. and De Luca, V. (eds) *Evolution of Metabolic Pathways.* Elsevier, Oxford, pp. 109–150.

Bohm, B.A. (1998) *Introduction to Flavonoids.* Harwood Academic Publishers, Amsterdam.

Davin, L.B. and Lewis, N.G. (1992) Phenylpropanoid metabolism: biosynthesis of monolignols, lignans, neolignans, lignins and suberins. In: Stafford, H.A. and Ibrahim, R.K. (eds) *Phenolic Metabolism in Plants.* Plenum Press, New York, pp. 325–375.

Dictionary of Natural Products (2003) Version 11.2. Chapman & Hall through CRC Press, Boca Raton, Florida, CD-ROM.

Forkmann, G. and Heller, W. (1999) Biosynthesis of flavonoids. In: Barton, D.H.R., Nakanishi, K. and Meth-Coon, O. (eds) *Comprehensive Natural Products Chemistry,* Vol. 1, *Polyketides and Other Secondary Metabolites Including Fatty Acids and Their Derivatives* (Sankawa, U., volume ed.). Elsevier Press, Oxford, pp. 713–748.

Geissman, T.A. and Crout, D.H.G. (1969) *Organic Chemistry of Secondary Plant Metabolism.* Freeman Cooper & Co., San Francisco, California.

Harborne, J.B. (ed.) (1989) Plant phenolics. In: Dey, P.M. and Harborne, J.B. (eds) *Methods in Plant Biochemistry,* Vol. 1. Academic Press, London.

Harborne, J.B. (ed.) (1996) *The Flavonoids: Advances in Research Since 1986.* Chapman & Hall, London.

Hrazdina, G. and Jensen, R.A. (1992) Spatial organization of enzymes in plant metabolic pathways. *Annual Review of Plant Physiology and Plant Molecular Biology* 43, 241–267.

Jarvis, B.B. (2000) The role of natural products in evolution. In: Romeo, J.T., Ibrahim, R., Varin, L. and De Luca, V. (eds) *Evolution of Metabolic Pathways.* Elsevier, Oxford, pp. 1–24.

Junghanns, K.T., Kneusel, R.E., Baumert, A., Maier, W., Gröger, D. and Matern, U. (1995) Molecular cloning and recombinant expression of acridone synthase from elicited *Ruta graveolens* L. cell suspension cultures. *Plant Molecular Biology* 27, 681–692.

Murray, R.D.H. (1997) Naturally occurring plant coumarins. *Progress in the Chemistry of Organic Natural Products* 72, 1–119.

Petersen, M., Stracke, D. and Matern, U. (1999) Biosynthesis of phenylpropanoids and related compounds. In: Wink, M. (ed.) *Biochemistry of Plant Secondary Metabolism.* Sheffield Academic Press, Sheffield, pp. 151–221.

Rohmer, M. (1999) A mevalonate-independent route to isopentenyl pyrophosphate. In: Cane, D.A. (ed.) *Comprehensive Natural Products: Isoprenoids, Including Carotenoids and Steroids,* Vol. 2. Pergamon Press, New York, pp. 45–67.

Rosenthal, G.A. and Berenbaum, M.R. (eds) (1991) *Herbivores: Their Interactions with Secondary Metabolites,* 2nd edn. Academic Press, San Diego, California.

Saito, K. and Murakoshi, I. (1998) Genes in alkaloid metabolism. In: Roberts, M.F. and Wink, M. (eds) *Alkaloids: Biochemistry, Ecology and Medical Applications.* Plenum Press, New York, pp. 147–157.

Schröder, J. (2000) The family of chalcone synthase-related proteins: functional diversity and evolution. In: Romeo, J.T., Ibrahim, R., Varin, L. and De Luca, V. (eds) *Evolution of Metabolic Pathways.* Elsevier, Oxford, pp. 55–89.

Schröder, J., Raiber, S., Berger, T., Schmidt, A., Schmidt, J., Soares-Sello, A.M., Bardshiri, E., Strack, D., Simpson, T.J., Veit, M. and Schröder, G. (1998) Plant polyketide synthases: a chalcone synthatase-type enzyme which performs a condensation reaction with methylmalonyl-CoA in the biosynthesis of C-methylated chalcones. *Biochemistry* 37, 8417–8425.

Stafford, H.A. (2000) The evolution of phenolics in plants. In: Romeo, J.T., Ibrahim, R., Varin, L. and De Luca, V. (eds) *Evolution of Metabolic Pathways.* Elsevier, Oxford, pp. 25–54.

Sultana, N., Sarker, S.D., Armstrong, J.A., Wilson, P.G. and Waterman, P.G. (2003) The coumarins of *Philotheca sensu lato*: distribution and systematic significance. *Biochemical Systematics and Ecology* 31, 681–691.

Tsantrizos, Y.S. and Yang, X.-S. (2000) Macrolide and polyether polyketides: biosynthesis and molecular diversity. In: Romeo, J.T., Ibrahim, R., Varin, L. and De Luca, V. (eds) *Evolution of Metabolic Pathways.* Elsevier, Oxford, pp. 91–107.

Waterman, P.G. (ed.) (1993) Alkaloids and sulphur compounds. In: Dey, P.M. and Harborne, J.B. (eds) *Methods in Plant Biochemistry,* Vol. 8. Academic Press, London.

Waterman, P.G. (1998) Alkaloid chemosystematics. In: Cordell, G.A. (ed.) *The Alkaloids: Chemistry and Biology,* Vol. 50. Academic Press, Boston, Massachusetts, pp. 537–565.

Waterman, P.G. and Grundon, M.F. (eds) (1983) *Chemistry and Chemical Taxonomy of the Rutales.* Academic Press, London.

Waterman, P.G. and Mole, S. (1994) *Analysis of Phenolic Plant Metabolites.* Blackwell Scientific Publications, Oxford.

Wink, M. (1998) Modes of action of alkaloids. In: Roberts, M.F. and Wink, M. (eds) *Alkaloids: Biochemistry, Ecology and Medical Applications.* Plenum Press, New York, pp. 301–326.

Wink, M. and Witte, L. (1983) Evidence for a widespread occurrence for the genes of quinolizidine biosynthesis. Induction of alkaloid accumulation in cell suspension cultures of alkaloid 'free' species. *FEBS Letters* 159, 196–200.

13 Ecological importance of species diversity

Carl Beierkuhnlein[1] and Anke Jentsch[2]

[1]*Universität Bayreuth, Lehrstuhl fur Biogeografie, D-95440 Bayreuth, Germany;*
[2]*UFZ Centre for Environmental Research Leipzig, Conservation Biology and Ecological Modelling, Permoserstr. 15, D-04318 Leipzig, Germany*

Introduction

Understanding the ecological importance of biodiversity

Understanding the ecological importance of biodiversity for ecosystem functioning and ecological services to mankind requires us to relate the diversity of ecosystem properties to the diversity of species performances in space, in time, in biotic interaction and under changing environmental conditions. Before discussing ecosystem functioning, we therefore explore three basic properties of ecological systems related to energy, matter and information, three fundamental aspects of biodiversity related to quantitative, qualitative and functional aspects, and knowledge on functional traits of species. We then review emerging theory on the role of biodiversity for ecosystem functioning, including functions such as productivity, stability, nutrient retention, resistance against invasion, and the temporal performance of communities. We further extend our scope to the benefits and services of biodiversity for human societies at large and discuss possible implications of losing biodiversity. Finally, we present prominent experimental methods, modelling and conceptual approaches in biodiversity science, thereby reviewing

the most important biodiversity hypotheses, which are still under debate. Concluding, we point to emerging challenges related to key functions, historical contingency, cross-scale and cross-system research, and the implications of spatio-temporal dynamics for the performance of biodiversity under changing environmental conditions.

Currently, an extensive and controversial debate is questioning the effects that are expected to follow the decline of plant species diversity (Mooney *et al.*, 1996; Grime, 1998; Kaiser, 2000; Schmid, 2002). This debate stimulated ecological theory and methodology (e.g. Risser, 1995; Wardle *et al.*, 1997; Lawton, 1999; Loreau, 2000; Bednekoff, 2001; Naeem and Wright, 2003). Initially, opposing standpoints were developed concerning functional implications of species richness (Huston, 1997; Hector *et al.*, 1999, 2000b). The progress in biodiversity research during recent years is thus a consequence of the engagement of various international research groups with differing approaches and perspectives. Scientists began to realize that no general unified mechanism can be found that could be applied to every ecosystem (or site), nor are individualistic restrictions per se responsible for patterns and processes. Today, paradigms are shifting (Loreau *et al.*, 2001;

Naeem *et al.*, 2002). Based on the critical analyses of data and experiments, new challenges and research issues are emerging in biodiversity science (Loreau *et al.*, 2001). In addition, the recent insights contribute to a better understanding of the potential effects of global changes for human society.

It is obvious that local and regional biodiversity is strongly influenced by human land use and its alteration in many landscapes (Machlis and Forester, 1996). The threat to biodiversity during the next century will be caused mainly by changes of land use. At the global scale, climate change, depositions of nutrients and toxic compounds, and invasive species will be less important (Sala *et al.*, 2000). This adds the human factor to the complexity of systems as humans act within ecosystems and control many functions directly. Anthropogenic action may influence biodiversity as well as key ecosystem functions. If we are interested in the effects of biodiversity loss on ecosystem functioning, direct and indirect effects of human influences will have to be considered.

Properties of ecological systems

In principle, there are always three different aspects or properties of ecological systems that are controlled or maintained by the assemblage and diversity of organisms: the flow and cycling of (i) energy, (ii) matter and (iii) information. In addition, storage and transformation occurs. Organisms are influencing these flows, transformations and storages in a non-stochastic, directed way. They regulate ecological processes and functions. This regulation cannot be predicted on the basis of physical laws or chemical processes only. The genetic information of species becomes ecologically effective as regards, for example, life-history traits, metabolisms and their plasticity under changing environmental conditions (Fig. 13.1).

As there is a limited range of ecological niches in any ecological system, species diversity is believed to be limited too (Cornell and Lawton, 1992). The diversity of coexisting species can probably be understood by considering their functional capabilities. Organisms have differentiated their functional traits and niche occupation during speciation (e.g. Cody, 1991). Their coexistence is a reflection of functional specialization and niche complementarity. Although redundancy of functions may occur in various species at a certain focus of interest, each species generally performs unique mechanisms and functions within an ecological system. Therefore, a correlation between species diversity and functional diversity is probable but is not necessarily a causal explanation (Tilman *et al.*, 1997b) (Fig. 13.2).

Still, the ecological implications of species diversity are more complex. Ecological systems and species assemblages are influenced by stochastic processes. Species combination and diversity is not deterministic and also not directly connected to a given environment. Additionally, plant species that contribute to the diversity and functioning of ecosystems differ in many aspects: for example in size, longevity and metabolisms. This indicates that the ecological importance of species diversity must be related to specific communities and ecosystems. For mankind, it has been shown early that there is a relationship between biodiversity and the rise of highly developed ancient cultures (Vavilov, 1935). For instance, the centres of old cultures and the origins of many crops are closely linked to the global 'hot spots' of biodiversity (Myers, 1988; Barthlott *et al.*, 1996). The decline of such cultures is very probably an effect of non-sustainable use of resources and biodiversity. Kim and Weaver (1994) even predict that the survival of mankind depends on the preservation of biodiversity.

The diversity of biodiversity

Initiated by Wilson's (1985) alert on the 'crisis of biodiversity' and the Rio Conference, intensive research on biodiversity topics emerged, followed up by an incredible number of publications ('The diversity of publications on diversity is overwhelming', van der Maarel, 1997). Public and political awareness occupied the theme as well.

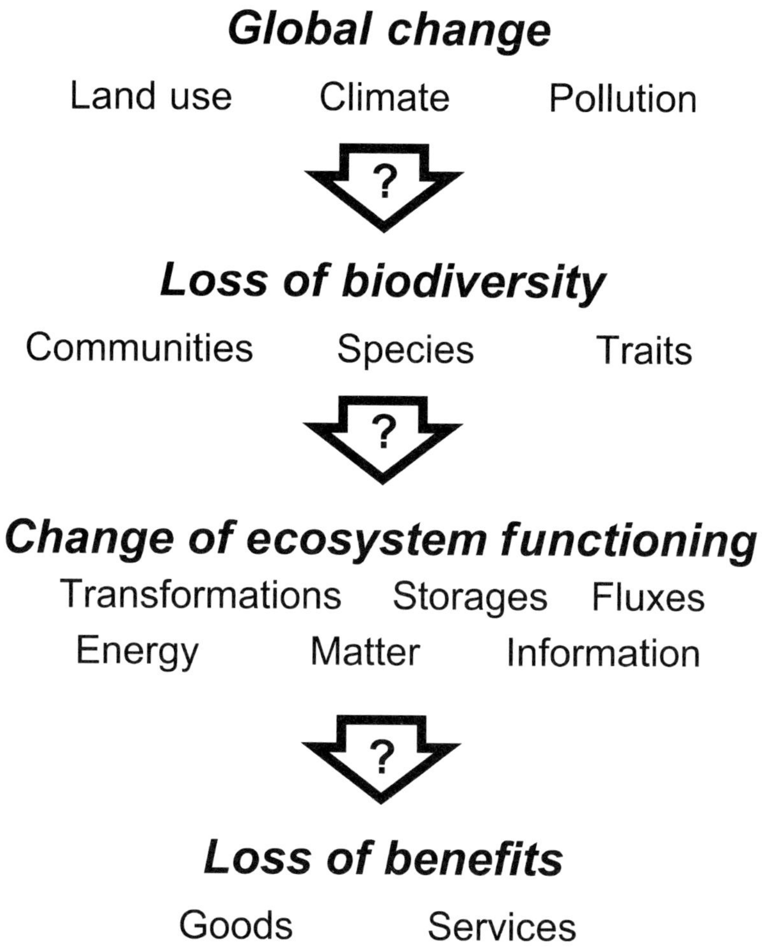

Global change

Land use   Climate   Pollution

Loss of biodiversity

Communities   Species   Traits

Change of ecosystem functioning

Transformations   Storages   Fluxes

Energy   Matter   Information

Loss of benefits

Goods   Services

Fig. 13.1. Hypothetical consequences of environmental and land-use changes. The subsequent loss of biodiversity is likely to be followed by functional changes and shifts of ecological complexity. Some of these changes result in a decline in human benefits. Note that there are many unclear connections and unsolved questions between these levels.

Looking closer, many research projects are continuing traditional approaches under the label of biodiversity just to gain funding. Perhaps more problematic is the lack of theory and concepts, which is a source of confusion and misinterpretation of results.

Different opinions and views of biodiversity research simply reflect the fact that the concept of biodiversity summarizes and integrates various aspects of biotic variability at different levels of organization (Bowman, 1993). Organisms are just one of these levels. Other levels are genes, populations, communities or ecosystems. Thus, species

diversity is just one part of biodiversity. Yet, it does not inform about abundance, dominance patterns or equitability.

Generally we can distinguish: (i) qualitative variability from (ii) quantitative richness of a community, an ecosystem or an area. In addition, different degrees of (iii) functional interactions create varying ecological complexity. With the focus on plants, this means that phytodiversity integrates the variability between plants, their number and their functional differences. Most attention is concentrated on the number of species, because this is easy to measure. However,

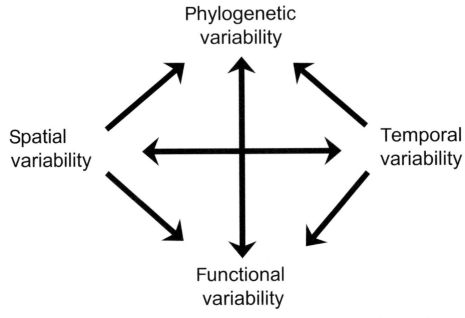

Fig. 13.2. Four aspects of biotic variability. Biodiversity occurs at all categories (spatial, temporal, phylogenetic and functional). Even if the performance of functional variability or diversity is influenced by spatio-temporal restrictions and reflects genetically fixed traits, it cannot merely be explained or predicted on the basis of one single criterion such as species diversity, but has its own quality.

we should keep in mind that taxonomic units such as species are just one possibility for the classification of plants. In this case, the types are based on phylogenetic relatedness. Other criteria could be applied as well, such as growth form or seasonality, and then other classes and units result. This in turn would influence the number of types to be counted.

There is no single index or value for all different aspects of biodiversity. Consequently, effects of biodiversity on ecosystem functioning have to be clearly related to the particular aspects of biodiversity considered in a study.

Functional traits and types

Based on approaches that concentrate on the functional response of species to certain environments (Grime, 1977), different functional groupings have been developed in vegetation science (Körner, 1993). However, guilds and functional groups have been gen-

erally more prominent in animal ecology (Hawkins and MacMahon, 1989). Another functional perspective in plant ecology is derived from population biology, where the regeneration of species is seen as a key factor for the maintenance of species diversity (Grubb, 1977). This approach concentrates only on the reaction of populations to functional processes.

The concept of plant functional types (PFTs) or functional groups deals explicitly with functional diversity (Smith *et al.*, 1993; Woodward and Cramer, 1996; Westoby and Leishman, 1997; Woodward and Kelly, 1997). This classification approach is based on functional traits (Walker *et al.*, 1999). In turn, the classification of individual species depends on those functional criteria applied. PFTs are very helpful if global forecasts of effects of climate change are undertaken, because they are both specific and coarse enough to show global patterns and processes (Smith *et al.*, 1993; Diaz and Cabido, 1997). It is not realistic to work at this level with species diversity. The same is true for changes in global biogeochemical cycles and land-use changes.

Functional attributes and traits focus on properties of plant organs or their metabolism. This approach is more flexible and clearly related to criteria such as fluxes of carbon or storage of water. Several functional species traits have been used to classify plants in ways that relate either effect or response to the environment (Noble and Slayter, 1980; Pavlovic, 1994). Now, the functional perspective concentrates on the effects of species, in particular of species diversity, *on* ecological functions (Lamont, 1995; Grime, 1997; Hector *et al.*, 1999). It is evident that the response to changes in the environment is likely to be shifts in the functional composition of a community (Aiguar *et al.*, 1996). If the 'importance' of plant diversity in a changing world is questioned, the concept of functional types is obviously successful (Boutin and Keddy, 1993; Chapin, 1993; Golluscio and Sala, 1993; Diaz, 1995; Box, 1996; Chapin *et al.*, 1996; Diaz and Cabido, 1997; Gitay and Noble, 1997; Diaz *et al.*, 1998). However, the numerical diversity of functional types (alpha-diversity) depends on the selection of certain key functions; it differs for the same real community with different criteria applied. It seems very much more concise to concentrate directly on specific functional traits (Leishman and Westoby, 1992; Solbrig, 1993).

Functional traits are properties of an organism that are considered to be important according to the response to or to the effect on the environment. Functional traits may be reflected in the morphology of plant organs (e.g. leaf size, seed structure) or in morphological capabilities (e.g. resprouting, clonal growth) (Kindscher and Wells, 1995). Other traits or attributes are related to the life cycles of plants. Their 'vital attributes' (Noble and Slayter, 1980) or 'life-history attributes' (McIntyre *et al.*, 1995) and the quantitative contribution of such strategies to communities help to predict their responses to disturbance events. The composition of temporal traits reflects the long-term disturbance regime at the community level (White and Jentsch, 2001). However,

important ecophysiological metabolic or mutualistic properties (C3/C4 grasses, nitrogen fixers) are not evident in many cases.

Functional traits can be defined not only by the optimum conditions for species responses, but also by the range of their tolerance and the shape of their response curves to a particular factor. Indeed, most species have plastic responses to the environment, and their role in, for example, post-disturbance recovery is a function not only of their optima but also of the competitive environment they encounter. For example, even shade-tolerant, slow-growing species respond to added light with accelerated growth, but at a slower rate than light-demanding species (Brokaw, 1985; White *et al.*, 1985). The problem is even more complex: the species of a particular ecosystem, and thus the range of responses in that ecosystem, have functional traits that were shaped by past exposures to environmental processes. Thus, there is a twofold historical contingency in responses of species diversity. First, in ecological time, only those species with access to the site can participate in recovery (this access can be influenced by prior disturbance) and, second, in evolutionary time, species adaptations reflect previous evolution. Both determine the diversity of functional responses within an ecosystem.

Whittaker's classification (1972) into alpha-, beta- and gamma-diversity can only be partly applied to this concept, because it does not consider functional diversity (Hooper *et al.*, 2002) or – in a systemic perspective – ecological complexity. Alpha- and gamma-diversity are just an index for the number of objects (species) in a certain subset. They depend on the scale of observation and the number of records. Beta-diversity may be seen as an index for qualitative differences between objects. However, it is mainly applied at the level of communities and then characterizes the resemblance or floristic distance or turn-over between samples. It can be applied to identify spatial heterogeneity and temporal trends. New genetic techniques would allow the calculation of the similarity or dissimilarity (contrast) between organisms. However this approach is as yet uncommon.

Ecosystem Functioning

Processes, mechanisms and functions

Functional aspects of biotic communities are characterized as 'ecosystem processes' (Odum, 1993), 'biogeochemical processes' (Schlesinger, 1991) or 'ecosystem functions' (Schulze and Mooney, 1993). We would like to make a distinction between biotic functions and processes. Processes are mechanisms such as photosynthesis, pollination or nitrogen fixation. Their properties do not depend on the object. In contrast, functions are a relation between processes and objects. We may find functions *of* an object according to a certain process, for example high capability of nitrogen fixation for nutrient retention in soils, or identify a function *for* an object via the same or another mechanism, for example nutrient retention in storage organs of plants.

As already mentioned, there is a difference between functions that affect an organism and those that are the effect of an organism. Some morphological properties of an organism may not easily be attributed to one or the other way of functional interaction. In plants, for instance, the reaction to a given environment (drought) may lead to certain growth. These structural properties, however, may also be genetically fixed and the occurrence of a certain species or ecotype will just indicate the competitiveness of certain functional traits under these site conditions. Functions may also result from different processes.

To ensure the persistence of ecological functions in plant communities in the face of disturbance, functional adaptations of species generally underlie two mechanisms of ecosystem response: complementarity and redundancy (Loreau and Hector, 2001). First, species have evolved a diverse spectrum of abilities relative to disturbance. After a particular disturbance, some species increase or invade, while others decrease or retreat (Vogl, 1974). Thus, ecosystem response is, in part, a result of niche complementarity. Second, when dominant species are primarily the ones affected by disturbances, other species may increase after a disturbance, even if their functional traits are similar to the previously dominant species. This has been expressed by the resilience hypothesis (Walker *et al.*, 1999). Dominant and minor species in the same functional groups are similar with respect to the contribution to ecosystem function, but they differ in their environmental requirements and tolerances and, thus, in their ability to respond to disturbances. Dominant and less dominant species switch in abundance under changing environmental conditions allowing functional stability. Thus, species diversity including functional redundancy is important in ensuring the persistence of ecosystem function under changing environmental conditions and in ensuring resilience in response to a disturbance. Moreover, apparently redundant species may operate on different spatial and temporal scales (Peterson *et al.*, 1998), thereby reinforcing function across scales.

Both complementarity and redundancy can be mechanisms that contribute to overall ecosystem stability. For example, Marks (1974) showed that fast-growing, early-successional trees are able to take up dissolved nitrogen after a disturbance, thus preventing nitrogen export to groundwater and streams. Vitousek's (1984) general theory of forest nutrient dynamics suggested that early-successional species immobilize limiting nutrients quickly after a disturbance.

Ecosystem functioning as a system property will be the integral of all different processes going on between the members of the community. Some of these functions (e.g. carbon cycling) may be relevant to objects (e.g. humans) outside the system (Reich *et al.*, 2001).

Initiated by DiCastri and Younès (1990) and then strongly supported by Chapin *et al.* (1992) and Schulze and Mooney (1993), functional aspects became a major focus of biodiversity research from the 1990s onwards (Baskin, 1994; Mooney *et al.*, 1995a,b; Chapin *et al.*, 1997, 1998; Tilman *et al.*, 1997c, 1998; Schläpfer *et al.*, 1999; Wall, 1999; Loreau *et al.*, 2001; Kinzig *et al.*, 2002; Mooney, 2002; Schmid *et al.*, 2002b). The identification and description of species and species diversity and its loss were the prominent concerns related to expected global changes (May, 1986, 1988, 1990; Soulé, 1991; Pimm *et al.*, 1995).

The functioning of ecosystems is hard to tackle scientifically. The perception of the functional diversity or complexity of ecological systems remains unclear and varies among authors (Franklin, 1988; Lawton, 1996; Martinez, 1996; Lavorel and Richardson, 1999). The connections between species diversity and ecological complexity are also controversial (Wilson, 1992; Lawton, 1994; Naeem *et al.*, 1994; Tilman and Downing, 1994; Tilman *et al.*, 1996, 1997a; Hooper, 1998; Hooper and Vitousek, 1998; Symstad *et al.*, 1998), even when functioning is related to evident parameters of diversity such as spatial vegetation structures (van der Maarel, 1986, 1988; Pacala and Deutschmann, 1996), plant species composition (Hooper and Vitousek, 1997; Tilman, 1997b) or dominance patterns (Grime, 1987; Smith and Knapp, 2003).

As animals depend directly on vegetation structure and composition, herbivores and other trophic groups have been correlated to plant species diversity (Asteraki *et al.*, 1995; Siemann *et al.*, 1998; Koricheva *et al.*, 2000). Further on, there are correlations between plant diversity and the diversity and functioning of soil bacteria and fungi (Spehn *et al.*, 2000b; Stephan *et al.*, 2000). There is evidence that plant species diversity positively affects key ecosystem processes such as decomposition via its influence on microbial functioning (Hector *et al.*, 2000a; Knops *et al.*, 2001; Mikola *et al.*, 2002). Another important type of interaction is direct mutualism between plants and fungi. Here as well, hints of positive correlations between plant diversity and fungal diversity are found, but methodological constraints have hindered further insights (van der Heijden and Cornelisson, 2002). Generally, the linkage between above-ground diversity and dynamics and below-ground processes and aspects (Wardle and van der Putten, 2002) is one of the important research gaps that has to be filled.

Temporal performance of diverse plant communities

It is assumed that species-rich communities will have the capability to react to a variety of events and disturbances, ensuring functions and ongoing dynamics despite disturbance. The probability that some members of a community will be able to cope with extremes increases with species richness. In the face of an increase in extreme events, which is expected during global climate change (IPCC, 2001), diverse communities that are adapted to an intensive disturbance regime might react flexibly to trends and events. With increasing diversity, plant species that are able to tolerate extremes are likely to occur. If environmental conditions swing back to former states in the future, such surviving species would be already on the spot.

Nevertheless, current biodiversity research is facing the dilemma that changing environmental conditions are accelerating, and that anthropogenic pressures on biodiversity are of global extent (Sala *et al.*, 2000). Obviously, the organismic potential to change its current location by large-distance migration or fast alteration of life-history cycle and growth form will offer survival mechanisms in the face of global change for a certain fraction of species (Higgins *et al.*, 2003). Still, the slow alteration of distribution areas (decades, centuries) will most likely cover only short distances, and evolutionary adaptation mechanisms will most likely take many generations (millennia). The expected spatio-temporal dynamics of global change clearly exceed such low-speed and short-range developments of most species. Hence, there is a sensitive threshold to the ongoing speed of change. As soon as species cannot cope with the speed or the spatial extent of environmental change, they are likely to become extinct. A decrease in local biodiversity is to be expected if the spatio-temporal mechanisms of migration, phenotypic plasticity and dispersal, meta-population dynamics or evolutionary development do not meet the scales of global change (Jentsch *et al.*, 2002; Jentsch and Beierkuhnlein, 2003). Consequently, ecosystem functioning may be permanently altered with respect to biotic feedback, material and energy cycles.

High species diversity is likely to go along with a diversity of ecological rhythms, which can also contribute to stability of ecosystem functioning. For instance, in high mountain ecosystems, a few dominating species reach very long life spans of several hundred to several thousand years. Most of them are trees, dwarf shrubs or clonal grasses such as *Betula nana*, *Pinus longaeva* or *Carex curvula*. Such long-lived, slow-growing species may neither be able to react to changing environmental conditions, nor die owing to some unfavourable decades or centuries. They represent ecological inertia in the face of altering conditions or competitive pressure by new species (Jentsch and Beierkuhnlein, 2003). This can mean both risk and potential for the future of biodiversity.

The risk of not being able to cope with changing environmental conditions by adaptation or migration is simply the fate of extinction (Ehrlich and Ehrlich, 1981). On the other hand, evolutionary inertia or the 'persistent niche' (Bond and Midgley, 2001) may provide temporal refuges in cyclically changing environments. The potential of enduring novel conditions via long-term survival is an option for 'a better future', in which conditions may become favourable again, although this strategy does not seem currently adequate. Species with very long lifespans exhibit genetic stability through time. They do not respond to novel conditions. When trends of alteration return to past conditions, their particular traits may be most successful and even ensure the persistence of these species through cyclic alterations. Populations of some tree species perform prolific resprouting after being cut or blown down.

Diversity and the stability of functions

One of the 'evergreen' topics of ecology is the relationship between diversity and stability of ecosystem functions (Tilman *et al.*, 1994, 2002a; Levine and D'Antonio, 1999; Loreau *et al.*, 2001, 2002; Tilman, 2001).

This debate has to be seen as a modern reflection of the idea of the balance of nature, which has been a general paradigm since the 19th century. Now it has shifted to whether and how plant diversity influences ecosystem functioning.

Stability includes the persistence of functions despite disturbance or despite change of environmental conditions. Examples are persistence of productivity, nutrient retention, carbon sequestration, air and water purification and slope stability, and the avoidance of erosion, desertification, and other forms of degradation. The constancy of species composition and abundance patterns, and finally of biodiversity, then reflects the functional continuity of an ecosystem in the face of disturbance impacts (Tilman, 1993; McIntyre *et al.*, 1995). McGrady-Steed *et al.* (1997) demonstrate the positive role of biodiversity for the control and predictability of ecosystems. Naeem (1998) points at the role of redundant species in changing ecosystems for maintaining their functioning ('reliability').

Based on the hypotheses from Connell and Orias (1964) about the expected feedback between species diversity and stability, a theoretical consideration of such mechanisms developed (e.g. Margalef, 1969; Goodman, 1975). May (1972) pointed early on at the restrictions according to the connection of the state of a system (diversity) and its ongoing processes (stability) and suggested replacing diversity by complexity. He postulated that with increasing complexity the variability of whole systems will be lower than the variability of species populations. Lehman and Tilman (2000) also found that diversity stabilizes the community but destabilizes individual populations.

To date much research has been directed towards the interrelation between stability and diversity, because this connection is both theoretically and practically attractive (McNaughton, 1977, 1978; Thierry, 1982; Kikkawa, 1986; Frank and McNaughton, 1991; Johnson *et al.*, 1996; Tilman, 1996; Doak *et al.*, 1998; Tilman *et al.*, 1998; Loreau and Behara, 1999; White and Jentsch, 2001). Decreases in population size as a consequence of resource partitioning in diverse stands may lead to higher sensitivity against stochastic fluctuations.

There is fundamental significance of multi-trophic dynamics for ecosystem processes such as stability, with primary producers in a key role (Raffaelli et al., 2002). Increasing the number of species without increasing the food-web linkages within an ecosystem is not likely to increase the stability (Leigh, 1965).

The purpose of species diversity for stability and maintenance of functions within ecosystems has been discussed in a community-based approach by Walker et al. (1999). This paper proposed that persistence in ecosystem function under changing environmental conditions and resilience against disturbance are ensured by functional similarities among dominant and minor species. According to the resilience hypothesis, major and minor species switch in abundance during times of stress or disturbance, thus maintaining ecosystem function. Abundant species contribute to ecosystem performance at any particular time (and are functionally dissimilar from each other). However, minor species contribute to ccosystem resilience during times of stress (Mulder et al., 2001) or disturbance (and are functionally similar to dominant species and could increase in abundance to maintain function if dominant species decline or disappear). Peterson et al. (1998) indicated that apparently redundant species operate at different scales and thus reinforce function across scales. It may be shown with the help of model ecosystems, that diversity–stability relationships are likely to occur (Doak et al., 1998).

The answer to the question of whether diversity and stability are related varies with the community or ecosystem that is dealt with. On the other hand, stability concepts differ. Grimm and Wissel (1997) identify four primary stability concepts (persistence, resistance, resilience and constancy). Diversity will influence such different qualities specifically.

Tilman and Downing (1994) demonstrate a linear relationship between species diversity and the recovery of grassland after severe drought. Givnish (1994) doubts the

higher stability of diverse stands because of their dependence on certain site conditions and nutrient availability. This indicates that in natural ecosystems both diversity and site effects will have to be considered. Huston and McBride (2002) also hint at the relative importance of both factors for the control of ecosystems. Wardle et al. (2000) suggest the importance of above-ground functional group richness and composition, which may dominate stability effects.

Species in turn are capable of changing their own environment. Mutualistic interrelationships between legumes and rhizobia strongly modify the nutrient status of a site (Spehn et al., 2002). Such species are indirect ecosystem engineers, such as termites or ants (Jones et al., 1994).

Species are idiosyncratic in their response to environmental constraints or disturbance regimes. Some species are keystone species that influence ecosystem dynamics more than others (Naeem et al., 2002). For instance, the fuel provided by a dominant understorey grass is critical to the fire regime, species diversity and pine regeneration in longleaf pine forests in the southeastern United States (Christensen, 1981).

Diversity and productivity

The effects of biodiversity on the productivity of stands are of crucial importance and are closely linked to economic perspectives (Tilman, 1999). Early studies found a positive correlation (Connel and Orias, 1964; Pianka, 1966; MacArthur, 1969). Others came to opposing conclusions (Margalef, 1969). It depends on the system whether effects of diversity per se or of certain parts of the community such as productive species emerge even if an increase in productivity is theoretically to be expected (Tilman et al., 1997b). Guo and Berry (1998) could not find clear effects of species number and biomass at different sites. Nevertheless, many studies detect a positive relationship between plant species richness and ecosystem processes, especially regarding aboveground primary production, and especially in species-poor communities (Schläpfer and

Schmid, 1999; Schwartz et al., 2000; Spehn et al., 2000a; Troumbis and Memtas, 2000; Bergamini et al., 2001; Hector, 2002; Schmid et al., 2002b; Tilman et al., 2002b). Not many studies are able to separate site effects that are controlling both biodiversity and productivity. There is also an interrelation between productivity and functional stability (Lehman and Tilman, 2000; Pfisterer and Schmid, 2002).

The probability of the occurrence of highly productive species grows with increasing species richness (Aarssen, 1997). Additionally species that promote nutrient availability, such as legumes (Spehn et al., 2002), or influence the ecosystems as ecosystem engineers in a positive way are likely to occur in species-rich stands. In Hooper's (1998) experiment biomass varied more within certain levels of diversity than between them. This also hints at the individual functional importance of certain species. Huston (1997) criticizes experimental approaches that ignore the possibility of 'hidden treatments' through the ecophysiological differences in species assemblages. Hooper and Vitousek (1998) find a better use of resources in species-rich stands. This reflects the paradigm of higher efficiency of diverse communities due to niche partitioning.

Not many approaches are able to separate such effects from mere biodiversity effects. Van Ruijven and Berendse (2003) find positive effects of plant species richness on the productivity of communities even in the absence of legumes.

In restoration ecology, it is of crucial importance to determine both the presence of particular functional traits in plant communities and the species diversity as a pool of complementary regeneration mechanisms for community assembly. For instance, diversity can affect initial productivity correlated with soil resources in several ways: the greater the diversity of present response groups to a disturbance, the higher the initial rates of establishment, growth and productivity. The greater the abruptness and magnitude of an increase in resources, the greater the initial selection for rapid colonization and the higher the initial growth rates, leading to critical uptake of soil ele-

ments that are otherwise vulnerable to leaching. The greater the productivity, the greater is the differentiation of successional roles and the greater the amount of successional turnover during assembly (White and Jentsch, 2004). This is reflected in changes in life-history traits: resource use efficiency, longevity and age at sexual maturity increase, while relative investment in reproduction decreases. As resources become immobilized in biomass and organic detritus, present diversity creates a filter for further establishment.

In this context, it is remarkable that the greater the resource supply in a diverse community, the greater the importance of disturbance to increase turnover by removal of inhibition (White and Jentsch, 2004). Whereas, the greater the stress or disturbance, the greater the importance of facilitation and mutualism within the species community (Temperton et al., 2004).

Nutrients, soil and relief

In the context of global climate change it is assumed that diverse ecosystems will have better capabilities to adapt to novel conditions and environmental constraints by shifting dominances (Peters and Lovejoy, 1992; Peters, 1994).

Following mechanical ground disturbances, the mineralization of nutrients would lead to nutrient leaching, as demonstrated for dry acidic grasslands in the lowland area of central Europe (Jentsch, 2004). If nutrients become available after disturbance, temporal aspects of species diversity increase in significance. Early colonists are able to store those resources rapidly. When resources become available after disturbance, such as in forest blowdowns, colonizing ability and growth rate are important and can have a lasting impact on ecosystem composition and structure. Rapid establishment supports rapid uptake of resources and stabilization of soil. Such mechanisms have been shown to be important in the tropics. However, generally, biodiversity and its temporal performance may play a decisive role in nutrient cycling.

Hooper and Vitousek (1998) could not prove lower leaching of nutrients in diverse stands. In contrast, Scherer-Lorenzen (1999) and Scherer-Lorenzen et al. (2003) found lower levels of nitrate in the leachate under grasslands when their species diversity was high, although plots with legumes showed higher nitrate values and the probability of the occurrence of legumes increased with species diversity. Even in non-fertilized plots without legumes, high concentrations of nitrate occurred due to atmospheric deposition and mineralization. This was only the case in species-poor communities. In rich stands, critical levels of nitrate in the leachate could not be measured.

On steep slopes, soil stability is an important property and also a service for the protection of human settlements and infrastructure. Soil stability is highly dependent on plant cover and rooting patterns. The more diverse the root growth forms, the less likely it is that extreme events will promote soil erosion. The loss of diversity could alter the sensitivity to soil erosion and slope stability in high mountains (Körner, 1999). This can also be perceived in terms of the insurance hypothesis (Yachi and Loreau, 1999).

Combating desertification and degradation is another of the most important international activities in the scope of sustainable development (UNEP, 1992). The problem is mainly caused by increasing human populations and overexploitation in developing countries. In semi-arid climates with high natural variability of drought and rainfall, sites could be negatively influenced by species loss. Species-rich plant communities are likely to be able to shift in abundance or dominance patterns and to preserve ecosystem functioning under stress. At higher scales, multi-patch vegetation patterns are likely to control the process of desertification (von Hardenberg et al., 2001). Spatial heterogeneity or evenness becomes effective for the maintenance of ecosystem functions in terms of promoting or hindering small-scale reactions to ecosystem changes (Wilsey and Potvin, 2000; Wilsey and Polley, 2002). However, this broad topic cannot be completely covered in this review.

Biodiversity and the invasibility of communities

Biodiversity is not constant in time. At the regional scale and within the temporal scales of ecosystems there are fluctuations. Dispersal and succession occur. Species composition may shift. Biodiversity is also influenced by invasive species and vice versa (Palmer and Maurer, 1997; Prieur-Richard and Lavorel, 2000). Due to the increasing connectivity of isolated habitats by anthropogenic vectors, species become introduced and thereby extend their former distribution. Competitive neophytes are initially adding to the flora of a region. As soon as they contribute to the local extinction of several less competitive indigenous species, negative effects on species diversity can follow, especially in biodiversity hot spots (Stohlgren et al., 1999).

High diversity was found to act as a barrier against or at least have a negative influence on ecological invasions and delay them in some communities (e.g. Tilman, 1997a; Crawley et al., 1999; Naeem et al., 2000; Prieur-Richard et al., 2000; Kennedy et al., 2002). Field experiments support the role of diversity in controlling invading plants (Knops et al., 1999; Hector et al., 2001a). However, there are also some contraindications in dynamic systems with high turnover and short-lived species (Robinson et al., 1995; Palmer and Maurer, 1997). Wardle (1999, 2001) is sceptical about findings that support diversity effects because the control of invasion can be species specific and then is related to diversity via the sampling effect.

Even if Dukes (2001) did not find an effect of species diversity in grassland microcosms on the establishment of alien Centaurea solstitialis, he observed a stronger suppressed growth of species-poor stands by this invasive species. This means that biodiversity did not prevent invasion but affected the stability of previous ecosystem properties. On longer timescales, this might produce negative feedback. Meiners et al. (2004) stress the fact that species-specific aspects and sampling effects are important and overlay diversity effects. Furthermore, it is important to understand species-specific interactions and the mechanisms of competition and mutualism that occur in a given community.

Stohlgren *et al.* (1999) and van Ruijven *et al.* (2003) point out that there is a scale dependence of diversity effects. Such mechanisms are likely to occur only at the community level. At larger scales, other factors (e.g. disturbances, heterogeneity of resource availability) are more decisive (see also Levine, 2000; Wardle, 2001).

However, also at the scale of the community, there is an influence of short-term disturbances that will affect invasibility (Rejmánek, 1989). The problem is that such disturbances and their effects have a short duration. Thus, it is absolutely necessary to look at spatio-temporal patterns. This is true not only for competition-free space but also for the performance of biodiversity that may differ over short distances and time steps (e.g. seasons). In some ecological zones, disturbances and temporal variability can foster invasions owing to environmental constraints. In the Mediterranean, temporal variability strongly controls the diversity of plant communities. The co-occurrence of therophytes, geophytes, dwarf-shrubs and bushes simply reflects the fact that there are temporal niches that are occupied. Even there, an effect of diversity on invasibility can be detected (Lavorel *et al.*, 1999). However, it is important to realize that disturbance may promote diversity. In many cases species-rich stands are frequently disturbed. The invasibility then is controlled by disturbances but these promote the initial diversity as well. This is the case in floodplains and riparian sites (McIntyre *et al.*, 1988; Planty-Tabacchi *et al.*, 1996).

The most sensitive phase is the establishment of invaders and thus the existence of competition-free safe sites. Such conditions can be delivered by disturbances (Burke and Grime, 1996), and this explains why dynamic systems are more prone to invasion than stable ecosystems. As we have seen for the Mediterranean vegetation, dynamic communities are often also rich in species. It will be important to separate effects of diversity and effects of temporal variability in invasion research. Diversity effects have to be differentiated into effects of species number per se and effects of functional diversity (Prieur-Richard and Lavorel, 2000), the latter being more likely to be important (Symstad, 2000).

Human Threats and Benefits

Crisis of biodiversity

The current species-extinction period is mainly caused by human impact, and is estimated to happen at a rate 1000 times greater than the natural rate of extinction (Primack, 1993). Recently, various global change scenarios have been developed that address the effects on biodiversity caused by atmospheric warming, altered precipitation patterns, land-use changes, increased fragmentation, urban expansion and other human activities (Sala *et al.*, 2000). Ongoing discussions among natural and social scientists (Jentsch *et al.*, 2003) have been further alerted by the last report of the Intergovernmental Panel on Climate Change (IPCC, 2001), stating an accelerated speed of change of environmental conditions. The upcoming IPCC report will emphasize the crucial role of extreme events for driving biodiversity patterns.

This 'crisis of biodiversity' as a consequence of human impacts has brought much attention and many repercussions in society, not only because of ethical responsibility and aesthetic interests. More than that, the fear of losing ecosystem functions and especially those that are of societal importance (Ehrlich and Wilson, 1991) characterize such functions as 'ecological services' (Daily, 1997; Daily *et al.*, 1997). This perception is the reason for political attention and societal awareness. Society is afraid that benefits could be lost that are delivered from nature for free. This would mean economic constraints as a follow-up of species decline (Jentsch *et al.*, 2003). The direct use of natural resources (plants) or the direct protection that is given by them (pure water, preservation of soil, protection against avalanches and landslides, etc.) is not the only concern; there is also the loss of potential benefits (e.g. pharmaceutical traits; Cragg and Newman, 2002) that have not yet been identified (Farnsworth, 1988). In this perspective biodiversity per se is regarded as a resource (Plotkin, 1988; Nader and Mateo, 1998).

Biodiversity, however, may also contribute to threats to human health and welfare (Dobson, 1995). Vectors may distribute toxic plants as well as diseases. Public awareness concentrates on human pathogenic microorganisms. Pathogenic microorganisms and insects that are distributed by trade and traffic can affect plants as well. Then species diversity and genetic variability within populations may contribute to the regulation of outbreaks of disease and the severity of such outbreaks (Mitchell et al., 2002). It is mainly because of such assumed capabilities, which are not easy to prove, that biodiversity has a positive image.

Goods, Services and Values

The benefits that society gains from biodiversity are differentiated into use values (extractive benefits) and non-use values (e.g. ethical, non-extractive benefits). Goods and services represent direct or indirect use values. Goods are directly related to an economic profit. Services represent the functioning of ecosystems (van Wilgen et al., 1996; Williams et al., 1996). Services of biodiversity include preservation and renewal of soil fertility, air and water purification, nutrient recycling, carbon uptake, waste detoxification and decomposition, moderation of disturbances and maintenance of genetic diversity for agricultural improvements, as well as control of agricultural pests and human diseases (e.g. Randall, 1994).

It is clear that the value of biodiversity to mankind has many aspects. The ethical value and the heritage that we must preserve for future generations represent a moral duty for society. The aesthetic values are evident as well (Heerwagen and Orians, 1993); some of them may be economically important. However, these values may only be identified and evaluated in the course of an inter-subjective participatory discussion. Biologists are no more competent in these fields than other groups of society.

Meanwhile, the socio-economic value of ecosystem services is widely acknowledged (Costanza et al., 1997). Biodiversity, especially functional biodiversity, is increasingly recognized as decisive for maintaining these services (Hooper and Vitousek, 1997; Hector et al., 2001b). Still, it remains a fundamental challenge to assign economic and ethical attributes to particular species, communities or to ecological functions, in order to propose conservation measures where the obtained benefits exceed the costs necessary for action (Jentsch et al., 2003). Nevertheless, there are some approaches to calculate services and goods derived from ecosystem functioning at a monetary scale (Huston, 1993; Buongiorno et al., 1994; Pearce and Moran, 1994; Montgomery and Pollack, 1996; Costanza et al., 1997; Pimentel et al., 1997; Rickleffs, 1997). As money is an efficient tool to make things comparable and to communicate the value of a subject, it may serve as a powerful argument for the preservation of biodiversity (Perrings, 1995). It may also contribute to finding solutions in conflicts between ecology and economy (Gowdy and Daniel, 1995).

Balmford et al. (2002) estimate, based on conservative assumptions and using a broad range of evaluation techniques such as hedonic pricing, contingent valuation and replacement cost methods, that there is a tremendous underinvestment in nature reserves. After their calculations and review, the benefit:cost ratio of reserve systems is around 100:1. Still, the value of biodiversity is most appreciated in a crisis, and in crises its value is extraordinary.

To determine monetary values of biodiversity that are not reflected in current market prices is an important task from an economic perspective (Jentsch et al., 2003). These values include: (i) non-consumptive use values, such as the benefits that species richness provides to tourism; (ii) indirect use values for ecosystem stability or functions, such as the provision of clean water; (iii) option values, such as future use in pharmaceuticals; (iv) existence; and (v) bequest values (Perrings, 1995; Costanza et al., 1997; Fromm, 2000; Heal, 2000; Dasgupta, 2001). A debate has evolved that elaborates the conditions under which economic valuation of biodiversity is sensible (Hampicke, 1999; Seidl and Gowdy, 1999; Nunes and van den Bergh, 2001). Limiting factors include the non-substitutability of the natural resource, the fact that societal preferences for the natural good to be valued cannot be represented by individual preferences, and that institutional structures significantly influence the results of the monetarization.

As biodiversity conservation typically causes costs at the local level while producing a global public good, special attention is paid to developing well-defined mechanisms for compensating local communities and land users (Ring, 2002). Hence, research on global biodiversity governance further includes the investigation of incentive structures and policy instruments (OECD, 1999; Barbier, 2000).

We can only briefly mention that, even if most authors focus on biodiversity-related economic evaluations on positive aspects of diversity and genetic resources and benefits (e.g. ten Kate and Laird, 2002), invasive non-indigenous species – which may add to species diversity – are causing enormous costs and hazards (e.g. Pimentel et al., 2000). Some invasive aliens such as giant hogweed (*Heracleum mantegazzianum*) in Central Europe (Pysek and Pysek, 1995) may even cause severe health problems.

If invasibility is reduced by biodiversity, as has been indicated above, then the risks that are related to non-indigenous plants will be reduced in species-rich stands. However, we have demonstrated that factors other than diversity decide whether a community is prone to invasion. In the case of giant hogweed, no effect of species diversity could be demonstrated but the disturbance regime was of crucial importance.

Heuristic Methods and Approaches

Theoretical considerations and diversity hypotheses

Theoretical considerations on the importance and mechanisms of species diversity are a challenging field in ecology (Naeem et al., 2002; Tilman and Lehman, 2002).

Vitousek and Hooper (1993) initially identified three major possible relationships between biological diversity and ecosystem-level biogeochemical functions: no effect, linear correlation and asymptotic approximation of a maximum level. The third was regarded as reflecting species redundancy by Lawton and Brown (1993). It has to be stressed that these concepts focus on 'species richness' only, even if graphical representations are often simplified to 'biodiversity' (Naeem et al., 2002). Functional diversity, which is not necessarily correlated to species richness, is rarely explicitly addressed (Wardle et al., 2000).

Since those initial theoretical concepts, many other views and hypotheses were published (review in Schläpfer and Schmid, 1999). Within the group of hypotheses that describes the occurrence of positive effects, perhaps the linear hypothesis (complementarity hypothesis) and the idiosyncratic hypothesis represent extreme positions. The first assumes that each species adds a comparable part to ecosystem processes. The reason for such a pattern is seen in the assumed complementarity of species (Hector, 1998; Hooper, 1998). We have already pointed out that, because of their ecological traits, species perform specifically. An equal contribution of each species to ecosystem functioning is unlikely.

Nevertheless, there is an indication that a close relationship between particular ecosystem processes and species diversity occurs. This does not imply general mechanisms. Species diversity might be relevant only within a given frame. Most experiments have been conducted within rather species-poor communities, and it is not surprising that within a set of just a few species a correlation of increasing number of species to functioning will be proven. The response of ecosystem functions to diversity is often controlled by restrictions of resource availability. Such restrictions will be effective in one case but not in another. If effective, increasing diversity would not influence the functioning of the system.

Biodiversity effects, such as higher biomass production with increasing species diversity in experimental grasslands (e.g.

Hector *et al.*, 1999), can occur per se as an effect of ecological complexity and functional complementarity (Hector, 1998) (Fig. 13.3). However, they can also be a result of probability. A critical perspective on biodiversity effects, the sampling hypothesis, says that with increasing numbers of species (or other objects) the probability arises that single powerful (e.g. productive) species will occur and contribute to the function of interest (Huston, 1997). Those species that are lacking in species-poor stands can be key species, which play decisive roles in diverse communities. This problem is hard to tackle. However, it seems possible to separate sampling and biodiversity effects (Loreau, 1998b).

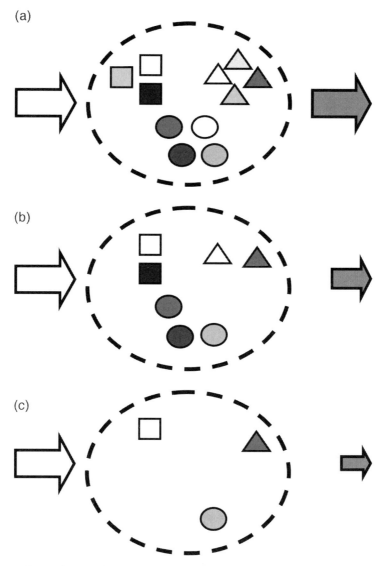

Fig. 13.3. According to the *complementarity, rivet* or *linear hypotheses,* the loss of species diversity is assumed to lead to a loss of functioning or output (dark arrows) of the community (dashed lines). This example simulates a loss from 11 (a) to 3 (c) species (objects) that belong to three functional groups (symbols). Every species contributes to the functioning of the community.

The rivet hypothesis (Ehrlich and Ehrlich, 1981) says that some species contribute additionally to the functioning of ecosystems. The hypothetic trajectory along a diversity gradient is then more stepwise than linear. However, it is inconsistently interpreted. Some views are close to the complementarity hypothesis rather than the keystone hypothesis.

A modification of the complementarity or linear theory is the redundancy hypothesis (Walker, 1992; Lawton and Brown, 1993; Gitay et al., 1996). This consideration predicts that the cumulative contribution of species in diverse communities will show an asymptotic distribution. When transformed to log scale, the data become linear. For a given diversity, each additional species contributes or adds in a different way to the functioning of this system. Just recording the number of species therefore ignores the individual response capabilities of species depending on the number of species that are already there. In any ecosystem, there are limits to the performance of certain functions. There are maximum values or thresholds that can be achieved. In less-diverse communities, any additional species is important, but after a certain threshold new species will not add remarkably to ecosystem functioning (Fig. 13.4). This hypothesis has been criticized from a nature conservation perspective because it implies that some species are unnecessary and their extinction would not cause negative effects.

The importance of redundancy effects depends on the dynamic trend of the community. The increase or decrease in functioning can be influenced by the direction of the development (Naeem et al., 2002). It matters whether species are added or lost. Non-linearity and hysteresis is possible and even likely to occur. The response of ecosystem functioning will differ for the same level of species diversity (Fig. 13.5). Another source of redundancy may be rareness. If organisms are rare or afford little space, they will not interact. Then, it is possible that the same ecological niche is occupied by different species. They will contribute to the species diversity of the community but not increase its complexity

('functional analogues' after Barbault et al., 1991). The taxonomic similarity between species (e.g. their assignment to a genus or family) does not necessarily hint at functional resemblance. Nevertheless, closely related species or representatives of one life form are quite often regarded to be functionally redundant. This is due to the fact that some morphological or ecological traits are restricted to a limited set of genetically related taxa.

The idiosyncratic hypothesis takes into account that the response of an additional species depends on the complexity of the community that is already established. Unpredictable interactions occur (Fig. 13.6). There is no clear linear or non-linear trend but more or less chaotic behaviour of the system. This cannot be explained by keystone species because here the functioning of a species is not regarded as independent. This theory does not state that there is no effect, but that the effects are individualistic and not to be predicted only by the number of species.

The unequal contribution of species to functioning is reflected by other hypotheses. The fundamental difference is that the following hypotheses assume that the contribution of a species is genetically fixed. Hence, its functional performance does not depend on the diversity of the community. Consequently, some species would be more efficient than others under certain site conditions.

If only a few or one single effective species occur, they can be regarded as 'keystone' or 'key species'. Functioning of the community will more or less exclusively depend on this species. In most cases, there will be more than one species that is strongly effective. With increasing species diversity the probability of the occurrence of effective species increases as well ('sampling effect') (Fig. 13.7).

Hector et al. (2002b) have tested the sampling effect hypothesis (Aarsen, 1997; Huston, 1997; Tilman et al., 1997b). Although diverse communities are strongly influenced by some dominant plant species, it could not be shown that species with highest biomass in monocultures were also most efficient in mixtures. Yields from mixtures

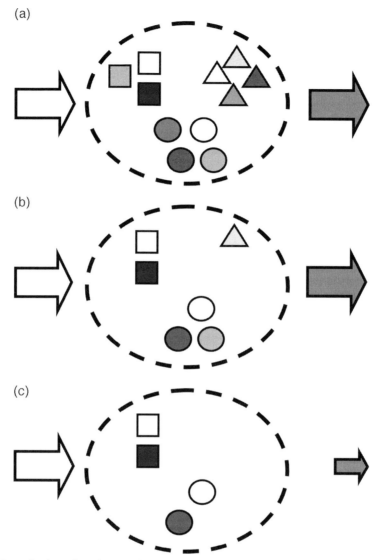

Fig. 13.4. The *redundancy hypothesis* predicts that there are redundant species within a functional group, which contribute only to species diversity but not remarkably to the functioning of a community (dashed lines) (a, b). Only if complete functional groups are lost (c) will reductions of functionality occur and the output of key functions (dark arrows) decline. Species diversity is reflected by the number of objects and functional diversity (functional groups) by different symbols.

were generally higher than the monoculture yield of dominant species within these mixtures. Pacala and Tilman (2002) support a shift from the sampling hypothesis to the complementarity hypothesis.

If there is a stronger effect of an increase in species diversity than expected from stochastic mixtures, 'overyielding' is detected. For instance, there are higher common values than might result from adding the values for single species derived from monocultures. In agricultural communities, it has also been found that certain combinations of species in polycultures had

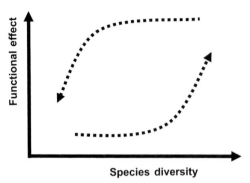

Fig. 13.5. The direction of species loss and gain is likely to cause different repercussions at the same level of species diversity. Hysteretic processes and non-linear response curves are mainly relevant in communities with low turnover rates and for species with high longevity. At the same level of diversity, saturated communities will then differ strongly from pioneer stages. Initial mechanisms such as the selection of functionally relevant species (e.g. in seed mixtures) will differ from extinction of key species. In communities with high turnover rates, both curves will be rather close. As ecosystem functions are not directly connected to species diversity but influenced more by ecological complexity, the redundancy of species is a matter of the direction of temporal processes such as invasion or extinction and of the time that is available to strengthen functional interactions such as competition.

higher biomass production than expected when reducing initial species diversity (Vandermeer *et al.*, 2002). However, Vandermeer *et al.* point out that overyielding does not necessarily imply that interspecific facilitation occurs. It could be related to species-specific capabilities in the use of resources.

Biodiversity is considered to be a potential resource that might become effective in the future. Rare species of today could play more important roles under the expected new environments of tomorrow. The different abilities of species to tolerate and react, to survive and to disperse are some sort of insurance against changing conditions in heterogeneous landscapes (Loreau *et al.*, 2003). They could become relevant even when no evident functional diversity *within* a group of plants can be detected under recent conditions. With increasing numbers

of species within such a group, the probability grows that the functions that are attributed to this group will be maintained in a changing environment (Chapin *et al.*, 1996). This means that redundancy under certain conditions will deliver the potential to react to new conditions (Fonseca and Ganade, 2001). The 'reliability' of communities increases with species diversity (Naeem and Li, 1997; Naeem, 1998). Based on such thoughts, Yachi and Loreau (1999) have developed the insurance hypothesis (Fig. 13.8).

The above-mentioned theories do not consider the influences of site conditions and especially of resource availability. Complementarity is most likely to occur when the participating species are not limited, for example by nutrient availability. At poor or dry sites, individuals of different species may exist in low abundances without any interaction. Adding or losing species will produce stochastic reactions within the limited frame of the environment. On the other hand, it can be observed that when resource availability is high, for example on fertilized or wet sites, only few specialists will be competitive and abundant. There, redundancy is common. In conclusion, we emphasize that the type of response depends very much on the availability of resources such as light, nutrients or water (Fig. 13.9).

Experimental Approaches

Removal experiments

It is strikingly simple to follow this approach and to exclude some species from formerly diverse communities in order to simulate the loss of plant species diversity. Most of the removal experiments in the literature are applied to animals and focus on food-web complexity. A critical point about removal experiments with plant communities is the impact on nutrient cycling. This is a problem even if no soil disturbances are affected or if no dead biomass or litter is left. After cutting a plant or destroying it with herbicides, its remaining root biomass

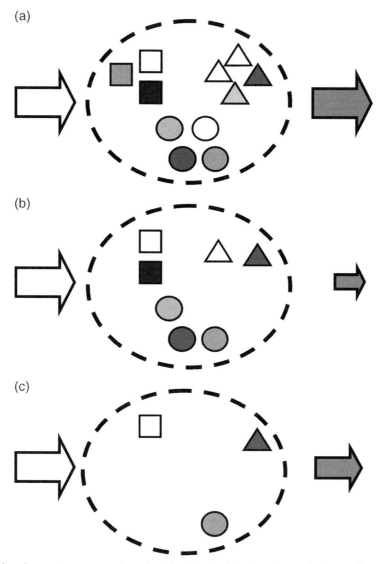

Fig. 13.6. The *idiosyncratic response hypothesis* implies that it is almost impossible to predict the effects of a decline of functional complexity that accompany species loss. Non-linear chaotic responses may occur. An initial decrease in the functioning (dark arrows) of a community (dashed lines) (a, b) can be followed by increases due to changing interactions (e.g. the removal of competitive but slow-growing species) during ongoing losses of species (objects) (c).

will be mineralized. This leads to temporarily enhanced nutrient values in the soil (Jentsch, 2004). In consequence, these nutrients will promote the remaining specimens from other species to be more productive. This effect then could be interpreted as a positive signal due to less competition whereas in fact it is a hidden fertilizing effect. Such processes are hard to avoid and even harder to quantify. This may explain why species-removal experiments have been widely neglected, whereas synthetic experiments have had a strong impact in the scientific community.

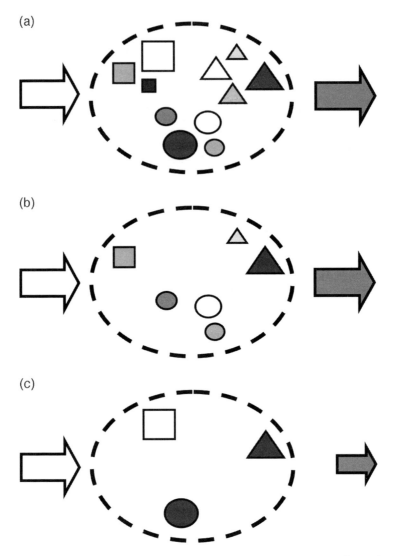

Fig. 13.7. The *sampling hypothesis* points at the lower probability of the occurrence of strongly influential species (represented by larger objects), according to the function of interest (e.g. biomass production) (dark arrows), with decreasing species diversity (a, b). It is difficult to avoid such effects in synthetic/additive experiments. Only some key species contribute substantially to the functioning of the community (dashed lines). If key species representing important functional groups (symbols) are preserved, there will be no negative effects of species loss to functioning (c).

Symstad and Tilman (2001) showed, in a 5-year removal experiment on abandoned agricultural fields, that there is a strong effect of the functional groups remaining in the community. The ability to occupy the space and to make use of the resources that have been required by the former competitors is unequally distributed across functional types. With this approach, traits have been identified that might become important in the course of species extinctions. Synthetic experiments with all species starting at the same time after the experiment has been installed are not able to simulate such spatio-temporal aspects.

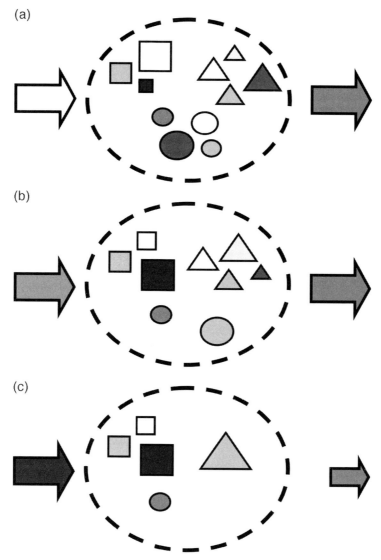

Fig. 13.8. The *insurance hypothesis* hints at the higher probability of flexible functional responses and adaptations to novel environments in species-rich communities (dashed lines). Shifts in abundance or dominance of species and in their relative contribution to the functioning may compensate for restrictions or the decline of sensitive species (a, b). If a negative threshold of diversity is surpassed, further changes in the environment will not be answered adequately (c).

Wootton and Downing (2003) point out that the results of species removal are highly idiosyncratic and therefore impossible to forecast. They suggest combining targeted species removals with general diversity manipulation. This approach could be helpful to find out which effect is related to key species (Mills *et al.*, 1993) and how biodiversity contributes via complex interactions between species to ecosystem functioning or buffers environmental extremes (Hughes *et al.*, 2002).

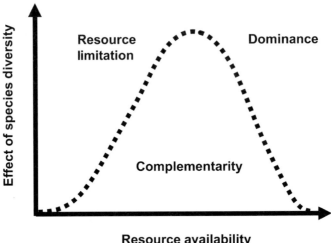

Fig. 13.9. Controversial findings from different experiments and field research can probably be explained by underlying effects of resource availability. An increase or loss of species diversity will hardly have any effect (e.g. on biomass production) if the community is strongly limited in resource supply (e.g. nutrients, water and energy). High resource availability on the other hand will support the predominance of a few or one specialized and competitive species (K species). Additional species will not be important. Only under intermediate conditions (for the ecological scale of the community) will complementarity be found and is species diversity likely to be functionally effective.

Synthetic experiments

During the search for interrelationships between biodiversity and ecosystem functions, experiments in simple model ecosystems were much supported. The reduction of complexity in these systems as well as their strongly controlled environmental conditions (see also Fukami *et al.*, 2001) would allow us to identify effects of changing variables such as species diversity (Schmid *et al.*, 2002a). In addition such experiments can be carried out within rather short timeframes. Their results can be validated by means of replicates. Furthermore, specific experimental designs can be repeated at different sites (Hector *et al.*, 2002a,c). This helps to identify generality behind individual experimental results.

One of the first influential projects in this field was the ECOTRON microcosm experiment (Naeem *et al.*, 1994, 1995, 1996; Hodgson *et al.*, 1998; Lawton *et al.*, 1998; Thompson and Hodgson, 1998). This approach was still very artificial. It was carried out in isolated chambers. However, it prepared the way for more natural field experiments.

Most experiments with plants focus on short-lived species (grasses, forbs) or are carried out on artificial substrates to reduce site heterogeneity and noise. The simplification of the approaches leads to a gap in the validation of gained results versus natural communities. Critical voices point at many other shortcomings and restrictions of such experiments (Grime, 1997; Huston, 1997; Fridley, 2002).

The heuristic value of experiments is clear. Functional consequences of biodiversity loss and causal effects can be detected – if just under the restricted conditions of an experiment. This helps to support or falsify hypotheses that have been developed theoretically. In nature many factors are interacting. It is impossible to separate them and to relate observed phenomena to selected site conditions directly.

Many recent experiments and manipulations that deal with biodiversity effects concentrate on plants because they are non-mobile and in many cases easy to control and to establish. In addition, they represent only one trophic level. Most of these research projects find positive correlations between species diversity of plants and key ecosystem functions (in most cases biomass production) (Hector *et al.*, 1999; Schläpfer and Schmid, 1999; Schwartz *et al.*, 2000; Hector, 2002; Schmid *et al.*, 2002b; Tilman *et al.*, 2002a,b).

In grasslands, two comprehensive field experiments had a strong impact. One experiment featured many replicates on one site in Minnesota, USA (Cedar Creek Experiment; Tilman *et al.*, 1996) and the other followed biogeographical gradients across Europe (Biodiversity and Ecosystem Processes in Terrestrial Herbaceous Ecosystems – BIODEPTH; Diemer *et al.*, 1997; Hector *et al.*, 1999). Grasslands do have the advantages of being spatially as well as economically important and they are easy to establish in a short time. Experiments have also been installed in forest communities (e.g. in Finland and in Germany). These will continue until stable conditions have developed, then the results will be published.

Due to the complexity of natural and anthropogenic ecosystems and landscapes, the monitoring of the loss of diversity is rather time consuming and will only deliver good results for selected examples and areas. There, it will be necessary to prove the generality of the results and to identify the causes of the decline, if it occurs at all. However, local species diversity may even increase because of new vectors and invading species at the same time as global extinctions occur, even in the same area, when rare species are lost.

Some important insights into ecological functions of biodiversity and driving factors for the decline of endangered species have been gained through the recent use of experimental designs at the landscape level (e.g. Caughley and Gunn, 1996). Results from experimental manipulations predict that high biodiversity will enhance ecosystem responses to elevated carbon dioxide and nitrogen deposition (Reich *et al.*, 2001; Catovsky *et al.*, 2002). Reich *et al.* (2001) show that current trait-based functional classifications alone might not be sufficient for understanding ecosystem responses to elevated carbon dioxide.

Nevertheless, experimentally proven correlations between biodiversity and ecosystem functions must be related to temporal and spatial scales (Oksanen, 1996; Rapson *et al.*, 1997; Bengtsson *et al.*, 2002; Levine *et al.*, 2002). Small-scale effects are not necessarily valid at larger scales (Waide *et al.*, 1999; Weiher, 1999; Chase and Leibold, 2002).

To apply a functional perspective and to identify the repercussions of changes in species diversity on the complexity of ecosystems will be almost impossible in most natural ecosystems. This is why reductionist models and experiments with defined conditions and environmental interactions became prominent during the past decade, when forecasts on the effects of species loss on key ecosystem functions were discussed.

Modelling approaches

Another promising heuristic approach for investigating the relationship between diversity and functioning is the application of ecological models (Loreau, 1998a). Models allow us to simulate interactions and multispecies diversity effects without being prone to the restrictions and noise of field investigations and experiments. In addition, they are not restricted to short-lived species. However, most models are still extremely simple and cannot cope with the reality of ecological complexity.

Doak *et al.* (1998) found that statistical averaging would result in greater stability of ecosystem functioning at high levels of diversity. Tilman *et al.* (1998) demonstrated that statistical averaging is not a necessary consequence of high diversity, but depends on the system that is investigated. The 'portfolio effect' may lead to a limited stabilization of the community due only to statistical grounds. The term is derived from economics, where experience shows that a diversified portfolio will be less endangered by stochastic market processes.

Lehman and Tilman (2000) compared different types of ecological models (mechanistic models, phenomenological models and statistical models). They showed that, even if the models perform differently, the general reactions of the simulated systems according to the stability of the communities are comparable: the variability of the entire communities decreased and the variability of the contributing populations increased.

Yachi and Loreau (1999) formulated the insurance hypothesis based on biodiversity models. Their model proves that the maintenance of key ecosystem functions as a reaction to temporal variability of the environment is more likely to occur in species-rich stands. Other ecological models that are dealing with insurance and related research questions have produced comparable results (Fonseca and Ganade, 2001; Petchey and Gaston, 2002).

Generally, ecological modelling can be used as a tool for integrating scientific results from various experimental and observational analyses as well as scenarios of changing environmental or socio-economic conditions. Spatially explicit grid-based models (De Angelis and Gross, 1992; Grimm, 1999) show that spatio-temporal correlations are a key to understanding system dynamics, their vulnerability or resilience (stability). There is growing evidence that such correlations are the currency to understand not only spatio-temporal dynamics of ecological systems, but also general mechanisms of biodiversity and species-specific functions and traits (Wiegand et al., 1999).

Recently developed ecological–economic models are promising techniques for the integration of social and natural sciences. They are pioneering approaches for economic assessments of different ecological management options (e.g. Frank and Ring, 1999; Johst et al., 2002). For instance, the modelling approach establishes the relationship between economic parameters of disturbance-management alternatives and the ecological effects on biodiversity properties.

Conclusion

Looking at certain key functions that are thought to be important, we do have to keep in mind that primary ecological factors such as water and nutrient cycling, energy in- and output and secondary or integrated abiotic constraints such as soils, relief and climate are strongly influencing ecosystem processes. The direct effect of such mechanisms will be much more important in many cases than biodiversity effects. In addition, direct human impact (e.g. pollution) may have consequences for biodiversity and ecosystem functioning. Cause and effect of changes in ecosystem functioning are then difficult to separate.

Nevertheless, plant species diversity plays a significant role for the control of ecosystem processes and overall functioning. In some cases, the effects will be related to complementarity of functional traits of species, in others just to the occurrence of key species (e.g. productive ones or ecosystem engineers). Today, the impact of biodiversity on ecosystem functioning can be neither predicted nor neglected.

Species are not similar. The historical and evolutionary background of each species may have a strong influence on the performance of entire ecosystems. Traits control the reaction pattern and metabolic capabilities of plant species. It is not possible to conclude general principles simply from the response of a given set of species. Mooney (2002) points at the fact that there is no simple solution to the controversial standpoints of whether the number of species or the variability of functional traits determines ecological functioning.

From a methodological point of view, it is absolutely necessary to link controlled but artificial experiments not only with models but also with standardized monitoring techniques. It is promising that the communication and exchange of results gained with different approaches will contribute to a better understanding of ecological mechanisms. In this field of cross-cutting research, enormous gaps still have to be filled.

Another challenge is the transfer or the validation of results across systems and communities. What has been found for the relationship between biodiversity and productivity in Central European grasslands (Minns et al., 2001; Hector, 2002) may be true for North American grasslands (Tilman et al., 2001, 2002a,b) but will be hard to apply to deciduous forests or even subtropical or tropical ecosystems.

Most of the diverse and threatened ecosystems of the world are poorly productive (e.g. the South African fynbos; Davis *et al.*, 1994). Other mechanisms and functional interrelations will be important in these communities and may be reduced or modified by biodiversity loss. The key functions (e.g. inflammability and proliferation of fire as a key disturbance for the maintenance of diversity) are largely to be identified. On the other hand, rather species-poor ecosystems such as mangroves might suffer severe functional restrictions with plant species losses (Field *et al.*, 1998).

The ecological importance of biodiversity can be subdivided into aspects that are relevant for ecosystem functioning and others that are, in addition, important to human society. Some aspects are exclusively relevant to humans (aesthetical and ethical values), but these have to be discussed at a broad societal level.

The paradox of depending on biodiversity and threatening it at the same time is one of the phenomena in complex human societies that are difficult to cope with. The human contribution to the processes that maintain and threaten biodiversity works at different scales. The regional diversity in Europe is to a major part dependent on anthropogenic disturbances and structures. If land use was stopped, biodiversity would be lost. Some species have even evolved with close dependence on land-use techniques and crops. At the global scale, land-use change today is the major driver for the irreversible loss of genetic variability (Sala *et al.*, 2000). This is due to the speed of transitions and to the technical and chemical intensity that is applied to fulfil social and economic requirements.

The same is true for the benefits that can be derived from biodiversity. Services at one spatial or temporal scale may be accompanied by a non-sustainable use of such benefits. The awareness of the risk of economic restrictions in connection with the loss of biodiversity might strongly support action to slow down or even stop this development.

Sustainable, long-term use and development is only possible if crucial ecological compartments and objects are maintained.

Biodiversity is a complex resource that is hard to define and to analyse with respect to its functional effects. However, there is strong support for the idea that it contributes to the maintenance of ecosystem functioning, which is fundamentally important for human beings. Some of the potential uses of biodiversity have not yet been discovered because of the large number of unknown species and our limited knowledge of the functional traits of plant species at the global scale.

Implicitly, the loss of biodiversity has been regarded as an indication of the loss of quality of life since the publication of Rachel Carson's book in 1962. In more recent decades this development has been perceived to be negative. However, it has actually speeded up in recent years. Short-term economic interests are more prominent and the survival of growing human populations in marginal habitats has to be ensured. Obviously, those socio-economic forces are powerful drivers for the loss of biodiversity. On the other hand, this ongoing loss is very likely to be followed up by violent negative feedback. Profits from the use of the global stock of biotic resources and ecosystems could be endangered in the future. Then, short-term individual economic gains would be followed by long-term societal economic losses.

Natural scientists tend to ignore normative social or economic values. In the case of biodiversity it would be foolish not to cooperate with socio-economic scientists in order to both identify the driving forces of extinctions and forecast and evaluate their effects. This chapter concentrates on the ecological part of the problem. Economists have developed methods to scrutinize the value that is given to natural subjects by people. One method is to ask for the 'willingness to pay'. However, this willingness is strongly influenced by knowledge, and knowledge on biodiversity is still fragmentary.

Ecological complexity, meaning the functional interactions between biota and their abiotic environment, can be seen as the most important aspect of biodiversity for human society as it controls the services and goods that can be derived from ecosystems. The biota does not exist in isolation from abiotic site conditions (extrinsic factors) but influences them and is influenced by them. This is crucially important for basic ecological research as well. The functional interactions between single elements of ecological systems such as organisms or soil or water bodies are still not completely understood. The web of interactions and reactions is woven at different levels of time (from osmotic responses to evolutionary patterns) and space (from cells to landscapes). Systemic features emerge from the non-stochastic interactions between species under specific site conditions and disturbance regimes resulting in comparable assemblages and structures. Such structures and their resource character from the human perspective are nothing less than the effect of ecological complexity. Thus, only a better understanding of ecological complexity will help us to save and manage biotic resources and functioning. Former experiences and knowledge have to be adapted to novel circumstances in the face of climate change.

References

Aarssen, L.W. (1997) High productivity in grassland ecosystems: effected by species diversity or productive species. *Oikos* 80, 183–184.

Aiguar, M.R., Paruelo, J.M., Sala, O.E. and Lauenroth, W.K. (1996) Ecosystem responses to changes in plant functional type composition: an example from Patagonian steppe. *Journal of Vegetation Science* 7, 381–390.

Asteraki, E.J., Hanks, C.B. and Clements, R.O. (1995) The influence of different types of grassland field margin on carabid beetle (*Coleoptera, Carabidae*) communities. *Agriculture, Ecosystems and Environment* 54, 195–202.

Balmford, A., Bruner, A., Cooper, P., Costanza, R., Farber, S., Green, R.S., Jenkins, M., Jefferiss, P., Jessamy, V., Madden, J., Munro, K., Myers, N., Naeem, S., Paavola, J., Rayment, M., Rosendo, S., Roughgarden, J., Trumper, K. and Turner, R.K. (2002) Economic reasons for conserving wild nature. *Science* 297, 950–953.

Barbault, R., Colwell, R.K., Dias, B., Hawksworth, D.L., Huston, M., Laserre, P., Stone, D. and Younes, T. (1991) Conceptual framework and research issues for species diversity at the community level. In: Solbrig, O.T. (ed.) *From Genes to Ecosystems: a Research Agenda for Biodiversity*. IUBS, Cambridge, Massachusetts, pp. 37–71.

Barbier, B.E. (2000) How to allocate biodiversity internationally? In: Siebert, H. (ed.) *The Economics of International Environmental Problems*. Mohr, Tübingen, pp. 79–106.

Barthlott, W., Lauer, W. and Placke, A. (1996) Global distribution of species diversity: towards a world map of phytodiversity. *Erdkunde* 50, 317–327.

Baskin, Y. (1994) Ecosystem function of biodiversity. *BioScience* 44, 657–660.

Bednekoff, P.A. (2001) Contributions – Commentary – A baseline theory of biodiversity and ecosystem function. *Bulletin of the Ecological Society of America* 82, 205.

Bengtsson, J., Engelhardt, K., Giller, P., Hobbie, S., Lawrence, D., Levine, J., Vilà, M. and Wolters, V. (2002) Slippin' and slidin' between the scales: the scaling components of biodiversity–ecosystem functioning relations. In: Loreau, M., Naeem, S. and Inchausti, P. (eds) *Biodiversity and Ecosystem Functioning*. Oxford University Press, Oxford, pp. 209–220.

Bergamini, A., Pauli, P., Peintinger, M. and Schmid, B. (2001) Relationships between productivity, number of shoots, and number of species in bryophytes and vascular plants. *Journal of Ecology* 89, 920–929.

Bond, W.J. and Midgley, J.J. (2001) Ecology of sprouting in woody plants: the persistence niche. *Trends in Ecology and Evolution* 16, 45–51.

Boutin, C. and Keddy, P.A. (1993) A functional classification of wetland plants. *Journal of Vegetation Science* 4, 591–600.

Bowman, D.M.J.S. (1993) Biodiversity: much more than biological inventory. *Biodiversity Letters* 1, 163.

Box, E.O. (1996) Plant functional types and climate at the global scale. *Journal of Vegetation Science* 7, 309–320.

Brokaw, N.V.L. (1985) Gap-phase regeneration in a tropical forest. *Ecology* 66, 682–687.

Buongiorno, J., Dahir, S., Lu, H.-C. and Lin, C.-R. (1994) Tree size diversity and economic returns in uneven-aged forest stands. *Forest Science* 40, 83–103.

Burke, M.J.W. and Grime, J.P. (1996) An experimental study of plant community invasibility. *Ecology* 77, 776–790.

Carson, R. (1962) *Silent Spring.* Houghton Mifflin, Boston, Massachusetts.

Catovsky, S., Bradford, M.A. and Hector, A. (2002) Biodiversity and ecosystem productivity: implications for carbon storage. *Oikos* 97, 443–448.

Caughley, G. and Gunn, A. (1996) *Conservation Biology in Theory and Practice.* Blackwell Science, Oxford.

Chapin, F.S. (1993) Functional role of growth forms in ecosystems and global processes. In: Ehleringer, J.R. and Field, C.B. (eds) *Scaling Physiological Processes: Leaf to Globe.* Academic Press, San Diego, California, pp. 287–312.

Chapin, F.S., Schulze, E.-D. and Mooney, H.A. (1992) Biodiversity and ecosystem processes. *Trends in Ecology and Evolution* 7, 107–108.

Chapin, F.S., Bret-Harte, M.S., Hobbie, S. and Zhong, H. (1996) Plant functional types as predictors of the transient response of arctic vegetation to global change. *Journal of Vegetation Science* 7, 347–357.

Chapin, F.S., Walker, B.H., Hobbs, R.J., Hooper, D.U., Lawton, J.H., Sala, O.E. and Tilman, D. (1997) Biotic control over the functioning of ecosystems. *Science* 277, 500–504.

Chapin, F.S., Sala, O.E., Burke, I.C., Grime, J.P., Hooper, D.U., Lauenroth, W.K., Lombard, A., Mooney, H.A., Mosier, A.R., Naeem, S., Pacala, S.W., Roy, J., Steffen, W.L. and Tilman, D. (1998) Ecosystem consequences of changing biodiversity. *BioScience* 48, 45–52.

Chase, J.M. and Leibold, M.A. (2002) Spatial scale dictates the productivity–diversity relationship. *Nature* 415, 427–430.

Christensen, N.L. (1981) Fire regimes in Southeastern ecosystems. In: Mooney, H.A., Bonnicksen, J.M., Christensen, N.L. and Reiners, W.A. (eds) *Fire Regimes and Ecosystem Properties.* USDA Forest Service general technical report WO-26. United States Department of Agriculture, Washington, DC, pp. 112–136.

Cody, M.L. (1991) Niche theory and plant growth form. *Vegetatio* 97, 39–55.

Connell, J.H. and Orias, E. (1964) The ecological regulation of species diversity. *American Naturalist* 98, 399–414.

Cornell, H.V. and Lawton, J.H. (1992) Species interactions, local and regional processes and limits to the richness of ecological communities: a theoretical perspective. *Journal of Animal Ecology* 61, 1–12.

Costanza, R., d'Arge, R., de Groot, R., Farber, S., Grasso, M., Hannon, B., Limburg, K., Naeem, S., O'Neill, R.V., Paruelo, J., Raskin, R.G., Sutton, P. and van den Belt, M. (1997) The value of the world's ecosystem services and natural capital. *Nature* 387, 253–260.

Cragg, G.M. and Newman, D.J. (2002) Chemical diversity: a function of biodiversity. *Trends in Pharmacological Sciences* 23, 404–405.

Crawley, M.J., Brown, S.L., Heard, M.S. and Edwards, G.R. (1999) Invasion-resistance in experimental grassland communities: species richness or species identity? *Ecology Letters* 2, 140–148.

Daily, G.C. (ed.) (1997) *Nature's Services.* Island Press, Washington, DC.

Daily, G.C., Alexander, S., Ehrlich, P.R., Goulder, L., Lubchenco, J., Matson, P.A., Mooney, H.A., Postel, S., Schneider, S.H., Tilman, D. and Woodwell, G.M. (1997) Ecosystem services: benefits supplied to human societies by natural ecosystems. *Issues in Ecology* 1, 1–16.

Dasgupta, P. (2001) Economic value of biodiversity, overview. In: Levin, S.A. (ed.) *Encyclopedia of Biodiversity,* Vol. 2. Academic Press, San Diego, California, pp. 291–303.

Davis, G.W., Midgley, G.F. and Hoffman, M.T. (1994) Linking biodiversity to ecosystem function – a challenge to Fynbos ecology. *South African Journal of Science* 90, 319–321.

De Angelis, D.L. and Gross, L.J. (1992) *Individual-based Models and Approaches in Ecology.* Chapman & Hall, London.

Diaz, S. (1995) Elevated CO_2 responsiveness, interactions at the community level, and plant functional types. *Journal of Biogeography* 22, 289–295.

Diaz, S. and Cabido, M. (1997) Plant functional types and ecosystem function in relation to global change: a multiscale approach. *Journal of Vegetation Science* 8, 463–474.

Diaz, S., Cabido, M. and Casanoves, F. (1998) Plant functional traits and environmental filters at a regional scale. *Journal of Vegetation Science* 9, 113–122.

DiCastri, F. and Younès, T. (1990) Ecosystem function of biological diversity. *Biology International Special Issue* 22, 1–20.

Diemer, M., Joshi, J., Schmid, B., Körner, C. and Spehn, E. (1997) An experimental protocol to assess the effects of plant diversity on ecosystem functioning utilised by a European research network. *Bulletin des Geobotanischen Instituts ETH Zürich* 63, 95–107.

Doak, D.F., Bigger, D., Harding, E.K., Marvier, M.A., O'Malley, R.E. and Thomson, D. (1998) The statistical inevitability of stability–diversity relationships in community ecology. *American Naturalist* 151, 264–276.

Dobson, A. (1995) Biodiversity and human health. *Trends in Ecology and Evolution* 10, 390–391.

Dukes, J.S. (2001) Biodiversity and invasibility in grassland microcosms. *Oecologia* 126, 563–568.

Ehrlich, P.R. and Ehrlich, A.H. (1981) *Extinction: the Causes and Consequences of the Disappearance of Species.* Random House, New York.

Ehrlich, P.R. and Wilson, E.O. (1991) Biodiversity studies: science and policy. *Science* 253, 758–762.

Farnsworth, N.R. (1988) Screening plants for new medicines. In: Wilson, E.O. and Peter, F.M. (eds) *BioDiversity.* National Academy Press, Washington, DC, pp. 83–97.

Field, C.B., Osborn, J.G., Hoffmann, L.L., Polsenberg, J.F., Ackerly, D.D., Berry, J.A., Bjorkman, O., Held, Z., Matson, P.A. and Mooney, H.A. (1998) Mangrove biodiversity and ecosystem function. *Global Ecology and Biogeography Letters* 7, 3–14.

Fonseca, C. and Ganade, G. (2001) Species functional redundancy, random extinctions and the stability of ecosystems. *Journal of Ecology* 89, 118–125.

Frank, D.A. and McNaughton, S.J. (1991) Stability increases with diversity in plant communities: empirical evidence from the 1988 Yellowstone drought. *Oikos* 62, 360–362.

Frank, K. and Ring, I. (1999) Model-based criteria for the effectiveness of conservation strategies – an evaluation of incentive programmes in Saxony, Germany. In: Ring, I., Klauer, B., Wätzold, F. and Månsson, B.Å. (eds) *Regional Sustainability. Applied Ecological Economics Bridging the Gap Between Natural and Social Sciences.* Physica-Verlag, Heidelberg, pp. 91–106.

Franklin, J.F. (1988) Structural and functional diversity in temperate forests. In: Wilson, E.O. and Peter, F.M. (eds) *BioDiversity.* National Academy Press, Washington, DC.

Fridley, J.D. (2002) Resource availability dominates and alters the relationship between species diversity and ecosystem productivity in experimental plant communities. *Oecologia* 132, 271–277.

Fromm, O. (2000) Ecological structure and functions of biodiversity as elements of its total economic value. *Environmental and Resource Economics* 16, 303–328.

Fukami, T., Naeem, S. and Wardle, D. (2001) On similarity among local communities in biodiversity experiments. *Oikos* 95, 340–348.

Gitay, H. and Noble, I.R. (1997) What are plant functional types and how should we seek them? In: Smith, T.M., Shugart, H.H. and Woodward, F.I. (eds) *Plant Functional Types: Their Relevance to Ecosystem Properties and Global Change.* Cambridge University Press, Cambridge, pp. 3–19.

Gitay, H., Wilson, J.B. and Lee, W.G. (1996) Species redundancy: a redundant concept? *Journal of Ecology* 84, 121–124.

Givnish, T.J. (1994) Does diversity beget stability? *Nature* 371, 113–114.

Golluscio, R.A. and Sala, O.E. (1993) Plant functional types and ecological strategies in Patagonian forbs. *Journal of Vegetation Science* 4, 839–846.

Goodman, D. (1975) The theory of diversity–stability relationships in ecology. *The Quarterly Review of Biology* 50, 237–266.

Gowdy, J.M. and Daniel, C.N. (1995) One world, one experiment: addressing the biodiversity – economics conflict. *Ecological Economics* 15, 181–192.

Grime, J.P. (1977) Evidence for the existence of three primary strategies in plants and its relevance to ecological and evolutionary theory. *American Naturalist* 111, 1169–1194.

Grime, J.P. (1987) Dominant and subordinate components of plant communities: implications for succession, stability and diversity. In: Gray, A.J., Edwards, P. and Crawley, M. (eds) *Colonisation, Succession and Stability.* Blackwell, Oxford, pp. 413–428.

Grime, J.P. (1997) Biodiversity and ecosystem function: the debate deepens. *Science* 277, 1260–1261.

Grime, J.P. (1998) Benefits of plant diversity to ecosystems: immediate, filter and founder effects. *Journal of Ecology* 86, 902–910.

Grimm, V. (1999) Ten years of individual-based modelling in ecology: what have we learned, and what could we learn in the future? *Ecological Modelling* 115, 129–148.

Grimm, V. and Wissel, C. (1997) Babel, or the ecological stability discussions: an inventory and analysis of terminology and a guide for avoiding confusion. *Oecologia* 109, 323–334.

Grubb, P.J. (1977) The maintenance of species richness in plant communities: the importance of the regeneration niche. *Biological Reviews of the Cambridge Philosophical Society* 52, 107–145.

Guo, Q. and Berry, W.L. (1998) Species richness and biomass: dissection of the hump-shaped relationships. *Ecology* 79, 2555–2559.

Hampicke, U. (1999) The limits to economic valuation of biodiversity. *International Journal of Social Economics* 26, 158–173.

Hawkins, C.P. and MacMahon, J.A. (1989) Guilds: the multiple meanings of a concept. *Annual Review of Entomology* 34, 423–451.

Heal, G. (2000) *Nature and the Marketplace: Capturing the Value of Ecosystem Services.* Island Press, Washington, DC.

Hector, A. (1998) The effect of diversity on productivity: detecting the role of species complementarity. *Oikos* 82, 597–599.

Hector, A. (2002) Biodiversity and the functioning of grassland ecosystems: multi-site comparisons. In: Kinzig, A.P., Pacala, S.W. and Tilman, D. (eds) *The Functional Consequences of Biodiversity.* Princeton University Press, Princeton, New Jersey, pp. 71–95.

Hector, A., Schmid, B., Beierkuhnlein, C., Caldeira, M.C., Diemer, M., Dimitrakopoulos, P.G., Finn, J., Freitas, H., Giller, P.S., Good, J., Harris, R., Högberg, P., Huss-Danell, K., Joshi, J., Jumpponen, A., Körner, C., Leadley, P.W., Loreau, M., Minns, A., Mulder, C.P.H., O'Donovan, G., Otway, S.J., Pereira, J.S., Prinz, A., Read, D.J., Scherer-Lorenzen, M., Schulze, E.-D., Siamantziouras, A.-S.D., Spehn, E., Terry, A.C., Troumbis, A.Y., Woodward, F.I., Yachi, S. and Lawton, J.H. (1999) Plant diversity and productivity of European grasslands. *Science* 286, 1123–1127.

Hector, A., Beale, A., Minns, A., Otway, S. and Lawton, J.H. (2000a) Consequences of loss of plant diversity for litter decomposition: mechanisms of litter quality and microenvironment. *Oikos* 90, 357–371.

Hector, A., Schmid, B., Beierkuhnlein, C., Caldeira, M.C., Diemer, M., Dimitrakopoulos, P.G., Finn, J., Freitas, H., Giller, P.S., Good, J., Harris, R., Högberg, P., Huss-Danell, K., Joshi, J., Jumpponen, A., Körner, C., Leadley, P.W., Loreau, M., Minns, A., Mulder, C.P.H., O'Donovan, G., Otway, S.J., Pereira, J.S., Prinz, A., Read, D.J., Scherer-Lorenzen, M., Schulze, E.-D., Siamantziouras, A.-S.D., Spehn, E., Terry, A.C., Troumbis, A.Y., Woodward, F.I., Yachi, S. and Lawton, J.H. (2000b) No consistent effect of plant diversity on productivity. *Science* 289, 1255.

Hector, A., Dobson, K., Minns, A., Bazeley-White, E. and Lawton, J.H. (2001a) Community diversity and invasion resistance: an experimental test in a grassland ecosystem and a review of comparable studies. *Ecological Research* 16, 819–831.

Hector, A., Joshi, J., Lawler, S., Spehn, E.M. and Wilby, A. (2001b) Conservation implications of the link between biodiversity and ecosystem functioning. *Oecologia* 129, 624–628.

Hector, A., Schmid, B., Beierkuhnlein, C., Caldeira, M.C., Diemer, M., Dimitrakopoulos, P.G., Finn, J., Freitas, H., Giller, P.S., Good, J., Harris, R., Högberg, P., Huss-Danell, K., Joshi, J., Jumpponen, A., Körner, C., Leadley, P.W., Loreau, M., Minns, A., Mulder, C.P.H., O'Donovan, G., Otway, S.J., Pereira, J.S., Prinz, A., Read, D.J., Scherer-Lorenzen, M., Schulze, E.-D., Siamantziouras, A.-S.D., Spehn, E.M., Terry, A.C., Troumbis, A.Y., Woodward, F.I., Yachi, S. and Lawton, J.H. (2002a) Biodiversity and the functioning of grassland ecosystems: multisite studies. In: Kinzig, A.P., Pacala, S.W. and Tilman, D. (eds) *Functional Consequences of Biodiversity: Experimental Progress and Theoretical Extensions.* Princeton University Press, Princeton, New Jersey, pp. 71–95.

Hector, A., Bazeley-White, E., Loreau, M., Otway, S. and Schmid, B. (2002b) Overyielding in grassland communities: testing the sampling effect hypothesis with replicated biodiversity experiments. *Ecology Letters* 5, 502–511.

Hector, A., Loreau, M., Schmid, B., Caldeira, M.C., Diemer, M., Dimitrakopoulos, P.G., Finn, J., Freitas, H., Giller, P.S., Good, J., Harris, R., Högberg, P., Huss-Danell, K., Joshi, J., Jumpponen, A., Körner, C., Leadley, P.W., Minns, A., Mulder, C.P.H., O'Donovan, G., Otway, S.J., Pereira, J.S., Prinz, A., Read, D.J., Scherer-Lorenzen, M., Schulze, E.-D., Siamantziouras, A.-S.D., Spehn, E., Terry, A.C., Troumbis, A.Y., Woodward, F.I., Yachi, S. and Lawton, J.H. (2002c) Biodiversity manipulation experiments: studies replicated at multiple sites. In: Loreau, M., Naeem, S. and Inchausti, P. (eds) *Biodiversity and Ecosystem Functioning.* Oxford University Press, Oxford, pp. 36–46.

Heerwagen, J.H. and Orians, G.H. (1993) Humans, habitats, and aesthetics. In: Kellert, S.R. and Wilson, E.O. (eds) *The Biophilia Hypothesis.* Island Press, Washington, DC, pp. 138–172.

Higgins, S.I., Clark, J.S., Nathan, R., Hovestadt, T., Schurr, F., Fragosoa, J.M.V., Aguiar, M.R., Ribbens, E. and Lavorel, S. (2003) Forecasting plant migration rates: managing uncertainty for risk assessment. *Journal of Ecology* 91, 341–347.

Hodgson, J.G., Thompson, K., Wilson, P.J. and Bogaard, A. (1998) Does biodiversity determine ecosystem function? The Ecotron experiment reconsidered. *Functional Ecology* 12, 843–848.

Hooper, D.U. (1998) The role of complementarity and competition in ecosystem responses to variation in plant diversity. *Ecology* 79, 704–719.

Hooper, D.U. and Vitousek, P.M. (1997) The effect of plant composition and diversity on ecosystem processes. *Science* 277, 1302–1305.

Hooper, D.U. and Vitousek, P.M. (1998) Effects of plant composition and diversity on nutrient cycling. *Ecological Monographs* 68, 121–149.

Hooper, D.U., Solan, M., Symstad, A., Díaz, S., Gessner, M.O., Buchmann, N., Degrange, V., Grime, P., Hulot, F., Mermillod-Blondin, F., Roy, J., Spehn, E. and van Peer, L. (2002) Species diversity, functional diversity, and ecosystem functioning. In: Loreau, M., Naeem, S. and Inchausti, P. (eds) *Biodiversity and Ecosystem Functioning*. Oxford University Press, Oxford, pp. 195–208.

Hughes, J.B., Ives, R. and Norberg, J. (2002) Do species interactions buffer environmental variation (in theory)? In: Loreau, M., Naeem, S. and Inchausti, P. (eds) *Biodiversity and Ecosystem Functioning*. Oxford University Press, Oxford, pp. 92–101.

Huston, M.A. (1993) Biological diversity, soils, and economics. *Science* 262, 1676–1680.

Huston, M.A. (1997) Hidden treatments in ecological experiments: re-evaluating the ecosystem function of biodiversity. *Oecologia* 110, 449–460.

Huston, M.A. and McBride, A.C. (2002) Evaluating the relative strengths of biotic versus abiotic controls on ecosystem processes. In: Loreau, M., Naeem, S. and Inchausti, P. (eds) *Biodiversity and Ecosystem Functioning*. Oxford University Press, Oxford, pp. 47–60.

IPCC (Intergovernmental Panel on Climate Change) (2001) *Climate Change 1999: the Scientific Basis*. Cambridge University Press, Cambridge.

Jentsch, A. (2004) Disturbance driven vegetation dynamics. Concepts from biogeography to community ecology, and experimental evidence from dry acidic grasslands in central Europe. *Dissertationes Botanicae* 384, 1–218.

Jentsch, A. and Beierkuhnlein, C. (2003) Global climate change and local disturbance regimes as interacting drivers for shifting altitudinal vegetation patterns. *Erdkunde* 57, 216–231.

Jentsch, A., Beierkuhnlein, C. and White, P. (2002) Scale, the dynamic stability of forest ecosystems, and the persistence of biodiversity. *Silva Fennica* 36, 393–400.

Jentsch, A., Wittmer, H., Jax, K., Ring, I. and Henle, K. (2003) Biodiversity – emerging issues for linking natural and social sciences. *Gaia* 2, 121–128.

Johnson, K.H., Vogt, K.A., Clark, H.J., Schmitz, O.J. and Vogt, D.J. (1996) Biodiversity and the productivity and stability of ecosystems. *Trends in Ecology and Evolution* 11, 372–377.

Johst, K., Drechsler, M. and Wätzold, F. (2002) An ecologic–economic modelling procedure to design effective and efficient compensation payments for the protection of species. *Ecological Economics* 41, 37–49.

Jones, C.G., Lawton, J.H. and Shachak, M. (1994) Organisms as ecosystem engineers. *Oikos* 69, 373–386.

Kaiser, C. (2000) Rift over biodiversity divides ecologists. *Science* 289, 1282–1283.

Kennedy, T.A., Naeem, S., Howe, K.M., Knops, J.M.H., Tilman, D. and Reich, P. (2002) Biodiversity as a barrier to ecological invasion. *Nature* 417, 636–638.

Kikkawa, J. (1986) Complexity, diversity, and stability. In: Kikkawa, J. and Anderson, D.J. (eds) *Community Ecology: Pattern and Process*. Blackwell, Melbourne, pp. 41–62.

Kim, K.C. and Weaver, R.D. (1994) Biodiversity and humanity: paradox and challenge. In: Kim, K.C. and Weaver, R.D. (eds) *Biodiversity and Landscapes*. Cambridge University Press, Cambridge, pp. 3–27.

Kindscher, K. and Wells, P.V. (1995) Prairie plant guilds: a multivariate analysis of prairie species based on ecological and morphological traits. *Vegetatio* 117, 29–50.

Kinzig, A.P., Pacala, S.W. and Tilman, D. (2002) *The Functional Consequences of Biodiversity: Empirical Progress and Theoretical Extensions*. Princeton University Press, Princeton, New Jersey.

Knops, J.M.H., Tilman, D., Haddad, N.M., Naeem, S., Mitchell, C.E., Haarstad, J., Ritchie, M.E., Howe, K.M., Reich, P.B., Siemann, E. and Groth, J. (1999) Effects of plant species richness on invasion dynamics, disease outbreaks, insect abundances and diversity. *Ecological Letters* 2, 286–293

Knops, J.M.H., Wedin, D. and Tilman, D. (2001) Biodiversity and decomposition in experimental grassland ecosystems. *Oecologia* 126, 429–433.

Koricheva, J., Mulder, C.P.H., Schmid, B., Joshi, J. and Huss-Danell, K. (2000) Numerical responses of different trophic groups of invertebrates to manipulations of plant diversity in grasslands. *Oecologia* 125, 271–282.

Körner, C. (1993) Scaling from species to vegetation: the usefulness of functional groups. In: Schulze, E.-D. and Mooney, H.A. (eds) *Biodiversity and Ecosystem Function*, Ecological Studies 99. Springer-Verlag, Berlin, pp. 117–140.

Körner, C. (1999) *Alpine Plant Life*. Springer, Berlin.

Lamont, B.B. (1995) Testing the effect of ecosystem composition/structure on its functioning. *Oikos* 74, 283–295.

Lavorel, S. and Richardson, D.M. (1999) Diversity, stability and conservation of Mediterranean-type ecosystems in a changing world: an introduction. *Diversity and Distributions* 5, 1–2.

Lavorel, S., Prieur-Richard, A.-H. and Grigulis, K. (1999) Invasibility and diversity of plant communities: from pattern to process. *Diversity and Distributions* 5, 41–49.

Lawton, J.H. (1994) What do species do in ecosystems? *Oikos* 71, 367–374.

Lawton, J.H. (1996) The role of species in ecosystems: aspects of ecological complexity and biological diversity. In: Abe, T., Levin, S.A. and Higashi, M. (eds) *Biodiversity: an Ecological Perspective*. Springer, New York, pp. 215–228.

Lawton, J.H. (1999) Biodiversity and ecosystem processes: theory, achievements and future directions. In: Kato, M. (ed.) *The Biology of Biodiversity*. Springer-Verlag, Tokyo, pp. 119–131.

Lawton, J.H. and Brown, V.K. (1993) Redundancy in ecosystems. In: Schulze, E.D. and Mooney, H.A. (eds) *Biodiversity and Ecosystem Function*, Ecological Studies 99. Springer, Berlin, pp. 255–270.

Lawton, J.H., Naeem, S., Thompson, L.J., Hector, A. and Crawley, M.J. (1998) Biodiversity and ecosystem function: getting the Ecotron experiment in its correct context. *Functional Ecology* 12, 848–852.

Lehman, C.L. and Tilman, D. (2000) Biodiversity, stability, and productivity in competitive communities. *The American Naturalist* 156, 534–552.

Leigh, E.G. (1965) On the relationship between productivity, biomass, diversity and stability of a community. *Proceedings of the National Academy of Sciences USA* 53, 777–783.

Leishman, M.R. and Westoby, M. (1992) Classifying plants into groups on the basis of associations of individual traits – evidence from Australian semi-arid woodlands. *Journal of Ecology* 80, 417–424.

Levine, J.M. (2000) Species diversity and biological invasions: relating local process and community pattern. *Science* 288, 852–854.

Levine, J.M. and D'Antonio, C.M. (1999) Elton revisited: a review of evidence linking diversity and stability. *Oikos* 87, 15–26.

Levine, J.M., Kennedy, T. and Naeem, S. (2002) Neighbourhood scale effects of species diversity on biological invasions and their relationship to community patterns. In: Loreau, M., Naeem, S. and Inchausti, P. (eds) *Biodiversity and Ecosystem Functioning*. Oxford University Press, Oxford, pp. 114–124.

Loreau, M. (1998a) Biodiversity and ecosystem functioning: a mechanistic model. *Proceedings of the National Academy of Sciences USA* 95, 5632–5636.

Loreau, M. (1998b) Separating sampling and other effects in biodiversity experiments. *Oikos* 82, 600–602.

Loreau, M. (2000) Biodiversity and ecosystem functioning: recent theoretical advances. *Oikos* 91, 3–17.

Loreau, M. and Behara, N. (1999) Phenotypic diversity and stability of ecosystem processes. *Theoretical Population Biology* 56, 29–47.

Loreau, M. and Hector, A. (2001) Partitioning selection and complementarity in biodiversity experiments. *Nature* 412, 72–76.

Loreau, M., Naeem, S., Inchausti, P., Bengtsson, J., Grime, J.P., Hector, A., Hooper, D.U., Huston, M.A., Raffaelli, D., Schmid, B., Tilman, D. and Wardle, D.A. (2001) Biodiversity and ecosystem functioning: current knowledge and future challenges. *Science* 294, 804–808.

Loreau, M., Downing, A., Emmerson, M., Gonzalez, A., Hughes, J., Inchausti, P., Joshi, J., Norberg, J. and Sala, O. (2002) A new look at the relationship between diversity and stability. In: Loreau, M., Naeem, S. and Inchausti, P. (eds) *Biodiversity and Ecosystem Functioning*. Oxford University Press, Oxford, pp. 79–91.

Loreau, M., Mouquet, N. and Gonzalez, A. (2003) Biodiversity as spatial insurance in heterogeneous landscapes. *Proceedings of the National Academy of Sciences USA* 100, 12765–12770.

MacArthur, R.H. (1969) Patterns of communities in the tropics. In: Lowe-McConnell, R.H. (ed.) *Speciation in Tropical Environments*. Academic Press, New York, pp. 19–30.

Machlis, G.E. and Forester, J.F. (1996) The relationship between socio-economic factors and the loss of biodiversity: first efforts at theoretical and quantitative models. In: Szaro, R.C. and Johnston, D.W. (eds) *Biodiversity in Managed Landscapes: Theory and Practice*. Oxford University Press, New York, pp. 121–146.

Margalef, R. (1969) Diversity and stability: a practical proposal and a model of interdependence. *Brookhaven Symposium in Biology* 22, 25–37.

Marks, P.L. (1974) The role of pin cherry (*Prunus pennsylvanica* L.) in the maintenance of stability in northern hardwood ecosystems. *Ecological Monographs* 44, 73–88.

Martinez, N.D. (1996) Defining and measuring functional aspects of biodiversity. In: Gaston, K.J. (ed.) *Biodiversity: a Biology of Numbers and Difference*. Blackwell, Oxford, pp. 114–148.

May, R.M. (1972) Will a large complex system be stable? *Nature* 238, 413–414.

May, R.M. (1986) How many species are there? *Nature* 324, 514–515.

May, R.M. (1988) How many species are there on earth? *Science* 324, 1441–1449.

May, R.M. (1990) How many species? *Philosophical Transactions of the Royal Society of London Series B* 330, 293–304.

McGrady-Steed, J., Harris, P.M. and Morin, P.J. (1997) Biodiversity regulates ecosystem predictability. *Nature* 390, 162–165.

McIntyre, S., Ladiges, P.Y. and Adams, G. (1988) Plant species-richness and invasion by exotics in relation to disturbance of wetland communities on the Riverine Plain, NSW. *Australian Journal of Ecology* 13, 361–373.

McIntyre, S., Lavorel, S. and Tremont, R.M. (1995) Plant life-history attributes: their relationship to disturbance response in herbaceous vegetation. *Journal of Ecology* 83, 31–44.

McNaughton, S.J. (1977) Diversity and stability of ecological communities: a comment on the role of empiricism in ecology. *American Naturalist* 111, 515–525.

McNaughton, S.J. (1978) Stability and diversity of ecological communities. *Nature* 274, 251–253.

Meiners, S.J., Cadenasso, M.L. and Pickett, S.T.A. (2004) Beyond biodiversity: individualistic controls of invasion in a self-assembled community. *Ecology Letters* 7, 121–126.

Mikola, J., Bardgett, R.D. and Hedlund, K. (2002) Biodiversity, ecosystem functioning and soil decomposer food webs. In: Loreau, M., Naeem, S. and Inchausti, P. (eds) *Biodiversity and Ecosystem Functioning*. Oxford University Press, Oxford, pp. 169–180.

Mills, L.S., Soulé, M.E. and Doak, D.F. (1993) The keystone-species concept in ecology and conservation. *BioScience* 43, 219–224.

Minns, A., Finn, J., Hector, A., Caldeira, M., Joshi, J., Palmborg, C., Schmid, B., Scherer-Lorenzen, M., Spehn, E., Troumbis, A., Beierkuhnlein, C., Dimitrakopoulos, P.G., Huss-Danell, K., Jumpponen, A., Loreau, M., Mulder, C.P.H., O'Donovan, G., Pereira, J.S., Schulze, E.-D., Terry, A.C. and Woodward, F.I. (2001) The functioning of European grassland ecosystems: potential benefits of biodiversity to agriculture. *Outlook on Agriculture* 30, 179–185.

Mitchell, C., Tilman, D. and Groth, J.V. (2002) Effects of grassland plant species diversity, abundance, and composition on foliar fungal disease. *Ecology* 83, 1713–1726.

Montgomery, C.A. and Pollack, R.A. (1996) Economics and biodiversity: weighing the benefits and costs of conservation. *Journal of Forestry* 94, 34–38.

Mooney, H.A. (2002) The debate on the role of biodiversity in ecosystem functioning. In: Loreau, M., Naeem, S. and Inchausti, P. (eds) *Biodiversity and Ecosystem Functioning*. Oxford University Press, Oxford, pp. 12–17.

Mooney, H.A., Lubchenko, J., Dirzo, R. and Sala, O.E. (1995a) Biodiversity and ecosystem functioning: ecosystem analyses. In: Heywood, V.H. and Watson, R.T. (eds) *Global Biodiversity Assessment*. Cambridge University Press, Cambridge, pp. 327–452.

Mooney, H.A., Lubchenko, J., Dirzo, R. and Sala, O.E. (1995b) Biodiversity and ecosystem functioning: basic principles. In: Heywood, V.H. and Watson, R.T. (eds) *Global Biodiversity Assessment*. Cambridge University Press, Cambridge, pp. 275–326.

Mooney, H.A., Cushman, J.H., Medina, E., Sala, O.E. and Schulze, E.-D. (eds) (1996) *Functional Roles of Biodiversity – a Global Perspective*. Wiley & Sons, Chichester.

Mulder, C.P.H., Uliassi, D.D. and Doak, D.F. (2001) Physical stress and diversity-productivity relationships: the role of positive species interactions. *Proceedings of the National Academy of Sciences USA* 98, 6704–6708.

Myers, N. (1988) Threatened biotas: 'hotspots' in tropical forests. *Environmentalist* 8, 1–20.

Nader, W. and Mateo, N. (1998) Biodiversity – resource for new products, development and self-reliance. In: Barthlott, W. and Winiger, M. (eds) *Biodiversity. A Challenge for Development Research and Policy*. Springer, Berlin, pp. 181–197.

Naeem, S. (1998) Species redundancy and ecosystem reliability. *Conservation Biology* 12, 39–45.

Naeem, S. and Li, S. (1997) Biodiversity enhances ecosystem reliability. *Nature* 390, 507–509.

Naeem, S. and Wright, J.P. (2003) Disentangling biodiversity effects on ecosystem functioning: deriving solutions to a seemingly insurmountable problem. *Ecology Letters* 6, 567–579.

Naeem, S., Thompson, L.J., Lawler, S.P., Lawton, J.H. and Woodfin, R.M. (1994) Declining biodiversity can alter the performance of ecosystems. *Nature* 368, 734–737.

Naeem, S., Thompson, L.J., Lawler, S.P., Lawton, J.H. and Woodfin, R.M. (1995) Empirical evidence that declining species diversity may alter the performance of terrestrial ecosystems. *Philosophical Transactions of the Royal Society of London Series B* 347, 249–262.

Naeem, S., Håkansson, K., Lawton, J.H., Crawley, M.J. and Thompson, L.J. (1996) Biodiversity and plant productivity in a model assemblage of plant species. *Oikos* 76, 259–264.

Naeem, S., Knops, J.M.H., Tilman, D., Howe, K.M., Kennedy, T. and Gale, S. (2000) Plant diversity increases resistance to invasion in the absence of covarying extrinsic factors. *Oikos* 91, 97–108.

Naeem, S., Loreau, M. and Inchausti, P. (2002) Biodiversity and ecosystem functioning: the emergence of a synthetic ecological framework. In: Loreau, M., Naeem, S. and Inchausti, P. (eds) *Biodiversity and Ecosystem Functioning*. Oxford University Press, Oxford, pp. 3–11.

Noble, I.R. and Slayter, R.O. (1980) The use of vital attributes to predict successional changes in plant communities subject to recurrent disturbances. *Vegetatio* 43, 5–21.

Nunes, P. and van den Bergh, J. (2001) Economic valuation of biodiversity: sense or nonsense? *Ecological Economics* 39, 203–222.

Odum, E.P. (1993) Ideas in ecology for the 1990s. *BioScience* 42, 542–545.

OECD (1999) *Handbook of Incentive Measures for Biodiversity. Design and Implementation*. OECD, Paris.

Oksanen, J. (1996) Is the humped relationship between species richness and biomass an artifact due to plot size? *Journal of Ecology* 84, 293–296.

Pacala, S.W. and Deutschmann, D.H. (1996) Details that matter: the spatial distribution of individual trees maintains forest ecosystem function. *Oikos* 74, 357–365.

Pacala, S. and Tilman, D. (2002) The transition from sampling to complementarity. In: Kinzig, A.P., Pacala, S.W. and Tilman, D. (eds) *The Functional Consequences of Biodiversity: Empirical Progress and Theoretical Extensions*. Princeton University Press, Princeton, New Jersey, pp. 151–166.

Palmer, M.W. and Maurer, T.A. (1997) Does diversity beget diversity? A case study of crops and weeds. *Journal of Vegetation Science* 8, 235–240.

Pavlovic, N.B. (1994) Disturbance-dependent persistence of rare plants: anthropogenic impacts and restoration implications. In: Boels, M.L. and Whelan, C. (eds) *Recovery and Restoration of Endangered Species*. Cambridge University Press, Cambridge, pp. 159–193.

Pearce, D. and Moran, D. (eds) (1994) *The Economic Value of Biodiversity*. Earthscan, London.

Perrings, C. (ed.) (1995) *Biodiversity Loss – Economic and Ecological Issues*. Cambridge University Press, Cambridge.

Petchey, O.L. and Gaston, K.J. (2002) Extinction and the loss of functional diversity. *Proceedings of the Royal Society of London Series B* 269, 1721–1727.

Peters, R.L. (1994) Conserving biological diversity in the face of climate change. In: Kim, K.C. and Weaver, R.D. (eds) *Biodiversity and Landscapes*. Cambridge University Press, Cambridge, pp. 105–132.

Peters, R.L. and Lovejoy, T.E. (eds) (1992) *Global Warming and Biological Diversity*. Yale University Press, New Haven, Connecticut.

Peterson, G., Allen, C.R. and Holling, C.S. (1998) Ecological resilience, biodiversity and scale. *Ecosystems* 1, 6–18.

Pfisterer, A.B. and Schmid, B. (2002) Diversity-dependent production can decrease the stability of ecosystem functioning. *Nature* 416, 84–86.

Pianka, E.R. (1966) Latitudinal gradients in species diversity: a review of concepts. *American Naturalist* 100, 33–46.

Pimentel, D., Wilson, C., McCullum, C., Huang, R., Dwen, P., Flack, J., Tran, Q., Saltman, T. and Cliff, B. (1997) Economic and environmental benefis of biodiversity. *BioScience* 47, 747–755.

Pimentel, D., Lach, L., Zuniga, R. and Morrison, D. (2000) Environmental and economic costs of nonindigenous species in the United States. *BioScience* 50, 54–65.

Pimm, S.L., Russell, G.J., Gittelman, J.L. and Brooks, T. (1995) The future of biodiversity. *Science* 269, 347–350.

Planty-Tabacchi, A.-M., Tabacchi, E., Naiman, R.J., DeFerrari, C. and Decamps, H. (1996) Invasibility of species-rich community in riparian zones. *Conservation Biology* 10, 598–607.

Plotkin, M.J. (1988) The outlook for new agricultural and industrial products from the tropics. In: Wilson, E.O. and Peter, F.M. (eds) *BioDiversity*. National Academy Press, Washington, DC, pp. 106–116.

Prieur-Richard, A.-H. and Lavorel, S. (2000) Invasions: the perspective of diverse plant communities. *Austral Ecology* 25, 1–7.

Prieur-Richard, A.-H., Lavorel, S., Grigulis, K. and Dos Santos, A. (2000) Plant community diversity and invasibility by exotics: invasion of Mediterranean old fields by *Conyza bonariensis* and *Conyza canadiensis*. *Ecology Letters* 3, 412–422.

Primack, R.B. (1993) *Essentials of Conservation Biology*. Sinauer Associates, Sunderland, Massachusetts.

Pysek, P. and Pysek, A. (1995) Invasion by *Heracleum mantegazzianum* in different habitats in the Czech Republic. *Journal of Vegetation Science* 6, 711–718.

Raffaelli, D., van der Putten, W.H., Persson, L., Wardle, D.A., Petchey, O.L., Koricheva, J., van der Heijden, M., Mikola, J. and Kennedy, T. (2002) Multi-trophic dynamics and ecosystem processes. In: Loreau, M., Naeem, S. and Inchausti, P. (eds) *Biodiversity and Ecosystem Functioning*. Oxford University Press, Oxford, pp. 147–154.

Randall, A. (1994) Thinking about the value of biodiversity. In: Kim, K.C. and Weaver, R.D. (eds) *Biodiversity and Landscapes*. Cambridge University Press, Cambridge, pp. 271–285.

Rapson, G.L., Thompson, K. and Hodgson, J.G. (1997) The humped relationship between species richness and biomass – testing its sensitivity to sample quadrat size. *Journal of Ecology* 85, 99–100.

Reich, P.B., Knops, J., Tilman, D., Craine, J., Ellsworth, D., Tjoelker, M., Lee, T., Wedin, D., Naeem, S., Bahauddin, D., Hendrey, G., Jose, S., Wrage, K., Goth, J. and Bengston, W. (2001) Plant diversity enhances ecosystem responses to elevated CO_2 and nitrogen deposition. *Nature* 410, 809–812.

Rejmánek, M. (1989) Invasibility of plant communities. In: Drake, J.A., Mooney, H.A., Di Castri, F., Groves, R.H., Kruger, F.J., Rejmánek, M. and Williamson, M. (eds) *Biological Invasions: a Global Perspective*. John Wiley & Sons, Chichester, UK, pp. 369–388.

Rickleffs, R.E. (1997) *The Economy of Nature*. Freeman, New York.

Ring, I. (2002) Ecological public functions and fiscal equalisation at the local level in Germany. *Ecological Economics* 42, 415–427.

Risser, P.G. (1995) Biodiversity and ecosystem function. *Conservation Biology* 9, 742–746.

Robinson, G.R., Quinn, J.F. and Stanton, M.L. (1995) Invasibility of experimental habitat islands in a California winter annual grassland. *Ecology* 76, 786–794.

Sala, O.E., Chapin, F.S. III, Armesto, J.J., Berlow, R., Bloomfield, J., Dirzo, R., Huber-Sanwald, E., Huenneke, L.F., Jackson, R.B., Kinzig, A., Leemans, R., Lodge, D., Mooney, H.A., Oesterheld, M., Poff, N.L., Sykes, M.T., Walker, B.H., Walker, M. and Wall, D.H. (2000) Global biodiversity scenarios for the year 2100. *Science* 287, 1770–1774.

Scherer-Lorenzen, M. (1999) Effects of plant diversity on ecosystem processes in experimental grassland communities. *Bayreuther Forum Ökologie* 75, 1–195.

Scherer-Lorenzen, M., Palmborg, C., Prinz, A. and Schulze, E.-D. (2003) The role of plant diversity and composition for nitrate leaching in grasslands. *Ecology* 84, 1539–1552.

Schläpfer, F. and Schmid, B. (1999) Ecosystem effects of biodiversity: a classification of hypotheses and exploration of empirical results. *Ecological Applications* 9, 893–912.

Schläpfer, F., Schmid, B. and Seidl, I. (1999) Expert estimates about effects of biodiversity on ecosystem processes and services. *Oikos* 84, 346–352.

Schlesinger, W.H. (ed.) (1991) *Biogeochemistry: an Analysis of Global Change*. Academic Press, San Diego, California.

Schmid, B. (2002) The species richness–productivity controversy. *Trends in Ecology and Evolution* 17, 113–114.

Schmid, B., Hector, A., Huston, M.A., Inchausti, P., Nijs, I., Leadley, P.W. and Tilman, D. (2002a) The design and analysis of biodiversity experiments. In: Loreau, M., Naeem, S. and Inchausti, P. (eds) *Biodiversity and Ecosystem Functioning*. Oxford University Press, Oxford, pp. 61–75.

Schmid, B., Joshi, J. and Schläpfer, F. (2002b) Empirical evidence for biodiversity–ecosystem functioning relationships. In: Kinzig, A.P., Pacala, S.W. and Tilman, D. (eds) The *Functional Consequences of Biodiversity: Empirical Progress and Theoretical Extensions*. Princeton University Press, Princeton, New Jersey, pp. 120–150.

Schulze, E.-D. and Mooney, H.A. (1993) Ecosystem function and biodiversity: a summary. In: Schulze, E.-D. and Mooney, H.A. (eds) *Biodiversity and Ecosystem Function*, Ecological Studies 99. Springer-Verlag, Berlin, pp. 497–510.

Schwartz, M.W., Brigham, C.A., Hoeksema, J.D., Lyons, K.G., Mills, M.H. and van Mantgem, P.J. (2000) Linking biodiversity to ecosystem function: implications for conservation ecology. *Oecologia* 122, 297–305.

Seidl, I. and Gowdy, J. (1999) Monetäre Bewertung von Biodiversität: Grundannahmen, Schritte, Probleme und Folgerungen. *GAIA* 8, 102–112.

Siemann, E., Tilman, D., Haarstad, J. and Ritchie, M. (1998) Experimental tests of the dependence of arthropod diversity on plant diversity. *American Naturalist* 152, 738–750.

Smith, M.D. and Knapp, A.K. (2003) Dominant species maintain ecosystem function with non-random species loss. *Ecology Letters* 6, 509–517.

Smith, T.M., Shugart, H.H., Woodward, F.I. and Burton, P.J. (1993) Plant functional types. In: Solomon, A.M. and Shugart, H.H. (eds) *Vegetation Dynamics and Global Change*. Chapman & Hall, New York, pp. 272–292.

Solbrig, O.T. (1993) Plant traits and adaptive strategies: their role in ecosystem function. In: Schulze, E.-D. and Mooney, H.A. (eds) *Biodiversity and Ecosystem Function*, Ecological Studies 99. Springer-Verlag, Berlin, pp. 97–116.

Soulé, M.E. (1991) Conservation: tactics for a constant crisis. *Science* 253, 744–750.

Spehn, E.M., Joshi, J., Schmid, B., Diemer, M. and Körner, C. (2000a) Aboveground resource use increases with plant species richness in experimental grassland ecosystems. *Functional Ecology* 14, 326–337.

Spehn, E., Joshi, J., Schmid, B., Alphei, J. and Körner, C. (2000b) Plant diversity effects on soil heterotrophic activity in experimental grassland ecosystems. *Plant and Soil* 224, 217–230.

Spehn, E.M., Scherer-Lorenzen, M., Schmid, B., Hector, A., Caldeira, M.C., Dimitrakopoulos, P.G., Finn, J.A., Jumpponen, A., O'Donnovan, G., Pereira, J.S., Schulze, E.-D., Troumbis, A.Y. and Körner, C. (2002) The role of legumes as a component of biodiversity in a cross-European study of grassland biomass nitrogen. *Oikos* 98, 205–218.

Stephan, A., Meyer, A.H. and Schmid, B. (2000) Plant diversity affects culturable soil bacteria in experimental grassland communities. *Journal of Ecology* 88, 988–998.

Stohlgren, T.J., Binkley, D., Chanog, G.W., Kalkhan, M.A., Schell, L.D., Bull, K.A., Otsuki, Y., Newman, G., Bashkin, M. and Son, Y. (1999) Exotic plant species invade hot spots of native plant diversity. *Ecological Monographs* 69, 25–46.

Symstad, A.J. (2000) A test of the effects of functional group richness and composition on grassland invasibility. *Ecology* 81, 99–109.

Symstad, A.J. and Tilman, D. (2001) Diversity loss, recruitment limitation and ecosystem functioning: lessons learned from a removal experiment. *Oikos* 92, 424–435.

Symstad, A.J., Tilman, D., Willson, J. and Knops, J. (1998) Species loss and ecosystem functioning: effects of species identity and community composition. *Oikos* 81, 389–397.

Temperton, V.M., Hobbs, R., Fattorini, M. and Halle, S. (eds) (2004) *Assembly Rules in Restoration Ecology – Bridging the Gap between Theory and Practice*. Island Press Books, Washington, DC, and Sinauer Associates, Sunderland, Massachusetts.

ten Kate, K. and Laird, S.A. (2002) *The Commercial Use of Biodiversity*. Earthscan, London.

Thierry, R.G. (1982) Environmental instability and community diversity. *Biological Reviews* 57, 671–710.

Thompson, K. and Hodgson, J.G. (1998) Biodiversity and ecosystem function: getting the Ecotron experiment in its correct context – response. *Functional Ecology* 12, 852–853.

Tilman, D. (1993) Community diversity and succession: the roles of competition, dispersal and habitat modification. In: Schulze, E.-D. and Mooney, H.A. (eds) *Biodiversity and Ecosystem Function*, Ecological Studies 99. Springer-Verlag, Berlin, pp. 327–344.

Tilman, D. (1996) Biodiversity: population versus ecosystem stability. *Ecology* 77, 350–363.

Tilman, D. (1997a) Community invasibility, recruitment limitation, and grassland biodiversity. *Ecology* 78, 81–92.

Tilman, D. (1997b) Distinguishing between the effects of species diversity and species composition. *Oikos* 80, 185–185.

Tilman, D. (1999) Diversity and production in European grasslands. *Science* 286, 1099–1100.

Tilman, D. (2001) Effects of diversity and composition on grassland stability and productivity. In: Press, M.C., Huntly, N.J. and Levin, S. (eds) *Ecology: Achievement and Challenge*. Blackwell Science, Oxford, pp. 183–207.

Tilman, D. and Downing, J.A. (1994) Biodiversity and stability in grasslands. *Nature* 367, 363–365.

Tilman, D. and Lehman, C. (2002) Biodiversity, composition, and ecosystem processes: theory and concepts. In: Kinzig, A., Pacala, S. and Tilman, D. (eds) *Functional Consequences of Biodiversity: Empirical Progress and Theoretical Extensions*. Princeton University Press, Princeton, New Jersey, pp. 9–41.

Tilman, D., Downing, J.A. and Wedin, D.A. (1994) Does diversity beget stability? *Nature* 371, 113–114.

Tilman, D., Wedin, D. and Knops, J. (1996) Productivity and sustainability influenced by biodiversity in grassland ecosystems. *Nature* 379, 718–720.

Tilman, D., Knops, J., Wedin, D., Reich, P., Ritchie, M. and Siemann, E. (1997a) The influence of functional diversity and composition on ecosystem processes. *Science* 277, 1300–1302.

Tilman, D., Lehman, C.L. and Thomson, K.T. (1997b) Plant diversity and ecosystem productivity: theoretical considerations. *Proceedings of the National Academy of Sciences USA* 94, 1857–1861.

Tilman, D., Naeem, S., Knops, J., Reich, P., Siemann, E., Wedin, D., Ritchie, M. and Lawton, J. (1997c) Biodiversity and ecosystem properties. *Science* 278, 1866–1867.

Tilman, D., Lehman, C.L. and Bristow, C.E. (1998) Diversity–stability relationships: statistical inevitability or ecological consequence? *The American Naturalist* 151, 277–282.

Tilman, D., Reich, P.B., Knops, J., Wedin, D., Mielke, T. and Lehman, C. (2001) Diversity and productivity in a long-term grassland experiment. *Science* 294, 843–845.

Tilman, D., Knops, J., Wedin, D. and Reich, P. (2002a) Experimental and observational studies of diversity, productivity and stability. In: Kinzig, A.P., Pacala, S.W. and Tilman, D. (eds) *The Functional Consequences of Biodiversity: Empirical Progress and Theoretical Extensions.* Princeton University Press, Princeton, New Jersey, pp. 42–70.

Tilmann, D., Knops, J., Wedin, D. and Reich, P. (2002b) Plant diversity and composition: effects on productivity and nutrient dynamics of experimental grasslands. In: Loreau, M., Naeem, S. and Inchausti, P. (eds) *Biodiversity and Ecosystem Functioning.* Oxford University Press, Oxford, pp. 21–35.

Troumbis, A. and Memtas, D. (2000) Observational evidence that diversity may increase productivity in Mediterranean shrublands. *Oecologia* 125, 101–108.

UNEP (1992) *World Atlas of Desertification.* UNEP, London.

van der Heijden, M.G.A. and Cornelissen, J.H.C. (2002) The critical role of plant–microbe interactions on biodiversity and ecosystem functioning: arbuscular mycorrhizal associations as an example. In: Loreau, M., Naeem, S. and Inchausti, P. (eds) *Biodiversity and Ecosystem Functioning.* Oxford University Press, Oxford, pp. 181–192.

van der Maarel, E. (1986) Vegetation structure and its relations with dynamics and diversity. *Journale Botanico Italiano* 120, 59–60.

van der Maarel, E. (1988) Diversity in plant communities in relation to structure and dynamics. In: During, H.J., Werger, M.J.A. and Willems, J.H. (eds) *Diversity and Pattern in Plant Communities.* SPB, The Hague, pp. 1–14.

van der Maarel, E. (1997) *Biodiversity: From Babel to Biosphere Management.* Opulus Press, Uppsala, Sweden.

van Ruijven, J. and Berendse, F. (2003) Positive effects of plant diversity on productivity in the absence of legumes. *Ecology Letters* 6, 170–175.

van Ruijven, J., De Deyn, G.B. and Berendse, F. (2003) Diversity reduces invasibility in experimental plant communities: the role of plant species. *Ecology Letters* 6, 910–918.

van Wilgen, B.W., Cowling, R.M. and Burgers, C.J. (1996) Valuation of ecosystem services. *BioScience* 46, 184–189.

Vandermeer, J., Lawrence, D., Symstad, A. and Hobbie, S. (2002) Effect of biodiversity on ecosystem functioning in managed ecosystems. In: Loreau, M., Naeem, S. and Inchausti, P. (eds) *Biodiversity and Ecosystem Functioning.* Oxford University Press, Oxford, pp. 221–233.

Vavilov, N.I. (1935) The phytogeographic basis of plant breeding. In: *The Origin, Variation, Immunity and Breeding of Cultivated Plants: Selected Writing of N.I. Vavilov.* (transl. by S. Chester) 13 (1949/50). Chronica Botanica Co., Waltham, Massachusetts, pp. 13–54.

Vitousek, P.M. (1984) A general theory of forest nutrient dynamics. In: Agren, G.I. (ed.) *State and Change of Forest Ecosystems – Indicators in Current Research,* Report 13. Swedish University of Agricultural Sciences, Uppsala, pp. 121–135.

Vitousek, P.M. and Hooper, D.U. (1993) Biological diversity and terrestrial ecosystem biogeochemistry. In: Schulze, E.-D. and Mooney, H.A. (eds) *Biodiversity and Ecosystem Function.* Springer-Verlag, Berlin, pp. 3–14.

Vogl, R.J. (1974) Effects of fire on grasslands. In: Kozlowski, T.T. and Ahlgren, C.E. (eds) *Fire and Ecosystems.* Academic Press, New York, pp. 139–194.

von Hardenberg, J., Meron, E., Shachak, M. and Zarmi, Y. (2001) Diversity of vegetation patterns and desertification. *Physical Review Letters* 87, 198101.

Waide, R.B., Willig, M.R., Steiner, C.F., Mittelbach, G., Gough, L., Dodson, S.I., Juday, G.P. and Parmenter, R. (1999) The relationship between productivity and species richness. *Annual Review of Ecology and Systematics* 30, 257–300.

Walker, B.H. (1992) Biodiversity and ecological redundancy. *Conservation Biology* 6, 18–23.

Walker, B.H., Kinzig, A. and Langridge, J. (1999) Plant attribute diversity, resilience, and ecosystem function: the nature and significance of dominant and minor species. *Ecosystems* 2, 95–113.

Wall, D.H. (1999) Biodiversity and ecosystem functioning. *BioScience* 49, 107–108.

Wardle, D.A. (1999) Is sampling effect a problem for experiments investigating biodiversity–ecosystem function relationships? *Oikos* 87, 403–407.

Wardle, D.A. (2001) Experimental demonstration that plant diversity reduces invasibility – evidence of a biological mechanism or a consequence of sampling effect? *Oikos* 95, 1–170.

Wardle, D.A. and van der Putten, W.H. (2002) Biodiversity, ecosystem functioning and above-ground–below-ground linkages. In: Loreau, M., Naeem, S. and Inchausti, P. (eds) *Biodiversity and Ecosystem Functioning.* Oxford University Press, Oxford, pp. 155–168.

Wardle, D.A., Bonner, K.I. and Nicholson, K.S. (1997) Biodiversity and plant litter: experimental evidence which does not support the view that enhanced species richness improves ecosystem function. *Oikos* 79, 247–258.

Wardle, D.A., Bonner, K.I. and Barker, G.M. (2000) Stability of ecosystem properties in response to above-ground functional group richness and composition. *Oikos* 89, 11–23.

Weiher, E. (1999) The combined effects of scale and productivity on species richness. *Journal of Ecology* 87, 1005–1011.

Westoby, M. and Leishman, M.R. (1997) Categorizing plant species into functional types. In: Smith, T.M., Shugart, H.H. and Woodward, F.I. (eds) *Plant Functional Types: Their Relevance to Ecosystem Properties and Global Change.* Cambridge University Press, Cambridge, pp. 104–121.

White, P.S. and Jentsch, A. (2001) The search for generality in studies of disturbance and ecosystem dynamics. *Progress in Botany* 62, 399–449.

White, P. and Jentsch, A. (2004) Disturbance, succession and community assembly in terrestrial plant communities. In: Temperton, V.M., Hobbs, R., Fattorini, M. and Halle, S. (eds) *Assembly Rules in Restoration Ecology – Bridging the Gap between Theory and Practise.* Island Press Books, Washington, DC, pp. 371–399.

White, P.S., MacKenzie, M.D. and Busing, R.T. (1985) A critique of overstory/understory comparisons based on transition probability analysis of an old growth spruce–fir stand in the Appalachians. *Vegetatio* 64, 37–45.

Whittaker, R.H. (1972) Evolution and measurement of species diversity. *Taxon* 12, 213–251.

Wiegand, T., Moloney, K.A., Naves, J. and Knauer, F.E. (1999) Finding the missing link between landscape structure and population dynamics: a spatially explicit perspective. *American Naturalist* 154, 605–627.

Williams, P.H., Humphries, C.J., Vane-Wright, R.I. and Gaston, K.J. (1996) Value in biodiversity, ecological services and consensus. *Trends in Ecology and Evolution* 11, 385.

Wilsey, B.J. and Polley, H.W. (2002) Reductions in grassland species evenness increase dicot seedling invasion and spittle bug infestation. *Ecology Letters* 5, 676–684.

Wilsey, B.J. and Potvin, C. (2000) Biodiversity and ecosystem functioning: the importance of species evenness in an old field. *Ecology* 81, 887–892.

Wilson, D.S. (1992) Complex interactions in meta-communities, with implications for biodiversity and higher levels of selection. *Ecology* 73, 1984–2000.

Wilson, E.O. (1985) The biological diversity crisis. *BioScience* 35, 700–706.

Woodward, F.I. and Cramer, W. (1996) Plant functional types and climatic changes: introduction. *Journal of Vegetation Science* 7, 306–308.

Woodward, F.I. and Kelly, C.K. (1997) Plant functional types: towards a definition by environmental constraints. In: Smith, T.M., Shugart, H.H. and Woodward, F.I. (eds) *Plant Functional Types: Their Relevance to Ecosystem Properties and Global Change.* Cambridge University Press, Cambridge, pp. 47–65.

Wootton, T.W. and Downing, A. (2003) Understanding the effects of reduced biodiversity: a comparison of two approaches. In: Kareiva, P. and Levin, S.A. (eds) *The Importance of Species: Perspectives on Expendability and Triage.* Princeton University Press, Princeton, New Jersey.

Yachi, S. and Loreau, M. (1999) Biodiversity and ecosystem productivity in a fluctuating environment: the insurance hypothesis. *Proceedings of the National Academy of Sciences USA* 96, 57–64.

14 Genomic diversity in nature and domestication

Eviatar Nevo

Institute of Evolution, University of Haifa, Mt Carmel, Haifa, Israel

Problems of Genomic Diversity in Nature and Domestication

Genomic diversity is the basis of evolutionary change in nature underlying the evolution of biodiversity at the level of genes, genomes, populations, species and ecosystems, and the evolution of genome–phenome diversity under environmental stress (Nevo, 2001a,b). It is the prime mover of molecular-genotypic and organismal-phenotypic evolution. Genetic diversity and selection drive the twin evolutionary processes of adaptation and speciation and the fast human evolutionary processes of domestication, agriculture and medicine.

Nevertheless, despite the cardinal role of genomic diversity in the evolutionary processes in nature and domestication, its evolutionary origin, dynamics and maintenance remain largely enigmatic, notwithstanding the dramatic discoveries of molecular biology of abundant diversity in nature. For example, what is the relative importance of deterministic (natural selection) and partly stochastic (mutation, migration, recombination, interspecific lateral gene transfer and random events) forces in driving the patterns, levels and dynamics of genomic diversity of both coding and non-coding regions, but especially of the latter? How much is genetic differentiation in nature and domestication determined by the mating system of selfing and outcrossing, hitch-hiking, ecological stresses, adaptive origin and maintenance, linkage disequilibria or epistasis? How much genomic change occurred during domestication evolution? Is genetic erosion threatening cultivars, their genetic diversity, the future of agriculture and the stable supply of food to a constantly increasing human population? If so, how should the genetic resources of the progenitors of cultivated plants be protected and utilized in crop improvement?

Evidence and Theory of Genetic Diversity in Nature

Substantial evidence derived from the analysis of protein and DNA genomic diversity and divergence in nature locally, regionally and globally indicate the following (Nevo, 2001a,b): (i) an abundant genotypic and phenotypic diversity exists in nature; (ii) the organization and evolution of molecular-genetic and organismal diversity in nature at all geographic scales are non-random and structured; (iii) genetic and genomic diversity display regularities across life and are positively correlated with, and partly predictable by, abiotic and biotic environmental heterogeneity and stress; and (iv) biodiversity evolution, even in small isolated populations, is primarily driven by natural selection including diversifying, balancing, cyclical and purifying selective regimes, interacting with, but ultimately overriding, the effects of mutation, migration and stochasticity (Nevo, 1978, 1988, 1998a,b, 2001a,b; Nevo *et al.*, 1984, 2002).

Evolution under Domestication

How does the aforementioned evidence and theory relate to one of the most dramatic and rapid evolutionary phenomena in nature, the domestication of plants and animals by humans, which is considered the most important human development in the past 13,000 years (Diamond, 2002)? How do agricultural practices create new selection pressures, in numerous traits, against those favoured by natural selection? How do cultivars and domesticates introduced by humans adapt to new environmental conditions? Diversity in characters associated with crop plant and animal domestication evolution is largely absent in the wild progenitors and much of it evolved under domestication, either unconsciously or consciously (Ladizinsky, 1998; Peng *et al.*, 2003). What is the taxonomic status of cultivated plants vis-à-vis their wild progenitors? Subspecies? New species? Did humans cause rapid speciation hundreds or thousands of years ago?

Domestication, as already noted by Darwin in his first chapter of 'The Origin' (1859), is mankind's most dramatic imitation of the evolutionary process of speciation. Darwin was so impressed by animal and plant domesticates that he regarded the cultivated derivatives more variable than their progenitors. Cultivated varieties for Darwin tend to be converted into new and distinct species. If so, human domestication of wild plants and animals during the last ten millennia is doubtlessly one of the best demonstrations of fast evolution in action. It is a process that has been condensed in a short time and advanced by *artificial* rather than *natural* selection. Luckily, the original plant and animal progenitors of many cultivars and domesticates still live in their natural environment and their local adaptations and genetic diversity resources are extractable for crop improvement. Plant and animal domestication revolutionized human cultural evolution and is the major factor underlying human civilization, but most emphatically, domestication is an extraordinary demonstration of active evolution of differential stages of speciation and adaptation to the human-created agricultural niche. So, if the current presumed predicament of domestication due to genetic erosion is verified by critical analysis, protecting this precious base of civilization becomes an urgent concern. This is particularly true in view of the rapid disappearance of centres of diversity because of human developments (Harlan, 1975a,b). It should be mentioned, however, that studies exist (e.g. Reeves *et al.*, 1999) indicating no quantitative changes in genetic diversity of UK cereal crops over the past 60–70 years.

Origin of Agriculture

A post-Pleistocene global rise in temperature following the ice age (i.e. climatic change) may have induced the expansion of economically important thermophilous plants and, in turn, promoted complex foraging and plant cultivation. The shift from foraging to steady production led to an incipient agriculture varying in time in various parts of the world. Vavilov (1926) proposed eight centres of origin for most world-cultivated plants (Fig. 1 in Harlan, 1975a; see also Fig. 2 in Diamond, 2002). Agricultural origins (following Harlan, 1975a) consist of centres and non-centres including the Near East centre, African non-centre, North Chinese centre, South-east Asia and South Pacific non-centres, Meso-American centre, and South American non-centre (Fig. 2 in Harlan, 1975a). In the Levant, agriculture developed out of an intensive specialized exploitation of plants and animals (Zohary, 1996). Natufian sedentism, followed by rapid population growth and resource stress induced by the expanding desert and coupled with available grinding technology, may have triggered plant domestication (Bar-Yosef, 1998).

The earliest signs of domestication found so far in the archaeological records (Smith, 1995; Bar-Yosef, 1998; Ladizinsky, 1998; Badr *et al.*, 2000; Lev-Yadun *et al.*, 2000; Zohary and Hopf, 2000; Diamond, 2002; Gopher *et al.*, 2002; Kislev, 2002; Salamini *et al.*, 2002) appear 10,000–12,000 years ago in the Fertile Crescent of the Near East, Central America and southern China, involving different crops and independent cradles of domestication. Each was founded on cereal staples whose reliability, yield and suitability for storage may have been crucial for early commitment to an agricultural way of life and the domestication of other plants. It is generally believed that plant domestication first took place in the Jordan Valley and in adjacent areas of the southern Levant (present-day Israel and Jordan). However, botanical, genetic and archaeological evidence point to a small core area within the Fertile Crescent – near the upper reaches of the Tigris and Euphrates Rivers, in present-day south-eastern Turkey/northern Syria – as the cradle of Old World agriculture (Bar-Yosef, 1998; Pasternak, 1998; Badr *et al.*, 2000; Lev-Yadun *et al.*, 2000; Zohary and Hopf, 2000; Salamini *et al.*, 2002). Further evidence is needed to clarify when and where wheat and barley domestication and agriculture originated, driving modern civilization. Was it mono- or polyphyletic in origin (Zohary, 1999)? For date estimates on microsatellite genetic analysis, which overlap the archaeological findings, see Ramachandran *et al.* (2005).

The genetic changes required for domestication to occur were relatively straightforward and rapid, most of them arising from direct and indirect selection for higher yield and quality traits. In domestication evolution, the relative importance of the evolutionary driving forces including hybridization, migration, drift and selection can best be assessed with a major emphasis on selection and hybridization. Considerable progress has been achieved in the field of the wild ancestry of Old World crops including wild cereals (Zohary and Hopf, 2000). The wild progenitors of most of these cultivated plants have now been satisfactorily identified by comparative morphology and genetic analysis. The

distribution and ecological ranges of the wild relatives have been established. Furthermore, comparisons between wild types and their cultivated counterparts have revealed the evolutionary changes brought about by domestication (Peng *et al.*, 2000, 2003) and taxonomic status of domesticates (Styles, 1986). For example, research in this field has enabled us to assess the relative importance of the evolutionary forces driving wheat evolution, hybridization, migration, drift and natural selection, interacting in generating the final wheat genotype. Our studies suggest that besides hybridization, which contributed to the tetraploidy of *Triticum dicoccoides*, natural selection played a significant role, and it has oriented wheat evolution primarily through the mechanisms of diversifying and balancing selection regimes (Nevo *et al.*, 2002).

The evolution of domestication can be considered in four contexts:

1. The evolution of new species of crop plants by humans through strong artificial selection leading to adaptive complexes fitting human demands. This provides fascinating insight into the evolutionary process (e.g. Hilton and Baut, 1998; Peng *et al.*, 2000, 2003).
2. The evolution of human civilization and the current consequences of population explosion and increasing world hunger in developing countries of the southern continents (Diamond, 2002).
3. How much genetic change is involved in domestication (e.g. Lauter and Doebly, 2002; Peng *et al.*, 2003)?
4. Did domestication involve a decline in genetic diversity threatening their future evolution?

This chapter primarily focuses on point 4.

Darwin's View on Variation under Domestication

Darwin first realized that evolutionary change is a two-step process: (i) the production of variation; and (ii) the sorting out of this variability by natural selection. This complementary of variation and selection is the essence of the evolutionary process both in nature and domestication. Darwin observed that the abundant variability in nature resulted not from major saltations but from the accumulation of small changes occurring at random with respect to environmental conditions. For Darwin, however, natural selection was the direction-giving force leading to adaptations. Moreover, Darwin claimed that a

> careful study of domesticated animals and cultivated plants would offer the best chance ... to gain a clear insight into the means of modification and co-adaptation ... our knowledge, imperfect though it be, of variation under domestication, afforded the best and safest clue to understanding the co-adaptations for organic beings to each other and to their physical conditions of life... Therefore, ... I shall devote the first chapter of the Origin to Variation Under Domestication... We shall see that a large amount of hereditary modification is at least possible and how great the power of man in accumulating by his selection successive slight variations. (Darwin, 1859, Introduction, phrases rearranged)

For Darwin, the

> vast diversity of the plants and animals which have been cultivated and which have varied during all ages under the most different climates and treatments are simply due to our domestic productions having been raised under conditions of life not so uniform as, and somewhat different from, those to which the parent species have been exposed under nature ... we may safely conclude that very many of the most strongly marked domestic varieties could not possibly live in a wild state. (Darwin, 1859)

Whereas the astounding varieties of cultivars and domesticates generated by humans across the globe are evident, the environmentally limited area of each cultivar and the vulnerability to biotic and abiotic stresses are notorious, and their turnover including extinction is extraordinary (Hutchinson and Weiss, 1999). Only under the favourable conditions of human agriculture of fertilizers, cultivation, spraying, dusting and so on can they survive under human-created artificial environments. Left alone, they will most probably competitively perish by local superiorly adapted wild species.

Our genetic insights into the structure and genomic architecture of the cultivars

compared with that of their wild progenitors clearly indicate the superiority under natural conditions of the latter over the former (Yamashita, 1965).

Domestication Evolution and its Predicaments

Domestication assembles from nature wild animals and plants and keeps them under cultivation. Domesticated quantitative trait loci defining the genetic changes accompanying utilization of plants by humans have been studied in rice, maize, sorghum, millet and wheat (Peng *et al.*, 2003, and references therein). Domesticates have been genetically changed by human selection either consciously or unconsciously. In general, domesticates lose some of their fitness components (reproductive and dispersal power as well as abiotic and biotic resistances to stress) adapting them to their natural habitats; hence, they largely cannot survive outside the agriculturally protected environment (see below).

Domesticates grow outside the natural distribution range of their progenitors and have changed morphologically, physiologically and phenologically, with a genetic change establishing a close and evolving symbiosis between humans and their domesticates in which humans protect and disperse the domesticates, and they provide us with food and other resources (Ladizinsky, 1998). Domestication is an evolutionary process in which new types are constantly being selected to meet human ecology, economy and culture, thereby distancing them genotypically and phenotypically from their wild progenitors and making them human-dependent, surviving only under cultivation.

Genetic Erosion

Humans protect crops (e.g. weed, pest and pathogen control) through cultivation, including ploughing, fertilizing, irrigating, spraying, and providing a more uniform environment for growth. Consequently, crops differentially and gradually lose, throughout domestication evolution, their natural genetic resistance against ecological abiotic and biotic stresses. This is reflected in their increasing vulnerability, susceptibility and loss of genetic diversity. Domestication resulted in the dramatic impoverishment of the gene pools of cultivars (Frankel and Bennett, 1970; Frankel and Hawkes, 1975; Harlan, 1975a,b; Frankel and Soule, 1981; Hawkes, 1983, 1991; IBPGR, 1985; Plucknett et al., 1987; Brown et al., 1989, 1990; Holden et al., 1993; Frankel et al., 1995; ICPGR, 1995; Gaut et al., 1997; Maxted et al., 1997; Valdes et al., 1997; Jones et al., 1998; Nevo, 1998b; Poncet et al., 1998; Hammer et al., 1999; Hoisington et al., 1999; Xu et al., 2000; Hernandez-Verdugo et al., 2001; Ellis et al., 2003). (See also Brush (1999) for a critical evaluation of the concept of genetic erosion of cultivars.) In contrast, wild plants and wild progenitors and relatives of crops have been under constant ecological stresses, both abiotic (climatic such as solar radiation, temperature, drought) and biotic (pathogens, pests, competitors). They have evolved throughout millions of years of evolution adaptive genetic strategies for fitness and survival. Moreover, generally, they are richer in genetic diversity of proteins and DNA than their cultivated derivatives (see below).

Therefore, wild relatives of crop plants constitute invaluable gene resources for crop improvement and sustain agricultural productivity (Vavilov, 1951; Nevo et al., 1979, 1986, 2002; Plucknett et al., 1987; Nevo, 1992; Tanksley and McCouch, 1997). The next agricultural revolution in the 21st century for increasing food production to cope with the exploding world population and increasing starvation in the southern continents will be primarily genetic. The best hope for crop improvement resides in the progenitors of cultivated plants that harbour rich genetic resources for tolerance against stressful abiotic and biotic stresses. Likewise, the progenitors harbour rich genetic resources for diverse agronomic traits, photosynthetic yield, high-quality proteins and amino acids, among others. For genetic resources of wild emmer wheat – the progenitor of most cultivated wheats – see Nevo (1983, 1988, 1995a) and Feldman and Sears (1981). It is this rich genetic resource that is so valuable to agriculture and should be tapped in future plant breeding for a variety of agronomically useful traits.

Agricultural Predicament and Plant Genetic Conservation

Human agriculture has revolutionized the planet and human evolution. However, this unmatched change in the planet and human societies has had the dramatic cost of environmental deterioration and loss of biodiversity through extinction (Nevo, 1995b), including the genetic erosion of domesticates. In an attempt to save the planet and humankind from the predicament of annihilation, massive directed conservation of biodiversity and genetic diversity is imperative.

Conservation of biodiversity in nature became more intensive during recent decades, in an attempt to alleviate the pending extinction of the biosphere by humans (Soule, 1986, 1987; Ehrlich and Wilson, 1991; Wilson, 1992, 2002; Fiedler and Jain, 1992). The salvation of biodiversity is very relevant to crop plants and their progenitors. Domestication resulted in a dramatic impoverishment of the gene pools of cultivars (see references above). Feeding a still growing world population (with >90% in developing countries) will require an astonishing increase in food production. Forecasts call for wheat to become the most important world cereal with maize close behind. Together these crops will account for 80% of developing countries' cereal import requirements. Access to a range of genetic diversity is critical to the success of a breeding programme. The global effort to assemble and use the genetic resources is enormous and genetic diversity in nature and gene banks is critical in the world's fight against hunger (Hoisington et al., 1999).

The accumulated information on biodiversity provides partial answers to the following questions:

1. How much and what types of genetic change were involved in domestication?
2. How fast was this change?
3. Can the crops be regarded as new species created by humans?
4. Could they survive without the ecological niche created by humans?

5. Can they be further improved to reinforce their qualitative and quantitative agricultural traits?

The following section focuses primarily on the first question in relation to genetic erosion of wheat and barley during domestication, based on genetic studies conducted thus far. In particular, I will give a brief overview of the research programme on wild progenitors of crop plants, wild cereals and wild lettuce conducted at the Institute of Evolution. This programme focuses on genetic diversity in nature and the decline under domestication, providing detailed evidence of genetic erosion in these crops and some others. This may bear directly on the available wild genetic resources that can be utilized to increase diversity and productivity of their cultivars.

Multidisciplinary Research Programme of Wild Cereals for Crop Improvement at the Institute of Evolution, University of Haifa, Israel

The following review leans heavily on a long-term multidisciplinary research programme of wild emmer wheat, *T. dicoccoides* (Nevo, 1986, 1989, 2001c; Nevo *et al.*, 2002), and wild barley, *Hordeum spontaneum* (Harlan and Zohary, 1966; Nevo, 1992), that has been conducted at the Institute of Evolution, University of Haifa, Israel, since 1975. The project is aimed at studying the ecological-genetic, genomic and domestication evolution of wild cereals, the progenitors of wheat and barley, across the Near East Fertile Crescent centre of origin and diversity. The multidisciplinary programme has focused on the patterning and maintenance of genetic diversity and divergence, genome organization and agronomically important traits in natural populations, attempting to highlight their adaptive evolutionary processes and potential genetic resources for crop improvement (Fig. 14.1). The programme has involved macrogeographic population genetics and ecology studies in the Near East Fertile Crescent in several countries: Israel, Turkey, Iran and Jordan (including 24 other countries added to our recent wild *Lactuca*

research programme); 675 natural populations and 15,500 genotypes have been collected in central, peripheral and marginal populations plus several Israeli microsites, which have been partly analysed genetically and agronomically (Table 14.1). In all, the programme has revealed a broad view of wild barley and wild emmer in a substantial part of the Fertile Crescent (Nevo *et al.*, 1986, 2002; Nevo and Beiles, 1989; Nevo, 1992). We have also conducted a series of microgeographic population genetics and ecology studies on allozymes and DNA polymorphisms of wild cereals in Israel. In addition, we have studied agronomic traits, disease resistance polymorphisms (Moseman *et al.*, 1984; Nevo *et al.*, 1985; Fetch *et al.*, 2000, 2003a,b; see list on wild cereals at http://research.haifa.ac.il~evolut), protein content, amino acid composition, herbicide resistance, photosynthetic yield, and drought and salt tolerance. A methodology has been developed, based on ecological factors and allozyme markers, for assessing wild cereal genetic resources, enabling us to screen single- and multiple-trait elite genotypes in nature (Nevo, 1987).

In our detailed study of domestication evolution in wheat, we studied wild emmer wheat, *T. dicoccoides*, which is the progenitor of modern tetraploid and hexaploid cultivated wheats (Nevo *et al.*, 2002). Our objective was to map domestication-related quantitative trait loci (QTL) in *T. dicoccoides* (Peng *et al.*, 2000, 2003). The studied traits included brittle rachis, heading date, plant height, grain size, yield and yield components. Our mapping population was derived from a cross between *T. dicoccoides* and *Triticum durum*. Approximately 70 domestication QTL effects were detected, non-randomly distributed among and along chromosomes. Seven domestication syndrome factors were proposed, each affecting 5–11 traits. We showed: (i) clustering and strong effects of some QTLs; (ii) remarkable genomic association of strong domestication-related QTLs with gene-rich regions; and (iii) unexpected predominance of QTL effects in the A genome. The A genome of wheat may have played a more important role than the B genome during domestica-

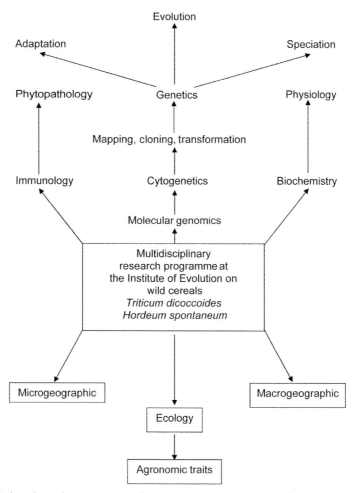

Fig. 14.1. Multidisciplinary long-term research programme of wild cereals at the Institute of Evolution, University of Haifa, Israel.

tion evolution. The cryptic beneficial alleles at specific QTL derived from *T. dicoccoides* may contribute to wheat and cereal improvement (Peng *et al.*, 2000, 2003, and Fig. 14.2). We conducted similar studies in wild barley, *H. spontaneum* (Guoxiong *et al.*, 2002).

Since 1994, we also embarked upon a world collection and analysis of wild lettuce species, primarily the progenitor of lettuce, *Lactuca serriola*, in collaboration with Richard Michelmore of Davis, California. Our current collection of wild lettuce includes 252 populations and 2876 genotypes from 21 countries (Sicard *et al.*, 1999; Kuang *et al.*, 2004a,b).

Genetic Resources of Wild Barley in the Near East: Structure, Evolution and Comparison with Cultivated Barley and Breeding

Genetic diversity and structure of populations of the wild progenitor of barley, *H. spontaneum*, from three countries (Israel, Turkey and Iran) in the Near East Fertile Crescent were compared and contrasted (Nevo *et al.*, 1986). The analysis was based on allozymic variation in proteins encoded by 27 shared loci in 2125 individuals representing 52 populations of wild barley. The results indicated

Table 14.1. Gene bank collections at the Institute of Evolution, University of Haifa, Israel, used for genetic and agronomic research programmes.

Species	Countries	Populations	Genotypes
Hordeum spontaneum (wild barley)	5	133[a]	3,393
Triticum dicoccoides (wild emmer wheat)	6[b]	32[c]	1,650
Aegilops species	3	220	6,600
Lactuca species (wild lettuce)	21	252	2,876
Avena (wild hexaploid oats)	1	38	985
Total	36 (24)[d]	675	15,504

[a]Including three populations of Tabigha, four populations of Newe Yaar and seven populations of 'Evolution Canyon' (see Nevo, 2001b).
[b]Including Iran, Syria and Iraq, which are represented by a few genotypes each.
[c]Excluding Iran, Syria and Iraq.
[d]Countries involved in the overall count (36) and the actual number of countries considering overlap of species (24).

that: (i) *H. spontaneum* in the Near East Fertile Crescent is very variable genetically; (ii) genetic differentiation of populations includes some clinal but primarily regional and local patterns often displaying sharp geographic differentiation over short distances; (iii) the average relative genetic differentiation (G_{st}) was 54% within populations, 39% among populations and 8% among the three countries; (iv) allele distribution is characterized by a high proportion of unique alleles (51%) and a high proportion of common alleles that are distributed either locally or sporadically; (v) discriminant analysis by allele frequencies successfully clustered wild barley of each of the three countries (96% correct classification); (vi) a substantial portion of the patterns of allozyme variation in the wild gene pool is significantly correlated with the environment and is predictable ecologically, chiefly by a combination of humidity and temperature variables; (vii) natural populations of wild barley are, on average, more variable than two composite crosses and landraces of cultivated barley (Fig. 14.2). The spatial patterns and environmental correlates and predictors of genetic variation of *H. spontaneum* in the Fertile Crescent indicate that genetic variation in wild barley populations is not only rich, but at least partly adaptive and predictable by ecology and allozyme markers. Consequently, conservation and utilization programmes should optimize sampling strategies by following ecological-genetic fac-

tors and molecular markers as effective predictive guidelines.

An intriguing question is: what is the pattern of DNA polymorphisms of *H. spontaneum*? Are they neutral as widely presumed, or do they display, like allozymes, ecological correlates and are they also subjected to natural selection? We elucidated this issue by studying microsatellite, amplified fragment length polymorphisms (AFLP) and ribosomal DNA (rDNA) diversities in wild cereals. To answer this question we analysed microsatellite molecular markers, also known as simple sequence repeats (SSRs). SSRs are short (1–6 bp), tandemly repeated DNA sequences and are highly polymorphic as a result of frequent variation in the number of times the core sequence is repeated (Li *et al.*, 1999).

Ecological-genomic Diversity of Microsatellites in Wild Barley, *H. spontaneum*, Populations in Israel and Jordan

Israel

Microsatellite diversity at 18 loci was analysed in 94 individual plants of ten wild barley, *H. spontaneum* (C. Koch) Thell., populations sampled from Israel across a southward transect of increasing aridity (Turpeinen *et al.*, 2001). Allelic distribution in populations was non-

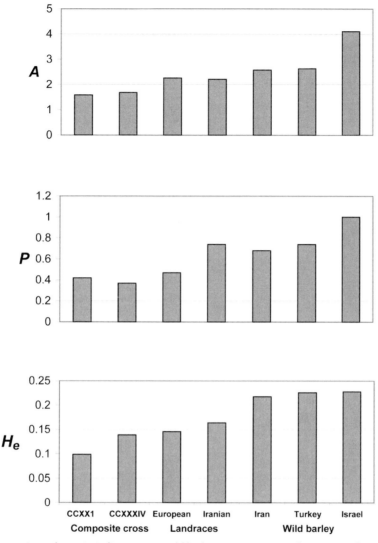

Fig. 14.2. Comparison of genetic indices among wild barley, *H. spontaneum* (from Iran, Turkey and Israel), and landraces (from Iran and Europe) and composite crosses of *H. vulgare*, based on 19 shared loci. Abbreviations: A = average number of alleles per locus; P = proportion of polymorphic loci per population; H_e = genic diversity. Twenty-nine accessions from the collection of Landbrugets Plantekulturs Sortssamling, Denmark, that includes European landraces, plus one from India and two from Egypt.

random. Estimates of mean gene diversity were highest in stressful arid-hot environments. Sixty-four per cent of the genetic variation was partitioned within populations and 36% between populations. Associations between ecogeographical variables and gene diversity, H_e, were established for nine microsatellite loci. By employing principal component analysis, we reduced the number of ecogeo-graphical variables to three principal components: water factors, temperature and geography. At three loci, stepwise multiple regression analysis significantly explained the gene diversity by a single principal component (water factors). Based on these observations, it was suggested that simple sequence repeats are at least partly subjected to natural selection. (For adaptive random amplified polymorphic DNA

(RAPD) variation in *H. spontaneum*, see also Volis *et al.*, 2001.) Similar patterns of *H. spontaneum*, in Israel, were obtained by AFLP analysis (Turpeinen *et al.*, 2003).

Jordan

We also analysed the ecological-genomic diversity of microsatellites of wild barley, *H. spontaneum*, at 18 loci in 306 individuals of 16 populations from Jordan across a southward transect of increasing aridity (Baek *et al.*, 2003). The 18 microsatellites revealed a total of 249 alleles, with an average of 13.8 alleles per locus (range 3–29), with non-random distribution. The proportion of polymorphic loci per population averaged 0.91 (range 0.83–1.00); gene diversity, H_e, averaged 0.512 (range 0.38–0.651). We compared the number of alleles of the 18 loci with those found in Israel populations by Turpeinen *et al.* (2001). Out of the 280 alleles, 138 (49.3%) were unique (i.e. occurred in only one of the countries). The percentage of unique alleles in Jordan and Israel populations was 43.0% and 17.9%, respectively, suggesting that Jordan together with Israel is an important centre of origin and diversity of wild barley (see also the 400 AFLP polymorphic loci analysis of Badr *et al.*, 2000). Estimates of mean gene diversity were highest in the populations collected near the Golan Heights, such as Shuni North, Shuni South and Jarash. Sixty-nine per cent of the microsatellite variation was partitioned within populations and 31% between populations. Associations between ecogeographical variables and gene diversity values were established for eight microsatellite loci. Remarkably, the cluster produced by SSR data of wild barley coincides with the result of the dendrogram of the *Spalax ehrenbergi* superspecies of blind subterranean mole rats in Jordan based on allozyme gene loci. The major soil type in the wild barley habitat of each ecological group was different. Stepwise multiple regression analysis indicated that the variance of gene diversity was explained by altitude (i.e. temperature; $R^2 = 0.362$, $P < 0.01$). These observations suggest that microsatellites are at least partly adaptive and subject to natural selection.

Near East Fertile Crescent

AFLP diversity in *H. spontaneum* in the Fertile Crescent was based on 39 genotypes, 268 AFLP loci, of which 204 proved to be polymorphic. Genotypes were grouped according to the area of origin. Shoot Na^+ content and carbon isotope (δ^{13}) reflecting drought resistance was associated with site of origin ecogeographic data (Pakniyat *et al.*, 1997).

Genetic Diversity of Wild Emmer Wheat in Israel and Turkey

Structure, evolution and application in breeding

Allozyme variation in the tetraploid wild emmer wheat, *T. dicoccoides*, the progenitor of all cultivated wheats, was studied for proteins encoded by 42 gene loci in 1815 plants representing 37 populations – 33 from Israel and four from Turkey – sampled in 33 localities from 1979 to 1987 (Nevo and Beiles, 1989). The results showed the following:

1. Six loci (14%) were monomorphic in all populations, 15 loci (36%) were locally polymorphic and 21 loci (50%) were regionally polymorphic. These results are similar to those obtained earlier on 12 Israeli populations (Nevo *et al.*, 1982). All polymorphic loci (except four) displayed high local levels of polymorphism ($\geqslant 10\%$).

2. The mean number of alleles per locus, A, was 1.252 (range: 1.050–1.634); the proportion of polymorphic loci per population averaged 0.220 (range: 0.050–0.415); gene diversity, H_e, averaged 0.059 (range: 0.002–0.119).

3. Altogether there were 119 alleles at the 42 putative loci tested, 114 (96%) of these in Israel.

4. Genetic differentiation was primarily regional and local, not clinal; 70% of the variant alleles were common ($\geqslant 10\%$) and not widespread, but rather localized or sporadic, displaying an 'archipelago' population genetics and ecology structure. The coefficients of genetic distance between populations were high and averaged $D = 0.134$ (range: 0.018–0.297), an indication of sharp genetic differentiation over short distances.

5. Discriminant analyses differentiated Israel into central and three marginal regions as well as those with different soil-type populations (Fig. 14.3).

6. Allozymic variation was 40% within and 60% between populations.

7. Gametic-phase disequilibria were abundant, their number being positively correlated ($r_s = 0.60$, $P<0.01$) with the humidity.

8. Multilocus organization (Brown *et al.*, 1980) was substantive and also positively correlated with humidity.

9. Allozyme diversity, overall and at single loci, was significantly correlated with, and partly predictable by, climatic and edaphic factors.

10. The distribution of the significant positive and negative values and the absence of autocorrelations in the correlogram revealed no similar geographic patterns across loci, eliminating migration as a prime factor of population genetic differentiation.

These results suggest that: (i) during the evolutionary history of wild emmer, diversifying natural selection, through climatic and edaphic factors, was a major agent of genetic structure and differentiation at both the single and multilocus levels; and (ii) wild emmer harbours large amounts of genetic diversity exploitable as genetic markers in sampling and abundant adaptive genetic resources utilizable for wheat improvement. In addition, we conducted DNA studies (RAPD and SSR) on wild emmer wheat (Fahima *et al.*, 1999, 2002).

RAPD Polymorphism of Wild Emmer Wheat Populations, *T. dicoccoides*, in Israel

Genetic diversity in RAPDs was studied in 110 genotypes of the tetraploid wild progenitor of wheat, *T. dicoccoides*, from 11 populations sampled in Israel and Turkey (Fahima *et al.*, 1999). Our results show a high level of diversity of RAPD markers in wild wheat populations in Israel. The ten primers used in this study amplified 59 scoreable RAPD loci of which 48 (81.4%) were polymorphic and 11 monomorphic. RAPD analysis was found to be highly effective in distinguishing genotypes of *T. dicoccoides* originating from diverse ecogeographical sites in Israel and Turkey, with 95.5% of the 100 genotypes correctly classified into sites of origin by discriminant analysis based on RAPD genotyping (Fig. 14.4a). However, interpopulation genetic distances showed no association with geographic distance between the population sites of origin, negating a simple isolation by distance model. Spatial autocorrelation of RAPD frequencies suggests that migration is not influential. Our present RAPD results are non-random and in agreement with the previously obtained allozyme patterns (Nevo and Beiles, 1989) although the genetic diversity values obtained with RAPDs are much higher than the allozyme values. Significant correlates of RAPD markers with various climatic and soil factors suggest that, as in the case of allozymes, natural selection causes adaptive RAPD ecogeographical differentiation, also in non-coding sequences. The results obtained suggest that RAPD markers are useful for the estimation of genetic diversity in wild material of *T. dicoccoides* and the identification of suitable parents for the development of mapping populations for tagging agronomically important traits derived from *T. dicoccoides*. Similar differentiation in *T. dicoccoides*, in Israel, was achieved by microsatellite analysis (Fahima *et al.*, 1999; Fig. 14.4b).

Microgeographic Critical Tests in Nature: Evaluation of Levels and Divergence of Genetic Diversity

Microsite ecological contrasts are excellent critical tests for evaluating the dynamics of genome and phenome evolution and assessing the relative importance for adaptation and speciation of the evolutionary forces causing differentiation (Nevo, 2001b). The latter involve mutation (in the broadest sense, including recombination), migration, chance and selection. At a microsite, mutation, which is usually considered a clockwise neutral process, is expected to be similar across the microsite. Migration, which operates for any organism at the microsite, even

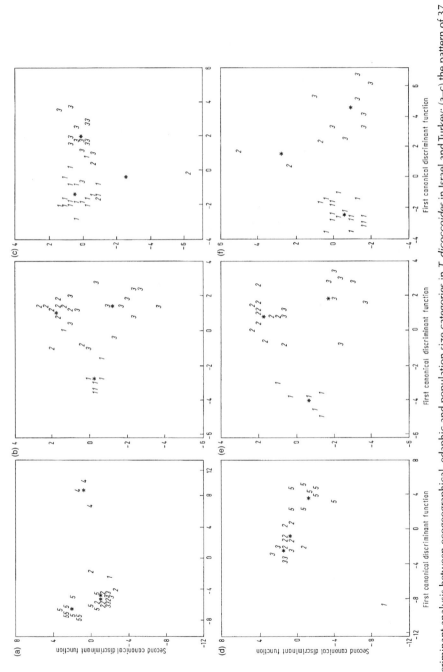

Fig. 14.3. Discriminant analysis between ecogeographical, edaphic and population size categories in *T. dicoccoides* in Israel and Turkey; (a–c) the pattern of 37 populations in Israel and Turkey; (d–f) the pattern of 29 populations from Israel (excluding the four micropopulations of Tabigha). (a) and (d) discriminate between four types of marginal and one type of central populations: *1* = cold steppe; *2* = west margin; *3* = east and south margin; *4* = Turkey; *5* = central population; * = centroid. (b) and (e) discriminate between small-, medium- and large-size populations: population size: *1* = small; *2* = medium; *3* = large; * = group centroid. (c) and (f) discriminate between populations growing on three soil types: *1* = terra rossa; *2* = rendzina; and *3* = basalt; * = group centroid (from Nevo and Beiles, 1989).

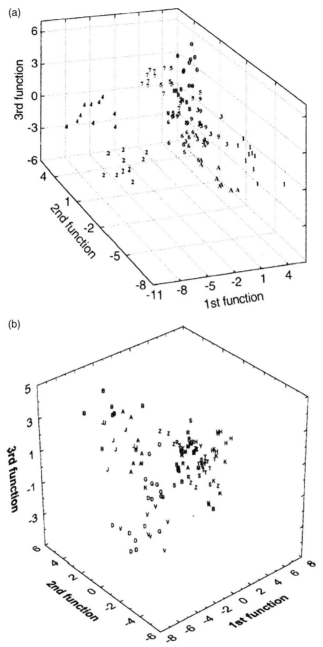

Fig. 14.4. (a) Plot of canonical discriminant functions 1, 2 and 3 based on the 48 polymorphic RAPD loci of *T. dicoccoides*. Symbols: *1* = Mt Hermon; *2* = Gamla; *3* = Rosh Pinna; *4* = Tabigha; *5* = Mt Gilboa; *6* = Mt Gerizim; *7* = Gitit; *8* = Kokhav Hashahar; *9* = Bat-Shelomo; *0* = Givat-Koach; *A* = Diyarbakir. The proportion of the eigen values of functions 1, 2 and 3 from the sum of the eigen values of the ten discriminant functions are 36.9%, 16% and 12%, respectively (Fahima *et al.*, 1999). (b) Plot of canonical discriminant functions 1, 2 and 3 based on ten polymorphic microsatellite loci of 15 populations of wild emmer wheat, *T. dicoccoides*, from Israel and Turkey. Symbols: *H* = Hermon; *Y* = Yehudiyya; *M* = Gamla; *R* = Rosh Pinna; *T* = Tabigha; *B* = Mt Gilboa; *Z* = Mt Gerizim; *G* = Gitit; *K* = Kokhav Hashahar; *J* = J'aba; *A* = Amirim; *B* = Bet-Oren; *S* = Bat-Shelomo; *V* = Givat-Koach; *D* = Diyarbakir (Fahima *et al.*, 2002).

in sessile organisms, is expected to homogenize allele frequencies. Stochasticity is not expected to result in repetitive, ecologically correlated patterns. Selection seems to be the only evolutionary force expected to result in repeated ecologically correlated and parallel patterns and contrasts.

In 1977, at the Institute of Evolution, we embarked on a series of microsite studies comparing diverse, sharply contrasting ecological alternative patterns including temperatures, aridity index, lithology, soil types, topography, and chemical and organic pollutants in marine organisms. The aforementioned studies demonstrated differential viability of allozyme and DNA genotypes where diversity and divergence were selected at a microscale or under critical empirically contrasting conditions and ecologies. Will the non-coding genome also display ecological correlates at regional and local scales? The answer is, emphatically, yes for outbreeding mammals (Nevo, 2001b) and inbreeding wild cereals (reviewed here).

Microscale Molecular Population Genetics of Wild Cereals at Four Israeli Microsites

We used three molecular marker systems that included allozymes, RAPDs and microsatellites (SSRs) to detect molecular diversity and divergence in three populations of wild emmer wheat (*T. dicoccoides*). These populations were from Ammiad, Tabigha and Yehudiyya microsites in northern Israel, and these microsites displayed topographic, edaphic and climatic ecological contrasts, respectively (references 29–33 in Nevo, 2001b). Likewise, we examined molecular diversity with allozymes, RAPD and SSR markers in wild barley, *H. spontaneum*, in the Tabigha microsite north of the Sea of Galilee and in Newe Yaar, Lower Galilee (Gupta *et al.*, 2002); the latter microsite represented a mosaic of microniches of sun, shade, rock, deep soil and their combinations. The three marker systems represented protein-coding (allozyme) and DNA noncoding and coding (most of RAPDs) regions plus short (most SSRs) and long (rDNAs)

repetitive DNA elements, hence providing comprehensive coverage of the wild wheat and barley genomes. At each microsite, we identified non-random divergence of allozyme, RAPD and SSR diversities. Extensive studies examined allozyme, RAPD and SSR and rDNA interslope diversity of *H. spontaneum* in 'Evolution Canyon' I in Mt Carmel, Israel, where a mesic 'European' slope faces the xeric 'African' slope at an average distance of 200 metres. Significant niche-specific (high frequency in niche type) and niche-unique (limited to a niche type) alleles and linkage disequilibria abounded, in all microsites in both wild cereals, barley and wheat, allowing classification into niches of either coded or non-coded markers (references 29–33 in Nevo, 2001b, and additional studies in 2002 and 2003; see our website http://research.haifa.ac.il/~evolut/). For adaptive RAPD variation in *H. spontaneum*, see also Volis *et al.* (2001, 2002).

In wild emmer wheat at Ammiad, the three marker systems used in wild wheat showed dramatically different levels of gene diversity (H_e) and genetic distance (D): SSR > RAPD > allozymes. The gene differentiation (G_{st}) order was allozymes > SSR > RAPD. Remarkably, the three marker systems revealed similar trends of diversity and divergence. All three molecular markers displayed non-random allele distributions, habitat-specific and habitat-unique alleles, and linkage disequilibria. The subpopulations in the drier habitats showed higher genetic diversities in the three marker systems. The genetic distances among the four subpopulations tended to increase with the difference in soil moisture after the early rain of the growing season. These results may suggest that ecological selection, probably through aridity stress, acts both on structural protein coding and on presumably partially regulatory noncoding DNA regions (SSR, RAPD and AFLP) resulting in microscale adaptive patterns. Similar microscale molecular (allozymes, RAPDs and SSRs) divergence was found in three populations of wild barley (*H. spontaneum*) at Tabigha, Newe Yaar (Owuor *et al.*, 2003) and for 'Evolution Canyon' (Owuor *et al.*, 1997). Ribosomal DNA diversity displayed dramatic differences in two microsites associated with soil and microclimatic divergence (Gupta *et al.*, 2002). A dramatic genetic separation in wild emmer wheat occurred in the shady and sunny microniches, separated by only a few metres, as shown in Fig. 14.5.

Genetic Diversity in Wild Barley versus Cultivated Barley

Allozymes

A comparison was made of total allozymic diversity of 19 shared loci of wild barley in Iran, Turkey and Israel with landraces of cultivated barley from Iran and a collection of European landraces. The comparison of two accessions from Egypt and one from India, and two composite crosses is shown in Table 14.2 and Fig. 14.2. Allozymic diversity was lowest in the composite crosses, intermediate in the landraces and highest in the wild germplasm in the order: Israel > Turkey > Iran (Fig. 14.2, Tables 14.2 and 14.3) for an analysis of three overall genetic indices, *A* (allelic diversity), *P* (genetic polymorphism) and H_e (genetic diversity) as well as for the analysis of diversity in each of the 19 gene loci (Table 14.3). These results appear to hold even if the world collection of cultivated barley, which presumably rep-

resents the allelic diversity maintained in the domestic gene pool, is compared. When this comparison is made for the four esterase loci, *Est 1–5*, the differences between the wild and cultivated gene pools in number of alleles are: *Est-1* (8,7); *Est-2* (17,12); *Est-4* (10,4); *Est-5* (6,6). These results emphatically demonstrate the potential role of the wild progenitor as well as that of the landraces as valuable genetic resources for plant breeding, and the depletion of allelic diversity in the cultivated germplasm. Importantly, a comparison between wild barley from Israel and two composite crosses of cultivated barley indicates more multilocus organization in the wild than in the cultivated germplasm (Brown *et al.*, 1980).

Microsatellite Comparisons in Wild and Cultivated Barley

Maroof *et al.* (1994) examined the extent of genetic variation in barley SSRs and studied the evolutionary dynamics of SSR alleles. SSR polymorphisms were resolved by the polymerase chain reaction of four pairs of primers. In total, 71 variants were observed in a sample of 207 accessions of wild and cultivated barley. Analyses of wheat–barley addition lines and barley-doubled haploids identified these vari-

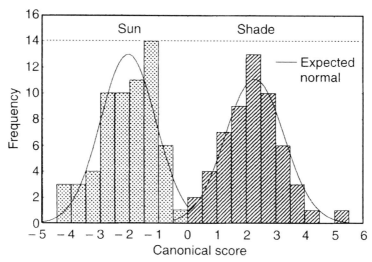

Fig. 14.5. Histogram of frequencies of canonical scores for wild emmer wheat, *T. dicoccoides,* and sunny niches in Yehudiyya according to 25 polymorphic RAPD loci (Li *et al.*,1999).

Table 14.2. Comparison of genetic indices among *H. spontaneum,* landraces and composite crosses of *H. vulgare* based on 19 shared loci (from Nevo *et al.,* 1986).

	N	A	P	H_e
H. vulgare				
Composite cross XXI (F_{17})	30	1.58	0.42	0.099
Composite cross XXXIV (F_4)	30	1.68	0.37	0.139
Landraces				
European[a]	178	2.26	0.47	0.146
Iranian	290	2.21	0.74	0.164
H. spontaneum				
Iran	309	2.58	0.68	0.218
Turkey	437	2.63	0.74	0.226
Israel	1179	4.11	1.00	0.228

[a]Twenty-nine accessions from the collection of Landrugets Plantekulturs Sortssamling, Denmark, which include European landraces, plus one from India and two from Egypt.

Table 14.3. Comparison of numbers of alleles and total diversity (H_t) among *H. spontaneum* and *H. vulgare* (from Nevo *et al.,* 1986).

\multicolumn{6} H. spontaneum (wild barley)						\multicolumn{6} H. vulgare (cultivated barley)						
Israel		Turkey		Iran		Near East		Iranian		European		
H_t	Allele	H_t	Allele	H_t	Allele	H_t	Allele	H_t	Allele	H_t	Allele	Locus
0.86	15	0.72	6	0.61	7	0.83	17	0.38	3	0.28	6	Est-2
0.66	4	0.63	5	0.48	4	0.73	8	0.57	4	0.47	4	Acph-3
0.70	7	0.63	4	0.62	4	0.70	10	0.45	3	0.58	4	Est-4
0.53	6	0.64	5	0.57	4	0.59	6	0.01	2	0.40	4	Est-5
0.54	4	0.48	2	0.29	3	0.47	4	0.50	3	0.01	2	Nadh-1
0.24	3	0.52	3	0.17	2	0.29	3	0.04	2	0.00	1	Adh-1
0.13	6	0.18	3	0.44	4	0.25	8	0.44	3	0.48	3	Est-1
0.03	2	0.28	2	0.30	2	0.17	3	0.00	1	0.00	1	Aat-1
0.13	3	0.17	4	0.22	4	0.16	5	0.19	3	0.00	1	Pgt
0.18	3	0.02	2	0.004	2	0.10	3	0.01	2	0.00	1	Pgm
0.11	3	0.04	2	0.11	3	0.10	5	0.00	1	0.00	1	Adh-2
0.01	2	0.01	2	0.34	2	0.09	3	0.15	2	0.00	1	Cat
0.08	4	0.02	3	0.004	2	0.05	7	0.00	1	0.00	1	Mdh-2
0.07	3	0.00	1	0.00	1	0.04	3	0.05	2	0.00	1	Gdh
0.03	4	0.00	1	0.00	1	0.02	4	0.02	2	0.04	2	Aat-2
0.02	3	0.00	1	0.00	1	0.02	3	0.21	3	0.46	4	Gpgd-2
0.01	2	0.00	1	0.00	1	0.01	2	0.00	1	0.00	1	Acph-1
0.01	3	0.00	1	0.00	1	0.00	3	0.00	1	0.07	6	Gpgd-1
0.01	2	0.002	2	0.00	1	0.00	2	0.11	2	0.00	1	Mdh-1

Twenty-nine accessions from the collection of Landbrugets Plantckulturs Sortssamling, Denmark, which include European landraces plus one Indian and two from Egypt.

ants (alleles) with four loci, each located on a different chromosome. The numbers of alleles detected at a locus corresponded to the number of nucleotide repeats in the microsatellite sequences. The numbers of alleles at two loci were 28 and 37. Three alleles were resolved by each of the other two loci. Allelic diversity was greater in wild than in cultivated barley and in surveys of two generations (F_8 and F_{53}) of *Composite Cross II* (Table 2 in Maroof *et al.*, 1994). An experimental population of cultivated barley showed that few of the alleles present in the 28 parents survived into generation F_{53} whereas some infrequent alleles reached high frequencies (Table 15.3 in Maroof *et al.*, 1994). Such changes in frequency indicate that the chromosomal segments marked by the SSR alleles are under the influence of natural selection. The SSR variants allow specific DNA sequences to be followed through generations. Thus, the great resolving power of SSR assays may provide clues regarding the precise targets of natural and human-directed selection (Maroof *et al.*, 1994). See also the comparison of rDNA in wild and cultivated barley in Maroof *et al.* (1984) and Gupta *et al.* (2002), and in wild and cultivated rice in Liu *et al.* (2002).

Using extensive chromosomally mapped multiallelic microsatellites, the SCRI (Scottish Crop Research Institute, www.scri.sari.ac.uk) showed, likewise, that allelic diversity in *H. spontaneum* far exceeds that of *Hordeum vulgare*. A major bottleneck associated with the domestication of barley (*H. vulgare*) from wild barley (*H. spontaneum*) has been identified (see Fig. 14.6 and other SCRI web figures). Large-scale sequencing projects can now compare and contrast thousands of gene sequences (*Ests*) in wild and cultivated germplasm.

Genetic Diversity of Wild Emmer versus Cultivated Wheat

In the regional wild wheat study (Fahima *et al.*, 2002) using 20 GWM (Gatersleben wheat microsatellites), 363 alleles were revealed among 135 wild wheat accessions, an average of 18 alleles per GWM. In a previous study, also using GWM markers, Plaschke *et al.* (1995) revealed a total of 142 alleles among 40 cultivated wheat lines, an average of 6.2 alleles per GWM (but see Huang *et al.*, 2002, where an extensive study showed high SSR diversity varying among A, B and D genomes). The results obtained in the wild emmer study demonstrate the high diversity in microsatellite sequences among *T. dicoccoides* accessions compared with the cultivated germplasm corroborating RAPDs (Fahima *et al.*, 1999) and allozyme markers (Nevo and Beiles, 1989). Clearly, genetic diversity was eroded across both the coding and non-coding genome during the domestication of major cereal crops, as was displayed by allozymes (Nevo *et al.*, 1986), RAPD and SSR loci, paralleling the genetic erosion shown earlier in wild barley proteins and DNA. This contributed to the susceptibility and vulnerability of the wheat cultivars to abiotic and biotic stresses.

The dendrogram presented in Fig. 15.4.6 in Nevo *et al.* (2002) demonstrates the ability of microsatellites developed from *Triticum aes-*

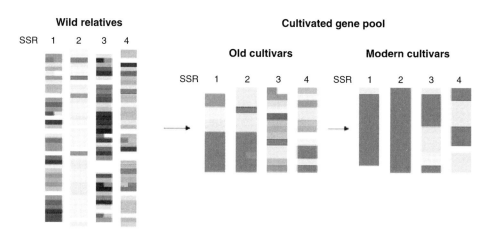

Fig. 14.6. Domestication: reduction of SSR alleles from wild germplasm to old and modern cultivars (from www.scri.sari.ac.uk).

tivum sequences to detect a large amount of genetic diversity in wild emmer wheat and to identify intergroup differences clearly identifying central vs. peripheral populations. All the examined *T. dicoccoides* populations were distinguishable by the GWM markers used, even among closely related populations originating from close geographic locations. Our results demonstrated that the SSR DNA polymorphism of wild emmer wheat was correlated with macro- and microscale ecogeographic factors. In particular, the Israeli populations exhibited high interpopulation and interregional polymorphism. These observations are consistent with previous results obtained with isozymes and different DNA markers for different collections of wild emmer wheat covering a much wider geographic range. They demonstrate geographic and genomic congruence and continuity from macro- to microscales and from coding (protein) to non-coding (presumably partly regulatory) DNA (Nevo *et al.*, 1982; Nevo and Beiles, 1989; Li *et al.*, 2002; and all our microsite studies in wild emmer; Li *et al.*, 1999; see citations in Nevo *et al.*, 2002, and also in http://research.haifa.ac.il/~evolut).

Molecular Diversity at the Major Cluster of Disease-resistance Genes in Cultivated and Wild *Lactuca* Species

We conducted the first study on diversity at the major resistance gene cluster of *Lactuca* species germplasm using molecular markers derived from resistance genes of the NBS–LRR type (nucleotide binding site (NBS) and a leucine-rich repeat region (LRR)) to downy mildew caused by the fungus *Bremia lactucae* (Sicard *et al.*, 1999). Large numbers of haplotypes were detected in three wild *Lactuca* species indicating the presence of numerous resistance genes in wild species compared with cultivated lettuce, *Lactuca sativa* (see Table 15.5 in Sicard *et al.*, 1999), exposing the distinct variation that was found within wild populations. The large number of wild haplotypes recommends the use of wild germplasm in protecting cultivated lettuce against pathogens. Additional in-depth studies of resistant gene candidates

(RGC2) in wild species confirmed the presence of numerous chimeric RGC2 genes in nature (Kuang *et al.*, 2004a,b) and the high AFLP polymorphism of the progenitor of cultivated lettuce, *Lactuca serriola* (Kuang *et al.*, 2004a,b).

Evidence on Domestication Genetics and Evolution of other Crops

Additional evidence on domestication genetics and evolution and the depletion of genetic diversity under domestication has been reported, generally, in:

● Plants (Tripp and van der Heide, 1996; Tanksley and McCouch, 1997; Hoisington *et al.*, 1999; Sandhu and Gill, 2002; Feuillet and Keller, 2002; www.gramene.org)
● Oilseed sunflower, *Helianthus annuus* (Burke *et al.*, 2002; Tang and Knapp, 2003)
● Barley, *H. vulgare* (Parzies *et al.*, 2000), and wild barley, landraces and cultivars (Lin *et al.*, 2001; SCRI at www.scri.sari.ac.uk; Neale *et al.*, 1988)
● Wheat, *T. aestivum* (Huang *et al.*, 2002)
● Maize, *Zea mays* (Matsuoka *et al.*, 2002; Vigouroux *et al.*, 2002 and Zhang *et al.*, 2002), and *Cucurbita* (Sanjur *et al.*, 2002)
● Rice, *Oryza sativa* (Morishima and Oka, 1970; Yang *et al.*, 1994; Liu *et al.*, 1996; Akimoto *et al.*, 1999; Xiong *et al.*, 1999; Zhu *et al.*, 2000)
● Cereals (Gale and Devos, 1998; Buckler *et al.*, 2001; Kellogg, 2001; Appels *et al.*, 2002)
● Diverse crops (Zohary, 1999)
● Melon, *Cucumis melo* (Christopher J. Duxler at www.genglab.ucdavis.edu/chris)
● Beans, *Phaseolus lunatus* (Mimura *et al.*, 2000; Beebe *et al.*, 2001; Fofana *et al.*, 2001)
● Cowpea, *Vigna unguiculata* (Coulibaly *et al.*, 2002)
● Soybeans (Maughan *et al.*, 1995, 1996) and millet landraces, *Pennisetum glaucum* (Bhattacharjee *et al.*, 2002)
● Potato, *Solanum tuberosum* (Ortiz and Huaman, 2001).

Conclusions and Prospects

Wheat and barley as model organisms in domestication

Wheat and barley are important model organisms for testing various aspects of evolutionary theory (both for speciation and adaptation) and a major source of human nutrition. Wheat speciation involves a polyploid series (2x, 4x, 6x). The origin of most wheat is wild emmer, *T. dicoccoides* (genome AABB). Wild barley is a diploid ($2n = 14$) and the progenitor of all barleys. In 1975, the Institute of Evolution at the University of Haifa established a long-term multidisciplinary research programme to study the genomics of wild cereal (http://research.haifa.ac.il/~evolut) in the origin and diversity centre of Old World agriculture in the Near East Fertile Crescent (Zohary and Hopf, 2000). The programme includes evolutionary ecological genetics and barley genomics and genetic mapping coupled with the exploration of genetic resources for wheat and barley improvement. Both aspects, the theoretical and the applied, have proved to be of great importance for studying evolutionary theory and cereal crop improvement.

Conclusion: Genetic Diversity of Wild Barley and Wild Emmer for Barley and Wheat Improvement

The present review of wild cereals supports the idea that the highest hope for future crop improvement lies in rationally and effectively exploring and exploiting the rich gene pool of the plant's wild relatives. What is the role that wild relatives can play in crop improvement? This question is of great importance in view of the dramatic reduction in genetic diversity and consequent increased vulnerability to abiotic and biotic stresses of some of the prime food crops for humans.

The molecular diversity and divergence of wild emmer wheat, *regionally* in the Near East Fertile Crescent and *locally* in four nat-

ural populations at Qazrin, Ammiad, Tabigha and Yehudiyya microsites (wild emmer wheat), and three natural populations of wild barley at Tabigha, Newe Yaar and 'Evolution Canyon' in northern Israel, display parallel ecological-genetic patterning. The *regional* and *local* results demonstrated significant spatial and temporal molecular divergence at the DNA and protein levels in the wild cereal populations and subpopulations. Specifically, these patterns revealed the following:

1. Significant genetic diversity and divergence exist at single-, two- and multilocus structures of allozymes, RAPD, AFLP, rDNA and SSRs both regionally within and between populations, but, most importantly, also locally, over very short distances of several to a few dozen metres in the six microsites.

2. The rich genetic patterns across coding (allozymes) and largely non-coding (RAPDs, AFLP and SSRs, but see Morgante *et al.*, 2002) genomic regions are correlated with, and predictable by, environmental stress (climatic, edaphic, biotic) and heterogeneity (the niche-width variation hypothesis), displaying significant niche-specific and -unique alleles and genotypes.

3. The genomic organization of wild cereals is non-random, heavily structured and, at least partly if not largely, adaptive. It defies explanation by genetic drift, neutrality, or near neutrality models as the primary driving forces of wild cereal molecular evolution. The only viable model to explain the genomic organization of wild cereals is natural selection, primarily diversifying, balancing and cyclical selection over space and time according to the two- or multiple-niche ecological models (see Nevo *et al.*, 2002). Spatial models are complemented by temporal models of genetic diversity and change (Kirzhner *et al.*, 1995, 1996, 1999). Natural selection may interact with mutation, migration and stochastic factors but it overrides them in orientating wild cereal evolutionary processes.

Based on mathematical modelling, we established that stabilizing selection with a cyclically moving optimum could efficiently protect polymorphism for linked loci, additively affecting the selected trait (Kirzhner et al., 1996; Korol et al., 1998). In particular, unequal gene action and/or dominance effects may lead to local polymorphism stability with substantial polymorphism attracting domain. Moreover, under strong cyclical selection, complex dynamic patterns were revealed including 'supercycles' (with periods comprising hundreds of environmental oscillation periods) and 'deterministic chaos' (Kirzhner et al., 1995, 1996, 1999; Korol et al., 1998). These patterns could substantiate polymorphism in natural populations of wild cereals and increase genetic diversity over long periods, thereby contributing to overcoming massive extinctions in natural populations (Nevo, 1995b).

Unique population genetic structure and centre of origin of wild emmer wheat and wild barley

Primarily, wild emmer wheat and, secondarily, wild barley have unique ecological-genetic structure. Emmer wheat central populations in the catchment area of the upper Jordan Valley, and wild barley populations in the Golan Heights, eastern Galilee and Jordanian Mountains, are massive and lush and represent their centre of origin and diversity. However, southwards in Israel and northwards into Turkey, wild emmer becomes fragmented into sporadic semi-isolated and isolated populations that are characterized by an archipelago genetic structure in which alleles are built up locally in high frequency, but are often missing in neighbouring localities. This phenomenon may even occur in the central continuous populations in which alternative fixation of up to eight alleles was described over tens to hundreds of metres in the Golan Heights between Qazrin and Yehudiyya (see Section 5.1 in Nevo et al., 2002).

The centre of origin and diversity of wild emmer and wild barley, the progenitors of most wheat and barleys, and that of other progenitors of cultivated plants is the Near East Fertile Crescent (Nesbitt and Samuel, 1998; Badr et al., 2000; Lev-Yadun et al.,

2000; Zohary and Hopf, 2000; Gopher et al., 2002; Salamini et al., 2002). Particularly in Israel with its extraordinary biotic and physical diversity, wild emmer (Nevo et al., 2002) and wild barley in the Near East (Nevo et al., 1986; Nevo, 1992) developed, both within and between populations, a wide range and rich adaptive diversity to multiple diseases, pests and ecological stresses over a long evolutionary history. Most importantly, this diversity is neither random nor neutral. In contrast, it displays at all levels, adaptive genetic diversity for biochemical, morphological and immunological characteristics which contribute to the species' ability to adapt to widely diverse climatic and edaphic conditions by diverse and complex fitness syndromes. The long-lasting co-evolution of wild emmer with parasites and the ecologically heterogeneous abiotic nature of Israel and the Near East Fertile Crescent led to the development of single multiallelic genes, multilocus structures and abiotic/biotic stress genomes locally and regionally co-adapted for both short- and long-term survival.

Genetic Resources for Cereal Improvement

Wild barley and wild emmer wheat are rich in genetic resources and represent the best hopes for enriching the genetically impoverished cultivars and advancing cereal improvement. These include abiotic (e.g. drought, cold, heat and salt) tolerances and biotic (viral, bacterial and fungal) and herbicide resistances, high quantity and quality storage proteins, hordeins (glutenins and gliadins), amylase and photosynthetic yield (all descriptions and citations in Nevo, 1992; Nevo et al., 2002). A small fraction of these resources have already been used for generating disease resistance cultivars in the USA and Europe. Most of these resources are as yet untapped and provide potentially precious sources for barley and wheat improvement (Nevo, 1989, 2001c). The current rich genetic map of T. dicoccoides with 549 molecular markers and 48 significant QTLs for 11 traits of agronomic importance (Peng et al., 2000, 2003), as well as the previously established association between molecular markers and disease resistance (Nevo 1987, 1992 and references therein; Fetch *et al.*, 2003a,b), permit the unravelling of beneficial alleles of candidate genes that are otherwise hidden. These beneficial alleles could be introduced into cultivated barley and wheat (simultaneously eliminating agronomically undesirable alleles) by using the strategy of marker-assisted selection. The genetic programmes of wild cereals conducted at the Institute of Evolution, University of Haifa, and elsewhere, confirmed that *H. spontaneum* and *T. dicoccoides* are very valuable wild germplasm resources for future barley and wheat improvement. This programme highlighted the evolution of wheat domestication (Peng *et al.*, 2003) and expedited the genomic analysis of wild cereal relatives. It can thus provide a solid basis for introgression or cloning and transformation of agronomically important genes and QTLs from the wild to the cultivated cereals and for advancing cereal improvement. This is particularly important in a world whose population is exploding, where hunger is prevalent, desertification and salinization are dramatically increasing, and water and fertile land resources are limited because of constant pollution and environmental degradation.

Prospects for Crop Improvement

What will be the next step in the research into wild cereals and other progenitors of cultivated plants in the genomic and post-genomic era in an attempt to improve crops? Conceptually, in-depth probing of comparative genome structure and function are the major challenges, in particular, the intimate relationship of the coding and non-coding genomes and the focus on genomic regulation. This may be particularly aided by the discovery that microsatellites are preferentially associated with non-repetitive DNA in plant genomes (Morgante *et al.*, 2002) and that the regulatory functions may be of great importance (Li *et al.*, 2002). Such studies will unravel genome evolution and highlight rich genetic potentials for wheat improvement residing in the progenitors

including *Hordeum, Triticum* and *Aegilops* species, and other Triticeae, the rich repertoire of R-genes in wild lettuce, and other diverse progenitors of cultivars. One example of QTL mapping as a basis for identifying candidate genes for wheat improvement is shown in Fig. 14.7. For a general review, see Mauricio (2001).

Specifically, we could divide the prospects somewhat arbitrarily into *theoretical* and *applied* perspectives (Nevo *et al.*, 2002). For a general review, see Kush (2001).

Theoretical Perspective

1. Highlighting the genetic structure, function, regulation and evolution at macro- and microgeographical scales of natural populations and the corresponding cultivars, bridging multilocus marker structures with fitness-related traits in order to get direct estimates of the adaptive fitness differentiation within and between populations (by using transplant experiments, mapping analysis, microarray methodology and selection-based mapping of fitness components and domesticated contributing genes over the entire life cycle).
2. Exposing the large-scale genome organization of the progenitors and their corresponding cultivars (wheat, soybeans, rice, maize, barley and others) using bacterial artificial chromosome (BAC) clones for molecular complete (*Arabidopsis* and rice) or partial sequence analysis of expressed sequence tags (ESTs), and open reading frames (ORFs) in the transcribed strand; molecular cytogenetic methods could complementarily probe the structure and interactions of the nuclear, mitochondrial and chloroplast genomes. Sequences upstream and downstream of selected ORFs (5′ and 3′ untranslated regions (UTRs), respectively) could be probed for functional regulatory polymorphisms and their function tested by genetic transformation.
3. Analysis of the progenitors' genetic system, or the 'transmission system', which determines the genetic flexibility of the species in diverse ecological contexts including:

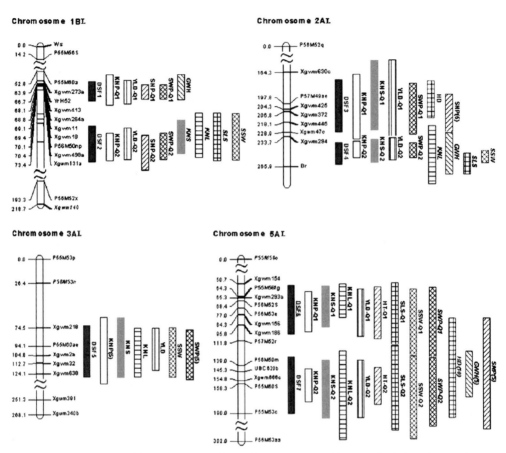

Fig. 14.7. Map locations of domestication syndrome factors (DSFs) and their involved QTLs in L version maps of wild emmer wheat, *T. dicoccoides*. (*Upper*) Short arms of chromosomes. (*Right*) DSFs and corresponding QTLs: ■, DSF; ▨, Kernal number/spike (KNS); ⊟, kernel number/spikelet (KNL); ⊡, grain yield/plant (YLD); ▨, plant height (HT); ⊞, spikelet number/spike (SLS); ⊠, single spike weight (SSW); ▨, spike weight/plant (SWP); ☐, kernel number/plant (KNP); ⊞, spike number/plant (HD); ◨, GWH; ▨, spike number/plant (SNP). The regular trait name represents a single QTL; the italic trait name represents a single QTL (Q2) detected by linked-QTL analysis; the regular trait name tailed with Q1 means the first QTL and tailed with Q2, the second QTL in a pair of linked QTLs. A tailed trait name (5) means that the QTL effect is not significant at the level of 5% of false discovery rate (FDR) but is significant at FDR 10%; (10) means that the effect is not significant at FDR 10% (from Peng *et al.*, 2003, where details can be found).

- Breeding system (reaction norm and genetic variation of the outcrossing rate).
- Mutation rate in different elements of the genome, both coding and non-coding.
- Recombination properties of the genome, their genetic and ecological control.
- Genomic distribution of structural genes, primarily abiotic and biotic stress genes and their regulatory function.
- Interface between ecological and genomic spatio-temporal dynamics and adaptive systems.
- Genome evolution in the polyploidization process.

Applied Perspective

1. Optimization of predictive sampling strategies based on ecological heterogeneity and molecular markers as guidelines, both regionally and locally, for *in situ* and *ex situ* conservation and utilization.

2. Genetic fine-mapping and dissection of

the collected unique wild genetic resources of agricultural importance by molecular markers including SSRs and single nucleotide polymorphisms (SNPs) and sequencing followed by transformation of the detected genes/alleles into elite cultivars via marker-assisted selection.

3. Molecular identification of adaptation genes based on integrated genomic strategies, and novel methodologies including genetic and physical mapping (Fig. 14.7), molecular markers, ESTs, mapping, cloning and sequencing, microarray expression analysis and genetic transformation of the defined target genes/alleles.

4. Comparative genetics/genomics of cereal plants aimed at deciphering the common specific ways of domestication evolution.

5. Critical comparative genetics/genomics of the progenitors across their ecogeographic ecological spectrum and their cultivars across diverse climates, soils and cultures and across coding and non-coding genomic regions to assess comparative diversities, and exploit the rich wild germplasm for crop improvement through genetic reinforcement. This could substantially optimize food production and meet the world's demands for food in a second expanded green revolution.

Acknowledgements

I thank my colleagues A. Beiles, A. Korol and G. Song for reading and commenting on the manuscript.

References

Akimoto, M., Shimamoto, Y. and Morishima, H. (1999) The extinction of genetic resources of Asian wild rice, *Oryza rufipogon* Griff.: a case study in Thailand. *Genetic Resources and Crop Evolution* 46, 419–425.

Appels, R., Francki, M. and Chibbar, R. (2002) Advances in cereal functional genomics. *Functional and Integrative Genomics* 3, 1–24.

Badr, A., Müller, K., Schafer-Pregl, R., El Rabey, H., Effgen, S., Ibrahim, H.H., Pozzi, C., Rohde, W. and Salamini, F. (2000) On the origin and domestication history of barley. *Molecular Biology Evolution* 17, 499–510.

Baek, H.J., Beharav, A. and Nevo, E. (2003) Ecological-genomic diversity of microsatellites in wild barley, *Hordeum spontaneum*, populations in Jordan. *Theoretical and Applied Genetics* 106, 397–410.

Bar-Yosef, O. (1998) The Natufian culture in the Levant, threshold of the origin of agriculture. *Evolutionary Anthropology* 6, 159–177.

Beebe, S., Rengifo, J., Gaitan, E., Duque, M.C. and Tohme, J. (2001) Diversity and origin of Andean landraces of common bean. *Crop Science* 41, 854–862.

Bhattacharjee, R., Bramel, P.J., Hash, C.T., Kolesnikova-Allen, M.A. and Khairwal, I.S. (2002) Assessment of genetic diversity within and between pearl millet landraces. *Theoretical and Applied Genetics* 105, 666–673.

Brown, A.H.D., Feldman, M.W. and Nevo, E. (1980) Multilocus structure of natural populations of *Hordeum spontaneum*. *Genetics* 96, 523–536.

Brown, A.H.D., Frankel, O.H., Marshall, D.R. and Williams, J.T. (eds) (1989) *The Use of Plant Genetic Resources*. Cambridge University Press, Cambridge.

Brown, A.H.D., Clegg, M.T., Kahler, A.L. and Weir, B.S. (1990) *Plant Population Genetics, Breeding, and Genetic Resources*. Sinauer Associates Inc., Sunderland, Massachusetts.

Brush, S. (1999) Genetic erosion of crop populations in centers of diversity: a revision. *FAO, Proceedings of the Technical Meeting on the Methodology of the FAO World Information and Early Warning System on Plant Genetic Resources, 21–23 June 1999*, Research Institute of Crop Production, Prague, Czech Republic.

Buckler, E., Thornsberry, J. and Kresovich, S. (2001) Molecular diversity, structure and domestication of grasses. *Genetic Resources, Cambridge* 77, 213–218.

Burke, J., Tang, S., Knapp, S. and Rieseberg, L. (2002) Genetic analysis of sunflower domestication. *Genetics* 161, 1257–1267.

Coulibaly, S., Pasquet, R., Papa, R. and Gepts, P. (2002) AFLP analysis of the phenetic organization and genetic diversity of *Vigna unguiculata* L. Walp. reveals extensive gene flow between wild and domesticated types. *Theoretical and Applied Genetics* 104, 358–366.

Darwin, C. (1859) *On the Origin of Species*. Atheneum, New York.

Diamond, J. (2002) Evolution, consequences, and future of plant and animal domestication. *Nature* 418, 700–706.

Ehrlich, P.R. and Wilson, E.O. (1991) Biodiversity studies: science and policy. *Science* 253, 758–762.

Ellis, R., Russell, J., Ransay, L., Waugh, R. and Powell, W. (2003) Barley domestication – *Hordeum spontaneum*, a source of new genes for crop improvement. SCRI at www.scri.sari.ac.uk

Fahima, T., Sun., G.L., Beharav, A., Krugman, T., Beiles, A. and Nevo, E. (1999) RAPD polymorphism of wild emmer wheat population, *Triticum dicoccoides*, in Israel. *Theoretical and Applied Genetics* 98, 434–447.

Fahima, T., Röder, M.S., Wendehake, K., Kirzhner, V.M. and Nevo, E. (2002) Microsatellite polymorphism in natural populations of wild emmer wheat, *Triticum dicoccoides*, in Israel. *Theoretical and Applied Genetics* 104, 17–29.

Feldman, M. and Sears, E.R. (1981) The wild gene resources of wheat. *Scientific American* 244, 102–112.

Fetch, T., Steffenson, B. and Nevo, E. (2000) Multiple disease resistance in *Hordeum spontaneum* accessions from Israel and Jordan. *8th International Barley Genetics Symposium, 22–27 October*, Adelaide, South Australia.

Fetch, T., Gold, J. and Nevo, E. (2003a) Evaluation of wild oat germplasm for stem rust resistance. *8th International Congress Plant Pathology, 2–8 February*, Christchurch, New Zealand (abstract).

Fetch, T.G. Steffenson, B.J. and Nevo E. (2003b) Diversity and sources of multiple disease resistance in *Hordeum spontaneum*. *Plant Disease* 87, 1439–1448.

Feuillet, C. and Keller, B. (2002) Comparative genomics in the grass family: molecular characterization of grass genome structure and evolution. *Annals of Botany* 89, 3–10.

Fiedler, P.L. and Jain, S.K. (eds) (1992) *Conservation Biology: the Theory and Practice of Nature Conservation Preservation and Management*. Chapman & Hall, New York.

Fofana, B., Jardin, P. and Baudoin, J. (2001) Genetic diversity in the lima bean (*Phaseolus lunatus* L.) as revealed by chloroplast DNA (cpDNA) variations. *Genetic Resources and Crop Evolution* 45, 437–445.

Frankel, O.H. and Bennett, E. (1970) *Genetic Resources in Plants – Their Exploration and Conservation*. Blackwell, Oxford.

Frankel, O.H. and Hawkes, J.G. (1975) *Crop Genetic Resources for Today and Tomorrow*. Cambridge University Press, Cambridge.

Frankel, O.H. and Soule, M.E. (1981) *Conservation and Evolution*. Cambridge University Press, Cambridge.

Frankel, O.H., Brown, A.H.D. and Burdon, J.J. (1995) *The Conservation of Plant Biodiversity*. Cambridge University Press, Cambridge.

Gale, M. and Devos, K. (1998) Comparative genetics in the grasses. *Proceedings of the National Academy of Sciences USA* 95, 1971–1974.

Gaut, B., Le Thierry d'Ennequin, M., Peek, A. and Sawkins, M. (1997) Maize as a model for the evolution of plant nuclear genomes. *Proceedings of the National Academy of Sciences USA* 97, 7008–7015.

Gopher, A., Abbo, S. and Lev-Yadun, S. (2002) The 'when', 'where' and the 'why' of the Neolithic revolution in the Levant. *Documenta Praehistorica* 28, 49–62.

Guoxiong, C., Krugman, T., Fahima, T., Korol, A.B. and Nevo, E. (2002) Comparative study of morphological and physiological traits related to drought resistance between xeric and mesic *Hordeum spontaneum* lines in Israel. *Barley Genetic Newsletter* 32, 22–33.

Gupta, P.K., Sharma, P.K., Balyan, H.S., Roy, J.K., Sharma, S., Beharav, A. and Nevo, E. (2002) Polymorphism at rDNA loci in barley and its relation with climatic variables. *Theoretical and Applied Genetics* 104, 473–481.

Hammer, K., Diederichsen, A. and Spahillari, M. (1999) Basic studies toward strategies for conservation of plant genetic resources. *FAO, Proceedings of the Technical Meeting on the Methodology of the FAO World Information and Early Warning System on Plant Genetic Resources, 21–23 June 1999*, Research Institute of Crop Production, Prague, Czech Republic.

Harlan, J.R. (1975a) *Crops and Man*. American Society of Agronomy, Crop Science Society of America, Madison, Wisconsin.

Harlan, J.R. (1975b) Our vanishing genetic resources. *Science* 188, 618–621.

Harlan, J.R. and Zohary, D. (1966) Distribution of wild wheats and barley. *Science* 153, 1074–1080.

Hawkes, J.G. (1983) *The Diversity of Crop Plants*. Harvard University Press, Cambridge, Massachusetts.

Hawkes, J.G. (1991) International workshop in dynamic in situ conservation of wild relatives of major cultivated plants: summary of final discussion and recommendations. *Israel Journal of Botany* 40, 529–536.

Hernandez-Verdugo, S., Luna-Reyes, R. and Oyama, K. (2001) Genetic structure and differentiation of wild
and domesticated populations of *Capsicum annuum* (Solanaceae) from Mexico. *Plant Systematics and
Evolution* 226, 129–142.

Hilton, H. and Baut, B. (1998) Speciation and domestication in maize and its wild relatives: evidence from
the globulin-1 gene. *Genetics* 150, 863–872.

Hoisington, D., Khairallah, M., Reeves, T., Ribaut, J.M., Skovmand, B., Taba, S. and Warburton, M. (1999)
Plant genetic resources: what can they contribute toward increased crop productivity? *Proceedings of
the National Academy of Sciences USA* 96, 5937–5943.

Holden, J., Peacock, J. and Williams, T. (1993) *Genes, Crops, and the Environment*. Cambridge University
Press, Cambridge.

Huang, X.Q., Börner, A., Röder, M.S. and Ganal, M.W. (2002) Assessing genetic diversity of wheat (*Triticum
aestivum* L.) germplasm using microsatellite markers. *Theoretical and Applied Genetics* 105, 699–707.

Hutchinson, C. and Weiss, E. (1999) Remote sensing contribution to an early warning system for genetic
erosion of agricultural crops. *FAO, Proceedings of the Technical Meeting on the Methodology of the
FAO World Information and Early Warning System on Plant Genetic Resources, 21–23 June 1999,*
Research Institute of Crop Production, Prague, Czech Republic.

IBPGR (1985) *Ecogeographical Surveying and in situ Conservation of Crop Relatives*. IBPGR Secretariat,
Rome.

ICPGR (1995) *Outline of the Report on the State of the World's Plant Genetic Resources*. Item 7 of the
Provisional Agenda, Sixth Session in Rome, 19–30 June 1995, Commission on Plant Genetic
Resources.

Jones, M.K., Allaby, R.G. and Brown, T.A. (1998) Wheat domestication. *Science* 279, 302–303.

Kellogg, E. (2001) Evolutionary history of the grasses. *Plant Physiology* 125, 1198–1205.

Khush, G. (2001) Green revolution: the way forward. *Nature Reviews Genetics* 2, 815–822.

Kirzhner, V.M., Korol, A., Turpeinen, T. and Nevo, E. (1995) Genetic supercycles caused by cyclical selection. *Proceedings of the National Academy of Sciences USA* 92, 7130–7133.

Kirzhner, V.M., Korol, A.B. and Nevo, E. (1996) Complex dynamics of multilocus systems subjected to cyclical selection. *Proceedings of the National Academy of Sciences USA* 93, 6532–6535.

Kirzhner, V.M., Korol, A.B. and Nevo, E. (1999) Abundant multilocus polymorphisms caused by genetic
interaction between species on trait-for-trait basis. *Journal of Theoretical Biology* 198, 61–70.

Kislev, M. (2002) Origin of annual crops by agro-evolution. *Israel Journal of Plant Sciences* 50, 85–88.

Korol, A.B., Kirzhner, V.M. and Nevo, E. (1998) Dynamics of recombination modifiers caused by cyclical
selection: interaction of forced and auto oscillations. *Genetic Resources* 72, 135–147.

Kuang, H., Woo, S.S., Meyers, B., Nevo, E. and Michelmore, R. (2004a) Multiple genetic processes result in
heterogeneous rates of evolution within the major cluster disease resistance genes in lettuce. *The Plant
Cell* 16 (in press).

Kuang, H., Ochoa, O., Nevo, E. and Michelmore, R. (2004b) The resistance gene Dm3 is infrequent in natural populations of *Lactuca serriola* due to deletions and gene conversions (in press).

Ladizinsky, G. (1998) *Plant Evolution Under Domestication*. Kluwer Academic Publishers, Dordrecht, The
Netherlands.

Lauter, N. and Doebly, J. (2002) Genetic variation for phenotypically invariant traits detected in teosinte:
implications for the evolution of novel forms. *Genetics* 160, 333–342.

Lev-Yadun, S., Gopher, A. and Abbo, S. (2000) The cradle of agriculture. *Science* 288, 1602–1603.

Li, Y.C., Fahima, T., Beiles, A., Korol, A.B. and Nevo, E. (1999) Microclimatic stress and adaptive DNA differentiation in wild emmer wheat (*Triticum dicoccoides*). *Theoretical and Applied Genetics* 98, 873–883.

Li, Y.C., Korol, A.B., Fahima, T., Beiles, A. and Nevo, E. (2002) Microsatellites: genomic distribution, putative
functions and mutational mechanisms: a review. *Molecular Ecology* 11, 2453–2465.

Lin, J., Brown, A. and Clegg, M. (2001) Heterogeneous geographic patterns of nucleotide sequence diversity
between two alcohol dehydrogenase genes in wild barley (*Hordeum vulgare* subspecies *spontaneum*).
Proceedings of the National Academy of Sciences USA 98, 531–536.

Liu, K.D., Yang, G.P., Zhu, S.H., Zhang, Q., Wang, X.M. and Saghai Maroof, M.A. (1996) Extraordinarily
polymorphic ribosomal DNA in wild and cultivated rice. *Genome* 39, 1109–1116.

Liu, Z., Sun, Q., Ni, Z., Nevo, E. and Yang, T. (2002) Molecular characterization of a novel powdery mildew
resistance gene Pm30 in wheat originating from wild emmer. *Euphytica* 123, 21–29.

Maroof, M.A., Soliman, K.M., Jorgensen, R.A. and Allard, R.W. (1984) Ribosomal DNA spacer-length polymorphisms in barley: Mendelian inheritance, chromosomal location and population dynamics. *Proceedings of the National Academy of Sciences USA* 81, 8014–8018.

Maroof, M.A., Biyashev, R.M., Yang, G.P., Zhang, Q. and Allard, R.W. (1994) Extraordinarily polymorphic microsatellite DNA in barley: species diversity, chromosomal locations, and population dynamics. *Proceedings of the National Academy of Sciences USA* 92, 5466–5470.

Matsuoka, Y., Vigouroux, Y., Goodman, M., Sanchez, J., Buckler, E. and Doebley, J. (2002) A single domestication for maize shown by multilocus microsatellite genotyping. *Proceedings of the National Academy of Sciences USA* 99, 6080–6084.

Maughan, P.J., Saghai Maroof, M.A. and Buss, G.R. (1995) Microsatellite and amplified sequence length polymorphisms in cultivated and wild soybean. *Genome* 38, 715–723.

Maughan, P.J., Saghai Maroof, M.A., Buss, G.R. and Huestis, G.M. (1996) Amplified fragment length polymorphism (AFLP) in soybean: species diversity, inheritance, and near-isogenic line analysis. *Theoretical and Applied Genetics* 93, 392–401.

Mauricio, R. (2001) Mapping quantitative trait loci in plants: uses and caveats for evolutionary biology. *Nature Reviews* 2, 370–381.

Maxted, J., Ford-Lloyd, B.V. and Hawkes, J.G. (eds) (1997) *Plant Genetic Conservation. The In Situ Approach.* Chapman & Hall, London, pp. 144–159.

Mimura, M., Yasuda, K. and Yamaguchi, H. (2000) RAPD variation in wild, weedy and cultivated azuki beans in Asia. *Genetic Resources and Crop Evolution* 47, 603–610.

Morgante, M., Hanafey, M. and Powell, W. (2002) Microsatellites are preferentially associated with non-repetitive DNA in plant genomes. *Genetics* 30, 194–200.

Morishima, H. and Oka, H. (1970) A survey of genetic variations in the populations of wild *Oryza* species and their cultivated relatives. *Japanese Journal of Genetics* 45, 371–385.

Moseman, J.G., Nevo, E., El Morshidy, M. and Zohary, D. (1984) Resistance of *Triticum dococcoides* to infection with *Erysiphe graminis tritici. Euphytica* 33, 41–47.

Neale, D., Saghai-Maroof, M., Allard, R., Zhang, Q. and Jorgensen, R. (1988) Chloroplast DNA diversity in populations of wild and cultivated barley. *Genetics* 120, 1105–1110.

Nesbitt, M. and Samuel, D. (1998) Wheat domestication: archaeological evidence. *Science* 279, 1433.

Nevo, E. (1978) Genetic variation in natural populations: patterns and theory. *Theoretical Population Biology* 13, 121–177.

Nevo, E. (1983) Genetic resources of wild emmer wheat: structure, evolution and application in breeding. In: Sakamoto, S. (ed.) *Proceedings of the 6th International Wheat Genetics Symposium.* Kyoto University, Kyoto, pp. 421–431.

Nevo, E. (1986) Genetic resources of wild cereals and crop improvement: Israel, a natural laboratory. *Israel Journal of Botany* 35, 255–278.

Nevo, E. (1987) Plant genetic resources: prediction by isozyme markers and ecology. In: Rattazi, M., Scandalios, J. and Whitt, G.S. (eds) *Isozymes: Current Topics in Biological Research. Agriculture, Physiology and Medicine* 16, 247–267.

Nevo, E. (1988) Genetic diversity in nature: patterns and theory. *Evolutionary Biology* 23, 217–247.

Nevo, E. (1989) Genetic resources of wild emmer wheat revisited: genetic evolution, conservation and utilization. In: Miller, T.E. and Koebner, R.M.D. (eds) *Proceedings of the Seventh International Wheat Genetics Symposium, 13–19 July 1998,* Institute of Plant Science Research, Cambridge, pp. 121–126.

Nevo, E. (1992) Origin, evolution, population genetics and resources for breeding of wild barley, *Hordeum spontaneum,* in the Fertile Crescent. In: Shewry, P.R. (ed.) *Barley: Genetics, Biochemistry, Molecular Biology and Biotechnology.* CAB International, Wallingford, UK, pp. 19–43.

Nevo, E. (1995a) Genetic resources of wild emmer, *Triticum dicoccoides,* for wheat improvement: News and Views. *Proceedings for the 8th International Wheat Genetic Symposium, 20–25 July 1993,* Beijing, pp. 79–87.

Nevo, E. (1995b) Evolution and extinction. In: Nierenberg, W.N. (ed.) *Encyclopedia of Environmental Biology,* Vol. 1. Academic Press, New York, pp. 717–745.

Nevo, E. (1998a) Molecular evolution and ecological stress at global, regional and local scales: the Israeli perspective. *Journal of Experimental Zoology* 282, 95–119.

Nevo, E. (1998b) Genetic diversity in wild cereals: regional and local studies and their bearing on conservation *ex-situ* and *in-situ. Genetic Resource and Crop Evolution* 45, 355–370.

Nevo, E. (2001a) Genetic diversity. In: Levin, S. (ed.) *Encyclopedia of Biodiversity,* Vol. 3. Academic Press, New York, pp. 195–213.

Nevo, E. (2001b) Evolution of genome–phenome diversity under environmental stress. *Proceedings of the National Academy of Sciences* USA 98, 6233–6240.

Nevo, E. (2001c) Genetic resources of wild emmer, *Triticum dicoccoides*, for wheat improvement in the third millennium. *Israel Journal of Plant Sciences* 49, 162.

Nevo, E. and Beiles, A. (1989) Genetic diversity of wild emmer wheat in Israel and Turkey: structure, evolution and application in breeding. *Theoretical and Applied Genetics* 77, 421–455.

Nevo, E., Zohary, D., Brown, A.H.D. and Haber, M. (1979) Genetic diversity and environmental associations of wild barley, *Hordeum spontaneum*, in Israel. *Evolution* 33, 815–833.

Nevo, E., Golenberg, E., Beiles, A., Brown, A.H.D. and Zohary, D. (1982) Genetic diversity and environmental associations of wild wheat, *Triticum dicoccoides*, in Israel. *Theoretical and Applied Genetics* 62, 241–254.

Nevo, E., Beiles, A. and Ben-Shlomo, R. (1984) The evolutionary significance of genetic diversity: ecological, demographic and life history correlates. *Lecture Notes in Biomathematics* 53, 13–213.

Nevo, E., Moseman, J.G., Beiles, A. and Zohary, D. (1985) Patterns of resistance of Israeli wild emmer wheat to pathogens. I. Predictive method by ecology and allozyme genotypes for powdery mildew and leaf rust. *Genetica* 67, 209–222.

Nevo, E., Beiles, A. and Zohary, D. (1986) Genetic resources of wild barley in the Near East: structure, evolution and application in breeding. *Biological Journal of the Linnean Society* 27, 355–380.

Nevo, E., Korol, A.B., Beiles, A. and Fahima, T. (2002) *Evolution of Wild Emmer Wheat Improvement. Population Genetics, Genetic Resources, and Genome Organization of Wheat's Progenitor*, Triticum dicoccoides. Springer-Verlag, Berlin, pp. 364.

Ortiz, R. and Huaman, Z. (2001) Allozyme polymorphisms in tetraploid potato gene pools and the effect on human selection. *Theoretical and Applied Genetics* 103, 792–796.

Owuor, E.D., Fahima, T., Beiles, A., Korol, A.B. and Nevo, E. (1997) Population genetics response to microsite ecological stress in wild barley *Hordeum spontaneum*. *Molecular Ecology* 6, 1177–1187.

Owuor, E.D., Beharav, A., Fahima, R., Kirzhner, V.M., Korol, A.B. and Nevo, E. (2003) Microscale ecological stress causes RAPD molecular selection in wild barley, Neve Yaar microsite, Israel. *Genetic Resources and Crop Evolution* 50, 213–224.

Pakniyat, H., Powell, W., Baird, E., Handley, L.L., Robinson, D., Scrimgeour, C.M., Nevo, E., Hackett, C.A., Caligari, P.D. and Forster, B.P. (1997) AFLP variation in wild barley (*Hordeum spontaneum* C. Koch) with reference to salt tolerance and associated ecogeography. *Genome* 40, 332–341.

Parzies, H., Spoor, W. and Ennos, R. (2000) Genetic diversity of barley landrace accessions (*Hordeum vulare* ssp. *vulgare*) conserved for different lengths of time in *ex situ* gene banks. *Heredity* 84, 476–486.

Pasternak, T. (1998) Investigations of botanical remains from Nevali Cori PPNB, Turkey: a short interim report. In: Damania, A.B., Valkoun, J., Wilcox, G. and Qualset, C.O. (eds) *The Origins of Agriculture and Crop Domestication*. ICARDA, Aleppo, Syria, pp. 170–176.

Peng, J.H., Korol, A.B., Fahima, T., Röder, M.S., Li, Y.C. and Nevo, E. (2000) Molecular genetic maps in wild emmer wheat, *Triticum dicoccoides*: genome-wide coverage, massive negative interference, and putative quasi-linkage. *Genome Research* 10, 1509–1531.

Peng, J.H., Ronin, Y.I., Fahima, T., Roder, M., Li, Y., Nevo, E. and Korol, A.B. (2003) Domestication quantitative trait loci in *Triticum dicoccoides*, the progenitor of wheat. *Proceedings of the National Academy of Sciences USA* 100, 2489–2494.

Plaschke, J., Ganal, M.W. and Röder, M.S. (1995) Detection of genetic diversity in closely related bread wheat using microsatellite markers. *Theoretical and Applied Genetics* 91, 1001–1007.

Plucknett, D.L., Smith, N.J., Williams, J.T. and Anishetty, N.M. (1987) *Gene Banks and the World's Food*. Princeton University Press, Princeton, New Jersey.

Poncet, V., Lamy, G., Enjalbert, J., Jolys, H., Sarr, A. and Robert, T. (1998) Genetic analysis of the domestication syndrome in pearl millet (*Pennisetum glaucum* L., Poaceae): inheritance of the major characters. *Heredity* 81, 648–658.

Ramachandran, S., Fahima, T., Krugman, T., Roder, M., Nevo, E. and Feldman, M. (2005) Domestication times of wheat and barley using microsatellite polymorphisms (in review).

Reeves, J.C., Law, J.R., Donini, P., Koebner, R. and Cooke, R. (1999) Changes over time in the genetic diversity of UK cereal crops. *FAO, Proceedings of the Technical Meeting on the Methodology of the FAO World Information and Early Warning System on Plant Genetic Resources, 21–23 June*, Research Institute of Crop Production, Prague, Czech Republic.

Salamini, F., Ozkan, H., Brandolini, A., Schafer-Pregl, R. and Martin, W. (2002) Genetics and geography of wild cereal domestication in the Near East. *Nature Reviews* 3, 429–441.

Sandhu, D. and Gill, K.S. (2002) Gene-containing regions of wheat and the other grass genomes. *Plant Physiology* 128, 803–811.

Sanjur, O., Piperno, D., Andres, T. and Wessel-Beaver, L. (2002) Phylogenetic relationships among domesticated and wild species of *Cucurbita* (Cucurbitaceae) inferred from a mitochondrial gene: implications for crop plant evolution and areas of origin. *Proceedings of the National Academy of Sciences USA* 99, 535–540.

Sicard, D., Woo, S.-S., Arroyo-Garcia, R., Ochoa, O., Nguyen, D., Korol, A.B., Nevo, E. and Michelmore, R. (1999) Molecular diversity at the major cluster of disease resistance genes in cultivated and wild *Lactuca* spp. *Theoretical and Applied Genetics* 99, 405–418.

Smith, B.D. (1995) *The Emergence of Agriculture.* Scientific American Library, New York.

Soule, M.E. (ed.) (1986) *Conservation Biology: the Science of Scarcity and Diversity.* Sinauer Association Inc., Sunderland, Massachusetts.

Soule, M.E. (1987) *Viable Populations for Conservation.* Cambridge University Press, Cambridge.

Styles, B.T. (ed.) (1986) *Infraspecific Classification of Wild and Cultivated Plants.* Systematics Association Special Volume No. 29, Clarendon Press, Oxford.

Tang, S. and Knapp, S. (2003) Microsatellites uncover extraordinary diversity in Native American land races and wild populations of cultivated sunflower. *Theoretical and Applied Genetics* 106, 990–1003.

Tanksley, S.D. and McCouch, S.R. (1997) Seed banks and molecular maps: unlocking genetic potential from the wild. *Science* 277, 1063–1066.

Tripp, R. and van der Heide, W. (1996) *The Erosion of Crop Genetic Diversity: Challenges, Strategies and Uncertainties.* Natural Resource Perspectives, No. 7. Overseas Development Institute, London.

Turpeinen, T., Tenhola, T., Manninen, O., Nevo, E. and Nissila, E. (2001) Microsatellite diversity associated with ecological factors in *Hordeum spontaneum* populations in Israel. *Molecular Ecology* 10, 1577–1591.

Turpeinen, T., Vanhala, V., Nevo, E. and Nissila, E. (2003) AFLP genetic polymorphism in wild barley (*Hordeum spotaneum*) populations in Israel. *Theoretical Applications of Genetics* 106, 1333–1339.

Valdes, B., Heywood, V.H., Raimondo, F. and Zohary, D. (eds) (1997) *Proceedings of the Workshop on Conservation of the Wild Relatives of European Cultivated Plants, 1–6 September 2003*, Palermo, Italy.

Vavilov, N.I. (1926) *Studies on the Origin of Cultivated Plants.* Institute of Botanique Applique et d'Amelioration des Plants, Leningrad.

Vavilov, N.I. (1951) The origin, variation and breeding of cultivated plants. *Cronica Botanica* 13, 1–364 (translated from Russian).

Vigouroux, Y., McMullen, M., Hittinger, C.T., Houchins, K., Schulz, L., Kresovich, S., Matsuoka, Y. and Doebley, J. (2002) Identifying genes of agronomic importance in maize by screening microsatellites for evidence of selection during domestication. *Proceedings of the National Academy of Sciences USA* 99, 9650–9655.

Volis, S., Yakubov, B., Shulgina, E., Ward, D., Zur, V. and Mendlinger, S. (2001) Tests for adaptive RAPD variation in population genetic structure of wild barley, *Hordeum spontaneum* Koch. *Biological Journal of the Linnean Society* 74, 289–303.

Volis, S., Mendlinger, S. and Ward, D. (2002) Differentiation in populations of *Hordeum spontaneum* along a gradient of environmental productivity and predictability: life history and local adaptation. *Biological Journal of the Linnean Society* 77, 479–490.

Wilson, E.O. (1992) *The Diversity of Life.* Harvard University Press, Cambridge, Massachusetts.

Wilson, E.O. (2002) *The Future of Life.* Alfred A. Knopf, New York.

Xiong, L., Liu, K., Dai, X., Xu, C. and Zhang, Q. (1999) Identification of genetic factors controlling domestication-related traits of rice using an F2 population of a cross between *Oryza sativa* and *O. rufipogon.* *Theoretical and Applied Genetics* 98, 243–251.

Xu, R., Tomooka, N., Vaughan, D. and Doi, K. (2000) The *Vigna angularis* complex: genetic variation and relationships revealed by RAPD analysis, and their implications for in situ conservation and domestication. *Genetic Resources* 47, 123–134.

Yamashita, K. (1965) *Cultivated Plants and Their Relatives.* Scientific Expedition to the Karakoram and Hindukush, 1955, Vol. I.

Yang, G.P., Saghai Maroof, M.A., Xu, C.G., Zhang, Q. and Biyashev, R.M. (1994) Comparative analysis of microsatellite DNA polymorphism in landraces and cultivars of rice. *Molecular and General Genetics* 245, 187–194.

Zhang, L., Peek, A., Dunams, D. and Gaut, B. (2002) Population genetics of duplicated disease-defense genes, hm1 and hm2, in maize (*Zea mays* ssp. *Mays* L.) and its wild ancestor (*Zea mays* ssp. *Parviglumis*). *Genetics* 162, 851–860.

Zhu, Y., Chen, H., Fan, J., Wang, Y., Li, Y., Chen, J., Fan, J.X., Yang, S., Hu, L., Leung, H., Mew, T., Teng, P., Wang, Z. and Mundt, C. (2000) Genetic diversity and disease control in rice. *Nature* 17, 718–722.

Zohary, D. (1996) The mode of domestication of the founder crops of Southwest Asian agriculture. In: Harris, D.R. (ed.) *The Origins and Spread of Agriculture and Pastoralism in Eurasia.* UCL Press, London, pp. 142–158.

Zohary, D. (1999) Monophyletic vs. polyphyletic origin of the crops on which agriculture was founded in the Near East. *Genetic Resources and Crop Evolution* 46, 133–142.

Zohary, D. and Hopf, J. (2000) *Domestication of Plants in the Old World,* 3rd edn. Oxford University Press, Oxford.

15 Conserving genetic diversity in plants of environmental, social or economic importance

Robert J. Henry

Centre for Plant Conservation Genetics, Southern Cross University, PO Box 157, Lismore, NSW 2480, Australia

Introduction

The case for conserving plant diversity was made in Chapter 1. This chapter further describes what needs to be conserved and how. Plants and plant-derived products are central features of human society (Table 15.1). Species that are used as sources of food, fibre, shelter or medicine may have added social significance beyond their direct practical or economic use. Diverse groups of plants are currently used by humans (Tables 15.2 and 15.3). Future uses of plants may involve using biotechnology to deliver vaccines or other high-value medicinal products or producing precursors for plastics or other industrial materials. Furthermore, biotechnology allows for the potential use of any or all plant species as a source of useful genes. This provides a convincing argument in support of efforts to conserve all higher plant diversity.

All plant species are of importance from an environmental perspective, in that loss of diversity in any species reduces plant diversity. However, most interest is directed towards plants that are rare or in danger of extinction in the short or medium term. This should not distract effort from the related issue of preserving diversity in more common or widespread species and especially the importance of conserving rare variants of more common species.

Rare and Endangered Species

Large numbers of plant species are considered to be rare or endangered. The security of wild plant populations is usually evaluated in the context of current land use. A species is threatened when extinction becomes a likely outcome if current land use is continued unaltered. Documentation of rare species continues to be a major challenge. Many rare species have probably become extinct before there was a chance to describe them. An example of an approach to the identification and classification of rare plants is given in Table 15.4. The development of a recovery plan for rare species is the next step beyond initial discovery and recognition of rarity. Implementation of recovery plans in most cases lags well behind these earlier stages of discovery, evaluation and planning.

Two approaches to plant conservation can be considered: *in situ* (Prance, 2004) and *ex situ* (Thormann, 2004). Conservation *in situ* aims to protect wild plant populations and ensure diversity is preserved and that evolution of the species can continue. This is the most common approach to conservation of environmentally important species. Conservation of plants in seed banks, or as living collections such as those in botanic gardens, may be necessary for critically

Table 15.1. Human uses of higher plants.

Category of use	Examples
Food	Cereal, pulses, fruit, vegetables, oilseeds and sugar
Beverage	Wine, beer, tea, coffee
Animal feed	Pastures, fodder
Fibre	Cotton, hemp and paper
Fuel	Firewood
Construction	Housing and furniture
Medicine	Pharmaceutical products, traditional medicines
Ornament	Cut flowers, pot plants, garden plants and turf grass
Industry	Ethanol for fuel, electricity
Environment	Environmental restoration, greenhouse gas sequestration
Other	Perfumes, cosmetics

Table 15.2. Groups of flowering plants defined by DNA analysis (see Chapter 2, Fig. 2.1, this volume).

Table 15.3. The uses of flowering plants (grouped as listed in Table 15.2) by humans. Based upon economic uses mainly as defined by Heywood (1978).

Amborellaceae	**Amborella** only New Caledonia
Nymphaeaceae	**Ornamental** water lilies
	Food (seeds and rhizomes)
	Cosmopolitan (fresh water)
Austrobaileyales	One genus, two species only in Australia, no known uses
Chloranthaceae	**Ornament** (*Chloranthus glaberi*)
	Beverage (*Chloranthus officinalis*)
	Medicine (*Hedyosmum brasiliense*)
Canellales	**Ornament**
	Food white cinnamon (*Canella winterana*)
	Medicine
Piperales	**Food** pepper (*Piper nigrum*)
	Beverage kava (*Piper methysticum*)
	Ornament
	Medicine
Laurales	**Food** avocado (*Persea americana*)
	Ornament cinnamon, bay leaves
	Construction
	Other perfume (*Doryphora sassafras*)
	Medicine
Magnoliales	**Food** nutmeg (*Myristica fragrans*), custard apple (*Annona*)
	Ornament *Magnolia*
	Construction
Alismatales	**Ornament**
	Food *Sagittaria sagittifolia* (*tubers*)
Asparagales	**Food** onions, garlic, leek, vanilla, asparagus
	Ornament *Gladiolus*, *Iris*, *Freesia*, daffodils, orchids
	Medicine
	Other saffron (*Crocus sativus*)
Dioscoreales	**Food** yams
	Medicine
Liliales	**Beverage** sarsaparilla
	Ornament *Lilium*, *Tulipa* (tulip)
	Mediine

Continued

Table 15.3. *Continued.*

Pandanales	**Food** (starchy fruits)
	Ornament *Pandanus*
	Other perfume, baskets
Dasypogonaceae	**Ornament** *Xanthorrhoea*
	Other varnishes
Arecales	**Food** coconuts, copra, dates, sago, palm oil
	Fibre coir, raffia
	Ornament
Poales	**Food** rice, wheat, maize, barley, sorghum, millet, sugarcane, bamboo, pineapple
	Animal feed pastures
	Ornament lawns, water plants
	Other baskets, brooms, thatching
Commelinales	**Ornament** wandering Jew, water hyacinth
Zingiberales	**Food** banana, ginger, cardamom, turmeric, arrowroot
	Fibre manila, hemp
	Ornament *Strelitzia, Canna*
	Other perfume
Ceratophyliales	**Other** protects fish in freshwater
Ranunculales	**Food** fruits
	Ornament *Clematis, Ranunculus* (buttercups)
	Medicine opium (*Papaver somniferum*)
Sabiaceae	**Ornament**
Proteales	**Food** *Macadamia integrifolia*
	Ornament *Banksia, Grevillea, Telopea, Protea, Leucadendron,* planes
	Construction timber
Buxaceae	**Ornament**
	Construction
Trochodendraceae	**Other** birdlime (seed to catch birds)
Gunnerales	**Ornament** *Gunnera*
	Other tanning and dyeing
Aextoxicaceae	
Berberidopsidaceae	**Ornament** (scared bamboo)
	Medicine
Dilleniaceae	**Ornament** *Hibbertia, Dillenia*
Caryophyliales	**Food** *Arraranthus*
	Ornament cockscombs (*Celosia cristata*), *Ptilotus*
Santalales	**Medicine**
Saxifragales	**Food** grapes (*Vitis vinifera*), gooseberries, currants (*Ribes*)
	Ornament *Hydrangeas, Kalanchoe*
	Construction timber
	Other perfume
Crossosomatales	**Ornament**
	Construction
	Medicine
Gereniales	**Ornament** *Geranium, Pelargonium*
	Construction timber
	Medicine
	Other perfume
Myrtales	**Food** cloves (*Syzgium aromatieun*), lilly pilly
	Ornament bottlebrushes, *Tibouchina, Fuschias*
	Construction *Eucalyptus*
	Medicine
	Other essential oils
Celastrales	**Beverage** Arabia tea (*Catha edulis*)
	Medicine
	Other essential oils

Continued

Table 15.3. *Continued.*

Malpighiales	**Food** cassava (*Manihot glaziovii*), passionfruit (*Passiflora*) **Ornament** Poinsettia (*Euphorbia*), violets **Industry** castor oil (*Ricinus communis*) **Other** rubber (*Hevea brasiliensis*)
Oxalidales	**Ornament** flycatcher plant (*Cephalotus follicularis*) **Construction** timber
Fabales	**Food** peas, beans, groundnut (peanut), soybean **Animal feed** clover, lucerne **Other** nitrogen fixing **Construction** timber **Ornament** *Acacia*
Rosales	**Food** fruits (apple, plum, pear, cherry, mulberries, fig, raspberries, strawberries) **Fibre** hemp (*Cannabis sativa*) **Ornament** roses **Beverage** hops (*Humulus lupulus*) **Construction** elms
Cucurbitales	**Food** cucumber, pumpkin, melon **Ornament** *Begonia*
Fagales	**Food** chestnut (*Castanea sativa*), walnut, pecan **Construction** beeches, oaks, birches ·
Brassicales	**Food** oilseed rape, mustard, vegetables (cabbage, cauliflower), papaya (*Carica papaya*) **Animal feed** fodder
Malvales	**Fibre** cotton (*Gossypium*) **Ornament** *Hibiscus* **Construction** **Other** chocolate
Sapindales	**Food** orange, lemon, lime, mango, cashew, pistachio, lychee (*Litchi chinensis*), maple sugar **Ornament** maples **Medicine** **Construction** mahoganies **Other** perfume, poison ivy
Cornales	**Ornament** dogwoods
Ericales	**Beverage** tea (*Camellia sinensis*) **Ornament** *Camellia* **Construction**
Garryales	**Medicine**
Gentianales	**Beverage** coffee (*Coffea*) **Ornament** *Gentian*, oleandas (*Nerium*), *Gardenia* **Medicine** quinine (*Cinchona*)
Lamiales	**Food** olives (*Olea europaca*)
Solanales	**Food** potato, aubergine, tomato, pepper, sweet potato **Ornament** morning glory (*Ipomoea purpurea*) **Medicine** **Other** tobacco
Aquifoliales	**Ornament** holly **Construction**
Apiales	**Food** carrot, celery, parsley, fennel, dill **Medicine** **Other** perfume
Asterales	**Food** sunflower, lettuce, chicory, Jerusalem artichoke **Ornament** *Dahlia, Gerbera*
Dipsacales	**Beverage** elderberry (wine) **Ornament** honeysuckles (*Lonicera*) **Medicine**

Table 15.4. Classification of rare and endangered plant species (Environment Protection and Biodiversity and Conservation Act, 1999, Australia).

Extinct – no reasonable doubt that the last member of the species has died
Extinct in the wild – species exists only in cultivation
Critically endangered – extremely high risk of extinction in the immediate future
Endangered – very high risk of extinction in the medium term
Vulnerable – high risk of extinction in the medium term
Conservation dependent – species is dependent on a specific conservation programme without which it could become vulnerable, endangered or critically endangered within 5 years

endangered species or in cases where the natural habitat is no longer available or suitable. This *ex situ* conservation is much more commonly used for species of economic importance such as the major crop species.

Species of Economic Importance

The sustainable production of food in agriculture requires the effective use of plant genetic resources (Henry, 2001). Increasing and even maintaining agricultural production is reliant upon continued genetic improvement of crop species. The available genetic resources for major crops are the key resource for global food supply. Humans rely on a very small number of plant species for a large proportion of their diets. Just three cereal species, wheat, maize and rice, may account for as much as half of all human food. Genetic improvement and the effective use of available genetic diversity in these species in essential to continued world food supply (Henry, 2004). Many other species are of importance to humans as a source of food or other products. This chapter will briefly review the diversity of genetic resources available for genetic improvement of higher plants of economic importance.

Wild populations

Plants in wild populations are an important source of genetic resources for economically important plants. The importance of wild populations may be reduced by the establishment of *ex situ* collections from these populations. However, the *in situ* conserva-

tion of many species will remain important. The relative security of wild populations and *ex situ* collections such as those provided by seed banks varies. Seed collections may be lost or destroyed. The wild populations should provide at the very least a backup for even the best-collected species. Conservation *in situ* should allow continued evolutionary development of the species in the longer term while seed collections or other *ex situ* collections represent the genetic resource at the time of collection.

Cultivated plants

The main source of genetic resources for many domesticated species is plants in crops on farms or in gardens. The domesticated gene pool can often be considered to be distinct from the wild gene pool. The increased adoption of higher performing plant varieties produced by modern plant breeding techniques may displace the production of older varieties or landraces resulting in loss of this genetic resource. This does not necessarily need to result in a loss of diversity in crop species in production if efforts are made to ensure improved varieties from genetically diverse sources are released for production. Private and public gardens represent the main source of germplasm for ornamentals and for many minor crops.

Genetic resource collections

The conservation of plant genetic resources *ex situ* may take different forms depending on the biology of the species. Seed collections are important for species with long-lived seeds. Some plants do not have seed that can be stored for any significant period of time. These must be maintained by vegetative propagation, usually as living collections in the field or glasshouse. Species that require clonal propagation to maintain genetic characteristics may also be stored as living collections even if their seed is suitable for storage. Very large numbers of genotypes are held in germplasm collections for the major species (Table 15.5).

Seed banks

Large collections of seed are stored for each of the major field crop species. These collections are deliberately duplicated in different collections in different countries to increase global security of the resource. This approach is only suitable for plants with long-lived seeds. DNA analysis tools are becoming increasingly important in managing these collections. Confirming sample identity and establishing relationships between accessions in collections allows more targeted efforts to conserve diversity and supports more effective utilization of these genetic resources in plant improvement (Nagamine, 2004). These benefits also apply to other types of genetic resource collection.

Table 15.5. Numbers of accessions held in germplasm collections of some major crop species. For more information see FAO (1996).

Species	Number of accessions in collections (Godwin, 2003)
Wheat	784,500
Barley	485,000
Rice	420,500
Maize	277,000
Oat	222,500
Sorghum	168,500
Soybean	174,500
Groundnut	81,000
Tomato	78,000
Apple	97,500
Grape	47,000

Other living collections

Species with seeds that are not suitable for storage or that require vegetative propagation to retain genetic integrity are usually maintained as living collections. Recalcitrant species have seeds that cannot be dried without loss of viability. These species require alternatives to seed storage for germplasm collections. Options include growth in the field, in a plant house or in tissue culture. Techniques such as cyropreservation may be used for some species. A range of different explants can be stored by cyropreservation. These include cell suspensions, calluses, shoot tips and embryos (Thormann, 2004).

Plants may be kept in tissue culture at low temperatures to reduce the cost of maintaining the cultures. Low temperatures allow a lower frequency of media change. Slow-growing collections are used for species such as banana, plantain, cassava, potato and sweet potato (Thormann, 2004).

Botanic gardens

Botanic gardens are widespread, usually public institutions that provide a repository usually including a very wide range of plant species. Some private gardens also contribute in this way but may provide a lower level of long-term security to the genetic resource. The added value of botanic gardens is usually associated with the documentation of the material in the collections (Makinson, 2004) and links to associated herbaria.

DNA banks

The storage of plant genomic DNA and even individual genes or gene libraries is an additional, more recent option for the conservation of plant genetic resources (Rice, 2004). This approach is not currently suitable as an alternative to other methods for conserving species because living plants cannot be recovered from the stored DNA. However, specific genes may be protected and reintroduced into plants. DNA banks for plants are currently being established and expanded in several key locations internationally.

Genetic resources available for different plant groups

The quantity (numbers of samples) and quality (diversity of samples and standard of storage conditions) of genetic resources available for plants varies. Some species are reliant completely on wild populations. Others have very large numbers of samples in seed banks (Table 15.5) and are distributed around the world.

Crop plants

Genetic resources available for major crop species are critical to world food security. A very small number of species account for most human food. Cereals are the stable food sources for most human diets (Henry, 2001). The very large collections of these species contrast with the relatively small genetic resource collections available for the minor regional crop species. Wild populations of crop species or their relatives remain very important genetic resources for most species. Protection of wild populations of crop species and important relatives has received comparatively little attention in many regions (Meilleur and Hodgkin, 2004). The gene pool for these species may be considered to include the primary gene pool of the species itself, the secondary gene pool of closely related species that can be used in breeding and a tertiary gene pool of more distant relatives that have genes that might be accessed using more novel techniques (Fig. 15.1). In many cases relatively little effort has been given to conservation of germplasm outside the primary gene pool. Many crop plants of economic importance are listed in Table 15.3.

Recently Meilleur and Hodgkin (2004) defined key objectives for advancing the conservation of crop wild relatives. The main recommendations included:

- Development of an agreed definition of crop wild relatives.
- Preparation of comprehensive lists of species.
- Identification and flagging of wild relatives of crops in existing germplasm databases.
- Expansion of data on these entries.
- Conducting targeted research to identify conservation needs.
- Greater coordination of conservation efforts.

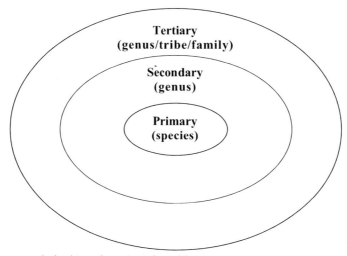

Fig. 15.1. The gene pool of cultivated species (adapted from Henry, 2000).

Ornamental plants

Ornamental plants have often been highly selected, and cultivated forms may appear very different from wild ancestors. In some cases the wild progenitors can no longer be identified. Often the apparent large phenotypic differences may be explained by relatively small genetic changes in loci contributing to key morphological traits. Gardens are important centres of conservation of plant diversity for ornamental and edible species. In some regions species diversity in gardens exceeds that in the wider environment. Ornamental plant conservation is subject to the vagaries of fashion in gardens. Very high values may be placed on some rare plants. The high price may assist in conservation of the species but often has a negative impact, generating collection pressure on wild genetic resources. The final outcome may be serious depletion of wild populations and if this is followed by loss of *ex situ* collections in gardens because of difficulty in cultivation or reduction in horticultural interest in the species, species diversity may be reduced and long-term survival of the species threatened.

Forest plants

Forest species are not yet well domesticated compared with crop plants (Campbell *et al.*, 2003). Most forest plantations use genetic resources that have been subjected to relatively little selection or genetic improvement. A relatively small number of plant species dominate forest planting worldwide. These include gymnosperms (specifically the pines (*Pinus*)) and angiosperms (poplars (*Populus*) and eucalypts (*Eucalyptus*)). The eucalypts are the most widely planted commercial forest species in the southern hemisphere and poplars are grown widely in northern regions. Other less widely available species have very high value (e.g. oaks (*Quercus robur* and *Quercus petrea*), walnut (*Jugulans regia*) and mahogany (*Swietinia humilis*)). Genetic resources for most forest species are mainly to be found *in situ* in wild populations and in forest plantations. Dedicated *ex situ* forest germplasm collections are relatively limited.

Future Opportunities in Plant Genetic Resources

Advances in plant genetics are continuing to expand the options for the recombination of genes and genomes to generate novel germplasm. Technologies for transfer of uncharacterized genes from distant wild relatives have been described (Abedinia *et al.*, 2000). Recombinant DNA technologies allow new variation of all types to be generated. Effective use of this capability is likely to continue to rely on an understanding of the arrangement and variation in plant genomes and on the processes of genome evolution.

Genomics has resulted in large amounts of genetic information becoming available for many important plant species. The expansion of DNA banks and rapid improvements in the efficiency of DNA analysis tools promise the potential to better characterize and utilize plant germplasm in the future. Plant diversity conservation will be greatly enhanced by these continuing technological developments (especially in molecular biology and bioinformatics), allowing very large amounts of biological and especially genetic data to be collected, stored and analysed.

References

Abedinia, M., Henry, R.J., Blakeney, A.B. and Lewin, L. (2000) Accessing genes in the tertiary gene pool of rice by direct introduction of total DNA from *Zizania palustris* (wild rice). *Plant Molecular Biology Reporter* 18, 133–138.

Campbell, M.M., Brunner, A.M., Jones, H.M. and Strauss, S.H. (2003) Forestry's fertile crescent: the application of biotechnology to forest trees. *Plant Biotechnology Journal* 1, 141–154.

FAO (1996) *The State of the World's Plant Genetic Resources for Food and Agriculture*. FAO, Rome.

Godwin, I. (2003) Plant germplasm collections as sources of useful genes. In: Newbury, H.J. (ed.) *Plant Molecular Breeding*. Blackwell, Oxford, pp. 134–151.

Henry, R.J. (2000) Technical advances in plant transformation providing opportunities to expand the cereal gene pool. In: O'Brien, L. and Henry, R.J. (eds) *Transgenic Cereals*. AACC, St Paul, Minnesota, pp. 252–276.

Henry, R.J. (2001) Exploiting cereal genetic resources. *Advances in Botanical Research* 34, 23–57.

Henry, R.J. (2004) Genetic improvement of cereals. *Cereal Foods World* 49, 122–129.

Heywood, V.H. (ed.) (1978) *Flowering Plants of the World*. Oxford University Press, Oxford.

Makinson, R. (2004) Botanic gardens and plant conservation. In: Henry, R.J. (ed.) *Plant Conservation*. Haworth Press, Binghampton, New York.

Meilleur, B.A. and Hodgkin, T. (2004) *In situ* conservation of crop wild relatives: status and trends. *Biodiversity and Conservation* 13, 663–684.

Nagamine, T. (2004) The role of genetic resources held in seedbanks. In: Henry, R.J. (ed.) *Plant Conservation*. Haworth Press, Binghampton, New York.

Prance, G. (2004) Strategies for *in situ* plant conservation. In: Henry, R.J. (ed.) *Plant Conservation*. Haworth Press, Binghampton, New York.

Rice, N. (2004) Conservation of plant genes and the role of DNA banks. In: Henry, R.J. (ed.) *Plant Conservation*. Haworth Press, Binghampton, New York.

Thormann, I. (2004) Techniques for *ex situ* plant conservation. In: Henry, R.J. (ed.) *Plant Conservation*. Haworth Press, Binghampton, New York.

Index

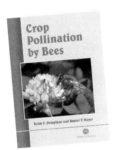
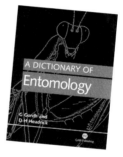
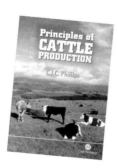

-